A GUIDE TO THE GEOGRAPHY OF PLINY THE ELDER

This is the first thorough English commentary on the geographical books of Pliny the Elder, written in the AD 70s. Pliny's account is the longest in Latin, and represents the geographical knowledge of that era, when the Roman Empire was the dominant force in the Mediterranean world. The work serves both cultural and ideological functions: much of it is topographical, but it also demonstrates the political need to express a geographical basis for the importance of the Roman state. In five books, Pliny covers the entire world as it was known in his era and includes some of the first information on the extremities of the inhabited region, including Scandinavia and the Baltic, eastern Asia, and sub-Saharan Africa. The commentary provides a detailed analysis of all the points Pliny raises: his sources, toponyms, and understanding of the place of the earth in the cosmos.

DUANE W. ROLLER is Professor Emeritus of Classics at The Ohio State University. He is the author of numerous books, including *The Geography of Strabo* (Cambridge, 2014), *A Historical and Topographical Guide to the Geography of Strabo* (Cambridge, 2018), *Cleopatra: A Biography* (2010), and *Empire of the Black Sea* (2020).

A GUIDE TO THE GEOGRAPHY OF PLINY THE ELDER

DUANE W. ROLLER

The Ohio State University

CAMBRIDGE
UNIVERSITY PRESS

CAMBRIDGE
UNIVERSITY PRESS

University Printing House, Cambridge CB2 8BS, United Kingdom

One Liberty Plaza, 20th Floor, New York, NY 10006, USA

477 Williamstown Road, Port Melbourne, VIC 3207, Australia

314–321, 3rd Floor, Plot 3, Splendor Forum, Jasola District Centre,
New Delhi – 110025, India

103 Penang Road, #05–06/07, Visioncrest Commercial, Singapore 238467

Cambridge University Press is part of the University of Cambridge.

It furthers the University's mission by disseminating knowledge in the pursuit of education, learning, and research at the highest international levels of excellence.

www.cambridge.org
Information on this title: www.cambridge.org/9781108481809
DOI: 10.1017/9781108693660

First published 2022

Printed in the United Kingdom by TJ Books Limited, Padstow Cornwall

A catalogue record for this publication is available from the British Library.

ISBN 978-1-108-48180-9 Hardback

In memory of Duane H. D. Roller (14 March 1920–24 August 1994), who inspired my interest in antiquity and Pliny

Contents

List of Maps page viii
Preface ix
List of Abbreviations xi

Introduction 1

Book 1: Pliny's List of Sources 15

Book 2: Cosmology 32

Book 3: Southern Europe 104

Book 4: The Greek Peninsula and Interior Europe 186

Book 5: Africa and Western Asia 247

Book 6: The Remainder of Asia 321

Bibliography 409
Index of Passages Cited 424
General Index 453

Maps

1 The Ancient World as Known to Pliny *page* 3
2 The Geographical Divisions of Books 3–6 of the *Natural History* 5
3 Places cited in Book 3 of the *Natural History*, according to chapters (courtesy of Ancient World Mapping Center). 105
4 Places cited in Book 4 of the *Natural History*, according to chapters (courtesy of Ancient World Mapping Center). 187
5 Places cited in Book 5 of the *Natural History*, according to chapters (courtesy of Ancient World Mapping Center). 248
6 Places cited in Book 6 of the *Natural History*, according to chapters (courtesy of Ancient World Mapping Center). 322

Preface

This guide is designed to provide the first modern English commentary on the geographical books of the *Natural History* of Pliny the Elder. It is based on the Latin text of the Budé edition, insofar as it is available (Books 5 and 6 are not yet complete), and the Tusculum edition, with some minor emendations by the present commentator.

The death of Pliny at the time of the eruption of Mt. Vesuvius on 24 August AD 79 is one of the best-known events of classical antiquity. Two months previously the emperor Vespasian (ruled AD 69–79) had died and was succeeded by his son Titus (ruled AD 79–81). The *Natural History* greatly reflects the era of these emperors, especially the former. When Pliny died he may still have been polishing the treatise, yet the geographical books (2–6) are complete in regard to their trajectory, and only additional editing would have been necessary. The work as a whole is the longest surviving work in Latin from antiquity, and, as is usual in such lengthy works, shows a process of layering of sources and internal inconsistency that can make a coherent analysis difficult. It should also be remembered that the five geographical books are not a complete work, but merely a limited portion of the totality of the *Natural History*. Moreover, Pliny was writing in Latin about a discipline that was essentially Greek, and this is reflected in his handling of toponyms and other proper nouns. He often vacillated between Greek and Latin, or would mix his forms between the two languages, which can make comprehension of his text problematic. For his geographical books, Pliny cited over a hundred sources and obviously used others. There are nearly six thousand toponyms; all that can be identified – and many cannot – appear on a map to be created by the Ancient World Mapping Center.

There has been no full commentary in English since the edition of John Bostock and H. T. Riley of 1855–7, whose notes are limited to comments at the bottom of the page and which is more valuable for its role in Victorian antiquarianism than as a modern conception of the *Natural History*. The sparse and problematic notes in the Loeb edition of Harris Rackham (1938–42) were produced in an era when the study of ancient geography was less favored than it is today, and ancient toponyms were often hardly understood. Many of the place names in the translation of that edition were changed to assumed modern ones, more often misleading than useful. The most thorough commentaries are those in the French Budé (started in 1950 and not yet finished) and the German Tusculum (1973–96; whose notes can be sporadic). The canonical edition of the text is the Teubner by Karl Mayhoff (1892–1909), but it has been superseded by the editions noted above. The most recent English translation of the geographical books is that by Brian Turner and Richard J. A. Talbert (Cambridge 2022).

The bibliography on Pliny's geography is enormous, and many of the relevant works are cited in the present volume. In addition, valuable research tools include the *Barrington Atlas* (*BA*), *Brill's New Pauly* (*BNP*), *The Encyclopedia of Ancient Natural Scientists* (*EANS*), and *Die Fragmente der Griechischen Historiker* (*FGrHist*). The present commentator has built on his experience in translating and creating commentaries for works such as Megasthenes' *Indika*, Eratosthenes' *Geography* (both of which were used extensively by Pliny), and especially the *Geography* of Strabo of Amaseia.

Enthusiastic thanks are due to the many who assisted in the completion of this volume, especially the libraries of the University of California at Berkeley and Stanford University, and the highly efficient interlibrary loan system of The Ohio State University, as well as David Braund, Stanley M. Burstein, Georgia M. Irby, Paul T. Keyser, Carolina López-Ruiz, Lee Patterson, Letitia K. Roller, Michael Sharp, and many others at Cambridge University Press, with special thanks to copy-editor Gill Cloke, Richard J. A. Talbert, Lisbet Thoresen, Brian Turner, and Wendy Watkins and the Ohio State University Center for Epigraphical and Palaeographical Studies. The primary work on the commentary was done in the author's study in Santa Fe.

Abbreviations

AAE:	*Arabian Archaeology and Epigraphy*
AAntHung:	*Acta antiqua Academiae Scientiarum Hungaricae*
AClass:	*Acta Classica*
AfrRom:	*L'Africa Romana*
AJA:	*American Journal of Archaeology*
AJPh:	*American Journal of Philology*
ANRW:	*Aufstieg und Niedergang der Römische Welt*
AntCl:	*L'Antiquité classique*
AOF:	*Archiv für Orientforschung*
AR:	*Archaeological Reports*
ARID:	*Analecta romana Instituti Danici*
BA:	*Barrington Atlas of the Greek and Roman World* (ed. Richard J. A. Talbert, Princeton 2000)
BAR-IS:	*British Archaeological Reports, International Series*
BASOR:	*Bulletin of the American Schools of Oriental Research*
BICS:	*Bulletin of the Institute of Classical Studies*
BNP:	*Brill's New Pauly*
Broughton:	T. Robert S. Broughton, *The Magistrates of the Roman Republic* (New York 1951–1952, 1986)
BSA:	*Annual of the British School at Athens*
CAH:	*The Cambridge Ancient History*
CIL:	*Corpus Inscriptionum Latinarum*
Cohen, *East:*	Getzel M. Cohen, *The Hellenistic Settlements in the East from Armenia and Mesopotamia to Bactria and India* (Berkeley, Calif. 2013)
Cohen, *Europe*:	Getzel M. Cohen, *The Hellenistic Settlements in Europe, the Islands, and Asia Minor* (Berkeley, Calif. 1995)
Cohen, *Syria:*	Getzel M. Cohen, *The Hellenistic Settlements in Syria, the Red Sea Basin, and North Africa* (Berkeley, Calif. 2006)

CP:	*Classical Philology*
CQ:	*The Classical Quarterly*
CR:	*The Classical Review*
CRAI:	*Comptes Rendus, Académie des Inscriptions et Belles-Lettres*
DHA:	*Dialogues d'histoire ancienne*
EANS:	*The Encyclopedia of Ancient Natural Scientists* (ed. Paul T. Keyser and Georgia L. Irby-Massie, London 2008)
FGrHist:	*Die Fragmente der Griechischen Historiker*
FRH:	*The Fragments of the Roman Historians* (ed. T. J. Cornell, Oxford 2013)
G&R:	*Greece and Rome*
GA:	*Geographia antiqua*
GJ:	*The Geographical Journal*
Graham:	Daniel W. Graham, *The Texts of Early Greek Philosophy* (Cambridge 2010)
GRBS:	*Greek, Roman, and Byzantine Studies*
HRR:	*Historicum romanorum reliquae* (ed. Hermann Peter, Leipzig 1870)
HSCP:	*Harvard Studies in Classical Philology*
IG:	*Inscriptiones Graecae*
ILS:	*Inscriptiones Latinae Selectae*
IrAnt:	*Iranica Antiqua*
JAH:	*Journal of African History*
JAS:	*Journal of the Royal Asiatic Society of Great Britain and Ireland*
JBL:	*Journal of Biblical Literature*
JESHO:	*Journal of the Economic and Social History of the Orient*
JHS:	*Journal of Hellenic Studies*
JNES:	*Journal of Near Eastern Studies*
JRS:	*Journal of Roman Studies*
MediterrAnt:	*Mediterraneo Antico*
NEAEHL:	*The New Encyclopedia of Archaeological Excavations in the Holy Land* (ed. Ephraim Stern, New York 1993)
NH:	Pliny, *Natural History*
OGIS:	*Orientis Graeci Inscriptiones Selectae*
OT:	*Orbis Terrarum*
PBSR:	*Papers of the British School at Rome*
PECS:	*Princeton Encyclopedia of Classical Sites*

PEQ:	*Palestine Exploration Quarterly*
PIR:	*Prosopographia Imperii Romani*
PP:	*La Parola del Passato*
RBi:	*Revue biblique*
RÉG:	*Revue des études grecques*
RÉL:	*Revue des études latines*
RELat:	*Revista de Estudios Latinos*
Richardson:	L. Richardson Jr., *A New Topographical Dictionary of Ancient Rome* (Baltimore, Md. 1992)
SE:	*Studi etruschi*
SEG:	*Supplementum Epigraphicum Graecum*
SIG:	*Sylloge Inscriptionum Graecarum*
SO:	*Symbolae Osloenses*
Tusculum:	*Plinius Naturkunde* (ed. Roderick König *et al.*, Munich 1973–2013)

Introduction

A Brief Sketch of the Life of Pliny the Elder

C. Plinius Secundus was born at Comum (modern Como) in late AD 23 or sometime the following year.[1] Comum, at the southern end of Lake Larius (modern Lago di Como), had been an obscure Celtic village until a century previously, when the Romans established a presence there.[2] Virtually nothing is known about Pliny's youth or education, but by AD 47 he had embarked on an equestrian military career in Germania under the command of Cn. Domitius Corbulo.[3] He had returned to Rome by the AD 50s, and seems to have remained relatively obscure during the principate of Nero. But with the accession of Vespasian – whom he already knew – in AD 69 he returned to public service and became a procurator (financial officer), with positions in various locations, including Narbonensis, Tarraconensis, Belgica, Africa, and perhaps elsewhere.[4] He became a close advisor to the emperor, conferring with him on a daily basis when in Rome.[5] He also practiced law. Eventually he became fleet commander at Misenum, the Roman naval base established by Augustus at the end of the long peninsula that forms the western side of the Bay of Naples.[6] His sister Plinia and her son, also C. Plinius Secundus, known today as Pliny the Younger, lived with him. It was here that they observed the eruption of Vesuvius on 24 August AD 79, two months after Vespasian had died and his son Titus (also an intimate of Pliny's) had come to the throne. The account of subsequent events reported by his nephew is well known.[7] Investigating

[1] He died at the age of 55 in August of AD 79: Pliny the Younger, *Letters* 3.5.8.
[2] Strabo, *Geography* 5.1.6. [3] Ronald Syme, "Pliny the Procurator," *HSCP* 73 (1969) 204–8.
[4] John F. Healy, *Pliny the Elder on Science and Technology* (Oxford 1999) 1–35.
[5] Pliny the Younger, *Letters* 3.5.7–9.
[6] Pliny the Younger, *Letters* 6.16.4; Suetonius, *Augustus* 49.1.
[7] Pliny the Younger, *Letters* 6.16, 20.

the eruption and attempting to offer assistance to those affected, Pliny collapsed on the beach at Stabiae, just southeast of the volcano. His body was found two days later.

In time Pliny the Younger provided his friend Q. Baebius Macer (the suffect consul of AD 103) with a list of his uncle's works.[8] It is in chronological order, beginning with *de iaculatione equestri* (*Throwing the Javelin From Horseback*), in one book, which was probably a technical manual for cavalry, produced during his years in Germania. Nothing more is known about it. This was followed by a two-book biography of P. Pomponius Secundus, the consul of AD 44, under whom Pliny served in Germania in AD 50–1, and who wrote tragedies and was a grammarian.[9] Pliny mentioned the biography in the *Natural History*.[10]

He also produced two lengthy historical treatises. The *German Wars*, in 20 books, examined the entire history of Roman military involvement in Germania, beginning with the campaigns of Drusus at the end of the first century BC. The work is astonishingly poorly preserved and only two fragments survive, but it was probably a major source for Tacitus when writing his *Germania*.[11]

There was also a general history, titled *A fine Aufidi Bassi* (*From the End of Aufidius Bassus*), in 31 books. Aufidius Bassus had written an account that seems to have extended from the 40s BC into the Julio-Claudian period.[12] Pliny's continuation would thus have dealt with the later years of that era. The handful of surviving fragments are all concerned with the reign of Nero, including its end; thus the work was completed after AD 68.[13]

Pliny also wrote on oratory and grammar: Pliny the Younger noted that the grammatical work was written during the last years of Nero, when other topics might not have been wise. And finally there was the *Natural History*, 37 encyclopedic books on virtually every conceivable topic of the natural world. It was a product of Pliny's last years, and essentially complete at his death in the summer of AD 79.

The Roman Encyclopedia and Imperial Ideology

It is common today to call Pliny's *Natural History* an "encyclopedia," but that concept did not truly exist until early modern times. The word in its current usage was derived from a Greek phrase, *enkyklios paideia*, roughly meaning "a general education." It seems to have originated in Hellenistic

[8] Pliny the Younger, *Letters* 3.5.3–6; *FRH* #80. [9] *PIR* P563. [10] *NH* 14.56.
[11] *FRH* #80, F2–3. [12] *FRH* #78. [13] *FRH* #80, F4–8.

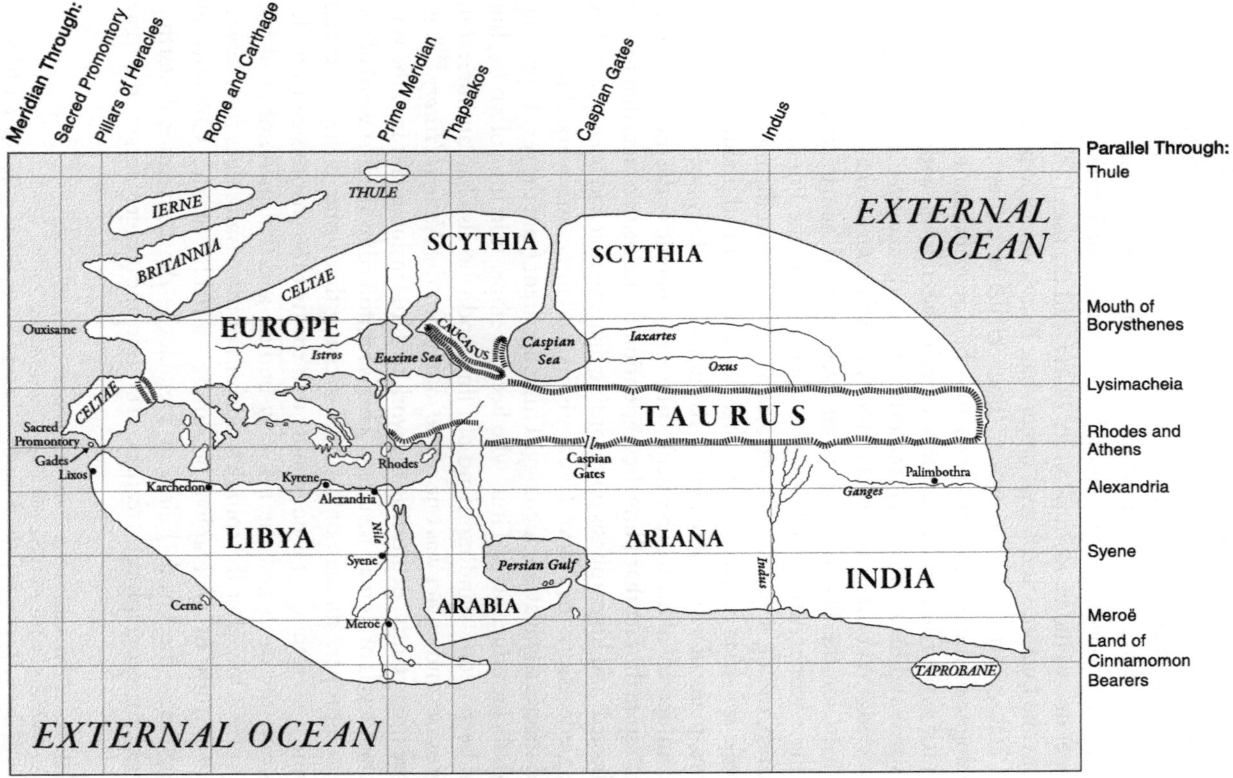

Map 1 The Ancient World as Known to Pliny

times, but was more common in the Roman Imperial period. Pliny himself used the phrase, but presented it in its Greek original and in Greek characters.[14] To him, it was essentially a curriculum of study, and hardly a way of describing the *Natural History* as a whole. That work was titled *Naturalis Historia*, and in a modern sense it could be called encyclopedic but not an encyclopedia. The common English translation is ambiguous; a more proper one would be *Research on the World*, and the use of "naturalis" tends to invoke Lucretius and Vergil rather than Pliny's catalogic predecessors.[15]

Catalogue treatises had existed since Hellenistic times: Callimachus' catalogue of the Alexandria library may be one of the earliest examples.[16] But it was the Romans who came to excel in the concept, with authors such as Varro and his *Antiquitates*, and other works that covered a vast range of knowledge. Yet the uniqueness of the *Natural History* is that it attempted, perhaps for the first time, to cover every part of human knowledge in a single work. Although Pliny was dismissive of the importance of his treatise, calling it "a lighter work," he placed it within the tradition of the Greek *enkyklios paideia*, noting that he had consulted about 2,000 volumes, many of which were rare.[17] His authorial modesty does not diminish the immensity of his achievement, truly encyclopedic in the modern sense.

Despite the fact that Pliny drew attention to the 20,000 facts that he had collected, and his emphasis on recording material that had been ignored by his predecessors, the *Natural History* is first and foremost a work of Roman, and specifically Flavian, ideology. This is apparent from the very beginning with its introduction to Titus, who became emperor in June of AD 79, but had been a long-standing friend of Pliny, and who also had an interest in curiosities and in fact was one of the sources for the *Natural History*.[18] But, more importantly, dedicating the work to Titus insinuated it into the very highest levels of Roman ideology. This is reinforced by the first word of the treatise proper, *mundus*, or "world," stressing that Rome is the world and Rome is everything. It is no wonder that the first 166 chapters concern the cosmos: the earth and the heavenly bodies, and the phenomena of the universe. Only after the complete analysis of the cosmos does the *Natural History* turn to geography, beginning, suitably, at the Columns of Hercules – the Greek Pillars of Herakles – the conventional starting point for geographical treatises. Yet lest the reader forget the true

[14] Dionysius of Halicarnassus, *On Literary Composition* 25; *NH Preface* 14.

[15] See, especially, Lucretius 1.25 etc. and Vergil, *Georgics* 2.490.

[16] Rudolf Blum, *Kallimachos: The Alexandrian Library and the Origins of Bibliography* (tr. Hans H. Wellisch, Madison 1991) 150–60.

[17] *NH Preface* 12, 17 (*levioris operae*). [18] *NH* 1.2, 2.89.

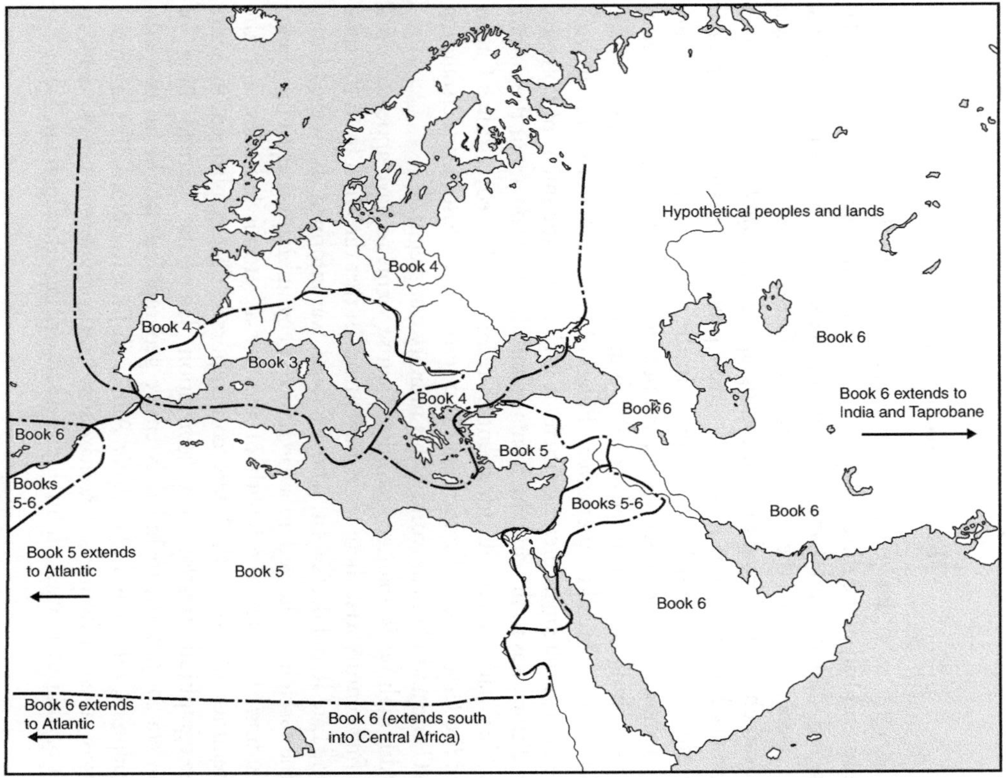

Map 2 The Geographical Divisions of Books 3–6 of the *Natural History*

orientation of the work, by the second sentence of the geographical portion the narrative has turned to the emperor Augustus and the expedition he commissioned to northern Europe.[19] This left no doubt that the focus of the treatise was to be Roman achievements: the cosmos, the world, and Rome are seen as virtually synonymous.

The Romans and Geography

Geography as a discipline was invented by Eratosthenes of Cyrene in the second half of the third century BC.[20] Although there had been speculation about the size and shape of the earth and the nature of its surface, places, and peoples since the beginnings of Greek intellectualism, it was Eratosthenes who created the word "geography" (*geographia*) and established it as a scholarly endeavor. Other Greek geographical treatises appeared thereafter, but the first Roman interest in the topic was largely for military and demographic purposes. At the fall of Carthage in 146 BC, P. Cornelius Scipio Aemilianus sent his advisor Polybius beyond the Columns or Pillars of Hercules or Herakles (hereafter, "Pillars of Hercules") in order to investigate the Carthaginian presence there. It is clear that the intent of Polybius' expeditions was to provide the Romans with information about the empire that they had conquered, not scientific research.[21] Moreover Julius Caesar used the *Geography* of Eratosthenes as a resource to assist in understanding the Celtic and German world of his military campaigns.[22]

Although no Roman geographical treatise seems to have appeared until the first century AD, there were movements in that direction from the time of the Late Republic. Caesar, who, as noted, was informed about the history of the discipline, showed an awareness of geographical issues in his works, including the tides, the length of the day at various latitudes, and the extremities of the earth.[23] His contemporary Cicero planned to write a geographical treatise.[24] A generation later, the learned king of Mauretania, Juba II, included extensive geographical data in his two ethnographies, *Libyka* and *On Arabia*. Although Juba was Numidian in origin, he was raised in Rome as a Roman aristocrat, yet wrote in Greek, demonstrating that there was still little interest in writing ethnologies in Latin.[25]

[19] *NH* 2.167. [20] Duane W. Roller, *Eratosthenes' Geography* (Princeton 2010) 1–2.
[21] Polybius 34.15.7–9. [22] Caesar, *Gallic War* 6.24.
[23] Duane W. Roller, *Ancient Geography* (London 2015) 149–50.
[24] Cicero, *Letters to Atticus* #24, 26 (April 59 BC).
[25] Duane W. Roller, *The World of Juba II and Kleopatra Selene: Royal Scholarship on Rome's African Frontier* (London 2003) 183–243; *FGrHist* #275.

M. Vipsanius Agrippa, the confidant of the emperor Augustus, was perhaps the first Roman to be considered an actual geographical scholar.[26] But even his work had a political end: it seems to have been a commentary on a public map that he created – probably completing a project begun by Julius Caesar – and as such it served the military and demographic purposes that were a standard feature of Roman geographical efforts.[27]

As Roman knowledge spread throughout the inhabited world, the need for legitimate geographical scholarship became more acute, even if it was still directed toward political needs. Greek scholars could not always be relied upon to express matters in ways that were politically correct to the Roman ruling elite. By the middle of the first century AD, Romans had explored the North Sea and perhaps penetrated the Baltic. They were aware of lands beyond India, perhaps as far as the Malay peninsula and eastern Asia. They also had gone into sub-Saharan Africa.[28] Thus it is perhaps no surprise that these efforts led to the first true Roman geographical treatise, the three-book *Chorographia* of Pomponius Mela.[29] It is the earliest surviving geographical work in Latin, and while brief, provides a new Latin vocabulary derived from long-existing Greek geographical terminology, as well as giving a Roman outlook to the discipline. Written in the generation before the *Natural History*, it established a Roman view of the inhabited world.

The *Natural History* and Geography

All of the geographical works cited in the previous section were known to Pliny when he began the *Natural History*, and he used them assiduously to create the five geographical books of the treatise. These present a geographical compendium that is less an analytical discussion than a catalogue of existing knowledge on the topic. That this was his intent

[26] Pliny (*NH* 3.17) wrote of his diligence in handling geographical material.

[27] The exact nature of Agrippa's map and the accompanying commentary remains one of the more complex and debatable issues in the study of ancient geographical scholarship: see Alfred Klotz, "Die geographischen commentarii des Agrippa und ihre Überreste," *Klio* 24 (1931) 38–58, 386–466; J. J. Tierney, "The Map of Agrippa," *Proceedings of the Royal Irish Academy* 63 (1962–4) 151–66; and Claude Nicolet, *Space, Geography, and Politics in the Early Roman Empire* (Ann Arbor, Mich. 1991) 95–122. One of the many undecided questions is how much Augustus contributed to the project after Agrippa's death in 12 BC: see Nicolet, *Space* 171–187.

[28] Roller, *Ancient Geography* 171–87.

[29] Georgia L. Irby, "Tracing the Orbis Terrarum From Tingentera," in *New Directions in the Study of Ancient Geography* (ed. Duane W. Roller, University Park, Pa. 2019) 103–34.

is made clear by his listing of the total pieces of information contained in each book (a technique not limited to the geographical books). Nevertheless, his role was to show how the entire world had been made known through the efforts of Rome, a concept following closely on the map of Agrippa and the *Chorographia* of Pomponius Mela. In fact, there is a certain map-like quality to the geography of the *Natural History*, manifested in sequences such as the lengthy lists of toponyms along the Nile. Pliny could see the map of Agrippa on a daily basis when in Rome, and used it, in part, as the basis for creating his own literary map. Unlike the *Geography* of Strabo, whose 17 books are the longest geographical work from antiquity, there is often little analysis in Pliny's text, and the lists of places and peoples predominate. Despite this, there is much of value in terms of description and exegesis, and Book 2 is a thorough study – however derivative – of cosmology.

Even though Pliny had such an emphasis on topographical lists, he was very much beholden to the tradition of geographical scholarship that he had inherited. There is a focus on coasts, in the tradition of the nautical sailing guides (*periploi*) that had been part of Greek culture since earliest times. But the Romans were not oriented as much to the coasts as were the Greeks, especially in Europe, where they had penetrated deep into the interior. Rivers and mountains defined these inland regions, and it is no accident that these are the first topographical features mentioned in the summaries that end the catalogues of Book 1. In the catalogue to Book 6 (the only one among the geographical books where the numbers are preserved), Pliny told his readers that he had recorded the names of 115 famous rivers and 38 famous mountains. Rivers, especially, were of significance to the Romans: Rome owed its origins to its control of a major crossing of the Tiber that affected travel throughout the Italian peninsula, and river cults were a major part of Roman ideology.[30] Hundreds of rivers are named in the *Natural History*, and the important ones, such as the Danube, Rhone, or Hiberus (Ebro), were the means by which Roman culture could be carried into the remote interior of the continents.

There were also mountains, the opposite of rivers. Rather than open up territory, they separated it. This was not a new idea: in the late fourth century BC, Dicaearchos of Messana had conceived of a vast mountain range that extended from Asia Minor to India and thus divided the world

[30] Trevor Murphy, *Pliny the Elder's Natural History: The Empire in the Encyclopedia* (Oxford 2004) 138–48.

east of the Mediterranean into a northern and southern portion.[31] Thus the world was both accessible and divisible, and mountains and rivers defined the interior of the continents. Rivers allowed access; mountains, to some extent, allowed control.

But another factor was Imperial ideology, which actually formed the basis of the Roman study of geography. Vespasian and Titus had brought peace to the Roman world. Previously it had been torn apart through civil war, not only in AD 68–9 (which ended with Vespasian acquiring the principate), but in the greater civil war of the first century BC. Now, however, everywhere was open, and many were able to go forth throughout the entire world (although Pliny at *NH* 2.117–18 rather tartly noted that such travels were for profit, not knowledge). There was no more piracy, which had been endemic in the Mediterranean until the early first century BC, and the world was no longer divided into hostile states. This new era of Vespasianic peace not only allowed Pliny to write the *Natural History*, but to devote its first books to revealing the world itself.

After the catalogues that make up Book 1, the *Natural History* begins in earnest. Its 37 books comprise one of the longest surviving works from classical antiquity. It was completed in the years just before Pliny's death in AD 79, and has survived intact, as can be seen by the reference of Pliny the Younger to its 37 books.[32] The work exists today in about 130 manuscripts, some of which are from late antiquity, but these are partial and do not contain the geographical books. Two manuscripts in Paris are the earliest that contain most of the treatise. Parisinus lat. 6795, from the ninth or tenth century, includes everything except Books 33–7 (the earliest manuscript of those books is Bambergensis M. V. 10, of the tenth century). Parisinus lat. 6797, of the thirteenth century, has the entire text except the very end, the last five chapters of Book 37. The earliest printed edition was by Johann of Speyer in Venice, from 1469.[33] Despite the many manuscripts, the text is in astonishingly poor shape, especially the geographical portions, due to the many unfamiliar and unique toponyms. Moreover, Pliny often used ethnyms to describe communities, since this was the format of the official Roman registry. All this can make it difficult to

[31] Paul T. Keyser, "The Geographical Work of Dikaiarchos," in *Dicaearchus of Messana: Text, Translation, and Discussion* (ed. William W. Fortenbaugh and Eckart Schütrumpf, New Brunswick, N.J. 2001) 353–72; see also Murphy, *Pliny* 148–54.

[32] Pliny the Younger, *Letters* 3.5.6.

[33] *BNP Supplement* 2: *Dictionary of Greek and Latin Authors and Texts*, 504–5.

understand the places involved.[34] He was in the awkward position of using the Latin language to write about a discipline that was still largely Greek.

It should be remembered that the five geographical books of the *Natural History* are not a free-standing work, but part of a lengthy composition that, in theory, included all the natural world, with topics such as zoology (Books 8–16), agriculture (Books 17–19), herbs and drugs (Books 20–32), and mineralogy (Books 33–7). Yet these broad discussions lead to others, such as the extensive section on art and artists (Books 34–6) or the biographical details about notable Romans (Book 7). Geographical material is scattered throughout, including the route to the frankincense region of southern Arabia (12.61–5) or information about the Trogodyte territory of East Africa (31.18). Pliny's sources were wide-ranging, with nearly five hundred cited in the catalogues of Book 1, and others noted in the text.

The geographical books begin with the single book (2) on cosmology and natural phenomena, and then continue with a circuit of the known world (3–6). The cosmology examines the nature of the earth and the various celestial bodies. There is also a discussion of celestial and terrestial phenomena. This is followed by a closer consideration of the surface of the earth, including the distinction between land and water, its inhabited portions, and the different lengths of day at various latitudes. Oceanic phenomena, such as its saltiness and the tides, are also included. The book closes with the dimensions of the earth.

Book 3 begins the geographical survey. Much of it is in *periplous* format, the style of the coastal sailing itinerary that had long existed in the Greek world.[35] It starts in southern Iberia and continues along the Mediterranean coast to Italy, including the western Mediterranean islands, the Adriatic, and the lands to its north. Book 4 moves from the Adriatic into the Greek peninsula and the islands of the Aegean, and north into the lands west of the Black Sea, but then curves west around through Europe, the coast of the North Sea, and into the Celtic lands, including the western Iberian peninsula and the Atlantic islands. Book 5 crosses over to Mauretania and then moves east to Egypt and north through the Levant and into Asia Minor, ending with Bithynia at its northwestern extremity. Book 6 begins with Pontus, just to the east of Bithynia, and includes the interior and remaining Black Sea coast of Asia Minor. Then it continues around the

[34] Nikos Kokkinos, "The City of 'Mariamme': An Unknown Herodian Connection?" *MediterrAnt* 5 (2002) 723–4.

[35] Kai Brodersen, "The Geographies of Pliny and His 'Ape' Solinus," in *Brill's Companion to Ancient Geography: The Inhabited World in Greek and Roman Tradition* (ed. Serena Bianchetti *et al.*, Leiden 2015) 300–2.

east side of the Black Sea into the Bosporanian territory on its north shore, and east toward the Caucasus and Caspian, and the presumed eastern coast of Asia, at India and Taprobane. The account then returns west across southern Asia through Mesopotamia and Arabia to the regions of the upper Nile, and back across Africa to Mauretania. It ends with the Fortunate Islands off the Atlantic coast, with a final discussion about terrestrial distances and the places on various latitude lines.

Pliny's geographical itinerary – from southern Iberia and back to the Atlantic coast of Mauretania just to the south – follows no known precedent.[36] A beginning at the western end of the inhabited earth (the Pillars of Hercules) was well established, but his technique of circling around (so that Iberia is discussed in two separate portions), rather than a more linear progression, is unusual.[37] Hecataeus of Miletus,[38] the author of the first Greek geographical treatise (his *Circuit of the Earth*, from around 500 BC), had established the concept of starting at the west end of the Mediterranean, and although exact details of his organization are not known, it seems that he created a great circular route from Iberia through coastal Europe, Asia Minor, southern Asia (as far as India), and back across Africa. Eratosthenes, however, reversed direction and began with India, heading west, and ending somewhere in the western Mediterranean, seemingly a more linear than circular route. This may have been due to the prominence of India in the century after Alexander the Great.

Yet later geographical treatises reverted to the west as a starting point, making a clockwise circuit: this was the pattern of Strabo, in the early first century AD (but whose *Geography* was not known to Pliny). Pomponius Mela, writing only a few years previous to Pliny, also began at the west end of the Mediterranean, but followed a counter-clockwise route around the sea, ending in Iberia. But then he made a second, clockwise circuit from Iberia around the perceived coasts of the External Ocean (and thus following the exterior of the inhabited earth), through Germania, Scythia, to India and then around the outside of Africa, ending again at the western outlet of the Mediterranean. Thus he created an itinerary of roughly two concentric circles, each going in the opposite direction.

Pliny, however, created his own scheme, breaking away from the ovoid circles and straight lines of his predecessors, using wandering interlocked

36 Benet Salway, "Putting the World in Order: Mapping in Roman Texts," in *Ancient Perspectives* (ed. Richard J. A. Talbert, Chicago 2012) 214–15.

37 Gerhard Winkler, "Geographie bei den Römern: Mela, Seneca, Plinius," in *Geographie und verwandte Wissenschaften* 2 (ed. Wolfgang Hübner and Germaine Aujac, Stuttgart 2000) 158.

38 *FGrHist* #1.

loops. Books 4–6 each begin where the previous one left off, but his itinerary places the focus on Europe, which is treated in its entirety before any consideration of Asia or Africa. This is to some extent a reversal of Greek geographical theory, and thus gives emphasis not only to the regions most recently important to Rome, but those particularly relevant to Pliny's own career.

Creating a Commentary on Pliny's Geography

A number of issues confront the commentator on the geographical books of the *Natural History*. As with any lengthy work written in classical antiquity, the difficulties of editing and proofreading a handwritten manuscript can have their effect on the author. Pliny often failed to reconcile divergent and contradictory sources, with the result that toponyms may appear in different forms in different citations. He made it perfectly clear that he consulted multiple copies of a text, and felt free to make his own editorial judgements.[39] This is a testimony to the varied nature of copies of a given work, but it also meant that Pliny's text would have its own vagaries.

Writing in Latin about a discipline that was historically Greek, he could be inconsistent in what he produced, especially in creating Latin forms of Greek proper nouns. For example in Book 4 he vacillated between the Greek "Isthmos" and the Latin "Isthmus." Needless to say this probably reflects his sources – their exact identity is not always apparent – but one could even argue that Pliny, totally bilingual, was hardly aware of which form he was using. Greek endings to latinized toponyms pervade the *Natural History*, and even hybrid forms, such as the Akran Iapygiam of *NH* 3.100, or the Mega Tichos and Graucomen of 6.179. Occasionally material even appears in Greek characters. Toponyms, whether latinized Greek, pure Greek, or pure Latin, may not appear in the nominative anywhere in the text, so determining that case, and whether the word is neuter or masculine, may be problematic when the nominative appears nowhere in extant Latin literature. As much as possible this commentary uses Pliny's actual forms, and attempts, when necessary, to create a proper nominative, relying on what Pliny would have used. To be sure some familiar names, such as Rome and Athens, are anglicized, and where Pliny's form is quite idiosyncratic, such as Delmatia and Zmyrna for Dalmatia and Smyrna, the more common form is also presented.

[39] *NH* 6.62, 170.

In addition to the issues of latinizing a Greek discipline, there are problems in understanding Pliny's language. The *Natural History* is a complex and confusing document, with many digressions that interrupt, even grammatically, the main thread of the narrative. To be sure, such issues are probably less of a problem in the geographical books than elsewhere, since lists of toponyms without further comment tend to be so common in Books 3–6. But the work as a whole has a sense of incompleteness and lack of polish, something that the author himself acknowledged.[40] Part of this may be authorial modesty (since he said that it also applied to his other writings), but one cannot deny that understanding the text is baffling at times.[41] The geographical portion can be merely catalogic, such as the list of populations around the Maeotis, or in southern Asia,[42] but there can be sudden digressions, which, to the modern reader, are some of the most fascinating material in the work. Examples include the expeditions of Hanno, Scipio Aemilianus, and Polybius in West Africa, or the processing of silk and the characteristics of the Silk People.[43]

The *Natural History* was a product of the era of the emperor Vespasian, who was mentioned a dozen times in the geographical books. As a whole, the work was dedicated to Vespasian's son, the future emperor Titus, mentioned five times. Nevertheless the text has some striking anachronisms, inevitable but often presented without explanation or clarification. For example, in Pliny's discussion of the Roman province of Africa, the context is the brief period in the 40s BC when the region was divided into two provinces.[44] This was the situation a century before Pliny wrote, yet he did not inform the reader of this, giving the impression that it was the current organization of Africa, despite the fact that he himself had served in the unified province.

Measurements are always an issue in consideration of geographical texts. For long distances Pliny relied on Roman miles, but also used the Greek stadion, Egyptian schoenus, and Persian parasang, as well as travel days. All of these except the Roman mile (1.48 km.) are variable, as are their formulas for conversion. Numerals may appear in the text without any units of measurement, or the wrong type of unit may be presented. As a convenience modern distances in kilometers are provided in the commentary, but none is a certain equivalent except those in Roman miles, yet

[40] *NH Preface* 28. [41] Murphy, *Pliny* 30–6. [42] *NH* 6.21, 48–51. [43] *NH* 5.8–9, 6.54–5, 88.
[44] *NH* 5.25.

even with these the measurement may have been converted from other units, not always correctly.

In creating a commentary on Pliny's geographical books, it has been necessary to take into account all these issues. Pliny's text can be confusing, even clumsy, and while producing a proper translation is not the matter at hand, a commentator must use the text in such a way as to make a meaningful analysis. There is always the problem of how much to say; one could write an article about each of the 893 sections or nearly 6,000 toponyms in the geographical narrative. Nevertheless the purpose of this commentary is to acquaint the reader with the overall pattern of the geographical books, which stand alone in Latin literature as the longest and most significant geographical treatise in that language and one of the few works of that discipline surviving from classical antiquity.

BOOK I

Pliny's List of Sources

Introduction

One of the features of the *Natural History* is the catalogue of topics and sources that comprises Book 1, a rare component of an ancient text. It provides the subjects of each book, followed by a list of the number of items within it. For Book 2 this is limited to "facts, investigations, and observations," but for books 3–6 there is the number of "towns and peoples, famous rivers, famous mountains, islands, towns and peoples that have perished," and then the summary, "facts, investigations, and observations." The actual numbers for Books 3 and 4 are missing. Only the number of islands (118) survives for Book 5, and Book 6 has the complete sequence of numbers.

The catalogues of topics speak for themselves and need not be discussed in the commentary. In the preface to the *Natural History*, which is in the form of a letter to Vespasian's son Titus, the future emperor, Pliny drew attention to the table of contents, recognizing its unusual quality.[1]

Following each catalogue of topics and the summary of items are lists of the sources that were used in that book, introduced by the phrase "ex auctoribus" ("from [these] authors"). First are Romans, and then "exter-nis" ("foreigners"). This second list casts a wide net, including anyone whom Pliny did not consider Roman; in fact some of the authors, such as Juba II of Mauretania, were of foreign origin but were Roman culturally, yet wrote in Greek. To some extent the texts reflect the order of use in the relevant book, but not inevitably, and, moreover, many are listed that are not cited by name in the subsequent text, although it can be possible to speculate when Pliny used a particular unnamed author. In addition, a number of sources appear in the text that are not listed in Book 2.

[1] *NH Preface* 33.

Nevertheless the index is a valuable tool for comprehending the nature of the *Natural History*.

In the catalogue for Books 2–6, 137 sources are listed (44 domestic and 93 foreign, although there is some duplication because of the use of variant names), covering a significant number of Greek and Latin authors up to Pliny's day. Many familiar ones appear, such as Livy, Pomponius Mela, and Seneca among the Latin writers, and Aristotle, Herodotus, and Thucydides among the "foreigners." Some expected sources are not mentioned, notably Homer (although cited in the text) and Strabo of Amaseia (almost certainly unknown to Pliny), but the lists are remarkably thorough, and extend into Pliny's own day with authors such as Pomponius Mela and the emperors Claudius and Titus. A number of those cited in the text but not in Book 1 may not have been consulted directly. In another peculiarity, Cicero is apparent as a source through much of Book 2 but is never named.

In one sense, Pliny's lists are derived from the library catalogues that had existed since Bronze Age Mesopotamia.[2] More relevant to Pliny, however, were those created for the great Hellenistic libraries. Callimachus of Cyrene, who worked in the Alexandria library in the third century BC, created his *Pinakes*, effectively a library catalogue and thus a list of books and their authors.[3] This work was certainly known to Pliny, but his own arrangement of authors follows no discernable pattern, and may reflect his actual use of the author in question, something difficult to determine because so many of them are not actually cited in the text. There may also have been chronological schemes, and vestiges of some appear, such as with the Roman authors of Book 4. Pliny cited at least three professional librarians among his sources (Callimachus, Crates, and Hyginus), and thus was probably well aware of the cataloguing process, and his own vast library probably had its own librarian and catalogue. Pliny also credited Valerius Soranus with being the first Roman to create such a list, as part of his *Epoptides* ("Initiates"), written in the early first century BC.[4] Like all scholars, Pliny made notes and extracts from his reading, and left 160 notebooks upon his death, which his nephew Pliny the Younger inherited.[5]

In the catalogue that follows, names are listed in the orthography of Pliny's Latin, although he was not immune to using Greek or foreign

[2] Lionel Casson, *Libraries in the Ancient World* (New Haven 2001) 4–9.

[3] Blum, *Kallimachos* 226–43.

[4] If this is the same person as the one mentioned at *NH* 3.65, his indiscretion about religious matters led to his downfall (Plutarch, *Roman Questions* 61).

[5] Pliny the Younger, *Letters* 3.5.17.

spellings on occasion. Names in **bold** type are the form used by Pliny; if the author's name is substantially different from that cited, it follows in parentheses.

Catalogue of Sources Listed for Books 2–6

Roman authors

Acta Triumphorum (Book 5), more properly the *Fasti Triumphales*, recorded the Roman triumphs from earliest times until the Augustan period. A copy, of which fragments survive, was placed on the Arch of Augustus in the Roman Forum. The reliability of the list has repeatedly been called into question, especially the earlier citations, which are believed to be a construct of early Roman history, but nonetheless they were a visible public document (R. M. Ogilvie and A. Drummond, "The Sources for Early Roman History," *CAH* 7.2 [1989] 17–21; Ettore Pais, *Fasti Triumphales Populi Romani* [Rome 1920]).

M. Agrippa (Marcus Vipsanius Agrippa, Books 3, 4, 5, 6), the confidant of Augustus, was cited regularly by Pliny. His geographical work was the basis of a map of the world that was displayed in the Porticus Vipsania in Rome, whose location is unknown (Richardson 319–20). The *Natural History* is the primary source for Agrippa's geographical scholarship (Klotz, "Geographischen Commentarii," 38–58, 386–466).

Antias (Valerius Antias, Books 2, 3) wrote a history of Rome from its origins to the early first century BC, which was produced during the period 80–60 BC (*FRH* #25). Over 70 fragments survive.

Arruntius (L. Arruntius, Books 3, 5, 6) is presumably the historian of the First Punic War (*FRH* #58), who was probably the consul of 22 BC, although his son, the consul of AD 6, is also a possibility. Whether father or son, the topic of his history is only known through a discussion and a few quotations by Seneca (*Letters* 114.17–19); Pliny did not cite him by name in the relevant books.

L. Ateius, Ateius Capito, Ateius Philologus (Books 3, 4). Pliny may have cited as many as three people with the name "Ateius," who are difficult to distinguish. L. Ateius was listed for Book 3; this may be the "Ateius Philologus" of Book 4, whose *praenomen* was Lucius (*FRH* #51). There was also an Ateius Capito (Book 4), who may be C. Ateius Capito, tribune in 55 BC (Broughton 2, 216), or perhaps the homonymous consul of AD 5. Any distinction between these personalities, and what works were used by Pliny, remains uncertain: the elder Ateius Capito is not known beyond his political career and the younger was a lawyer (*PIR* A1062). Ateius Philologus is more distinct: he was a freedman from Athens and an associate of Sallust and Asinius Pollio, who wrote a summary Roman history.

Divine Augustus (Books 3, 4), the Roman emperor (*FRH* #60), wrote a number of works that could have been used by Pliny, including an autobiography, the extant *Res gestae*, and various letters, edicts, and decrees.

Caecina (Aulus Caecina, Book 2) was an Etruscan scholar (his family was of that origin) and a close friend of Cicero's (Cicero, *Letters to His Friends* 6.7–9).

Caelius Antipater (Coelius Antipater, Book 2, and almost certainly the "Caelius" of Book 3) was the first Latin historian to limit his work to a specific period, in this case the Second Punic War (*FRH* #15). He lived near the end of the second century BC. Nearly 70 fragments of his work exist, cited by a variety of historians.

Cato the Censor (M. Porcius Cato, Books 3, 4), who lived from 239 to 149 or 148 BC, was the first to write a history in Latin prose. He was also a major political activist. His *Origines*, of which over 150 fragments survive, was a selective study of Roman history from its beginning, and was widely used by later writers.

Claudius Caesar, the emperor (Books 5, 6), lived from 10 BC to AD 54 (*FRH* #75). He was a polymath who wrote widely on history and antiquarianism (Duane W. Roller, *Scholarly Kings: The Writings of Juba II of Mauretania, Archelaos of Kappadokia, Herod the Great and the Emperor Claudius* [Chicago 2004] 185–95). Of particular interest to Pliny would have been his Roman history, which covered the Augustan and early Julio-Claudian periods. His Greek-language works on the Etruscans and Carthaginians might also have been relevant to the composition of the *Natural History*. Yet little of Claudius' work survives (eight fragments at most), with the *Natural History* the primary source.

Cornelius Nepos (Books 2, 3, 4, 5, 6), who lived throughout much of the first century BC (*FRH* #45), was used regularly by Pliny, cited in each of the geographical books. He was a pioneer in the genre of Latin biography, and some of his entries survive. He also wrote widely on other topics, including a universal history (the first in Latin), famous for its citation by Catullus at the beginning of his first poem, as well as a geography, which is hardly known (*NH* 2.169–70, 3.127).

Curio the Elder (Book 3) is not easily identified, since no person of that name appears in the appropriate book. The most likely is C. Scribonius Curio, the consul of 76 BC, who had an active military and political career.

He was proconsul of Macedonia in 75–72 BC (Broughton 2, 92–3, 99) and campaigned in Thrace as far as the Danube. Pliny may have used his report of this expedition (*NH* 3.147–9).

Domitius Corbulo (Cn. Domitius Corbulo, Books 5, 6) was an important military leader during the period of Claudius and Nero, with commands in Asia Minor and Syria (*FRH* #82); he was also Pliny's commanding officer in Germania. He published an account of at least the eastern portion of his career, probably shortly before his death in AD 67; as such, he was one of Pliny's most recent sources (*NH* 6.23). Yet only three fragments survive, two of which are from the *Natural History* (NH 2.180, 5.83).

Fabianus (Papirius Fabianus, Book 2) was a noted natural philosopher of the era of Augustus and Tiberius (Arnaud Zucker, "Papirius Fabianus," *EANS* 610–11). He wrote extensively on scientific topics, and was used repeatedly by Pliny, who valued his research highly (*NH* 36.125). His writings on meteorology and oceanic phenomena were of relevance to Pliny's geographical material (*NH* 2.121, 224).

Fabricius Tuscus (Books 3, 4, 6) cannot be identified. He was not cited by name in any of the three books in which his material appears. The *cognomen* Tuscus suggests an Etruscan origin. One possibility is C. Fabricius Tuscus, equestrian prefect in Germania during the last years of Augustus (Ségolène Demougin, *Prosopographie des chevaliers romains Julio-Claudiens* [Paris 1992], no. 216).

Gellianus (Book 3) is generally believed to be Gnaeus Gellius (cited in the index to Book 7), a historian of the second century BC (*FRH* #14; Pliny, *Histoire naturelle* 3 [ed. Hubert Zehnacker, Paris 2004] 223–4). He wrote a lengthy (at least 97 books) history of Rome from its origins to his own era, with nearly 40 fragments surviving. Pliny cited Gellianus for the early history of central Italy (*NH* 3.108), and Gellius for various matters connected with the origins of civilization (*NH* 7.192–8).

Hyginus (C. Julius Hyginus, Books 3, 4, 5, 6) was the librarian of the emperor Augustus (*FRH* #63). He was a prolific author, with works on the ethnology of Italy, the history of religion, and agriculture, as well as literary commentaries. Of particular interest to Pliny would have been his geographical treatise (Christophe Cusset and Philip Thibodeau, "C. Iulius Hyginus," *EANS* 454–5).

Julius Aquila (Book 2) cannot be certainly identified. He wrote on Etruscan matters, which do not seem to be predictable expertise of known people with this name. These include the prefect of Egypt in AD 10–11 (Demougin, *Prosopographie*, no. 120), and his possible relative, who led an expedition into the Cimmerian Bosporus in AD 49 (Demougin, *Prosopographie*, no. 533). Pliny's source may have been a member of this family.

Licinius Mucianus (C. Licinius Mucianus, Books 2, 3, 4, 5, 6), was consul in AD 64, 70, and 72 and a close associate of the emperor Vespasian. After a lengthy military career he turned to scholarship and wrote, among other things, on geographical marvels. He was cited 32 times by Pliny, especially on the topic of unusual springs (Paul T. Keyser, "Licinius Mucianus," *EANS* 507–8; Thomas Francis Caldwell, "The Career of Licinius Mucianus," [M. A. thesis, University of Melbourne, 2015] 27–49).

T. Livius (Livy, Books 2, 3), was the well-known historian of the Augustan period, cited occasionally in the *Natural History*.

T. Livius the Son (Books 5, 6) cannot be identified, but was presumably a relative of the historian.

Nigidius (P. Nigidius Figulus, Book 6) was an important politician and scholar of the first half of the first century BC: Cicero (*Timaeus* 1) spoke highly of his diligence (Bruno Centrone, "P. Nigidius Figulus," *EANS* 572–3).

L. Piso (Books 2, 3) was probably the consul of 133 BC, L. Calpurnius Piso Frugi (*FRH* #9). He wrote a history of Rome, which survives in nearly 50 fragments. To be sure, L. Calpurnius Piso is one of the most common names of the late Republic and into the early Empire, and it is not certain that Pliny's two geographical citations of "L. Piso" (*NH* 2.140, 3.131) refer to the same person.

Pomponius Mela (Books 3, 4, 5, 6) is the only extant Latin geographer previous to Pliny. His three-book *Chorographia*, composed during the reign of Claudius, is an important testimonium to the study of ancient geography, and was used regularly by Pliny.

Sebosus (Sebosus Statius, Books 2, 3, 5, 6) wrote a little-known geographical treatise whose material extended from the Atlantic to India (*NH* 6.183, 201–2, 9.46). He probably lived in the Julio-Claudian period (Paul T. Keyser, "Sebosus Statius," *EANS* 729).

Seneca (L. Annaeus Seneca the Younger, Book 6), the polymath and advisor to the emperor Nero, included among his interests a study of rivers (Daryn Lehoux, "L. Annaeus Seneca," *EANS* 84–5). His *Natural Questions* were a probable source for the cosmological material of Book 2, but he was not cited by name in that book.

Sergius Plautus (Book 2) may be the little-known Stoic author cited by Quintilian (*Institutes* 10.1.124, etc.; *PIR* S378), but his identity is uncertain.

Suetonius Paulinus (C. Suetonius Paulinus, Book 5) was the first Roman officer to cross the Atlas mountains and to penetrate into tropical West Africa, during a propraetorian command in AD 42 (*PIR* S694). He published a report of his expedition (*NH* 5.14–15), some of the earliest material in Latin about the strange world of the tropics. He was later governor of Britannia during the revolt of Boudica.

Sulpicius Gallus (C. Sulpicius Gallus or Galus, Book 2) was a little-known writer on astronomy from the early second century BC (*NH* 2.53; Bruno Centrone, "C. Sulpicius Gallus" *EANS* 768).

Tarquitius (Book 2) was an Etruscan scholar, probably the same one mentioned by Macrobius (*Saturnalia* 3.7.2), who translated an Etruscan book of portents.

Imperator Titus Caesar (Book 2), who became emperor in June of AD 79, two months before Pliny's death, wrote a poem in AD 76 about a comet which appeared in that year (*NH* 2.89).

Q. Tubero (Q. Aelius Tubero, Book 2) was primarily a legal scholar. He fought alongside Cn. Pompeius Magnus (Pompey the Great) at Pharsalus and then devoted his life to law and history (*FRH* #38). His history was 14 or more books long, from the earliest times to at least the second century BC.

Tullius Tiro (M. Tullius Tiro, Book 2), the freedman and confidant of Cicero (*FRH* #46), was most noted for a biography of his patron and editing some of his speeches and letters. Exactly what use Pliny made of his writings is unknown.

Turranius Gracilis (Book 3), prefect of Egypt in the Augustan period, wrote on geography and agriculture. Pliny's several citations of his work (*NH* 3.3, 9.11, 18.75) all relate to Iberia, his home region (Paul T. Keyser, "Turranius Gracilis," *EANS* 820).

Valerianus (Book 3), cited for the Roman destruction of the city of the Vidicinians (in Picenum), may be the Cornelius Valerianus cited in Books 8, 10, 14, 15, or the name is possibly an error for Valerius Antias (*FRH* #25, F70; *supra*, p. 17).

M. Varro (M. Terentius Varro, Books 2, 3, 4, 5, 6), the prolific polymath of the first century BC, was quoted extensively by Pliny. His works on the Latin language and agriculture survive; of particular interest to Pliny would have been his extensive geographical and meteorological writings (Philip Thibodeau, "M. Terentius Varro of Reate," *EANS* 774–8).

Varro Atacinus (P. Terentius Varro, Books 3, 4, 5, 6), of the first century BC and known as "Atacinus" from the Atax River at Narbo, his home, wrote a *Chorographia* in three books, which was didactic in tone, and an

Ephemeris, on weather forecasting, which influenced Vergil in his *Georgics* (Georgia L. Irby-Massie, "P. Terentius Varro of Narbo, 'Atax'," *EANS* 774).

Verrius Flaccus (M. Verrius Flaccus, Book 3) was an important grammarian and linguist during the second half of the first century BC and into the following century. He was the teacher of the grandsons of Augustus, C. and L. Caesar (Suetonius, *Grammarians* 17; Karl-Ludwig Elvers, "Verrius," *BNP* 15 [2010] 323–5).

L. Vetus (Books 3, 4, 5, 6) is probably L. Antistius Vetus, consul in AD 55 and proconsul of Asia in AD 64–5 (*PIR* A612). He also served in Germania. Nothing is known about his writings.

"Foreign" authors

Agathocles (Agathokles, Books 4, 5, 6) cannot easily be identified. In addition to the tyrant of Syracuse in the fourth century BC, there was an Agathocles from Atrax, active probably in the third century BC, who wrote *On Fishes* (Paul T. Keyser and Georgia L. Irby-Massie, "Agathokles," *EANS* 43–4), and one from Chios, who was the author of a treatise on agriculture (Philip Thibodeau, "Agathokles of Khios" *EANS* 44). An Agathocles from Kyzikos, of uncertain Hellenistic date, wrote about his native city (*FGrHist* #472). All these have been suggested to be Pliny's source, and since no Agathocles is mentioned by name in the text of Books 4–6, it is difficult to determine Pliny's use of him.

Aglaosthenes (or Aglosthenes, Book 4), from Naxos, lived in the fourth or third century BC and was the author of a poem called the *Naxila*, about the island and its relationship with Zeus (*FGrHist* #499).

Alexander the Great (Book 6) was mentioned regularly by Pliny, but it is difficult to determine how he might have been used as a literary source. It is possible that Pliny was merely citing quotations from the king that appeared in the accounts of his career. Yet a reference to "Itineraries of Alexander the Great" (*NH* 6.45) may refer to published documents, and, moreover, papers were found at the time of his death which later became available to scholars (E. Badian, "A King's Notebooks," *HSCP* 72 [1968] 183–204).

Alexander Polyhistor (Books 3, 4, 5, 6), from Miletus, obtained his surname because of his polymathic abilities (*FGrHist* #273). He was brought to Rome in the 80s BC and became a prolific writer, dying after 40 BC. Nearly 150 fragments of his works survive, heavily tilted toward geography and ethnography. Pliny used him regularly.

Amometus (Book 6), probably from Cyrene, where the name is common, lived in the third century BC and wrote on geographical paradoxes and fantasies (Paul T. Keyser, "Amometos (of Kurene?)," *EANS* 67).

Anaximander (Books 2, 4), may be the scholar from Miletus, who lived in the early sixth century BC and was the second of the Ionian monists. He drew the first map of the world, and as a natural scientist was interested in geographical speculation (Roller, *Ancient Geography* 27–9). There was also a later Anaximander, also from Miletus (*FGrHist* #9), of the fourth century BC, who is obscure but may be Pliny's source for the entry at *NH* 4.58.

Anticlides (Antikleides, Book 4), an Athenian, lived around 300 BC, and wrote a history of Alexander the Great and on various antiquarian topics, including a study of the mythological history of Delos (*FGrHist* #140).

Antigenes (Book 5) cannot be accurately identified, but may be the officer of Alexander the Great who was also prominent in the early era of the Successors (Waldemar Heckel, *Who's Who in the Age of Alexander the Great* [Malden 2006] 30–1).

Apollodorus (Books 4, 6), is probably the scholar from Artemita, active in the early first century BC. His *Parthika* may have been one of Pliny's sources for Central Asia (Georgia L. Irby-Massie, "Apollodoros of Artemita," *EANS* 106–7). Another possibility is the geographical writer from Athens, also of the first century BC (*FGrHist* #244); the sheer number of literary personalities with this name makes certainty impossible.

Archimedes (Book 2) is the scientist and engineer of the third century BC from Syracuse. He was a major personality in the history of mathematics (Reviel Netz, "Archimedes of Surakousai," *EANS* 125–8).

Aristarchus of Sicyon (Book 5) is little known, but at some time after 60 BC he wrote a description of Greece that extended north to Dacia (Paul T. Keyser, "Aristarkhos of Sikuon," *EANS* 133).

Aristides (Aristeides, Book 4), perhaps from Cnidus and active in the fourth century BC, wrote on the Aegean islands (*NH* 4.64, 70; Paul T. Keyser, "Aristeides (of Knidos?)," *EANS* 135).

Aristocreon (Aristokreon, Books 5, 6), probably from the early third century BC, wrote an *Aithiopika* (*NH* 5.59, 6.183). He may have come from Cyprus or Cos (Georgia L. Irby-Massie, "Aristokreon," *EANS* 138).

Aristocritus (Aristokritos, Books 4, 5) is obscure (*FGrHist* #493). He may have been from Miletus, which he wrote about. He also provided Pliny with information about Samos (*NH* 5.135). Since he was quoted by Parthenios, the poet of the early first century BC, he must be contemporary to or predate that time, but nothing more is known.

Aristotle (Books 2, 5), the polymath of the fourth century BC from Stageira in Macedonia, was cited occasionally by Pliny.

Artemidorus (Books 2, 3, 4, 5, 6), from Ephesus and active at the end of the second century BC, wrote a geography in 11 books, probably the longest work on its topic up to that time. The work was used regularly by Pliny. He cited "Artemidorus the Ephesian" in the catalogue to Book 2, and just "Artemidorus" for the remaining books, but it is probable that all the references are to the scholar from Ephesus (R. Stiehle, "Der Geograph Artemidoros von Ephesos," *Philologus* 11 [1856] 193–244).

Astynomus (Books 4, 5) wrote on islands, probably before 100 BC (Paul T. Keyser, "Astunomos," *EANS* 175).

Baeton (Baiton, Books 5, 6) was one of those keeping the record of the itinerary of Alexander the Great, perhaps in company with Diognetus (*NH* 6.61), and published it with additional ethnographic and scientific data. He may have returned to India for further research after the expedition (*NH* 6.69). His name suggests that he might have been Egyptian (Paul T. Keyser, "Baiton," *EANS* 186).

Basilis (Book 6) wrote an *Indika*, and perhaps other works on the eastern and southern portions of the known world. He probably lived in the third or second century BC, and is otherwise unknown (Georgia L. Irby-Massie, "Basilis," *EANS* 190–1).

Bion (Book 6), from Soloi, wrote a lengthy work on Aethiopia, as well as on agricultural topics. He probably lived in the early third century BC (Philip Thibodeau, "Bion of Soloi," *EANS* 193)

Callicrates (Kallikrates, Books 3, 5) cannot be identified. Whether he was the famous Athenian architect of the fifth century BC remains speculative.

Callidemus (Kallidemos, Book 4), wrote about Euboea (*NH* 4.64), but is only known through Pliny's citation and the parallel one by Solinus (11.15).

Callimachus (Kallimachos, Books 4, 5, 6), the polymath from Cyrene, was active in the third century BC. He was said to have written over 800 works; those most relevant to Pliny were his scientific and geographical treatises (Jan Bollansée *et al.*, "Kallimakhos of Kurene," *EANS* 463).

Calliphanes (Kalliphanes, Books 3, 5) is unknown except for Pliny's citations in the catalogue and a single reference to the Machlyes, a hermaphroditic people in sub-Saharan Africa (*NH* 7.15). He was probably of Hellenistic date.

Cleobulus (Kleoboulos, Books 4–5) was a little-known geographer probably of Late Hellenistic or Early Imperial date.

Clitarchus (Kleitarchos, Book 6), from Alexandria, wrote a history of Alexander's eastern journey at the end of the fourth century BC (although he does not seem to have been a member of the expedition). His material was widely used by later Alexander historians (*FGrHist* #137; Heckel, *Who's Who* 86).

Coeranus the Philosopher (Koiranos, Book 2) was a scholar of the era of Claudius and Nero, who seems to have written on cosmology (Paul T. Keyser, "Koiranos," *EANS* 485).

Crates the Grammarian (Krates, Book 4), from Mallos in Cilicia, who lived in the second century BC, was primarily a grammarian, but his Homeric studies led him to geographical issues and the construction of the first world globe, which was on display at Pergamum, where he was librarian and ambassador to Rome (Daniela Dueck, "Krates of Mallos," *EANS* 490).

Critodemus (Kritodemos, Book 2) was a Greco-Egyptian writer on astrology, active perhaps in the third century BC. He was said to have been a student of Berossos, the Babylonian historian, and thus to have had access to Babylonian material (*NH* 7.193; Francesca Rochberg, "Kritodemos," *EANS* 493).

Ctesias (Ktesias, Book 2) was from Cnidus and active around 400 BC. He was physician to the Persian king Artaxerxes II, and wrote a history of Persia and an account of India. His writings tend toward the fantastic and marvellous, and his Persian history had many suspect accounts (*FGrHist* #688; Philip Kaplan, "Ktesias of Knidos," *EANS* 496).

Dalion (Book 6) was the author of an *Aithiopika*, written during the early Ptolemaic period. He was probably the first Greek to go up the Nile far beyond Meroë (*NH* 6.183; *FGrHist* #666; Georgia L. Irby-Massie, "Dalion (Geog.)," *EANS* 222).

Damastes (Books 4, 5, 6), from Sigeion, wrote in the later part of the fifth century BC, and was one of the early Greek ethnographers (*FGrHist* #5).

Democritus (Demokritos, Book 2) was presumably the natural philosopher from Abdera, of the late fifth and early fourth centuries BC, noted for his development of the atomic theory. There was, however, another Democritus who wrote on lapidary topics (*NH* 1.37). Whichever person was meant, he was cited by Pliny at *NH* 2.14 because of his views on the gods.

Demodamas (Book 6), from Miletus, was a military leader in the early Seleucid period, who went far into Central Asia, crossing the Iaxartes (Paul T. Keyser, "Demodamas of Miletos," *EANS* 233).

Dicaearchus (Dikaiarchos, Books 2, 4, 5, 6) was from Messana and one of Aristotle's students. He was noted for his innovative geographical research, measuring the heights of mountains and creating a baseline of latitude across the inhabited world (Keyser, "Geographical Work" 353–72).

Diodorus of Syracuse (Books 3, 5) cannot be identified with certainty, since he is only mentioned in the index and not in the text. He is probably the historian of the first century BC, known popularly today as Diodorus of Sicily, whose lengthy history, the *Bibliotheke*, is partially extant. But this Diodorus was actually from Agyrion, well to the northwest of Syracuse.

Diognetus (Diognetos, Book 6), perhaps from one of the towns named Erythrai, recorded the route of Alexander the Great, perhaps in cooperation with Baeton (*NH* 6.61). He is otherwise unknown (*FGrHist* #120).

Dionysius (Books 4, 5) cannot be identified, as there are many scholars with the name. The most likely is the familiar historian and ethnographer from Halicarnassus, known for his extant *Roman Antiquities* and other works, active in the first century BC. A reference by Pliny to a Dionysius from Charax (at the head of the Persian Gulf, *NH* 6.141) is certainly an error for Isidorus.

Dosiades (Book 4) was a historian of Crete, active in the early third century BC (*FGrHist* #458).

Ephorus (Books 4, 5, 6), from Cyme in Aeolis, was an important historian of the fourth century BC, who also made significant advances in geographical theory (*FGrHist* #70). His history was in 30 books, of which over 200 fragments survive.

Epigenes (Book 2) is probably either the scholar from Byzantium of Late Hellenistic or Augustan date, who wrote on astrology and had access to Babylonian sources (*NH* 7.160; Francesca Rochberg, "Epigenes of Buzantion," *EANS* 290), or the agrarian writer from Rhodes of uncertain Hellenistic date (Philip Thibodeau, "Epigenes of Rhodes," *EANS* 291).

Eratosthenes (Books 2, 4, 5, 6), the polymath from Cyrene, spent his academic career in Alexandria. He was the inventor of the discipline of geography and also calculated the circumference of the earth. He was active in the second half of the third century BC (Roller, *Eratosthenes*).

Euclides (Euclid, Book 2), from Alexandria, was the famous mathematician of the early Ptolemaic period, noted for his development of the discipline of geometry.

Eudoxus (Books 2, 4, 5, 6), from Cnidus, lived in the early fourth century BC. He was a mathematician and astronomer, but also had a particular interest in geography. His *Circuit of the Earth* was the longest geographical work to date, and was notable for its consideration of the dimensions and size of the inhabited world (Roller, *Ancient Geography* 75–6). The citation at *NH* 6.198 may actually refer to the adventurer from Cyzicus, of the late second century BC.

Hanno (Book 5), a Carthaginian aristocrat, made a voyage around 500 BC along the West African coast as far as the tropics. A summary of his report was translated into Greek at an early date (Duane W. Roller, *Through the Pillars of Herakles* [New York 2006] 29–43).

Hecataeus (Hekataios, Books 4, 5, 6), from Miletus and active around 500 BC, wrote a *Circuit of the Earth*, the first work of its genre, based in part on his extensive travels. Nearly 400 fragments survive, and it was used regularly by later writers (*FGrHist* #1). But the citation at *NH* 6.55 is certainly Hecataeus of Abdera, of the late fourth century BC, who wrote on fantasy geography (*FGrHist* #264).

Hellanicus (Hellanikos, Books 4, 5, 6), from Lesbos, was a mythographer and ethnographer of the fifth century BC. He was also interested in chronology (*FGrHist* #4).

Heraclides (Herakleides, Book 4), cited by Pliny for a detail about the Aegean island of Melos (*NH* 4.70), is probably the geographer of the third century BC known as "Kritikos," but this is not certain (Daniela Dueck, "Herakleides 'Kritikos'," *EANS* 366).

Herodotus (Books 2, 5), is the familiar logographic historian of the fifth century BC, from Halicarnassus.

Himilco (Himilkon, Book 5), a Carthaginian, travelled around 500 BC into the Atlantic, from Iberia north to the British Isles. Details of his voyage are scantily preserved, although a written report seems to have survived into the Roman period (Roller, *Through the Pillars* 27–9).

Hipparchus (Books 2, 5, 6), from Nicaea, was a mathematician of the latter second century BC. He also had astrological interests (*NH* 2.95), and wrote a critique of Eratosthenes' *Geography* from a mathematical point of view, which survives in about 60 fragments and was cited occasionally by Pliny (D. R. Dicks, *The Geographical Fragments of Hipparchus* [London 1960]).

Isidorus the Characene (Books 2, 3, 4, 5, 6, cited only as "Isidorus" in Books 3–6) was from the town of Charax, at the head of the Persian Gulf (*FGrHist* #781). His *Parthian Stations*, an itinerary from Mesopotamia to India, is extant. He was commissioned by the emperor Augustus to write on the geography of the East as advance preparation for Gaius Caesar's extensive travels of 2 BC–AD 4. The "Dionysius" from Charax at *NH* 6.141 is an error for Isidorus.

King Juba (Books 5, 6), the scholarly king of Mauretania during the period of Augustus and Tiberius, wrote extensively on Roman history, the ethnography of his kingdom, Africa, and the route east to India (*FGrHist* #275; Roller, *World of Juba*).

Megasthenes (Books 5, 6), active during the generation after Alexander the Great, spent time at the court of Chandragupta, the Mauryan king, at Pataliputra in India, and wrote an *Indika*, an important ethnography of the region (*FGrHist* #715).

Menaechmus (Menaichmos, Book 4), from Sikyon, was an Alexander historian who also wrote about his native region, as well as on artistic topics. He was active in the late fourth century BC (*FGrHist* #131).

Metrodorus (Books 3, 4, 5, 6), from Scepsis in the Troad, was an intimate of Mithridates VI of Pontus, and thus active in the early first century BC. He wrote a variety of works, including a history and a geography, the latter perhaps to assist Mithridates in his grandiose plans. He was also famous for devising a unique mnemonic system (*FGrHist* #184).

Myrsilus (Books 4, 5), from Methymna and active in the first half of the third century BC, wrote a *Lesbiaka* and a treatise on historical and natural paradoxes (*FGrHist* #477).

Nearchus (Book 6), from Crete, was a member of Alexander's expedition and was the commander of the fleet down the Indus as well as the one that returned from its mouth to the Persian Gulf. He wrote an account of the voyages (*FGrHist* #133).

Nechepsus (Nechepso, Book 2) was the pseudonym of the author of a group of Greek texts on astrology emanating from Hellenistic Egypt and probably collated in the second century BC. The name is associated with another Egyptian pseudonym, Petosiris, also cited by Pliny (Ernest Riess, *Nechepsonis et Petosiridis Fragmenta Magica* [Bonn 1890]; Francesca Rochberg, "Petosiris, or Nekhepso-Petosiris," *EANS* 637–8).

Nymphodorus (Books 3, 5), from Syracuse and living perhaps around 200 BC, wrote on Sicily, and also a geographical work covering at least Asia. Both seem to have emphasized the marvelous (*FGrHist* #572).

Onesicritus (Onesikritos, Books 2, 6), from Astypalaea, was a member of Alexander's expedition. He was the king's liaison with the Brahman community at Taxila, and then the chief pilot of the sea voyage from India to the Persian Gulf. He wrote on the philosophy and culture of India, and also an account of the sea voyage (*FGrHist* #134).

Panaetius (Panaitios, Book 5, 6), from Lindos on Rhodes, may have been the scholar who was the head of the Athenian Stoa in the late second century BC, but since he is not cited in the text of the *Natural History*, this is uncertain (Daryn Lehoux, "Panaitios of Rhodes (Lindos)," *EANS* 608).

Patrocles (Patrokles, Book 6), a military leader and explorer of the early Seleucid era, was the first Greek to make a reconnaissance of the Caspian Sea. His report is not extant (*FGrHist* #712).

Petosiris (Book 2), along with Nechepsus, is one of the names given to the author of works on astrology that emanated from Hellenistic Egypt (Rochberg, "Petosiris," *EANS* 637–8).

Philemon (Book 4), dating to the early first century AD, wrote a geographical treatise on northern Europe, perhaps with a focus on the amber trade (Paul T. Keyser, "Philemon," *EANS* 645).

Philistides of Mallos (Book 4), known only from two citations in the *Natural History* (4.58, 120), wrote on islands, perhaps in the early first century BC.

Philonides (Books 4, 5), from Chersonesus on Crete, was a professional runner (*hemerodromos*), and was one of those recording the itinerary of Alexander the Great (*FGrHist* #121).

Polybius (Books 4, 5, 6), from Megalopolis in the Peloponnesus, was the political leader, explorer, and polymath of the second century BC. He included extensive geographical data in his history, and led an expedition to tropical West Africa. His geographical works exist only in fragments.

Posidonius (Poseidonios, Books 2, 4, 5, 6), from Apamea in Syria, was a polymath of the late second and early first centuries BC. He was a significant contributor to geography, meteorology, and cosmology. His works are all lost, but survive in nearly 300 fragments (L. Edelstein and I. G. Kidd, *Posidonius 1: The Fragments* [second edition, Cambridge 1989]).

Pyrrander (Book 5) cannot be identified, but may be the Hellenistic author (Pyrandros) of a *Peloponnesiaka* who also wrote on myth (*FGrHist* #504).

The Pythagoreans (Book 2) were the followers of Pythagoras of Croton, and were active from the fifth century BC into the Early Hellenistic period. In addition, there was a later body of material attributed to them. Pliny did not mention any of them by name except for Pythagoras himself (*NH* 2.83–4, 191).

Pytheas (Books 2, 4) was a Greek explorer from Massilia (Greek Massalia), dating to the late fourth century BC and noted for being the first Greek to explore the British Isles and to penetrate the Arctic, reporting on the phenomena of high latitudes. His treatise, *On the Ocean*, is lost but survives in a number of fragments (Christina Horst Roseman, *Pytheas of Massalia, On the Ocean* [Chicago 1994]).

Serapion (Books 2, 4, 5), from the early first century BC, was from Antioch in Syria, and wrote on geography. Pliny distinguished between Serapion the *gnomicus* (an expert on sundials) and Serapion of Antioch, but these are probably the same person (Paul T. Keyser, "Serapion of Antioch," *EANS* 733).

Simonides the Younger (Book 6) was a Ptolemaic envoy who spent five years in Meroë and wrote an *Aithiopika*, cited only by Pliny (*NH* 6.183; *FGrHist* #669).

Sosigenes (Book 2) is probably the astronomer from Alexandria who assisted Julius Caesar with his reform of the calendar (*NH* 18.210–12; Alan C. Bowen, "Sosigenes (I)," *EANS* 752).

Sotades (Book 5) cannot be identified. He may be the Athenian author on mystery cults, of uncertain date, but this is speculative (*FGrHist* #358).

Staphylus (Book 4, 5) was a historian of uncertain Hellenistic date (perhaps from before 150 BC), who wrote a number of ethnologies (*FGrHist* #269).

Theophrastus (Book 3), of Eresos, was the student and successor of Aristotle. Remembered today largely for his botanical works, he was a prolific polymath.

Theopompus (Books 2, 3) from Chios, was a major historian of the fourth century BC, who wrote a general history of the Greek world as well as a treatise on the environment of Philip II of Macedonia. Over 400 fragments of his works survive (*FGrHist* #115).

Thrasyllus (Book 2) is probably the writer on rivers, from Mendes, perhaps dating to the Julio-Claudian period (*FGrHist* #622), but this is not certain. Whether he is identical with the mathematician who became the astrologer of Tiberius cannot be proven (Harold Tarrant, "Thrasullus, Ti. Claudius (of Mendes?)," *EANS* 806–7).

Thucydides (Books 3, 4) was the familiar historian of the Peloponnesian War, from the late fifth century BC.

Timaeus (Timaios, Books 2, 4, 6), from Tauromenium, was the historian of the late fourth and early third century BC, who included a geographical component in his work (*FGrHist* #566).

Timaeus the Mathematician (Book 5) is probably the astrological writer of the Late Hellenistic or Julio-Claudian period, who also wrote on the flooding of the Nile (Alexander Jones, "Timaios (Astrol.)," *EANS* 810).

Timagenes (Book 3), from Alexandria, became a close associate of Augustus (although they eventually had a falling out), and wrote a universal history (*FGrHist* #88).

Timosthenes (Books 4, 5, 6), from Rhodes, was the naval chief of staff of Ptolemy II and wrote *On Harbors*, a nautical guide for Ptolemaic seamen

(Duane W. Roller, "Timosthenes of Rhodes," in *New Directions in the Study of Ancient Geography* [ed. Duane W. Roller, University Park, Pa. 2019] 56–79).

Xenagoras (Books 4, 5), perhaps from Herakleia Pontika, wrote *On Islands* and a chronological treatise at some time during the Hellenistic period (*FGrHist* #240).

Xenophon of Lampsacus (Books 3, 4, 5, 6) was a geographer who wrote on Syria and the coasts beyond the Mediterranean. He probably lived in the first century BC (Daniela Dueck, "Xenophon of Lampsakos," *EANS* 840–1).

Literary Sources Not Mentioned in Book 1

There are ten authors who are mentioned in Book 2 but who do not appear in the lists of Book 1. Homer was also noted in the other books. In some cases the failure of these authors to appear in Book 1 probably means that Pliny did not consult them directly, but through Hellenistic syntheses of earlier scholarship, especially on the topic of Greek philosophy. In some cases their actual writings may not have been available. Yet it is highly unlikely that he did not have copies of Homer and Plato in his library. Following is a list of the ten authors.

Anaxagoras of Clazomenae (*NH* 2.149–50) was the philosopher and natural scientist of the fifth century BC, who spent most of his career in Athens.

Anaximenes of Miletus (*NH* 2.187) was the third of the Ionian monists, active in the sixth century BC, and was noted for a physical theory that saw air as the basis of the origin of the cosmos.

Cidenas (Kidennas or Kidennu) of Babylon (*NH* 2.38–9), probably from 150 BC or later, was an astronomer, largely concerned with the nature of the moon (Francesca Rochberg, "Kidennas (Kidinnu) of Babylon," *EANS* 475).

Cleostratus (Kleostratos) of Tenedos (*NH* 2.31), from the late sixth century BC, was an early astronomical and calendrical theorist.

Homer, the epic poet, was cited several times.

Pherecydes of Syros (*NH* 2.191) was allegedly the teacher of Pythagoras and was active in the mid-sixth century BC and an early cosmologist.

Pindar (*NH* 2.54) was the lyric poet of the late sixth and fifth centuries BC.

Plato (*NH* 2.205) was mentioned only once, in reference to Atlantis.

Stesichorus (*NH* 2.54) was the lyric poet of the early sixth century BC.

Thales of Miletus (*NH* 2.53), of the early sixth century BC, was the founder of the Greek intellectual tradition and a hydraulic engineer.

BOOK 2

Cosmology

Book 2 of the *Natural History*, the opening book of the text proper, is not strictly geographical, but its discussion of the cosmos is a fitting introduction for the geographical examination that follows. There is an emphasis on the divinity (*deus*) and its relevance to Roman life, yet this is a divinity that has little or no concern for human affairs. This lays the groundwork for the sense of utility that underlies much of the *Natural History*, more apparent, perhaps, in practical sections such as those on agriculture and zoology than geography.

Stoic views pervade Pliny's cosmology, such as the bonding together of the universe, the role of the moon, and the inevitability of death (*NH* 2.5, 43, 175; Mary Beagon, *Roman Nature: The Thought of Pliny the Elder* [Oxford 1992] 26–33). Yet Stoicism is less apparent, and indeed less relevant, in the purely geographical portions of the treatise, and in fact Pliny's philosophical views were quite eclectic. The book begins with a consideration of the earth itself, but in a cosmic sense (1–153), and with a focus on the heavenly bodies (28–101; Alan C. Bowen, "Hellenistic Astronomy," in *The Oxford Handbook of Science and Medicine in the Classical World* [ed. Paul T. Keyser with John Scarborough, Oxford 2018] 305–10). Then there is a lengthy section on weather (105–53): since the planets were said to affect the weather, this follows naturally from his cosmic discussions.

The remainder of the book is somewhat more geographical (154–248), with a discussion of the nature of the earth and the natural phenomena affecting it. This includes issues such as the water on its surface (166–70, 201–6, 212–34), the location of the earth in the cosmos as well as the seasons (176–88), earthquakes (191–200), and the seas and fire (212–48). Again, there is a focus on utility to the Roman people. As is common throughout the *Natural History*, there are many other topics and cultural anecdotes scattered throughout. Although Pliny listed over forty sources for Book 2, from Herodotus to ones of his own era, most apparent are Cicero (especially *de natura deorum*), Seneca's *Natural Questions*, and Aristotle's *Meteorologika*.

The Heavens

2.1–2. The solemn opening of the *Natural History* has a religious tone, seeing the cosmos (or universe) as divine. It is similar to a passage in Plato's *Timaios* (28b), where there is also a difficulty about what the cosmos (or, to Plato, the *ouranos*) should be called, a question whose answer was avoided by both Plato and Pliny. Pliny chose *mundus* – normally "world" – rather than the broader terms used by Plato, relying more on Cicero's translation of the dialogue. This was also the word used by Pomponius Mela (1.1). Although in some ways *mundus* is more limiting than the Greek words, it is ambiguous and also has cosmic overtones, as noted by Lucretius (4.134–6), as well as Vergil with his "clarissima mundi lumina" ("brightest lights of the world," *Georgics* 1.5–6; Mireille Courrént, "*Innumerabiles mundos*... Pline l'Ancien et l'esprit de découverte (*Histoire Naturelle* 2, 1–3)," *Euphrosyne* 27 [1999] 51–64). With *mundus*, Pliny has not only aligned himself with Roman views on cosmology, but used the word in order to look beyond his discussion of the cosmos to the world itself, the essential subject of the *Natural History*. Another influence may have been Posidonius' critique of the *Timaios* (F85).

The *mundus* is seen as divine, and has all the specific attributes of divinity. This is not the finite world of the Stoics (Diogenes Laertios 7.140), but a more Pythagorean one, reflecting the treatise preserved under the name of Ocellus Lucanus and probably of late Hellenistic date, which described the universe as eternal and indestructible (Bruno Centrone, "Ocellus Lucanus," *EANS* 586). Paradoxically it is infinite but appears to be finite.

As an aside, Pliny noted that what was beyond the *mundus* was not of human concern and not even of human comprehension, since everything was under the sky (*caelum*). This difficult issue was discussed slightly more thoroughly at *NH* 2.4.

2.3. Without mentioning specifics, Pliny expressed his indignation at those who attempted to determine the size of the universe and the distance between its component parts, as well as the idea that there could be multiple universes. Such thoughts had been expressed by Aristotle (*On the Heavens* 1.9.278b), although Pliny was somewhat more polemical. This issue was considered further at *NH* 2.83–8. The phrase "natura omnium artifex" is similar to Cicero, *de natura deorum* 2.58, and the repeated use of "rerum natura" is a memory of Lucretius.

2.4. Pliny's disdain for investigating what is beyond the universe was founded, in part, on the belief that the universe itself is so imperfectly known that such inquiries are foolish. There is also a hint of the universe as a living being, an idea put forth by Plato (*Timaios* 30ab), and originating in Ionian monistic philosophy, especially with Anaximander (Paul T. Keyser, "The Kozy Kosmos of Early Cosmology," in *New Directions in the Study of Ancient Geography* [ed. Duane W. Roller, University Park, Pa. 2019] 30–1). The section concludes Pliny's discussion about the ultimate nature of the universe.

2.5. Although Pliny's terminology is somewhat ambiguous, it is clear that here (and through 2.7) he was describing the universe as a whole, not merely the earth. This is despite the use of the word "orbis," which can also mean the earth, used by Pomponius Mela (1.1) as the first word of his treatise. Much of Pliny's account would apply to the earth itself (Cicero, *de natura deorum* 2.115–16). The universe – as manifested to humans by the half-sphere visible above the earth and the celestial bodies – is a perfect sphere; this is obvious to the viewer. It is all bonded together, as Cicero noted (*de natura deorum* 2.115), in a shape most suitable for rotation. Such a universe was popularized by Posidonius (F8) and was a Stoic belief. In addition to the universe, the earth itself is spherical, as are the sun and moon (Posidonius F117, 122). Thus all three were analogies to the universe itself. Posidonius was probably Pliny's immediate source, but this concept originated, in part, with the Pythagoreans and their spherical earth (Diogenes Laertios 8.48). The spherical universe could be validated empirically by a viewer on the surface of the earth, who would see an arc rising above in every direction.

2.6. The motion of the sun was also seen as proof of the spherical universe and its rotation (which is at an unknown speed). The concept of the "harmony of the spheres" – that the rotation of the universe and its components makes a harmonic sound – was attributed to the Pythagoreans (Censorinus, *de die natali* 13.5; Hippolytos, *Refutations* 1.2.2). Aristotle, who rejected the idea, nevertheless described it in detail, but also noted as part of the theory that humanity would have become conditioned to the sounds and would not hear them (Aristotle, *On the Heavens* 2.9.290b), something alluded to by Pliny. See further, *NH* 2.84.

2.7. The celestial sphere is not smooth and featureless but has representations of living beings on it, a sculptural analogy of the constellations that was explored further at *NH* 2.8–9. This section is peculiar, not only because of its reference to the surface of an egg but the suggestion that creatures on the earth are produced by fertilization from the celestial ones, something more folk wisdom than cosmology, with Orphic overtones. There is also a rejection of the Platonic perfection of the cosmos, whose surface was smooth (Plato, *Timaios* 33bc). The Bear (Ursa Major) and Bull (Taurus) are both prominent constellations, known since earliest times, but it is difficult to understand why Deltoton – shaped like the letter delta – was also cited. This constellation, today known as Triangulum and lying between Andromeda and Perseus, is faint and obscure. Although the region around the celestial pole has few stars, Pliny may have been alluding to the density of the Milky Way, which in Greek antiquity was generally known merely as the Milk (Gala), and was discussed in detail by Aristotle (*Meteorologika* 1.8).

2.8. A linguistic digression considers various words for the cosmos. *Ornamentum* is Pliny's translation of Greek *kosmos*; the Greek word had been used since the beginning of Greek literature to mean "ornament," such as when Hera adorned herself in order to distract Zeus (Homer, *Iliad* 14.187). The word only came to have an astronomical meaning with the Pythagoreans, perhaps first so used by Parmenides (Diogenes Laertios 8.48). *Mundus* also has the meaning of "ornament," and by the second century BC it seems to have had its astronomical definition (Varro, *de lingua latina* 7.14 [quoting Accius]). This was eventually popularized by Lucretius (4.134) and Vergil (*Georgics* 1.5). The word *caelum*, generally translated as "sky" or "heavens," seems to have originated from a concept that it was engraved, with the celestial bodies positioned on the cosmic sphere in the manner of gold or silver art work (Varro, *de lingua latina* 5.18). This is Pliny's first mention (outside of Book 1) of one of his major sources, the polymath M. Terentius Varro, of the first century BC, who was cited regularly throughout the geographical sections of the *Natural History* (Klaus Gunther Sallmann, *Die Geographie des älteren Plinius in ihrem Verhältnis zu Varro: Versuch einer Quellenanalyse* [Berlin 1971]).

2.9. The discussion of the celestial sphere concludes with a passing reference to the zodiac, the 12 constellations that mark the path of the sun through the heavens. A full catalogue had been provided by Aratos of Soloi in the early third century BC (*Phainomena* 544–52; B. L. van der Waerden, "History of the Zodiac," *AOF* 16 [1952–3] 216–30). The final statement returns to *NH* 2.1 and reminds the reader of the eternal nature of the universe.

The Four Elements

2.10. The concept of four elements is attributed to Empedokles of Akragas, active in the fifth century BC. He called them "roots" (*rhizomata*), and described them in divine terms: heat (Pliny's fire) was Zeus, air was Hera, earth was Aidoneus (Hades), and water was Nestis, probably an obscure nature divinity (Empedokles F26–7 Graham). Pliny's source for the suspension of the earth in the middle of the cosmos may be Cicero (*de natura deorum* 2.116), whose thoughts seem to pervade these sections of the *Natural History*. Each of the four elements is discussed in greater detail later in Book 2.

2.11. Pliny continued to rely on Cicero's *de natura deorum*, which is more detailed in describing the relationship between the four elements (Cicero, *de natura deorum* 2.84). There is an eternal cycle of change, from earth to water, to air, and to aether (this roughly equivalent to Pliny's fire), and then returning to earth. This creates a balance between the parts of the cosmos. Since the earth is at the center, it provides stability and equilibrium. The concept of the ever-changing cycle of the four elements is Pythagorean, as well as the belief that the earth is both at the center of the universe and is connected to everything (Aristotle, *On the Heavens* 2.13.293; Ovid, *Metamorphoses* 15.237–51).

2.12. There is a lengthy discussion (through *NH* 2.52, with digressions) of the seven wandering stars (*errantia* in Pliny's Latin, derived from the Greek *planetai*). The movements of the planets were a constant concern of ancient astronomical scholarship. Although the word *planetai* in its astronomical sense is not documented until the fifth century BC, in the context of Socrates' teachings (Xenophon, *Memorabilia* 4.7.5), planetary theory was another Pythagorean endeavor (Aristotle, *Meteorologika* 1.6.342b). There was some disagreement as to whether the sun and moon should be considered planets, as Pliny thought; Aratos, on the other hand, excluded them (*Phainomena* 454–9). Saturn (Greek Kronos) was the outermost planet known in antiquity and thus was considered the highest (*NH* 2.32); Uranus, although faintly visible to the naked eye, was not identified as a planet until the eighteenth century.

2.13. The first wandering star Pliny discussed was the sun. As the brightest object in the heavens, it had unique qualities and dominated any discussion of the cosmos. Its importance was demonstrated by its role in determining the seasons and the change of the year; there were also psychological ramifications, since it affected people with its light (or lack of it).

Its power was such that it had animate qualities, being able to see and hear everything. This was a concept as old as Homer: Agamemnon prayed to the all-seeing and all-hearing sun before the combat between Alexander (Paris) and Menelaus (Homer, *Iliad* 3.275–7). Pliny's phrase is a direct translation of Homer's words, and his eulogistic comments may also reflect the writings of Kleanthes of Assos, who considered the sun to be the Hearth of the Cosmos and the Ruler of the Cosmos (Plutarch, *Concerning the Face that Appears in the Globe of the Moon* 6.923a; Diogenes Laertios 7.139).

Divinity

2.14. There is a long digression (through *NH* 2.27) about the nature of divinity (Thomas Köves-Zulauf, "Plinius d. Ä. und die römische Religion," *ANRW* 2.16.1 [1978] 194–6). The Latin word is *deus*, but to translate this as "god" is limiting, as Pliny himself made clear at *NH* 2.15. The tone through *NH* 2.17 is sceptical, and again based to some extent on Cicero, especially *de natura deorum* 2.60–5. It also brings to mind the outlook of Lucretius. Pliny stressed that *deus* was something separate from the world or universe itself (although *mundus* must be supplied in the text: see Pliny, *Histoire naturelle* 2 [ed. Jean Beaujeu, Paris 1950] 126–7). *Deus* was defined by a number of non-corporeal aspects that were more mental than physical. But Pliny rejected any suggestion that there were innumerable gods, a concept originating with Xenophanes of Colophon (F29–42 Graham), who criticized the tendency to attribute human qualities to the gods. The list of divinities (Pudicitia, Concordia, Mens, Spes, Honor, Clementia, and Fides) is also a catalogue of temples in the city of Rome in Pliny's time; the only one visible today is the Temple of Concordia, whose remains are conspicuous on the northeast slope of the Capitoline Hill and which was a notable art museum (Richardson 98–9). The context of the reference to Democritus, assuming that it is the atomist of the fifth century BC, is unknown, yet he was noted for his varying views on the nature of the gods (Cicero, *de natura deorum* 1.120–1).

2.15. The limitations of humanity have required innumerable gods, each relating to a particular human need. Although Pliny cited no source, the passage is an echo of Xenophanes' famous statements about each ethnic group having gods in their own image, and how animals would also have gods like themselves (Xenophanes F32–3 Graham). This naturally leads to the conclusion, expressed at *NH* 2.16, that there are more gods than humans.

2.16. Continuing the theme of innumerable gods, despite his rejection of the idea, Pliny, in a passage taken almost verbatim from Cicero's *de natura deorum* (3.63), cited several examples of gods whose utility was extremely limited. The shrine to Febris (Fever), on the Palatine, was one of three to the divinity in the city of Rome. It was probably merely an ancient altar but cannot be located today (Richardson 149–50).

The shrine of Orbona, mentioned only here in extant literature (the text of Cicero has a lacuna), was seemingly dedicated to parents who had lost a child. Its exact location is unknown, but it was near the shrine of the Lares, on the summit of the Via Sacra near the Arch of Titus (Richardson 277–8). The third shrine, to Mala Fortuna, was on the Esquiline near the top of the Clivus Suburanus (Richardson 156). All three of these shrines commemorated rather dismal situations, and citation of them supports Pliny's negative tone about the multiplicity of gods and the existence of one for every human condition.

The concept of innumerable gods also reached a personal level, something that would run counter to the universal nature of divinity. Such divine qualities were described satirically by Petronius (16–18; see also Seneca, *Letter* 110.1). Gods in the form of animals were believed to be a stock feature of certain peoples other than Greeks and Romans. Syrians and Egyptians were frequently cited, with the latter notorious for deifying every animal (Cicero, *de natura deorum* 1.81–2, 101; 3.39). Pliny may also have had in mind the Egyptian trait of using garlic and onions when swearing oaths (*NH* 19.101).

2.17. Adopting a somewhat more sarcastic tone, Pliny catalogued a number of traits attributed to the gods that tended to make them more human (see also Cicero, *de natura deorum* 1.83). Xenophanes had criticized their anthropomorphic qualities as well as their penchant for adultery (Xenophanes F29–30 Graham). An additional cause for concern was their fixed ages, whether old like Saturn or youthful like Apollo and Bacchus. There were also other problematic characteristics, such as winged like Mercury, lame like Vulcan, or born from eggs like the Dioscuri. The dark-skinned divinity may be Erebus, who, as the consort of Nyx (Night), was associated with the darkness (Hesiod, *Theogony* 123–5), and the one who lives and dies could be Proserpina (Tusculum 2, 218). Divinities of theft and crime are more difficult to identify, but the former may be a reference to Mercury (Hermes) and his theft of the cattle of Apollo (Homeric *Hymn to Hermes* 68–141). A god of crime is perhaps Dolus, one of the children of Erebus and Nyx (Cicero, *de natura deorum* 3.44). Yet the specific divinities are hardly of importance in the overall thrust of Pliny's argument.

2.18. Changing his tone, Pliny turned to a more eulogistic view of divinity, presenting it in an allegorical sense as the best in human nature, with the implication that it would be revealed in people who act for the common good. Benefactors of humanity were considered divine, and in some cases, such as Hercules and Aesculapius, actually became divine themselves (Cicero, *de natura deorum* 2.62). As Seneca noted, such benefactions are far more important than any glory received from war (Seneca, *de clementia* 1.26.5). Pliny placed this within a contemporary political context by referring to Vespasian – enthusiastically called the "the greatest leader [*rector*] of all time" – and his sons Titus and Domitian. *Rector* can have divine overtones (Cicero, *de finibus* 4.11; Seneca, *Letter* 16.4), and was also applied to the heroes of the past (Vergil, *Aeneid* 9.173). By using it to describe Vespasian, Pliny was pointedly seeing him as the person who was guiding the Roman state back to normalcy after the civil war of AD 68–9, and thus was himself on a path to the divine.

2.19. Reinforcing the idea that considering humans divine is a way of rewarding them for their services, Pliny adopted a view similar to that of Euhemeros of Messene (*FGrHist* #63), who believed that many gods were originally human and had achieved their divine status as a reward for their benefactions to humanity (Diodorus 6.1–2). He lived around 300 BC and is best known for his fantasy geographical writings, described in a treatise titled *The Sacred Record*, in which he also set forth his ideas about mortals achieving divinity.

2.20–1. Pliny emphasized that it is unreasonable to assume that the divine power has any interest in human affairs (if it did have such concerns, it would be depressing). This leads to the matter of foreign cults, which were common in Rome by Pliny's day. The ways of their adherents remained perplexing to the Romans: many of them seemed rigidly and excessively governed by rituals. In Pliny's time, rings with figures of Egyptian gods were common, especially representations of Harpokration, the child of Isis, whose images had become popular in Ptolemaic times (*NH* 33.41; Françoise Dunand and Christiane Zivie-Coche, *Gods and Men in Egypt, 3000* BCE *to 395* CE [tr. David Lorton, Ithaca 2004] 273–4).

The cults first came to Rome by the late third century BC. Cybele, the Magna Mater, arrived from Phrgyia in 204 BC and was housed in a temple at the west corner of the Palatine, whose remains are still visible (Livy 29.37.2, 36.36.3–5; Richardson 242–3). Isis and Serapis were seen in the city by the first century BC (Richardson 211–13). There were several local synagogues by the Augustan period (Duane W. Roller, *The Building Program of Herod the Great* [Berkeley 1998] 75), and Christians were notoriously visible during the era of Nero (Tacitus, *Annals* 15.44). Belief in worship of monsters was a rather harsh view of the foreign religions, presumably based on the animal cults prevalent in Egypt and elsewhere, which could involve creatures alien to the Romans, such as crocodiles.

Dietary restrictions were seen to be common, perhaps most conspicuously among the Jews.

2.22. The cult of Fortuna was ancient, perhaps known in Rome since the regal period, when, it was said, King Servius Tullius had popularized it, building numerous shrines to various aspects of the goddess (Plutarch, *Roman Questions* 74). Perhaps originally an Italian women's goddess, she evolved into a protective divinity, best known today through her massive and spectacular temple complex at Praeneste, just south of Rome (Fritz Graf, "Fortuna," *BNP* 5 [2004] 505–9). She had over thirty shrines in Rome, most notably a temple in the Forum Boarium, parts of which are still visible (Richardson 155–8). The idea that the fickleness of fortune could be a governing principle of the cosmos was anathema to the Stoics and their ordered universe, yet the numerous shrines demonstrate her popularity in Roman society (Juvenal 13.86–9; see also Tacitus, *Annals* 6.22). In explaining the activities of Fortuna, Pliny used a bilateral accounting method (*utramque paginam*; B. R. Rees, "Pliny, *N. H.* ii.22," *CR* 8 [1958] 213–15).

2.23. Astrology was also popular in the Roman world of the first century AD. It had ancient Mesopotamian origins, and became an element of propaganda in the Roman Imperial period. The basic theory of astrology was that future events could be predicted by the motion and position of heavenly bodies, especially at one's birth. There were both intense supporters and critics of the concept; the emperor Tiberius was perhaps its most famous adherent, and the process could easily be manipulated (Tacitus, *Annals* 6.21) The *Astronomica* of Manilius, from the early first century AD, is the earliest extant treatise on the topic, but it is not known whether Pliny was familiar with it. Pliny sarcastically noted the essential flaw in astrological theory: it assumes that all the future is determined at the moment of birth.

2.24–5. Similar to astrology, but affecting people throughout their lives, were portents, prophecies, and omens. These could range from the spectacular, such as weather phenomena, to the trivial, such as a sneeze or misstep. Again these beliefs conflicted with the ordered Stoic universe, although one could not deny that minor incidents might have extensive unexpected consequences. Cicero summarized well the absurdity of too much reliance on omens and their frequent misinterpretation, as well as how this might cause havoc to daily life (Cicero, *de divinatione* 2.40–1). Lightning was perhaps the most conspicuous omen, and would be interpreted in many ways, often conveniently (Seneca, *Natural Questions* 2.32–50). The incident regarding Augustus and his boot lacks context, but presumably was early in his career, perhaps connected with events around

Perusia in 41 BC (Suetonius, *Augustus* 14, 92). Reliance on such omens was seen as an essential proof of human weakness, with the peculiar analogy that animals could be better off than humans because they have no such concerns.

2.26–7. Pliny concluded his digression on the nature of divinity by describing its limitations, yet admitting that in some ways divinity was connected to human affairs, and that there were matters peculiar to humanity that not even the gods could control. The positive comment about suicide brings to mind Seneca, who had committed the deed in AD 65, only a few years before Pliny completed the *Natural History*. Seneca had expressed similar views about divinity, which he called Jupiter, seeing him as "the controller [*rector*] and guardian of the universe" who did not do everything but who made certain that everything could happen. Jupiter was not always present for everyone but gave the signal (*signum*) and was the force (*vis*) for all things (Seneca, *Natural Questions* 2.45–6). Thus divinity, Seneca's Jupiter, reveals the power of Nature (*naturae potentia*).

The Wandering Stars

2.28–9. After the digression on divinity (since *NH* 2.14), Pliny returned to astronomical concerns (see *NH* 2.7–9), with a discussion of the wandering stars (through *NH* 2.43). Nevertheless he could not resist another comment on how stars have nothing to do with human affairs. Falling stars, or meteorites, were popularly associated with a death: Pliny included a brief scientific explanation of the astronomical phenomenon. In antiquity they were most commonly called "shooting stars" and were caused, it was believed, by combustion in the lower levels of the heavens (Aristotle, *Meteorologika* 1.4). The image of a stream of oil from a lamp is especially precise. Observation of shooting stars is recorded from the beginnings of Greek literature, such as the comparison of their luminous trail with the movement of Athena from Olympus to the earth (Homer, *Iliad* 4.75–8). They continued to inspire poets into the Roman period (Aratos, *Phainomena* 926–32; Vergil, *Georgics* 1.365–7).

2.30. Since shooting stars are transient, Pliny felt it necessary to remind his readers that the other heavenly bodies are eternal, and are all considered a sort of cosmic glue that holds the universe together. Because they are always visible – within the limits of their cycles – they were of great influence and have been studied extensively; Pliny's analysis was reserved for later (*NH* 18.207–17). Rather he proceeded to a discussion of the zodiac and the wandering stars (the zodiac had been mentioned in passing at *NH* 2.9). The examination that follows is totally geocentric; there is no hint of the heliocentric ideas that existed in antiquity.

2.31. Anaximander of Miletus was the second of the Ionian monists, said to be the student of his compatriot Thales. A brilliant and innovative scholar, his endeavors included inquiry into much of the natural world (Daniel W. Graham, "Anaximandros of Miletos," *EANS* 75–6). His assumed date of Ol. 58 (548–545 BC) is reasonable, but probably late in his career. This is the only extant reference to his discovery of the obliquity of the zodiac, or indeed anything to do with the zodiac, and such thought on his part is dubious, since despite his broad range of knowledge it is anomalous to the trajectory of his work, and the zodiac does not seem to have been defined even in Babylonia until the fifth century BC (Dirk L. Couprie *et al.*, *Anaximander in Context: New Studies in the Origins of Greek Philosophy* [Albany 2003] 174–5).

Cleostratus (Kleostratos) was an astronomical writer of probably the late sixth century BC, little known today. He was the author of an *Astrologia* (J. K. Fotheringham, "Cleostratus," *JHS* 39 [1919] 164–84). Pliny's passage is difficult to interpret, for it is certainly not the case that Cleostratus was the first Greek to identify constellations, since some were known to Homer (*Odyssey* 5.270–5), but Cleostratus may have published a catalogue of them. He is not cited in Pliny's list of sources, so probably was not consulted directly. Whether he knew about the zodiac is problematic – like Anaximander he seems too early – and definition of it is usually credited to Oinopides of Chios in the fifth century BC (Couprie *et al.*, *Anaximander* 174). It may be that Cleostratus was the first to define Aries and Sagittarius, but even this is questionable.

The mythological giant Atlas was believed to hold up the celestial sphere. He was positioned somewhere in the far west, where the toponyms Atlas and Atlantic still exist (Hesiod, *Theogony* 517–19; Aeschylus, *Prometheus Bound* 349–52).

2.32–3. Saturn (Greek Kronos or Phainon, Babylonian Ninurta) is the faintest of the planets and was thus believed to be the farthest and the coldest (*NH* 2.34). Its stated period of 30 years is accurate; the modern figure is 29.458 years (Mark R. Chartrand, *National Audubon Society Field Guide to the Night Sky* [New York 1991] 658). The issue of the opposite motion of the planets was discussed in detail by Aristotle in a complex and obscure argument. This reverse motion was believed, to some extent, to cause the winds and thus to keep the cosmos from settling into a sluggish mass (Aristotle, *On the Heavens* 2.3; Cicero, *de natura deorum* 2.51–2).

Seneca also made a brief reference to the phenomenon, calling it *occursu contrario* (*Natural Questions* 2.11.2). This is actually a simplification of the apparent complex motion of the planets, but it is a good descriptive phrase, since they appear to move from west to east relative to the fixed stars (Harry M. Hine, *An Edition With Commentary of Seneca, Natural Questions, Book Two* [New York 1981] 221–4).

2.34. Because of the distance of Saturn, it was believed to be frozen and cold. Jupiter (Greek Zeus or Phaethon, Babylonian Marduk) was said to have a cycle of 12 years (the modern figure is 11.8623 years: Chartrand, *National Audubon Society* 655). It was thought to be in a balanced position between cold Saturn and hot Mars, and thus was temperate, an analogy for the nature of Italy, located between the cold north and hot south (Vitruvius 6.1.11).

Mars (Greek Ares or Herakles – hence Pliny's Hercules – Babylonian Nergal) was also called Pyroeis, the "Fiery One," from its reddish tinge and assumed closeness to the sun (Cicero, *de natura deorum* 2.53). Its stated period of about two years is unusually inaccurate; it is actually 686.98 days (Chartrand, *National Audubon Society* 650).

2.35. The sun was said to have an orbit between Mars and Venus: in other words, located where the earth is in the heliocentric system. Pliny's figures are based on the Julian calendar, which had been established in 46 BC by Julius Caesar with the assistance of Sosigenes of Alexandria, whom Caesar may have met when he was involved with placing Cleopatra VII on the throne (*NH* 18.210–12). Sosigenes also studied the orbit of Mercury (*NH* 2.39). The Julian calendar – still in use today, although with modifications – adjusted the assumed 360 degrees (days) of the solar orbit to create a year of 365¼ days with an intercalary (leap) day every four years. To some extent it was based on an Egyptian model (Appian, *Civil War* 2.154; Sacha Stern, *Calendars in Antiquity* [Oxford 2012] 211–14).

2.36–8. Venus (Greek Aphrodite, Babylonian Ishtar) was also known as Phosphoros, "the Bringer of Light," due to its vivid appearance at sunrise ahead of the sun. The Romans refined this somewhat and called it Lucifer (i. e. Phosphoros) in morning and Vesper (Evening) when it was after the sun (Cicero, *de natura deorum* 2.53).

Pythagoras of Samos, who was active in the late sixth and early fifth centuries BC, mostly in southern Italy, was a charismatic teacher who cannot easily be separated from the Pythagorean school. He was generally believed to have been the first to realize that the morning and evening stars were the same, although this discovery is also attributed to Parmenides (Diogenes Laertios 9.23), and there were Egyptian and Mesopotamian precedents (Micah T. Ross, "Hellenistic Egyptian Planetary Theory," in *Hellenistic Astronomy: the Science in Its Contexts* [ed. Alan C. Bowen and Francesca Rochberg, Leiden 2020] 163).

Pliny is the first extant author to mention that Venus casts a shadow. The planet is the third brightest object in the sky (magnitude -4), and its shadow is elusive and faint. Because of its unusual nature it not only had many names but powerful qualities not associated with the other planets. Its connection to Aphrodite meant that it was responsible for life and love, as was the goddess herself (Lucretius 1.1–20).

The reported cycle of 348 days is far from the modern figure of 224.701 days. The Timaeus cited is not the historian from Sicily but a little-known astrological writer of uncertain late Hellenistic date, who was cited three times by Pliny (also *NH* 5.55, 16.82) and wrote on a variety of topics including the flooding of the Nile (Jones, "Timaios," *EANS* 810).

2.39–40. Mercury (Greek Hermes, Apollo, or Stilbon ["Glittering One"]; Babylonian Nabu), the lowest of the five conventional planets, was similar to Venus in its characteristics, since it also was believed to be between the earth and sun, and never be more than 22 degrees (actually 27) from the latter. Its period of 87.969 days is far less than Pliny thought, but its name was due to its swift motion.

Cidenas (Kidennas or Kidennu) was a Babylonian of late Hellenistic times who wrote about the planets (Strabo, *Geography* 16.1.6; Rochberg, "Kidennas," *EANS* 475). Sosigenes was the Alexandrian astronomer who assisted Julius Caesar in creating his calendar (Appian, *Civil War* 2.154).

The concept of the Great Year (Magnus Annus), the length of the interval necessary for all the planets to return to the same position, is Babylonian in origin. It was known to Greeks by the time of Herakleitos of Ephesus in the late sixth century BC (the discovery is also attributed to Oinopides (Henry Mendell, "Oinopides of Khios" *EANS* 587). Herakleitos considered it to be 18,000 years (F97 Graham), and it was discussed in detail by Cicero (*de natura deorum* 2.51–2).

2.41–3. There is a lengthy section about the final planet, the moon (Luna, Greek Selene). It was the most problematic of the ancient planets because of its phases and the variable height of its orbit. The moon was believed to exist in order to counteract darkness, but its phases and visible surface – unique to the planets – were more matters of curiosity than enigmas. Study of it and its peculiarities began with Thales of Miletus (F25 Graham). Plutarch's essay *Concerning the Face that Appears in the Globe of the Moon* summarizes the state of lunar scholarship shortly after Pliny's time.

Endymion was the beloved of Selene, the moon goddess. Their relationship was perhaps first treated by Sappho (F199), and became popular in Hellenistic times (Theokritos 20.37–9). His eternal sleep was a popular subject in Roman Imperial sarcophagal art (Diana E. E. Kleiner, *Roman Sculpture* [New Haven 1992] 304). There is a faint tradition – documented no earlier than Pliny – that he was the first astronomer and calculated the lunar phases (Nonnos 41.379–81). A Stoic digression contrasts the moon as a source of light (and indeed enlightenment) with human proclivities toward violence.

2.44–6. The figures cited by Pliny for the lunar orbit are accurate, and are contrasted with the time for the longest orbit, that of Saturn. The moon as a source of knowledge was suggested at *NH* 2.43; Cicero believed that the study of the heavens (not merely the moon) was a revelatory experience that inspired the origins of Greek intellectualism (Cicero, *Tusculan Disputations* 5.68–9).

There is a catalogue of seven things that can be learned from the moon, beginning with the concept of 12 months to a year, an exceedingly ancient Mesopotamian idea (E. J. Bickerman, *Chronology of the Ancient World* [second edition, Ithaca 1980] 16–21). The fact that the light of the moon was a reflection of that of the sun was established by Anaxagoras in the fifth century BC: Plato considered it a recent theory (*Kratylos* 24). Much of the following argumentation concerns the variability of its light; the exuding of moisture from the moon is discussed in greater detail at *NH* 2.223. Its visibly rough topography – Pliny called the irregularities *maculae*, or spots – was said to be produced by dirt from the earth due to the transference of moisture, but this only happens around the time of the full moon, since the rough surface is not visible except at that time. Pliny concluded his discussion of the moon with a look ahead to what he considered to be the greatest marvel in nature: eclipses (*NH* 2.47–50).

2.47. Lunar and solar eclipses, which had long been an object of fear and superstition, had been recorded in Mesopotamia since at least the second millennium BC. The prediction of the year of an eclipse of the sun by Thales, in the early sixth century BC, is the first record of Greek intellectual interest in the phenomena (Herodotus 1.74; *NH* 2.53); by the second century BC it was possible to predict them over a span of 600 years (*NH* 2.53). A detailed description of the nature of eclipses was provided by Kleomedes (see especially his 2.6). The conical shape of the shadow was a curiosity, and Pliny felt it necessary to use household terminology for an explanation: the shadow was similar to an inverted *turbo*, a child's toy, which was essentially like a modern spinning top (Vergil, *Aeneid* 7.378–81).

Its pointed end helped to explain the nature of the lunar eclipse. It was also worth noting that only the moon was eclipsed in this way.

2.48. It seems that Pliny attributed the tapering of the shadow to an inability to see it at a great distance. He is the only extant source to place the division between the air and aether at the level of the earth's shadow (Pliny, *Histoire naturelle* 2, 141). The clarity of the aether above the moon helps to explain why the stars are only visible at night (it does not seem to have been recognized that Venus can be seen, on occasion, in the daytime).

In theory there could be a lunar eclipse every month, but because of the erratic orbit of the moon one occurs only two to five times a year (Chartrand, *National Audubon Society* 56–8). Moreover the observed experience of antiquity regarding solar eclipses was limited to what was visible in the Mediterranean world and places known to it, so the data are incomplete.

2.49–50. The idea that the moon was larger than the earth was proposed by Anaximander of Miletus (F25 Graham); it was a belief essentially obsolete by Pliny's day, but may have continued in the popular imagination. There were various calculations of the relative size of the moon and the earth, beginning with Aristarchus of Samos, who wrote in his extant *On the Sizes and Distances of the Sun and the Moon* that the moon was $\frac{9}{25}$ the size of the earth (the actual figure is $\frac{5}{18}$). Other than Pliny, and whatever source he was using, there were essentially no further assumptions that the moon was larger than the earth, except for a strange reference attributed to Posidonius (F122) which is probably an error in the late antique transmission of the data (I. G. Kidd, *Posidonius* 2: *The Commentary* [Cambridge 1988] 472). But the idea that the moon was larger was persistent enough into late Hellenistic times – at least in some circles – that Kleomedes felt the need to refute it (Kleomedes 2.3).

There was little doubt that the sun was immense in size. Although Anaximander had said the sun and moon were of equal dimensions (F23 Graham), by Hellenistic times it was recognized that the sun was far larger. The first known source to conclude this was Eudoxus of Cnidus, in the fourth century BC, who estimated that the diameter of the sun was nine times that of the moon (Archimedes, *Sand Reckoner* 1.9).

Pliny, however, provided a number of assumed proofs of the size of the sun, generally more popular wisdom than mathematical data. Solar shadows are parallel, much like a long row of trees: Pliny was probably thinking of the poplar trees that lined Roman roads, some of which are still visible today in northern Italy (Kleomedes [2.1], perhaps less familiar with Italy, substituted a colonnade).

The second proof of the size of the sun is somewhat more scientific: the fact that its shadows, or lack of them, can be variable, depending on the time of day and of the year. The source here may be Posidonius (F208), who used solar shadows as a way of defining latitude, but Pliny's account is somewhat garbled.

The final example returns to folk wisdom. It was said that the rising sun appeared to be wider than Mt. Ida. Presumably this is the mountain in the Troad (although Mt. Ida on Crete also had peculiar characteristics: Solinus 11.7). The nature of the sunrise at Mt. Ida in the Troad was first reported by Diodorus (17.7.4–7), writing in the first century BC but recording an incident from the era of Alexander the Great, although it is not clear whether the anecdote about the sunrise belongs to that period. According to Diodorus, just before sunrise the light of the sun appeared like fiery flames three *plethra* (about 300 feet or 90 m.) across (see also Pomponius Mela 1.94–5). This is not the same as Pliny's report that it was wider than Ida itself, and needless to say such accounts depend on the perspective of the viewer, but nevertheless Mt. Ida seemed to contribute to strange solar phenomena. Comparing the size of the sun to terrestrial geography was a known technique: Anaxagoras had said that it was larger than the Peloponnesus (F37 Graham).

2.51–2. Pliny returned to the matter of shadows and how they reflect the size of the object producing them (see the diagram, Tusculum 2, 221). The conical shadow of a lunar eclipse demonstrates that the sun is larger than the earth. There is also a passing reference to how the varying orbit of the sun produces the seasons. It was originally believed that localities near the equator were burned by the sun, but the idea that the tropics were hot was disputed by Eratosthenes (*Geography* F45) in the third century BC and then more definitively by Polybius in the following century, when he learned that the tropical west African coast was mountainous. He reported on his journey into these regions in a treatise titled *On the Inhabited Parts of the Earth Under the Equator* (Geminos 16.32). The concept of the hot tropics originated with Parmenides (F43 Graham), who divided the earth into climate zones, calling the tropical one the "burned [*diakekaumene*]." This theory was refined by Posidonius (F49) and Strabo (*Geography* 2.2.2).

2.53. C. Sulpicius Gallus (or Galus) had an active political and military career in the first half of the second century BC. Most notably, he was military tribune at the battle of Pydna in 168 BC, which brought the Macedonian kingdom to an end. He became consul two years later. At Pydna he was said to have announced that there would be a lunar eclipse on the following day and that it was not something to be feared. This encouraged the Roman soldiers to victory, and became a paradigm for the value of scientific thinking instead of superstition (Philip Thibodeau, "Traditionalism and Originality in Roman Science," in *The Oxford Handbook of Science and Medicine in the Classical World* [ed. Paul T. Keyser with John Scarborough, Oxford 2018] 597–8). But the Macedonians saw the eclipse as foretelling the death of their king, Perseus (Polybius 29.16; Livy 44.37.5–9). Sulpicius Gallus' astronomical work was used by Pliny (*NH* 1.2) and was also known to Cicero (*de senectute* 49–50), but no fragments survive.

It was generally accepted that Thales of Miletus was the first Greek to investigate eclipses, during the reign of Alyattes of Lydia and famously described by Herodotus (1.74). Pliny's date would be equivalent to 585/4 BC; the synchronism with the founding of Rome may mean that Pliny's source was Sulpicius Gallus' treatise.

Pliny has been discussing eclipses, and seemingly relying on material from Hipparchus, but given the limited geographical area available for observation in antiquity, any universal catalogue for the sun would have been impossible, and there is no other evidence for one. He probably did not consult Hipparchus directly, and while his eulogistic comments about the earlier scholar are well taken, they may reflect the view of Posidonius. Presumably Pliny was referring to Hipparchus' solar and lunar tables, which recorded their position on any particular date (Dicks, *Geographical Fragments* 41–2, 108–10). Nevertheless Pliny cited Hipparchus regularly, even if indirectly.

2.54. Ancient literature is replete with examples of the fear and hysteria caused by eclipses (see the list, Pliny, *Histoire naturelle* 2, 144). Stesichorus was active in the early sixth century BC; nearly three hundred fragments of his poetry survive. The reference is his F271, also noted by Plutarch (*Concerning the Face that Appears in the Globe of the Moon* 19.931e). Pindar, the poet of the early fifth century BC, described the effects of a contemporary solar eclipse in his *Paian* 9 (F52 k). Twenty lines survive, which recount how the eclipse might have been a portent of an unprecedented disaster, war, or crop failure, or weather-related incidents that would lead to the extinction of the human race, a surprisingly modern point of view.

Nicias (Nikias), the fleet commander of the ill-advised Athenian expedition to Sicily in 413 BC, used an eclipse of the moon to delay further progress of the invasion. He had no competent advisor regarding the meaning of the eclipse; his astrologer, Stilbides, had recently died. This delay was said to be a contributing factor to the Athenians' eventual defeat, and came to be compared to the measured response of Sulpicius Gallus (Thucydides 7.50; Quintilian, *Institutes* 1.10.47–8; Plutarch, *Nikias* 23).

2.55–7. After a brief and textually uncertain passage seemingly reminding readers that it was best to determine their own fate, a final discussion of eclipses takes a more scientific tone. The period of 223 months, the so-called *saros* cycle, is Babylonian in origin. The word, a hellenized version of an Akkadian term, originally meant a period of 3,600 years (Abydenos [*FGrHist* #685] F2), but in early modern times was applied to the lunar cycle of 223 months (James Evans, *The History and Practice of Ancient Astronomy* [Oxford 1998] 316). Solar eclipses only occur at the new moon and lunar eclipses only at the full moon, but Pliny cautiously noted that many eclipses were not visible because of clouds (rather obvious) or, more seriously, the curvature of the earth.

Continuing his summary of details from Hipparchus' researches, Pliny made another eulogistic comment about him, praising his insight (*sagacitas*). Peculiarly, he reported that the shadow of the earth during a lunar eclipse could move either west to east or east to west (actually it is only west to east); it is difficult to imagine Hipparchus making this error, and it may be a confusion by Pliny about the different motions of lunar and solar eclipses (Kleomedes 2.6; Alan C. Bowen and Robert B. Todd, *Cleomedes' Lectures on Astronomy* [Berkeley 2004] 155).

The consulship of Vespasian and Titus cited by Pliny was in AD 72; the two eclipses were actually in March AD 71 (Theodor Von Oppolzer, *Canon der Finsternisse* [Vienna 1887] 124, 344). Father and son had also been consuls in AD 70, but this had been the son's first holding of the office. The manuscripts of the *Natural History* vary somewhat, but assuming that the date of the eclipses is correct, the report is probably an error by Pliny (Bickerman, *Chronology* 155; Pliny, *Histoire naturelle* 2, 145–6).

2.58. The phases of the moon were first studied by Anaxagoras of Clazomenae in the fifth century BC (Anaxagoras F45 Graham). But the detailed calculations reported by Pliny are from a later source, perhaps Hipparchus, but converted into Roman terminology, using Latin numerology with the word *semuncia* (a twenty-fourth). It was established that although the moon appears larger, it is actually the smallest heavenly body, and is also the closest to the earth (Kleomedes 2.5). Additionally it was important to realize that the stars did not cease to shine in the daytime (as shown by their appearance during solar eclipses, somewhat of a contradiction to *NH* 2.48). The idea that they could be seen from the bottom of a well is folk wisdom and hardly true, although one is reminded of the anecdote that Thales of Miletus fell into a well while observing the stars (Plato, *Theaitetos* 174a).

2.59–61. A lengthy discussion about the planets (through *NH* 2.84) focuses on their positions and movements. Again, the primary source may be Hipparchus, although he is not mentioned by name in these sections.

Planetary motion is difficult to explain within a geocentric system. Apollonios of Perge, around 200 BC, seems to have been the first to attempt a theoretical structure (Evans, *History* 337–41). The theories continued to be modified throughout antiquity and then with the adoption of the heliocentric system in the Renaissance. Yet Pliny's material is not theoretical but a collection of data, beginning with the position of the three superior planets – those "supra solem", or above the sun – Mars, Jupiter, and Saturn, and concluding with the two lower ones, Mercury and Venus. Despite the thoroughness of Pliny's information about the planets and their movements, it is problematic: as one recent author reported, "when it is not just wrong, [it is] either very confused or maddeningly confusing" (Alan C. Bowen, "Some Early Hypotheses in Greco-Roman Astronomy," in *Hellenistic Astronomy: the Science in Its Contexts* [ed. Alan C. Bowen and Francesca Rochberg, Leiden 2020] 102–10, with numerous diagrams). As always, a major problem was the lack of Latin terminology for the phenomena he was attempting to describe. The commentary that follows is not an attempt to determine the "correctness" of Pliny's arguments but to analyze them as presented.

2.62–7. The next topic is a discussion of the changes in appearance of the planets. They can be higher or lower, brighter or dimmer, and differing in color. Acknowledgement of the great ideas of one's predecessors, even if they were in error, was a common Stoic trait and a feature found elsewhere in the *Natural History* (e. g. 23.112; see also Seneca, *Natural Questions* 6.5).

Several reasons were advanced for planetary motions. Pliny used the Greek word *apsis* (presented in Greek characters; his other citations are in Latin), which originally meant a loop, as in a snare (Homer, *Iliad* 5.487). It came to be a more generic term, defining something circular, and then entered astronomical terminology. This was a small planetary orbit that was different from its general movement through the cosmos. By Roman times the common term for this was *epikyklos* ("epicycle"), as used by Plutarch (*On the Generation of the Soul in the Timaios* 31; see the diagram in Evans, *History* 212).

Pliny used the *apsides* as the basis of his arguments, stating that they were the reason for changes in the appearance of the planets. Also affecting them (as seen from earth) was the slant of the zodiac – which lies over the habitable part of the earth – since the planets could wander beyond its limits. Yet Pliny realized that no explanation, or set of explanations, fully explained planetary motions, and all the data needed to be integrated with great perception (*subtilitas immensa*) into a working planetary theory.

2.68–71. Pliny discussed the movement of the three outer planets. He may have been the first to latinize the Greek word *ekleiptikos,* translated into English as "ecliptic." According to his theory, the planets move more quickly when nearest to the earth (at rising and setting), and less so when at their highest point.

The idea that the movement of the planets was affected by the sun was a common belief, suggested by Vitruvius (9.1.11), although he was dubious about the matter. It may have been an Egyptian theory: Julius Caesar was said to have learned about it from the Egyptian astronomer Acoreus (Lucan 10.201–3). Pliny bolstered his argument by stating that such movements and effects could not easily be seen.

In his transition to the discussion of the two inner planets, Pliny made the strange statement that he was the first to explain their motions. This seems unlikely, and Pliny may have copied something from his unnamed source. Eudoxus of Cnidus, in the fourth century BC, was the first to discuss planetary motion in detail (Aristotle, *Metaphysics* 12.8; Evans, *History* 305–10). Pliny may also have been familiar with the recension of Eudoxus' theories by Aratos of Soloi in the third century BC (see especially Aratos, *Phainomena* 454–61).

2.72–6. Pliny next considered the movements of the two inner plants, Mercury and Venus, which are never far from the sun and are seen to move back and forth near it (Pliny, *Histoire naturelle* 2, 159–64). No source was provided, but it probably continued to be Hipparchus or Posidonius. The speed of the motion of the two planets was reasonably seen as proof of their closeness to the earth (Evans, *History* 299–301). Again, Pliny's account is a collection of data rather than any theorization; in fact, he noted that explanations eluded the (unnamed) theorists.

2.77–8. Pliny concluded his discussion of planetary motion with some scattered comments, reminding his readers that all he has said was according to secretive natural laws. The emphasis is on the relation of the planets to various constellations of the zodiac, and the fact that they are invisible for extended periods, as much as half a year for Mars and Saturn.

Although the material so far has been limited to scientific data, with a small amount of theorization, Pliny digressed to provide a mythological anecdote. Lynceus was the son of the king of Messene, and was involved in many events of Greek mythology, including the voyage of the *Argo* and the Calydonian Boar Hunt. His tomb was still visible at Sparta in the second century AD (Apollonios 1.151–5; Ovid, *Metamorphoses* 8.304; Apollodorus, *Bibliotheke* 3.10.3; Pausanias 3.13.1). Most relevant to Pliny was his sharp vision, which even allowed him to see through solid objects. This characteristic was already proverbial by the fourth century BC (Aristophanes, *Ploutos* 210). The astronomical abilities of his eyesight were suggested by Apollonios (4.1477–80) and Kleomedes (2.1), but Pliny is the only source to provide any detail.

2.79–80. The appearance of the planets is due to the constellations that they happen to be in and the effect of the nearby stars, a point of view more popular than scientific. Although Pliny called this effect planetary "coloring" (*colores*), it was more a matter of their visual aspect, from the brightness of Jupiter to the softness of the moon. The use of the name Vesper for Venus – not in the immediately previous sections – suggests a different source, one more attuned to folk wisdom than astronomical data. The relative positions of the sun, moon, and the five planets also affected how they appeared. Phases of the moon were attributed to variations in the impact of the light of the sun.

2.81. There is a brief digression on the variations of the solar orbits that cause the seasons, attributed to the slant of the zodiac. Despite the fact that the universe is always divided equally between what is above and what is below the earth, this slant causes the unequal length of the days. Thales of Miletus was said to have been the first to observe these phenomena (F24 Graham).

Determination of the moment of the solstice – when it was believed that the sun stood still (*solstitium*) – as well as the equinoxes, was a regular endeavor in the Greco-Roman world, usually performed by measuring a sequence of solar heights by a gnomon (Evans, *History* 205–7).

2.82. Another digression discusses thunderbolts (*fulmina*). This connects with the fuller discussion of the topic at *NH* 2.135–46 as well as 2.191. The source here may be Berossos, the Babylonian historian of the third century BC, whom Pliny used elsewhere (*NH* 1.7). Here he advanced a theory that they came from the three upper planets and were similar to the extrusions from a crackling fire. In a somewhat circular argument, he mixed the meteorological and mythological, connecting Jupiter the planet with Jupiter (Zeus) the god as the source of thunderbolts. The idea that the god hurled thunderbolts is as old as Homer (*Iliad* 8.133–4), but it was Anaximander who first suggested a scientific explanation, that thunder was caused by wind (F31 Graham = Seneca, *Natural Questions* 2.18). Pliny's theory that material was cast off the outer planets (attributed to unnamed scholars who assiduously studied the heavens, perhaps the Etruscans) was from Anaxagoras (F50 Graham). Yet it was rejected by Seneca, using the argument that fire goes up, not down, and in order for the theory to be valid thunder should happen far more frequently (Seneca, *Natural Questions* 2.13.1). But Pliny made an attempt to combine the theories by suggesting that moisture and disturbance in the atmosphere were also contributing factors.

2.83. The earliest calculations of planetary distances were attributed to the Pythagoreans, as Pliny noted; he probably obtained his data from the treatise of Sulpicius Gallus (*NH* 2.53). One suspects that there had been earlier speculation: Anaximander had considered the relative size of the sun, earth, and moon (F22–5 Graham), although there is no evidence of any consideration of distances. The Pythagorean figures were probably intuitive rather the result of any calculations: 126,000 stadia would be about 1,575 miles, or 2,500 km., with appropriately farther distances to the sun and the zodiac. It was only Aristarchus of Samos, in the third century BC, who actually made geometric calculations, which were recorded in his surviving treatise *On the Sizes and Distances of the Sun and Moon* (Evans, *History* 67–71). Pliny (and thus probably Sulpicius Gallus) showed no knowledge of Aristarchus' material, preferring to rely on earlier speculation.

2.84. The Pythagorean interest in harmony and music extended to comprehension of the universe, and thus planetary distances were expressed in musical terms. The extent of the universe – from the earth to the zodiac – consisted of seven tones; the concept was called "the harmony of everything," a phrase that Pliny expressed in Greek characters. The subtleties of this are complex: the basic meaning of *harmonia* is a "joint" or "fastener," perhaps originally a shipbuilding term. It was first used when Odysseus fitted together the planks to build the raft that he used to leave Kalypso (Homer, *Odyssey* 5.246–8). Thus the universe is constructed like a well-built ship, an apt analogy for the constantly seafaring Greeks. As relating to the cosmos, the phrase is Pythagorean (Jesús Luque Moreno, "Plinio, *nat.* 2.20 (§ 84): *De sideribus musica*," *RELat* 14 [2014] 53–86). Yet it took on an extended life, especially as the title of Johannes Kepler's great work, *Harmonices Mundi* (Linz 1619).

Pliny used the Greek word *phthongos* for the modal constructs of the planets; the word essentially means "sound" or a human voice (Homer, *Iliad* 5.234), but came to have a musical meaning. The Dorian and Phrygian *phthongoi* varied in antiquity (John G. Landels, *Music in*

Ancient Greece and Rome [London 1999] 87, 98–9). Pliny believed the use of modes in regard to the planets was more entertaining than serious.

2.85. The Greek stadion was a variable measurement. Many scholars have been unaware of this and have thus made erroneous conversions of the unit to modern distances, something especially pernicious in the understanding of ancient geography (Aubrey Diller, "The Ancient Measurements of the Earth," *Isis* 40 [1949] 6–9). The stadion was the most common – although by no means universal – measurement of distance in the Greek world, first mentioned in extant literature in the fifth century BC (Herodotus 1.26, etc.). Pliny regularly had a conversion factor of eight stadia to a mile, as here, but other sources – some of which were used by Pliny – had different equivalents: Polybius, for example, had 8⅓ stadia to the mile (Polybius 34.12.2a-8).

At this point Pliny was using measurements supplied by Posidonius (F120), although there were also unnamed sources ("plures") who disagreed (the clouds are either as high as 40 or 9,000 stadia). The text is uncertain, either reading "minus" or "non minus" 40 stadia, with the former more probable (Pliny, *Histoire naturelle* 2, 173–5). Pliny did not understand Posidonius' figures – a distance from the moon seems irrelevant – and he suggested that the measurements were not based on geometry. To be sure, the figures are obviously round numbers, and Pliny's reference to geometric calculation may refer to Hipparchus, although Pliny believed that figures so derived would also be estimates.

2.86–7. Pliny's figures make little sense. No source was cited, and the final conclusion, that the moon is halfway between the sun and earth, is not only bizarre but contradicts the Pythagorean distances of *NH* 2.83–4. It is possible that Pliny was actually too vague in his terminology, and his "in medio" merely means "in between." Yet, in summation, he was rather dismissive of such calculations, seeing them not only as a useless endeavor (something already suggested at 2.85) but an improper application of geometry, which, as its name implies, originated in terrestrial measurement. His reference to a *perpendiculum*, or a plumb line, basically a carpentry tool, supports this (Vitruvius 8.5.1).

Comets and Shooting Stars

2.88. Petosiris and Nechepso are pseudonyms for the authors of a body of Egyptian astrological material from the second century BC. Pliny seems to have had access to a version of the collection, or at least a summary of it (Riess, *Nechepsonis* F2; Rochberg, "Petosiris," *EANS* 637–8). It may have become popular in Rome by the early first century AD (Pliny, *Histoire naturelle* 2, 174–5). Pliny was contemptuous of the calculations.

2.89–91. The next topic is comets, called *kometes* in Greek (latinized by Pliny to *cometes*), and *crinites* in Latin, with both terms meaning essentially "long-haired" or "hairy." Comets that had an especially long tail were *pogonaios* ("bearded," latinized by Pliny; Aristotle, *Meteorologika* 1.7). Comets were first discussed by Xenophanes of Colophon in the sixth century BC (F70 Graham), and further analyzed by Anaxagoras (F3 Graham = *NH* 2.149–50; F37 Graham). There were several different types, identified by descriptive names. Pliny's immediate source is Roman rather than Greek, even though the names are all Greek (invariably latinized by Pliny) except for the *hirci*, which may be from the Samnite word for "wolf," perhaps applied due to its shagginess (Strabo, *Geography* 5.4.12).

The poem by the emperor Titus, written during his fifth consulship (AD 76), one of the latest datable references in the *Natural History*, is otherwise unknown, although Pliny praised his abilities as a poet in the preface (5). Even though the date of the poem is thus documented, the comet may have been a number of years earlier, perhaps the one reported by Josephus to have lasted for a year in AD 70 at the end of the Judaean war (Josephus, *Jewish War* 6.289; Burkhard Tautz, "Ein Komet im Jahre 76 n. Chr.?" *Gymnasium* 111 [2004] 15–23).

The horned comet that appeared at the time of the Battle of Salamis (summer 480 BC) is also otherwise undocumented, as is the comet of Ol. 108 (346 BC). The comet that appeared for only seven days was shortly after the death of Julius Caesar in 44 BC (see further, *NH* 2.94); the one lasting for 180 days was early in the principate of Nero (Seneca, *Natural Questions* 7.21.3–4).

Aristotle (*Meterologika* 1.6.343b) disagreed with those who believed that comets were due to planetary conjunctions, perhaps especially having in mind Democritus of Abdera. Yet Aristotle did not precisely write that many comets have been seen at the same time, only that they were frequent. The occurrence of comets at the south pole is a strange statement, since no one from the Greco-Roman world had even been far south of the equator, and it is probably merely an inference based on Aristotle's belief that comets are more frequent outside the tropics.

Pliny was using a popular source at this point, rather than the more scientific ones of previously: this is shown by the matter of the south pole, that comets in that region have no rays, and the mention of Typhon. He may have been quoting from the astrological material attributed to

Petosiris and Nechepso (*NH* 2.88). In the hellenized version of Egyptian myth, Typhon was one of the divine kings of Egypt and was eventually equated with Seth (Herodotus 2.144; Plutarch, *On Isis and Osiris* 41). The violent comets that appeared in Aethiopia and Egypt – which were terrible, fiery, twisted, and grim – would be appropriate for Seth, who had volcanic and meteorological aspects.

2.92–4. A further discussion of comets lists those known at Rome in the century and a half previous to Pliny. The source is unknown and may be popular memory that Pliny himself collected. It is certainly not true that comets cannot appear in the west, and this is either a piece of folk wisdom or a misunderstanding of Aristotle's statement (*Meteorologika* 1.6.343b) that comets fade away without setting (although oddly in the same passage Aristotle mentioned a comet that appeared in the west during the archonship of Asteios, or 373/2 BC).

That comets appear during periods of warfare or other times of difficulty was a common belief. The Octavius mentioned was the consul of 87 BC, who was killed while in office, inaugurating a reign of terror and the annulling of the reforms of L. Cornelius Sulla (Cicero, *de natura deorum* 2.14; Broughton 2, 45–6). This may have been an appearance of Halley's comet (Pliny, *Histoire naturelle* 2, 179). The war between Cn. Pompeius Magnus (Pompey) and Julius Caesar was in 49 and 48 BC; the comet, described with intensity by Lucan (1.524–9), was probably visible early in the conflict. The emperor Claudius was poisoned by his wife Agrippina and succeeded by her son Nero in AD 54; the contemporary comet was seen to foretell his death (Suetonius, *Claudius* 46). There were at least two comets during the reign of Nero, including the one that lasted for six months, the longest ever recorded (*NH* 2.90; Seneca, *Natural Questions* 7.21; Tacitus, *Annals* 15.47).

After some astrological comments about the significance of the location and shape of comets, Pliny turned to an account of their importance to the emperor Augustus (Octavian). The one in question, which came to be named the Sidus Iulium, appeared during games that Octavian celebrated in the late summer of 44 BC, either (as Pliny reported) those for Venus Genetrix, or those in July of that year for the Victory of Caesar (Kenneth Scott, "The *Sidus Iulium* and the Apotheosis of Caesar," *CP* 36 [1941] 257–72). The comet quickly became part of the accepted mythology of the apotheosis of Caesar, and statues, coins, and gems, showing the late Dictator crowned with a star, soon appeared (Seneca, *Natural Questions* 7.17.2; Plutarch, *Caesar* 69.3; Suetonius, *Divine Julius* 88; Dio 45.7.1–2). The Augustan poets also took up the cause (Horace, *Odes* 1.12.46–8; Vergil,

Eclogue 9.47; Vergil, *Aeneid* 8.681). According to Dio the temple in question would be that of Venus Genetrix in the Forum Iulium, but Pliny reported that it was the Temple of the Divine Julius in the Forum Romanum, and it is possible that there were appropriate statues in both locations. To be sure, Pliny's statement that a comet was worshipped in a temple is somewhat of an exaggeration.

Of particular interest is the quotation from Augustus' autobiography, which was 13 books long and went to the end of the Cantabrian War in 25 BC (*FRH* #60, T1). A maximum of 19 fragments can be identified, and this is the longest that is unquestionably verbatim. As Pliny noted, Augustus was astute in using the memory of Caesar to advance his own interests, even only a few months after the Ides of March, although it is quite possible that this was a later construct of the early years of the future Princeps (Nandini B. Pandey, "Caesar's Comet, the Julian Star, and the Invention of Augustus," *TAPA* 143 [2013] 405–49).

2.95. Eulogistic comments about Hipparchus begin a discussion (through *NH* 2.101) about other celestial phenomena related to the stars. Hipparchus' new star was presumably a nova, and cannot otherwise be identified, but it inspired him to create a catalogue of over a thousand stars in order to demonstrate that the stars did change and move (Dicks, *Geographical Fragments* 110). He also developed instruments to aid in his observations. A basic sighting tube, known as a *dioptra*, could be used in both directions to determine the positions of celestial bodies relative to one another. Hipparchus seems to have refined it, and probably used other instruments. He was also credited with inventing the technique of stereographic projection (Evans, *History* 34–5, 155, 247–8).

2.96. What Pliny called *faces* (the singular is *fax*, meaning "torch") are meteorites (Seneca, *Natural Questions* 1.1.5–6). The names for them are Greek: *lampades* (or *lampadias*, the diminutive) are torches and *bolides* are missiles, which describe their characteristics (Aristotle, *Meteorologika* 1.5). The gladiatorial games of Germanicus cannot be identified; they were probably when his career was at its peak, during the early years of Tiberius, perhaps around the time of his triumph in May of AD 17 (Tacitus, *Annals* 2.41.2).

Mutina (modern Modena), in northern Italy, was the location of the engagement between Marcus Antonius and the forces of Decimus Iunius Brutus and Octavian. Taking place from December 44 BC into the following spring, it was an important event in the legitimization of the aspirations of the future Augustus (Livy, *Summary* 117–18; Appian, *Civil War* 3.49–76; Dio 46.29–38).

There were also "beams," *dokoi* in Greek (Pliny used Greek characters), or linear bolts of light, and chasms, where the light was against a dark background, giving the appearance of depth (Aristotle, *Meteorologika* 1.5.342b). The historical reference is to a naval battle off Cnidus in 394 BC, when the Spartans were defeated by the Persians, ending Spartan sea power (Xenophon, *Hellenika* 4.3.11–14; Diodorus 14.83.4–7).

2.97. A meteorite with the color of blood was believed to be the most pernicious. The instance cited was in 350 BC, just before Philip II of Macedonia moved into Chalcidice, events leading to the subsequent destruction of Olynthos (Diodorus 16.53–5).

After this lengthy discussion of the connections between unexpected heavenly phenomena and major events, Pliny reverted to a more philosophical tone, noting that meteorites and comets are rare, and that misfortunes do not occur because of circumstances in the heavens.

2.98. After the philosophical digression, Pliny continued his catalogue of celestial events at particular times. The phenomena were discussed by Aristotle (*Meteorologika* 3.2–4), including coronas, spikes, and rings, which can be visible at times around heavenly bodies. Octavian arrived in Italy in April 44 BC and made his first public appearance in Rome in early May (Cicero, *Letters to Atticus* #359, 374). The consulship of L. Opimus and Q. Fabius was in 121 BC, that of C. Porcius Cato and M'. Acilius Balbus in 114 BC, and that of L. Julius Caesar and P. Rutilius in 90 BC. The solar phenomena associated with these years are not mentioned elsewhere.

The lengthy solar event cited would have been from sometime in late 44 BC into the following spring, although Dio placed it after the creation of the triumvirate late in 43 BC (Dio 45.17). Whatever happened, it is clear from Dio and Plutarch (*Caesar* 69) that it was not a solar eclipse but a weakening of sunlight, accompanied by a general lowering of temperatures and crop failures. This seems more a weather event – and a solar eclipse lasting more than a few hours is impossible – such as the harsh winter of 1880–1 in the North American Great Plains.

2.99. The "numerous suns" is the phenomenon known as the parhelion, first identified by Aristotle (*Meteorologika* 3.2.372a); Pliny's account is a close paraphrase of Aristotle's description. Sp. Postumius Albinus and Q. Mucius Scaevola were consuls in 174 BC, Q. Marcius Rex and M. Porcius Cato in 118 BC, Marcus Antonius and P. Cornelius Dolabella in 44 BC (the latter assumed the office after Caesar's assassination), and M. Aemilius Lepidus and L. Munatius Plancus in 42 BC. The consulship of the emperor Claudius and Ser. Cornelius Orfitus was in AD 51. The lunar counterpart of a parhelion is a *paraselene*, which is not an ancient word but one created in early modern times. The consulship of Cn. Domitius Ahenobarbus and C. Fannius was in 122 BC.

2.100. The final discussion of strange celestial phenomena begins with the so-called "night suns," not obviously explainable but perhaps an intensive glow in the sky caused by the sun or full moon before rising or after setting. C. Caecilius Metellus and Cn. Papirius Carbo were consuls in 113 BC. The burning shield (*clipeus*), perhaps similar to the amber-colored *discus* of *NH* 2.89, may be an especially large meteorite. The one that was seen during the consulship of L. Valerius Flaccus and C. Marius was in 100 BC (Seneca, *Natural Questions* 1.1.15). The falling spark (*scintilla*) during the consulship of Cn. Octavius and C. Scribonius Curio (76 BC) was probably a similar occurrence. Which of the several people named Silanus was cited by Pliny is not certain, but most probably it was M. Junius Silanus, who was proconsul of Asia in AD 54, where, presumably, he saw the *scintilla* (*PIR* I533). The stars that seem to move around the sky are probably also meteorites; the phenomenon was described in greater detail by Vergil and was thought to be connected with the winds (Vergil, *Georgics* 1.365–9).

2.101. The sparks on the javelins of soldiers and the extremities of ships are called St. Elmo's fire today, a type of electrical discharge. The phenomenon was long associated with the Dioscuri (Alkaios F34). On land, it afflicted the soldiers in Sicily during the war with Hannibal (Livy 22.1.8) as well as Caesar's troops in Africa (*de bello africo* 47). If there was only a single affected point, the phenomenon was called Helena, perhaps related to the mythological queen of Sparta, although the evidence is contradictory and obscure (Euripides, *Orestes* 1635–7; Sosibios [*FGrHist* #595] F20).

Weather

2.102. Having completed his discussion of the stars, planets, and other celestial phenomena, Pliny's next topic was the weather (through 2.153). He made a distinction between *caelum*, the traditional Latin word for the sky, and *aer*, the air (one might anachronistically say "atmosphere"), a word of Greek origin. To some extent Pliny's material parallels that of Posidonius (F120), already mentioned at *NH* 2.85, and stresses that weather is formed in the region between the earth and the moon. There is also the suggestion (perhaps obtained from Cicero, *de natura deorum* 2.56) that because of the instability of weather, it is limited to the lower regions, which thus does not negate any concept of the purity of the heavens. According to Posidonius this turbulent weather zone was limited to the lowest 40 stadia (about five miles, or eight km.).

2.103–4. Weather was attributed to pressure from the stars, with a cycle of downward movement, weakening, and then a return upward, all generated by the greater motion of the cosmos (Cicero, *de natura deorum* 2.117; Seneca, *Natural Questions* 2.4.1). Wind, needless to say, was an important element of weather. The cyclic movement and the role of the winds had been established by Aristotle (*Meteorologika* 1.9, 13).

2.105–6. It was believed that the stars exuded moisture, and this was the cause of precipitation, which was also connected to the annual cycle (Cicero, *de divinatione* 2.89; Seneca, *Natural Questions* 2.11.2). The moisture that was discharged depended on the nature and motion of the stars and planets themselves: folk wisdom believed that rainfall was more prevalent when Saturn was visible.

Popular linguistics connected the constellation Suculae with rainfall. Suculae is the Latin name (no longer used) for the Hyades, a group of stars that form the face of Taurus, including bright Aldebaran (α Tauri), the eye of the bull. The Hyades were the sisters of Hyas, and were placed in the sky after their brother's death while hunting in Libya (Ovid, *Fasti* 5.165–82). Clearly this has nothing to do with rain or pigs, and such associations were due to a Roman mistranslation of the name, believing that Hyades came from *hys*, or "pig," which led to the Roman name Suculae. Tiro, the secretary of Cicero, was quite contemptuous of the early Roman inability to understand Greek, yet he compounded the error by ignoring the Greek myth of the Hyades and insisting that the name of the stars was from Greek *hyein*, "to rain" (Aulus Gellius 13.9). This etymology and the popular wisdom resulting from it entered the literature (Vergil, *Aeneid* 1.744, 3.516).

The Haedi are the Kids, today Capella (α Aurigae) and nearby ζ Aurigae; these were the goats that nurtured the infant Zeus (Chartrand, *National Audubon Society* 443–5). Arcturus (α Boötis) is the Guardian of the Bear. It was not so much the rising and setting of these stars that were important, but their especially bright aspect: yellow-orange Arcturus, the fourth brightest star, was said to produce violent weather. Both Capella and Arcturus were important for navigation, which may have interested Pliny (Vergil, *Georgics* 1.204–7): Boötes, in which Arcturus lies, was one of the constellations that Kalypso cited in order to tell Odysseus the way home (Homer, *Odyssey* 5.272).

2.107. Canicula is Sirius (α Canis Majoris), the Dog Star. Since it was the brightest star, it was believed to have special properties: its Greek name,

Scirios (Hesiod, *Works and Days* 587, 609) means the Scorcher, and its rising in summer marked the beginning of hot weather (Chartrand, *National Audubon Society* 458–60). This was also believed to have a pernicious effect on wild animals: the oryx was allegedly the first animal to be affected, although this was disputed. Nevertheless there was a rich lore about animals and the rising of Sirius, including the predilection of dogs to catch rabies at this time of year, perhaps merely a factor of the hot weather (*NH* 8.152; Aelian, *On the Characteristics of Animals* 7.8; Plutarch, *On the Cleverness of Animals* 21.974 f).

2.108–9. The influence of stars on the weather was, to some extent, a circular argument, but it was astrologically easy to make the connection (Pliny discussed this in greater detail at *NH* 18.311–12). Seasonal changes of plants were seen as part of these phenomena, which Pliny presented as practical information in an astrological context. The turning of the leaves of the olive and certain other plants was an indicator that the solstice had passed, useful to farmers who did not have the tools to measure its actual time (Theophrastus, *Research on Plants* 1.10.1). *Puleium* is an aromatic herb, probably a type of mint and perhaps modern pennyroyal (Andrew Dalby, *Food In the Ancient World from A to Z* [London 2003] 254).

The heliotrope was the most visible proof that the sun could influence plants. Its ability to follow the sun even on cloudy days gave its movements (and the strength of the sun) an astrological interpretation: the sun did not even have to be visible to exert its power.

The discussion closes with a list of animals said to react to celestial phenomena, especially the moon. The *sorex* (shrew) and the *formica* (ant) were visible everywhere and thus could be closely observed, with their actions easily attributed to the heavens (*NH* 11.109).

2.110. A polemic follows, suggesting that human awareness of such data would be advantageous. The matter of eye disease in beasts of burden and its connection with lunar phases is another piece of folk wisdom (see also *NH* 11.149), used as a demonstration of the dangers of not understanding the influence of celestial bodies. The defense of ignorance is peculiar, since only the planets and the brightest stars exerted much influence on humans.

Hipparchus was the first to count the stars, with an enumeration of 1,080. Ptolemy had 1,022 (Pliny, *Histoire naturelle* 2, 190–1); the source for Pliny's 1,600 is unknown. Under the best conditions about 5,500 can be seen today, but, as Pliny noted, the ancient counts ignored the faintest ones. The 72 constellations is a suspiciously interesting number, perhaps Babylonian in origin, and would depend on how a constellation is defined (Tusculum 2, 231).

The Vergiliae are the Pleiades (the meaning of the Latin name is uncertain), which are the tail of Taurus, and the Suculae (see *NH* 2.106) are in his head. Boötes and the Septemtriones or Septentriones (Ursa Major, the Great Bear: see *NH* 2.178) extend across the northern sky.

2.111–13. The explanations of unpredictable weather phenomena are more scientific than astrological, beginning with an account of the causes of precipitation taken from Aristotle's *Meteorologika* (1.9). The earth produces moisture, and this eventually condenses into clouds; the sun can still be seen through clouds from an underwater perspective.

Several reasons were advanced for thunder and storms: the first is that fire falls from the stars into clouds, generating heat and producing thunder if there is wind resistance or the fires collide with one another. This was the theory of Anaxagoras, who also noted that the flashing of the fire was lightning (F50–1 Graham). Other explanations for thunder are that clouds blocked the moisture rising from the earth, resulting in the sound, or that it was caused by friction, or when clouds collide. It was also possible that there was fire trapped in the clouds, according to Empedokles of Akragas (Aristotle, *Meteorologika* 2.9.369b). But Pliny saw thunder and lightning as random phenomena and thus of no interest in the overall understanding of nature.

2.114. The next topic is the winds (though *NH* 2.130); a major source was probably Seneca (*Natural Questions* 5.1–18). In order to begin his discussion, Pliny listed several explanations for them, most of which relate to the presence of water. Aristotle commented that some believed all winds were one (*Meteorologika* 1.13.349a), but this idea was generally rejected, and Pliny, without being specific, realized that winds were diverse both in their characteristics and origin. There is no obvious reason for him to have singled out the Altanus wind from the many known by name; Vitruvius in his catalogue of over twenty winds placed it at the southwest (Vitruvius 1.6.10), but to Pliny it was a land wind, which seems to be reflected in the name.

2.115. It was obvious that winds were generated by topography and were similar to echoes. The cave of Senta, in Dalmatia, is otherwise unknown. Caves are notorious for inhalation and exhalation, yet the details of the phenomenon described are not easily explained. The rock in Cyrenaica is actually at the famous oracle of Ammon, where a cliff sacred to the south wind had the characteristics mentioned (Pomponius Mela 1.39). The matter of the containers of liquids is difficult to understand – how could they be observed if closed and in the dark? – and may have something to do with evaporation.

2.116. Pliny made a distinction between *flatus*, or gusts, and *ventus*, the winds. The same difference was implied in the Greek words *aurai* and *anemoi*. Gusts were more localized, whereas winds were more universal, and would be defined by universal rules generally attributed to cosmic motions, so they could be considered in terms of natural law. Some gusts, as the name *flatus* implies, were said to be similar to flatulence in humans, although this was disputed by Seneca, as well as the exhalation from caves and other openings in the earth (Seneca, *Natural Questions* 5.4–5; Lucan 10.247–8).

2.117–18. The unnamed 20 authors on the winds may have come from a handbook on the topic, perhaps a Stoic treatise. Anaximander was the first to consider the winds, believing that they were currents of air whose moisture and fine parts were affected by the sun (F32–3 Graham). These thoughts were refined by Anaximenes (F29–31 Graham). Many others considered the nature of the winds; the earliest extant detailed discussion is in Aristotle's *Meteorologika* (2.6).

Yet before actual examination of the topic of the winds, Pliny digressed to consider the remarkable nature of human intellectual achievements, even in perilous times. This may also be from a Stoic handbook, as the first comments are anachronistic to Pliny's day. The belief that there was discord throughout the world seems better suited to the first century BC than Pliny's own era (the brief Roman civil war of AD 68–9 was hardly worthy of such a strong statement). Moreover, piracy, which had been a serious problem in the Mediterranean during Hellenistic times, had greatly diminished by the Augustan period, due to intense Roman efforts (Philip de Souza, *Piracy in the Greco-Roman World* [Cambridge 1999] 195–200); as Pliny noted, the seas and coasts were open to everyone, another achievement of Roman supremacy. Eulogistic comments about the emperor Vespasian are surely Pliny's addition, yet in fact the emperor did play a role in the suppression of the vestiges of piracy, which had to some extent reasserted itself during the Judaean rebellion of AD 66–70 (Josephus, *Jewish War* 3.416; de Souza, *Piracy* 208–10). But Pliny, referring to the Indian trade, then turned his argument into a criticism of how a desire for profit had superseded one for knowledge, yet with the ironic suggestion that this intense mercantilism required knowledge of the winds. There had been Greek travel on the Indian Ocean since the late sixth century BC, when Skylax of Caryanda went from the mouth of the Indus to East Africa, commissioned by Darius I of Persia (*FGrHist* #709; Herodotus 4.44). Nearly two centuries later, Nearchus and Onesicritus had gone from the Indus to the Persian Gulf, as part of the expedition of

Alexander the Great (*FGrHist* #133, 134). All these may have encountered unusual winds on the Ocean (Matthew Adam Cobb, "From the Ptolemies to Augustus," in *The Indian Ocean Trade in Antiquity* [ed. Matthew Adam Cobb, Leiden 2018] 23–4), but there was no detailed knowledge of the monsoon until the late second century BC, when sophisticated understanding of it opened trade with India (*Periplous of the Erythraian Sea* 57).

2.119–21. Knowledge of winds was indeed an essential factor in Greek seamanship from earliest times. Odysseus was the first to be affected by them, both negatively and positively. Greek literature and history is replete with stories of pernicious or useful winds. Kolaios of Samos was blown all the way across the Mediterranean into the Atlantic in 630 BC, with the allegedly serendipitous result of opening trade with the wealthy regions of southwest Iberia (Herodotus 4.152). Posidonius, on a voyage from that district to Italy, was kept at sea for three months due to contrary winds, being blown all around the western Mediterranean (Posidonius T22 = Strabo, *Geography* 3.2.5).

The earliest recorded Greek comment on the topic is by Homer, who knew of four winds, essentially corresponding to the cardinal directions (*Odyssey* 5.295–6, etc.). As seamanship became more complex, more winds were added to general knowledge. Aristotle catalogued eight (*Meteorologika* 2.4–5), and Timosthenes of Rhodes, in the early third century BC, increased this to 12 (Timosthenes F6 Roller). By the first century BC, over 20 were known to Vitruvius (1.6.10). Yet for practical use, according to Pliny, eight were sufficient, essentially those familiar to Aristotle.

Pliny provided the Latin names for the winds, as well as the original Greek (latinized). His catalogue is in the form of the device today called a wind rose (a modern term, not used in antiquity), presenting the winds in a circuit, probably something invented by Timosthenes (F6–8 Roller). This begins with the Subsolanus (Greek Apeliotes) in the east, and proceeds clockwise around to the Aquilo (Greek Boreas) in the northeast. The names came from several sources, and were either mythological, ethnic, or topographical. Some of the winds existed anywhere in the Mediterranean world, but others were localized or depended on the viewer's perspective, or could even have different names in different places. For example, the Sciron wind in Attica, similar to the Argestes (from the west-northwest), and which is depicted on the Tower of the Winds in Athens, was named after a mythological personality from Megara, west-northwest of Athens, and thus would only be meaningful in that region, yet the same wind was called the Olympias in northern Greece around Mt. Olympus.

Although wind theory was a Greek concept, and to some extent originated from conditions in the Aegean and surrounding areas, there were also winds peculiar to Italy and the western Mediterranean. The *circius* was a ferocious wind in Narbonensis that cut across the sea to Italy but did not penetrate into the Gallic interior: this is surely seamen's evidence.

2.122. Winds had seasonal characteristics, and this was important to sailors. The reference to seamanship that introduces this section suggests that Pliny's source may be a sailing manual such as that by Timosthenes. Sailing in antiquity was seasonal. Hesiod reported the details of the Greek sailing season, which lasted for six to eight months, generally from March or April into late October (Hesiod, *Works and Days* 618–94; James Beresford, *The Ancient Sailing Season* [Leiden 2013] 10–12). In Pliny's time, it started earlier – in February, when winter began to moderate – which may reflect the larger ships of Hellenistic and Roman times.

Hirundo is a generic name for swallows or martins; the Greek wind name, *Chelidoniae*, refers to the same class of birds (W. Geoffrey Arnott, *Birds in the Ancient World From A to Z* [Abingdon 2007] 28–9). They return to temperate latitudes in late February or March and are considered locally to mark the beginning of spring. The more general name for the wind, *Ornithias* ("Bird Winds"), also reflected this event and implied a southerly wind (Aristotle, *Meteorologika* 2.5.362a).

2.123–5. The Vergiliae rise in May and mark the beginning of summer, whose hottest part is contemporary with the rise of Canicula (Sirius) in mid-August. At *NH* 18.269–71, Pliny provided a detailed account of the rising of various constellations, making it clear that he was using the Julian calendar established by Caesar in 46 BC. These dates of rising differ from those at *NH* 2.123, showing that Pliny was at times using pre-Roman data, perhaps Aristotle (*Meteorologika* 2.5), although it is not obvious whether *prodromi* ("Precursors") is the name of a wind or simply a description of the Aquilones. Nevertheless the *prodromi* could be hazardous; Cicero was relieved not to encounter them in July of 44 BC off the southwest coast of Italy (Cicero, *Letters to Atticus* #414).

Autumn was marked by the rising of Arcturus in mid-September. The setting of the Vergiliae in mid-November was the beginning of winter. Although this created a short autumn, it was perhaps reasonable in the temperate Mediterranean world. It was well known that one could not prepare too early for winter: as Hesiod wrote, "it will not always be summer" (*Works and Days* 503). Again the activities of birds marked the season, with the birthing of the *alcyons* (or halcyons), which gave rise to the phrase "halcyon days," still in use to refer to a period of serenity or calmness.

The ancient sailing season traditionally closed when the Pleiades set (Hesiod, *Works and Days* 618–23), although Pliny, as before, implied a longer season, and noted that there was pressure to keep sailing well into winter, since this was possible due to diminished piracy. In his day, however, it was greed, not piracy, that prolonged the season, made possible by improvements in ship construction, especially for merchant vessels (Beresford, *Ancient Sailing Season* 138).

2.126–9. Pliny changed sources, largely moving away from material oriented around seamanship to a simple catalogue of the characteristics of various winds. Most of these – dry, snowy, or warm – are issues related to weather and climate that accompany the seasons or depend on whatever wind is blowing. Although the passage ultimately derives from Aristotle's *Meteorologika* (2.6), the references to Hispania, Asia, and Africa show that the material has been adapted for a Roman audience. There is the passing comment that ships going in different directions with the same wind might collide, an interesting reference to tacking, and reflecting Pliny's practical knowledge of seamanship (Lionel Casson, *Ships and Seamanship in the Ancient World* [Princeton 1971] 273–4). At the end of the passage Pliny adopted a more philosophical mode: the gender of numbers was a Pythagorean concept (Walter Burkert, *Lore and Science in Ancient Pythagoreanism* [tr. Edwin L. Minar, Jr., Cambridge, Mass. 1972] 33–4).

2.130. Eudoxus of Cnidus, active in the fourth century BC, may have learned about the cosmic quadrennial cycle while in Egypt (Seneca, *Natural Questions* 7.3.2; Pliny, *Histoire naturelle* 2, 210). The cycle would have begun in the summer of every leap year (Pliny used the Roman term, "intercalary year").

2.131. The next topic (through 2.150) is things falling suddenly from the sky, especially thunderbolts, but including other phenomena. Again there is a navigational context, since some were especially pernicious to ships. Sudden gusts had already been attributed by Pliny to exhalation from the earth and the effects of the clouds (*NH* 2.111–12). The *ecnephias* (Pliny's latinized Greek), "from the clouds," was not so much a cloudburst in the modern sense but more of a sudden microburst of wind, which might or might not produce rain (Aristotle, *Meteorologika* 2.6.365a; Seneca, *Natural Questions* 5.12.1). The *typhon*, basically a whirlwind (*vortex*), took its name from the giant of the same name (Typhon or Typhoeus), who was cast into Tartaros by Zeus and was responsible for violent weather (Hesiod, *Theogony* 820–52).

2.132–4. Returning to nautical data, Pliny described waterspouts and their effect on ships. They existed in several forms. The common term for the phenomenon was *prester*, a Greek word that had become naturalized in Latin. The word seems to be a generic term for any kind of whirlwind, whether on land or sea, but those at sea – waterspouts – were the most dangerous (Aristotle, *Meteorologika* 3.1; Lucretius 6.423–50). Other terms were *turbo* (something twisting), *columna* (something vertical), and *aulon* (a flute); the last term was rare in this context in both Greek and Latin.

2.135–7. The frequency of thunderbolts (*fulmina*) varied according to location and weather. Pliny's account has a Roman outlook. Generally, thunder was more prevalent in spring and autumn, but did not occur where it was very cold (Scythia, the general term for the far north) or hot (Egypt). It was believed that cold extinguished the fire and if there was heat, it could not condense. Thunderbolts were said to be common in Italy, especially on the southern slopes of the Apennines, where they occurred all year, allegedly because of the temperate climate.

They could come in a variety of types, and there were many anecdotes from the time of Aristotle (*Meteorologika* 3.1) and later about the bolts causing the melting of objects or the evaporation of liquids. Lucretius wrote how they could pass through walls, melt metals, and dry up wine (Lucretius 6.225–38; see also Seneca, *Natural Questions* 2.31.1).

The Marcia cited was probably the paternal grandmother of Julius Caesar, who was descended from the Roman kings (Suetonius, *Divine Julius* 6.1); thus Pliny styled her as "princeps Romanorum." To be sure, there were many prominent women with the name in the late Republic, but this Marcia is the most likely. The anecdote was mentioned only by Pliny.

The revolt of Q. Sergius Catilina was in 63 BC, during Cicero's consulship. The poem that he wrote about the events makes note of the violent weather at that time, seen as proof of the disaster affecting the Roman state (Cicero, *de divinatione* 1.18–19). M. Herennius was a *decurio*, or city councillor, of Pompeii; Cicero, without naming him, also noted the event in his poem.

2.138–9. There is a digression on Etruscan beliefs about thunderbolts, and how the data were partially assimilated by the Romans. The *Etrusca disciplina* – "Etruscan system" – pervaded certain aspects of Roman life, from architecture to political institutions, but nowhere so much as in religion. This is one of the most thorough ancient discussions of *fulmina* (see also Seneca, *Natural Questions* 2.39–51; John of Lydia, *On Portents* 43–52). Although Pliny attributed the material to Etruscan sources,

Babylonian concepts, probably obtained from Berossos, lie behind it. Pliny used Berossos elsewhere for celestial matters (*NH* 17.19.3; Stefan Weinstock, "Libri Fvlgvrales," *PBSR* 19 (1951) 122–53).

Many Romans looked at the Etruscan influence with a certain diffidence, despite extensive popular knowledge of their lore (Cicero, *de divinatione* 1.35, 2.42–3; Seneca, *Natural Questions* 2.32). Summanus was an Etruscan divinity associated with Jupiter, and his statue was on the pediment of that god's Capitoline temple. In the early third century BC, due to a lightning strike, it was thrown into the Tiber, and Summanus then received his own temple on the slopes of the Aventine. Because he was associated with night lightning, the temple was outside the *pomerium*, the official boundary of the city. No remains are known (Richardson 373–4).

Lightning could strike in several different ways, and even rebound from the earth (rather than come from within the earth, as the Etruscans believed). It was thought that it originated from Saturn or Mars, and lightning that occurred when setting up a household could affect that person's entire life. Volsinii (modern Orvieto) was one of the 12 major Etruscan cities, but it is unknown when it was destroyed by a thunderbolt. The rich tombs around the city testify to its wealth.

The source of this Etruscan-oriented material was not specified by Pliny, but latinized Etruscan writings were among his sources for Book 2. Three authors are listed who wrote on *Etrusca disciplina*: Caecinus, who was a friend of Cicero's (*Letters to Friends* 6.7–8), Julius Aquila, who cannot be easily identified, and Tarquitius, who translated an Etruscan book of portents (Macrobius, *Saturnalia* 3.7.2). All these authors were Romans who knew Etruscan material and gave it a Roman recension.

2.140–1. The unattributed *Annals* at the beginning of the passage are probably those of L. Calpurnius Piso, noted below (*FRH* #9, F12, 15a). He was active in the mid-second century BC. The monster called Volta (or Olta, depending on the manuscripts), whose depredations were brought to an end by the king of Clusium, Porsina (or Porsenna), is not mentioned elsewhere; the event occurred at the end of the sixth century BC.

Numa was the second king of Rome, who was knowledgeable about Etrusco-Roman ritual and was able to summon thunderbolts. However, Tullus Hostilius, the sixth king, was evidently not so well informed and was struck down. This is one of many stories about his death; the thunderbolt tale may be an attempt to cover up an assassination (Dionysius of Halicarnassus, *Roman Antiquities* 3.35; *FRH* #3, 200–2).

The various shrines to Jupiter were scattered around Rome. There were two temples to Jupiter Stator; Pliny was probably referring to the earlier one, from the third century BC, whose foundations may have been identified near the Arch of Titus and which was famous in Roman history since it was where the Senate met during the Catalina episode (Cicero, *in Catilinam* 1.11). Jupiter Tonans had a shrine on the Capitoline built by Augustus after his narrow escape from lightning in Cantabria in 26 BC; whether some remains on the Capitoline belong to this temple is debatable. Jupiter Feretrius was also on the Capitoline; the temple was said to date from the time of Romulus, but there are no identifiable remains. The shrine of Jupiter Elicius ("Calling Down") was merely an altar, on the Aventine; the attribute seems to be associated with rainfall (Richardson 218–29, 225–7). In his final comments, Pliny admitted that everyone had their own opinion about the role of ritual, a somewhat more balanced statement than previous ones about religion.

2.142. Pliny considered the fact that lightning was seen before thunder, since light travelled faster than sound. This was well known: Aristotle provided the example of the difference between seeing and hearing rowing on a ship far away, and Lucretius the matter of a tree being cut down at a distance (Aristotle, *Meteorologika* 2.9.369b; Lucretius 6.167–70).

For the sun to rise on the left side requires a perspective to the south, which was the normal direction of a Roman augur (Varro, *de lingua latina* 7.7). The Romans differed from Greeks and others in that they considered the left auspicious (Cicero, *de divinatione* 2.82), despite the eventual meaning of the word *sinistra.*

2.143–4. The sixteen parts of the Etruscan sky are somewhat analogous to a wind rose (see *NH* 2.119–21). There were four primary divisions (approximating the cardinal directions), and then 12 subdivisions (Dionysius of Halicarnassus, *Roman Antiquities* 2.5). They are represented on many Etruscan artifacts, most notably the bronze liver found near Piacenza in 1877 and now in the local museum (L. B. Van der Meer, *The Bronze Liver of Piacenza: Analysis of a Polytheistic Structure* [Amsterdam 1987]). To the Etruscans, the most auspicious thunderbolts came from the northeast and returned in that direction.

L. Cornelius Sulla, in his conflict with C. Marius in 88 BC, dreamed that a goddess (whose identity is uncertain) appeared and gave him a thunderbolt, which he used to vanquish his enemies (Plutarch, *Sulla* 9). It is not possible to identify which of the several shrines to Juno in Rome is meant, and there may have been a confusion in Pliny's account. The Temple of Juno Lacina, on the Cispian hill, was struck in 190 BC (Livy 37.3.2), and that of Juno Regina at the Circus Flaminius in 156 BC (Julius Obsequens 16), but there is no other record of a strike during the consulship of M. Aemilius Scaurus in 115 BC. Pliny did not explain the context of the event and why it was related to restrictions on reporting thunderbolts.

2.145. There are various popular anecdotes about the effect of lightning on humans. No source was provided. It was believed that people could not always be killed by lightning – and indeed such is rare – and this was seen as proof of the superiority of humans.

2.146. Further stories about the properties of lightning include various ideas about how people can protect themselves from being struck. Some of these tales are associated with Augustus and Tiberius: the former always carried a sealskin (that of the Mediterranean monk seal, *Monachus monachus*) with him and retreated underground at the sign of a storm, because of his narrow escape in Cantabria when a slave next to him was killed by lightning (Suetonius, *Augustus* 29, 90). Tiberius always wore a laurel wreath during stormy weather (Suetonius, *Tiberius* 69). The incident that occurred at a locality between Tarracina in southern Latium and the Temple of Feronia (at the Lucus Feroniae, a few miles to the northwest) is otherwise unknown, but perhaps was during the Marsic (Social) War of the early first century BC.

2.147. Following is a list of strange items precipitated from the heavens. Some of these incidents were recorded in the *Acta*, the various official Roman records. There is no obvious explanation for any of them except popular rumor and unusual interpretations of celestial phenomena. The consulship of M'. Acilius Balbus and C. Porcius Cato was in 114 BC, and that of P. Volumnius Amintinus and Ser. Sulpicius Camerinus in 461 BC. This was a year of unusual portents, including a talking cow, and the events were long remembered (Livy 3.10.6; Dionysius of Halicarnassus, *Roman Antiquities* 10.2; Valerius Maximus 1.6.5). One wonders if the rain of flesh was in a battlefield context.

The death of M. Licinius Crassus in 53 BC on his ill-advised Parthian expedition is one of the better-known events of the late Republic, although any connection between it and a rain of iron in Lucania (perhaps due to vulcanism) is tenuous at best, given the high casualty rate of the expedition. L. Aemilius Paullus and C. Claudius Marcellus were consuls in 50 BC. T. Annius Milo cannot be identified, unless he was the praetor of 55 BC who was killed at Cosa in 48 BC, requiring a slight adjustment to Pliny's narrative (Broughton 2, 201). Compsa (modern Conza) is in interior Campania east of Naples.

2.148. The Cimbrian Wars were the engagements prosecuted by C. Marius at the end of the second century BC against northern peoples in what is now southeastern France and the Maritime Alps. It would hardly be surprising that battle sounds would be heard during the engagements, perhaps appearing to be in the sky due to echoes in the mountain passes. Marius' third consulship was in 103 BC. Ameria (modern Amelia) and Tuder (modern Todi) are both in Umbria. The portents were among many that appeared during Marius' campaign, but not necessarily in his presence (Plutarch, *Marius* 17).

2.149–50. The story of Anaxagoras and his rock, an event of 467 BC (see also Diogenes Laertios 2.10), is told in its greatest detail by Pliny. His emphasis is less on the event than the prediction, which would be the least reliable part of the tale. It would perhaps be more accurate to report that, from a scientific point of view, Anaxagoras, using "his knowledge of writings about the heavens," was able to predict the possibility of such an event (Graham 314). Like the eclipse "prediction" of Thales (Herodotus 1.74), the account evolved into something more than it was. Even though it is certain that there was a meteorite, it is not known when Anaxagoras made his statement, but he was responsible for the idea that the sun was a stone (F9 Graham).

Nevertheless, the meteorite became a tourist attraction, still visible in the first century AD. The Aegos (Aigos) River is a small stream in the Thracian Chersonesus that empties into the Hellespont; the primary town on the river is Aigospotamoi (near modern Cumalı), which may be where the stone was on display. At some time it came to be believed that it was a portent of the Athenian defeat there in 405 BC (Plutarch, *Lysandros* 12.1–2).

There were several other places where large meteorites were on display: the one at Abydos (at modern Maltepe in the Troad) may actually be that of Anaxagoras. Even though it was dubiously identified as another stone whose fall had been predicted by him, it seems odd that he should mention two meteorites falling a short distance from one another. Cassandria (Kassandreia), or Potidaea, is in the Chalcidice of eastern Macedonia. The city had existed since the beginning of the sixth century BC but was renamed Kassandreia in 316 BC (Diodorus 19.52); presumably the meteorite was after that date, and it is unlikely that the town was established because of the event since it was merely a typical example of city foundation or renaming in the world of the Successors of Alexander the Great.

The Vocontii lived in the mountains east of the Rhone (their major city was Vasio, modern Vaison-la-Romaine); Pliny was probably in the region when he was procurator in Narbonensis during the reign of Vespasian (Syme, "Pliny" 208–18).

2.151. Rainbows (Latin "arci") were recognized more as optical phenomena than meteorological ones: Pliny made it clear that there was nothing portentous about them. The optical explanation – light broken up – was described in detail by Aristotle (*Meteorologika* 3.2.372a, 3.4.375a), who included the corona around the full moon as a type of rainbow, but without color because it occurs at night. Yet Pliny was probably more influenced by the lengthy discussion of the phenomenon by Seneca, which is one of the first major topics in his *Natural Questions* (1.3–14), and which cites both Aristotle and the Stoics.

2.152–3. In his final comments about the atmosphere, Pliny listed a number of forms of precipitation in a straightforward catalogue. It provides the origins of hail, snow, mist, and dew, with much of the material derived from Aristotle (*Meteorologika* 1.10–12).

There are several localities where abnormal precipitation occurs; presumably these reports are more anecdotal than scientific. Africa presumably refers to the Roman province (essentially modern Tunisia), where Pliny had been procurator. Locri (Lokroi) was a Greek settlement on the east side of the heel of Italy (Strabo, *Geography* 6.1.7). Lake Velinus was north of Rome; some remnants of the lake survive along the modern Velino River between Rieti and Terni. The circumstances of Syracuse were well known, and it was especially noted for its mild winters (Seneca, *Consolation to Marcia* 17.4–5). With these comments Pliny brought his discussion of meteorology and cosmology to an end and turned his efforts for the rest of Book 2 to the earth itself.

The Earth

2.154–9. Pliny began his discussion with a highly eulogistic passage about the earth as the nurturing mother from birth until one returns to her through burial. The abundance of the earth is a familiar theme that Pliny discussed further elsewhere (*NH* 18.1, 33.2–3). It was an ancient idea that pervaded Greco-Roman thought, perhaps expressed most eloquently by Lucretius (2.991–8).

There is a distinct contrast between the violence of the heavens and the benign quality of the earth, which produces colorful flora and fauna, and even precious gems unlike anywhere else in the cosmos (*NH* 2.207). To be sure, there are poisonous animals and plants, but the earth provides medicinal remedies for them. In fact, it was believed that some animals would die of remorse after biting a human (*NH* 29.74). Pliny was dismissive of the noxious effects of poisons, noting that human misuse of them was hardly the fault of the earth.

His account digressed to a broader statement about natural resources, which in his day were obtained merely for pleasure (*deliciae*) rather than nourishment (*alimentum*). Pliny expressed indignation at mining – all for a ring to put on the finger – which would dislocate any underground creatures; Seneca noted that such activities had been part of human endeavor since the beginning (Seneca, *Natural Questions* 5.15). Oddly, Pliny wrote that poisonous flora and fauna in fact inhibited mining, but the avaricious search for wealth led to criminal activities and warfare. Yet despite this the earth was still nurturing the human race.

2.160. The first topic about the earth was the matter of its shape. Pliny unequivocally said that it was spherical, a Pythagorean idea perhaps originally published by Parmenides around the beginning of the fifth century BC (Diogenes Laertios 8.48, 9.21) and first documented in extant literature by Plato (*Phaidon* 58.108d-110b). Although the idea of a spherical earth was slow in gaining acceptance – even as late as the third century BC there were still doubts (Eratosthenes, *Geography* F25) – by Roman times it was generally agreed upon. Yet it was obvious that the earth was not a perfect sphere – mountains thousands of feet high were known – but these irregularities were trivial given its overall size (Eratosthenes, *Geography* F30). Nevertheless its sphericity was not the same as that of the heavens – it was the difference between convexity and concavity – and the rotation of the celestial sphere forces the earth into its own sphere and thus it is held together (Cicero, *de natura deorum* 2.98, 115; Manilius 1.206–14).

2.161. With the agreement that the earth was a sphere, the question arose as to why people on the opposite side would not fall off. They were considered *antipous* (having their "feet opposite"), a term that (in the plural, *antipodes*) eventually became a hypothetical toponym (Roller, *Ancient Geography* 74–5). Moreover, as soon as the size of the earth was determined by Eratosthenes in the late third century BC, it became obvious that the known part of the world was only a small portion of its entirety, limited to less than half the circumference of the northern hemisphere. This gave rise to Pliny's second question, whether people lived all around the earth. He did not answer the first question and his answer to the second is rather confused: it is by no means clear why the irregular surface – like a pine cone – should prove that people lived all the way around the earth.

2.162. Another question was why the earth remains suspended and does not fall. Again this was not really answered beyond the suggestion that it was its nature not to fall and, moreover, there was nowhere else for it to go.

Dicaearchus (Dikaiarchos) of Messana in Sicily, active in the late fourth century BC, was the first to create a method for determining the height of mountains. This was an issue that had long intrigued Greeks, but previous efforts were merely guesswork. His research was under royal patronage, but no specifics are provided. Only three short fragments of his treatise survive, but it is clear that he used an optical method developed by Euclid (Keyser, "Geographical Work" 353–61). His altitude for Pelion in Thessaly was reported by Pliny at 1,250 *passus*, perhaps from 10 stadia (about 1,850 m; the actual height is 1,547 m.). This is about 20% too high, and all Dicaearchus' known calculations have a similar error. The extravagant figure reported by Pliny for the Alps, much higher than any such amount in antiquity, is probably a textual error for 5,000 *passus*, again excessive but closer to an accurate figure. The Alps were certainly well enough known in Pliny's day to realize the problem with 50,000 *passus*.

2.163. As an introduction to his discussion of the External Sea, Pliny considered the nature of droplets of water, which are globular in form. Archimedes, in the late third century BC, had determined that the surface of any body of water was curved like the surface of the earth (*On Floating Bodies* 1, *Proposition* 2; Strabo, *Geography* 1.3.11), and this assisted in solving the problem of the External Ocean on the spherical earth.

2.164–5. Since the beginning of seamanship, it had been known that a vessel going out to sea would sink below the horizon, and conversely, land would appear to rise up as one approached it. This showed that the earth was curved, but such empirical data from seamen probably did not directly influence the Pythagorean idea of a spherical earth, which was based more on ideal concepts of number and harmony. But once it was accepted that the earth was spherical, it was necessary to account for the Ocean and explain why it would not fall off the earth. Archimedes' curved surface of all bodies of water was a partial answer, but not totally satisfactory. Thus it was suggested, rather obscurely, that all water moved from a higher to a lower point (using the analogy of rivers). This showed that water would seek the lowest point and would be pressed down, and thus would reflect the curvature of the earth (Aristotle, *On the Heavens* 2.4; *Meteorologika* 2.2; A. J. Fridh, "Les théories de l'Océan chez Pline l'Ancien," in *Göteborgs Kungl. Vetenskaps- och Vitterhets-Samhälles Handlingar* 4 [1952] 1–22).

2.166. Water and earth bond together and intermingle, thereby creating a stable world. Even when water comes to the surface at high points and appears in the manner of a siphon – such as springs and artesian wells – it returns to its own level and eventually the sea, producing an equilibrium so that the seas do not rise despite the constant draining of rivers into them.

Water was reported below the surface at great depth – as much as 300 feet – so rainwater was not thought to be the only source of the earth's moisture (Seneca, *Natural Questions* 3.3–7).

2.167. Considering the External Ocean, Pliny provided a summary of the exploration of its coasts (through *NH* 2.170). Gades (Gadeira, modern Cádiz) was an ancient Phoenician trading post, lying on the Hispanic coast a short distance outside the Pillars of Hercules. It had been reached by Greeks in 630 BC (Herodotus 4.152) and served as the contact point between the Mediterranean and Atlantic worlds, where trading voyages could be sent both north and south, along the Hispanic, Gallic, and African coasts. Carthaginians had explored this region as early as 500 BC (*NH* 2.169), and Greeks and then Romans followed.

The Septentrionales (or Septemtrionales) Ocean is the modern North Sea; the name was applied by Drusus, the brother of the emperor Tiberius, during his expedition late in 12 BC (Suetonius, *Claudius* 1.2). Pliny's notice is the earliest extant citation of the name. Tiberius himself was in the region in AD 5 (Velleius 2.106; Dio 55.28.5), and Drusus' son Germanicus followed in AD 13 (Strabo, *Geography* 7.1.3–4). Which, if any, of these expeditions is the one that Augustus commissioned to sail around Germania is not certain. Augustus reported that his fleet had sailed from the mouth of the Rhenus (Rhine) to the Cimbrian territory, where no Roman had gone before (Augustus, *Res gestae* 26). The Promontory of the Cimbrians is probably the northern tip of Jutland (Skagens Odde); from there the fleet eventually entered the Baltic (Pliny's "immense sea"), or at least heard about it. How far they went is not known; by the mid-first century AD the Romans knew about the Baltic coast as far as the Vistula, but some of this information may have been gathered from overland expeditions (*NH* 37.45; Pomponius Mela 3.30–2), and in fact a number of the details are probably as early as the expedition of Pytheas of Massilia in the late fourth century BC (Roller, *Through the Pillars* 121–3). The Romans certainly learned something about the eastern interior ("the Scythian expanse") and were aware of the frozen and wet nature of the territory, which Pliny used as proof that the External Ocean extended around to the north (D. W. L. van Son, "Zur Deutung von Plinius N. H. II 167," *Mnemosyne* 15 [1962] 146–52).

The account then examines the Ocean in the east. It was believed that the coast extended from eastern India to the Caspian Sea. The Caspian had been known to the Greek world since the expedition of Cyrus the Great of Persia in 530 BC (Herodotus 1.204); Alexander the Great and those reporting on his expedition put forth the idea that it was connected to the Ocean

(Quintus Curtius 6.4.18–19; Arrian, *Anabasis* 3.29.2, 5.26.1–2, 7.16). Seleucus I commissioned one of his officers, Patrocles, to explore the sea; he reported that it was in fact an inlet of the Ocean, a conclusion based on a mixture of preconceptions and misinterpretation of the vast mouths of rivers emptying into its north end, especially the Rha, or modern Volga (*FGrHist* #712; Roller, *Ancient Geography* 114–17). This led to the assumption that the coast of the Ocean did run from the Caspian to India, a point of view that had become somewhat obsolete in Pliny's day as more had been learned about eastern Asia beyond the Himalayas (*Periplous of the Erythraian Sea* 64). The suggestion that the Caspian and this part of the Ocean be named after the first two Seleucid kings may be from Patrocles' report.

2.168. The exploration of the Caspian shores was part of Patrocles' expedition, but Pliny was rather optimistic in believing that the entire northern coast of the inhabited world had been explored. Nothing was known in his day about the Ocean from the Baltic to eastern India, although some erroneous assumptions had been made (*NH* 2.167).

The Maeotic Marsh is the modern Sea of Azov. It had been known since at least the fifth century BC as the farthest extremity of the Mediterranean system and the division between Europe and Asia (Herodotus 4.45; *Airs, Waters, and Places* 13). At its interior end was the Greek outpost of Tanais (at modern Nedvigovka), located at the mouth of the river of the same name (the modern Don). It lay at the end of a major trade route to the interior (Strabo, *Geography* 11.2.3). This should have provided proof that the Maeotis was nowhere near the Ocean, but from the time of Alexander the Great it was believed otherwise, largely due to the manipulations of northern topography in order to enhance his reputation (Roller, *Ancient Geography* 102–4).

The next consideration is the southern part of the known world, from Gades around Africa to Arabia. Pliny's source for much of this material is the two treatises of Juba II of Mauretania, *Libyka* and *On Arabia* (*FGrHist* #275). To be sure, Alexander the Great never explored the coasts of Africa, and Pliny's description is somewhat sloppy. Nevertheless Alexander, in the last months of his life, sent out several expeditions along the coast of the Arabian peninsula (Roller, *Ancient Geography* 100–1). Pliny's Arabian Sea is the modern Red Sea, which had been explored by a certain Anaxikrates (Eratosthenes, *Geography* F95; Strabo, *Geography* 16.4.4).

The figureheads from Hispanian shipwrecks remain a curious point. The source is Juba's *On Arabia*; he was a member of Gaius Caesar's Arabian expedition, which was in the field for several years from late 2

BC (*NH* 6.141, 32.10; Roller, *World of Juba* 212–26). It is improbable that Hispanic shipwrecks reached Arabia, but a century previously Eudoxus of Cyzicus (see *NH* 2.169) had reported that one from Gades was found on the East African coast (Strabo, *Geography* 2.3.4). Pliny's report is probably a somewhat garbled account of this event. Regardless, the story of the shipwrecks remains inscrutable. It is remotely possible that ships from the far-ranging trading community of Gades reached East Africa, but it is more likely that it was information used by Eudoxus to demonstrate that a circumnavigation of Africa was possible.

2.169. The expedition of Hanno along the West African coast took place around 500 BC. A summary of his report survives in a Greek translation, and there are other notices of his cruise from as early as the fifth century BC (Herodotus 4.196; Pseudo-Skylax 112; Strabo, *Geography* 1.3.2; Jerker Blomqvist, *The Date and Origin of the Greek Version of Hanno's Periplus* [Lund 1979]). He went from Carthage into the West African tropics. Although the ultimate purpose of his voyage is unknown, it is unlikely that he intended to go as far as Arabia. Himilco, his contemporary, went north, but details of his journey remain obscure. Pliny's comment is the earliest extant, and he seems to have had access to an actual report (*NH* 1.5). According to the *Ora maritima* of Avienus, a geographical poem written in the fourth century AD but including much earlier material, Himilco reached Brittany, and perhaps the British Isles, Ireland, and the Azores (Avienus 117, 383, 412–15).

Eudoxus, from Cyzicus in the Propontis, was active from the late second century BC. The primary source for his adventures is the *Geography* of Strabo (2.3.4, from Posidonius F49), supplemented by Pliny's comment and some confused details by Pomponius Mela (3.90–5), both of whom relied on the historian Cornelius Nepos, from the first century BC (*FRH* #45). After being involved in establishing the route to India, Eudoxus had a dispute with King Ptolemy IX Lathyrus and went to Gades to outfit an expedition to reach India by means of circumnavigating Africa. He set forth, but was lost at sea, perhaps in the early 90s BC. L. Caelius (more commonly Coelius) Antipater wrote in the early first century BC (*FRH* #15), and his reference may refer to Eudoxus (with the use of "Aethiopia" as a generic term for sub-Saharan Africa, as established by Ephorus [F30a]), but this cannot be proven.

2.170. Nepos was also the source for the peculiar tale of the Indians reaching western Europe. Q. Caecilius Metellus Celer and L. Afranius were consuls in 60 BC, but Metellus Celer had had a command in Cisalpine Gaul two years previous (Broughton 2, 176), not, as Pliny had it, after his consulship. The only other report of the incident, from Pomponius Mela

(3.45), records that the Indians were presented by the king of the Boians, not the Suebians. The Boians were a central European people (Strabo, *Geography* 5.1.6) and the Suebians were east of the Rhine (Caesar, *Gallic War* 1.37), both rather far from coastal Gaul. Pomponius Mela and Pliny both implied that the Indians had come into the known world by sea, using the familiar image of being driven off course by storms (Pliny's *navigantes tempestatibus)*, but this need not mean that their sea voyage was a long one, although it has even been suggested that they were from North America, rather improbably.

It is improbable, indeed essentially impossible, that Indians made it into central Europe, and the people may have been a remote European ethnic group speaking no known language, perhaps from the northern Baltic. It may also be that Metellus Celer did encounter Indians, but elsewhere, perhaps when he was a legate in Asia in 66 BC (Dio 36.54.2–4). Yet the confused tale evolved at the expense of the facts into an example of proof of the possibility of coming from India to Europe by a northern sea route (Klaus Tausend, "Inder in Germanien," *OT* 5 [1999] 115–25; Alexander V. Podossinov, "The Indians in Northern Europe? On the Ancient Roman Notion of the Configuration of Eurasia," in *The Periphery of the Classical World in Ancient Geography and Cartography* [ed. Alexander V. Podossinov, Leuven 2014] 133–45).

Implicit in Pliny's statement about the Ocean encircling the inhabited world is the idea that there might be another part to it, in other words, another land mass, which he had already suggested at *NH* 2.161–2. This was perfectly reasonable, since it was understood that the known world was only a small portion of the surface of the earth. The idea of another continent across the Atlantic had been put forth as early as the fourth century BC (Strabo, *Geography* 2.3.6, 2.5.13). Later, Seneca famously prophesied that some day the Ocean would no longer be the limit and that new worlds would be discovered (*Medea* 375–9). Pliny did not explore the topic, and actually suggested that such thoughts were due to the failure of people to be satisfied with what they had.

2.171. Pliny believed that the inhabited part of the earth was as much as half of its total surface. Actually it was far less, as Eratosthenes had established (Eratosthenes, *Geography* F30; Strabo, *Geography* 2.5.5–6). The inhabited region was limited to the northern hemisphere, and Eratosthenes provided figures to demonstrate it was only a small portion of it.

2.172. The concept of zones originated with Parmenides (F43 Graham), who believed that there were five, with the northern and southern in constant cold and darkness and the middle one eternally hot because it was closest to the sun. Only the two temperate zones were habitable, but it was impossible to go from one to the other because of the intervening hot zone; this vaguely supports the idea that the southern temperate zone might be inhabited. Pliny's name here for the northern (Arctic) zone, Triones Septem (the Seven Oxen), is more poetic than cosmological (Vergil, *Aeneid* 1.744; Ovid, *Metamorphoses* 2.171); his name for the southern, Austrinus, merely means "southern." The term "antarctic" may not have been available to him (Geminos 5.16). Thus astronomical conditions caused much of the earth to be uninhabitable, and it was assumed that the same could be said for the Ocean.

2.173. Pliny pointed out two places where the Ocean, having almost animistic qualities, penetrated the land and created narrow isthmuses, which were seen both as a threat to the inhabited world and a primeval conflict between land and sea (Beagon, *Roman Nature* 159–61). The Arabian Bay (Red Sea) was said to lie only 115 miles (about 165 km.) from the Egyptian Sea (the part of the Mediterranean immediately north of Egypt). This is far more than the direct distance (about 120 km.), but it may reflect a route through the maze of canals between the two seas. The Caspian Sea (believed to connect with the Ocean) was only 375 miles (about 525 km.) from the Pontic (Black) Sea, an essentially accurate figure (see also *NH* 6.31). Although Pliny did not mention it, it had been known since the sixth century BC that the earth was not unchanging (Herodotus 2.12; see also Xenophanes, F59 Graham).

2.174–5. Because of rivers, wetlands, mountains, gorges, and other deserted lands, much of the landmass of the earth is uninhabitable. This is a reasonable statement, but it led Pliny to further comments about the madness (*furores*) of its inhabitants: the earth was also the place where civil wars occurred, which, ironically, made it less crowded. People appropriated the land of others and evicted those on it. To some extent, this is a comment on the agrarian issues that affected the late Republic, and the tendency of the rich to drive the poor off their land in order to create large estates, a situation that first became particularly noticeable at the time of the land reform measures of Tiberius Gracchus in the second half of the second century BC (Luuk de Ligt, "The Economy: Agrarian Change During the Second Century," in *A Companion to the Roman Republic* [ed. Nathan Rosenstein and Robert Morstein-Marx, Chichester 2007] 590–605). Pliny, however, placed this within a Stoic paradigm of the uselessness of it all because of the inevitability of death.

2.176. Pliny was totally committed to a geocentric system of the universe, arguing that the equal days and nights at the equinox proved this.

Moreover, through use of a *dioptra*, the sighting instrument probably invented by Euclid (*Optics* 19; Evans, *History* 34–5), it could be shown that at the equinox sunrise and sunset were at the same latitude, which was not the case at the solstices. Obviously Pliny's proofs were less reasonable than he believed, but provided a seemingly valid explanation for locating the earth at the center of the universe.

2.177. The seasons were created by the north–south movement of the sun. The line marking the summer solstice was where the sun "turned," or began to move south: hence the word "tropic" (Greek *tropos* or "turning"), a word in use astronomically since at least the fifth century BC (Herodotus 2.19). In antiquity, the zodiac sign associated with the summer solstice was Cancer (hence "Tropic of Cancer"), and the winter solstice and tropic were connected to Capricorn. Today the sun at the summer solstice is moving from Taurus to Gemini, and in Sagittarius at the winter one. Midway between the two tropics is the equator; the celestial equator runs through the zodiac.

2.178. Trogodytice is the region along the west coast of the southern Red Sea, between it and the Nile (*NH* 6.169). The Septentriones (Seven Oxen) are the constellation Ursa Major (the Great Bear). As one of the most prominent and earliest-identified features of the heavens, it guided Odysseus home (Homer, *Odyssey* 5.273). But as one went south it began to drop below the horizon, a matter of interest.

Canopus (α Carinae) is the second brightest star, but visible only south of 38° north latitude (that of southern Italy). It was first identified by Eudoxus of Cnidus in the fourth century BC at his observatory on Cnidus, and then studied by Posidonius when he lived in Gades (Strabo, *Geography* 2.5.14; Posidonius F204–5). It was important for navigation, named after Kanobos, the pilot of Menelaus.

The Crinis Berenices (Coma Berenices, the Lock of Berenice) is a small, dim constellation near the hindquarters of Leo (Chartrand, *National Audubon Society* 489–92). It was identified by the astronomer Konon and named in honor of Berenice II of Egypt, who had dedicated a lock of her hair to Aphrodite after the safe return of her husband, Ptolemy III, from war in 246 BC. The circumstances are best documented by Catullus' translation of a poem by Callimachus (Catullus 66). Canopus and the Coma Berenice were at times mentioned together as celestial features recently discovered (Strabo, *Geography* 1.1.6). But Pliny's citation of the latter is peculiar: it is much farther north than Canopus and is visible throughout Italy. The Throne of Caesar is otherwise unknown, although it may be the comet that appeared at the death of Julius Caesar (*NH* 2.94;

Pliny, *Histoire naturelle* 2, 233). The passage is confusing because the antecedent of "ibi" ("there") – where the stars are conspicuous – is unclear.

A list of places, catalogued from south to north, demonstrates the visibility of Canopus, from Alexandria, where it is elevated, to Pontus (presumably the Roman province on the south coast of the Black Sea), where it cannot be seen. Then there is a return south, showing the relative altitude of the Septentriones, from high in the sky in Pontus to limited visibility at Meroë (which lies near the summer tropic). This list must be later than Eudoxus because of the citation of Alexandria, and perhaps was from Hipparchus or Posidonius.

2.179. After noting that the data presented are largely due to the reports of sailors, Pliny stated that the changes in elevation of celestial bodies are due to the curvature of the earth. It was believed that the earth curved up to the north and down to the south (Diodorus 1.40.5; Vergil, *Georgics* 1.240–6). The passage suggests that people live south of the equator, where the earth slopes downward (*devexitas*), yet few if any people from the Greco-Roman world had gone that far south. The most likely person to have reported on the equatorial regions, and perhaps what lay to the south, was Polybius, who went into the West African tropics after the fall of Carthage and published a treatise on his discoveries (Polybius 34.1.7; *NH* 5.9–10).

2.180. The differing times of eclipses at various points was the basis of determining longitude, something realized by Hipparchus in the second century BC (Hipparchus, *Against the Geography of Eratosthenes,* F11). Pliny gave two examples of such data (although only the earlier could have been known to Hipparchus). Arbela (at modern Erbil in northern Mesopotamia) was near where Alexander the Great defeated Darius III of Persia on 1 October 331 BC (Plutarch, *Alexander* 31–3; Arrian, *Anabasis* 3.8–15). This seems to have been the only such example known to Hipparchus, and although the difference in lunar eclipse times between Arbela and Sicily does not fit the relative longitude (Dicks, *Geographical Fragments* 121–2). the data became important in understanding longitude as well as the sphericity of the earth (Kleomedes 1.5.30; Ptolemy, *Geographical Guide* 1.4; Bowen and Todd, *Cleomedes' Lectures* 67).

On 30 April AD 59, when C. Vipstanus Apronianus and C. Fonteius Capito were consuls, Cn. Domitius Corbulo was on campaign in Armenia because of Parthian interference in the Armenian kingship (Tacitus, *Annals* 13.8, etc.), and was able to observe a solar eclipse. In Italy it occurred at about the time of Nero's murder of his mother Agrippina and was seen to be somehow connected with that event (Tacitus, *Annals* 14.12). The difference in time between Armenia and Campania is remarkably accurate (Pliny, *Histoire naturelle* 2, 234).

2.181. There are several anecdotes about the different lengths of day in various places, all of which are unrealistic. The Towers of Hannibal were signaling stations, a regular feature of ancient warfare, but those mentioned in Africa and Hispania have not been located. The exact context and position of the signal fires in Asia used in warfare against pirates are not specified. The fires would have to be many miles apart – probably too distant from one another to be seen – and the story may be a popular soldier's tale based on misinformation: the time difference would be nine hours. Yet one is reminded of the account of signal fires in Aeschylus' *Agamemnon* (281–316).

Philonides, from Crete, was a *hemerodromos*, or day runner, for Alexander, and also a bematist, or recorder of distances. He published some of his data but only a few references survive (*FGrHist* #121). There was also a statue of him at Olympia that was still visible in the second century AD (Pausanias 6.16.5; Heckel, *Who's Who* 216). The distance from Sicyon west to Elis (across the northern edge of the Peloponnesus) was reported at 1,200 stadia (about 150 miles or 240 km.), and Philonides' timings were comparable to those of other known runners (*NH* 7.84). The context is not recorded, but he made the journey regularly, and it probably was a part of his life separate from that with Alexander. Both towns are coastal, yet it is more likely that he would have cut inland, creating a longer stretch of downhill running on the return. Pliny's account is not totally clear, but it seems that the westbound run was nine hours and the eastbound 15. It was a common yet untrue belief that the movement of the sun accelerated motion, with the westbound journey faster than the eastbound.

2.182. Timekeeping in antiquity was dependent on the length of the day, which was always divided into 12 hours, and which would vary as the latitude changed. An ancient hour was between 45 and 75 modern minutes (Bickerman, *Chronology* 15). This difference would be visible over an extent of 300–500 stadia (about 35–60 miles, or 60–100 km.). Use of a gnomon at noon, especially at the solstice, would detect the variances in solar shadows, as Pliny showed: going from Egypt to Rome and north to Ancona and Venetia (the region at the head of the Adriatic), the shadow measurement went from half the length of the gnomon to its actual length (the datum for Ancona is wrong: the shadow should be shorter). The idea of using a gnomon to measure the length of day (and thus latitude) was probably developed by Eratosthenes (*Geography* F54). Pliny preferred the vague term *horoscopium* – a timepiece – as an instrument to make the calculations.

2.183. Those who went up the Nile to Syene (modern Aswan, at the First Cataract, at 24° north latitude) were aware of the changes resulting from being at low latitudes. Syene was just north of the Tropic of Cancer and thus had no shadows at the solstice; this became the basis of Eratosthenes' calculations of the size of the earth (Eratosthenes, *Geography* F41–3). It is improbable that the famous well at Syene was constructed merely for observing the sun at the solstice. In fact, despite popular opinion, it was probably not used by Eratosthenes: it is not mentioned in any extant source before the Roman period, by which time it had become a tourist attraction (Strabo, *Geography* 17.1.48).

Onesicritus was the pilot for Alexander's voyage on the Indus and back by sea to the Persian Gulf (*FGrHist* #134, F9). The insertion of his datum seems somewhat out of place, but may in fact be from Eratosthenes' *Geography*, although the members of Alexander's expedition, who went as far south as 25° north latitude, would also have observed the astronomical changes. The Hypasis (modern Beas) River is well north of the latitude of Syene, but Onesicritus would have been at its latitude at the mouth of the Indus.

Berenice (Berenike) Trogodytice was founded by Ptolemy II at the Red Sea end of the route from the Nile, and was the major port of embarkation for the India trade (Strabo, *Geography* 16.4.5; *NH* 6.168). Ptolemais, 4,820 stadia (about 600 miles or 960 km.) farther down the coast, was founded at about the same time as a staging point for elephant hunts. It cannot exactly be located today, but would have been well within the tropics, and thus at certain times of the year shadows would have been toward the south (Strabo, *Geography* 16.4.7; Cohen, *Syria* 341–3).

2.184–5. Meroë was the main city of the Meroitic kingdom, and had been known to Greeks since the time of Herodotus (2.29). It was considered the southernmost major city in the world. Its distance from Syene, and also the local astronomical data, may have been provided by Philon, who was sent there by Ptolemy II in the second century BC. He wrote a treatise on his observations (*FGrHist* #670), which was used by Eratosthenes (*Geography* F40; Strabo, *Geography* 2.1.20; *NH* 37.108).

The Indian mountain named Maleus, in the territory of the Oretians (at *NH* 6.69 it is located in the land of the Suarians) would be well into southern India, beyond regions reached by Alexander and his people. Information on it may have come from Megasthenes, who was at the Mauryan court in Pataliputra around 300 BC, and received information from all over India (*FGrHist* #715, F7b). Maleus cannot be identified, and its presumed southern location may mean that the name was confused with

Malaia, the central mountain of Sri Lanka. The Oretians lay west of the mouth of the Indus (Diodorus 17.105.1; Arrian, *Anabasis* 6.21–2), but the material on Maleus lacks any clarity and is beyond comprehension (Klaus Karttunen, *India and the Hellenistic World* [Helsinki 1997] 127–8).

Patalis was at the mouth of the Indus, hardly at the presumed latitude of Maleus. Alexander and his expedition resided there in the summer of 325 BC prepatory to the journey back west. Some of Pliny's material continues to be from Onesicritus (F10), who used the word *askios* to describe a land without shadows; Pliny latinized this into *ascius*. But it is probable that all Pliny's material in these sections came through Eratosthenes (*Geography* F41), mentioned decisively at the end.

2.186. Eratosthenes established a prime meridian based on the length of day at various points from Syene north to the Black Sea (Eratosthenes, *Geography* F59–60). Pliny presented a revised and romanized version of this, eliminating all the Greek locations north of Alexandria, and substituting Italia and Britannia. Yet his list is vague, given the north–south extent of Italia and no indication of where his length of day was determined. Britannia, while a meaningful toponym to the Romans, also extends over a broad latitude, and the unspecified datum point was probably originally obtained from Pytheas of Massilia (*NH* 2.187), the explorer of the late fourth century BC who calculated several instances of the length of day in the British Isles (Strabo, *Geography* 2.5.8). Moreover, "Britannia" is a Roman term; Pytheas used "Prettanike." The phrases "bright nights in summer" and "narrow course of light" (*angusto lucis ambitu*, perhaps the *aurora borealis*) may also be from Pytheas, since he documented various examples of unusual Arctic celestial phenomena. Although it could be established theoretically that there would be six months of continuous daylight and six months of continuous darkness at high latitudes, Pytheas was probably the first person from the Mediterranean world to experience the phenomenon (Hipparchus, *Against the Geography of Eratosthenes* F57–8; Strabo, *Geography* 2.1.18, 2.5.42; Roller, *Through the Pillars* 71–2).

2.187. Pytheas' most famous discovery was Thule (Thoule), whose location is still greatly debated (T18a): his journey was described in his treatise, *On the Ocean.* His six-day journey supports the most probable location for Thule, Iceland (Pytheas F2; Strabo, *Geography* 1.4.2; Roller, *Through the Pillars* 57–91).

Mona is modern Anglesey and Camalodunum is modern Colchester. The distance of 200 miles (about 300 km.) is shorter than the modern one, but the Roman road system would have provided a more direct route than is available today. Mona, at approximately 53° north latitude, would have had completely different hours of sunlight than Thule, which cannot have been farther south than 60° north. Information about Mona, however inaccurate, probably came from the expedition of C. Suetonius Paulinus in AD 59–60, whose material was known to Pliny (*NH* 5.14–15; Tacitus, *Annals* 14.29).

Anaximenes was the third of the Ionian monists. Pliny credited him with having been the first to use the gnomon (Pliny used a Greek word, *gnomonikos*, but in latinized form yet with a Greek ending) for astronomical purposes. Anaximenes was also said to have used a *horologium* (or *horoskopeion*), an instrument to mark the hours, as well as the *sciothericon* (presented in latinized Greek; see Strabo, *Geography* 2.5.24), which was also a means of measuring shadows, perhaps little different from a gnomon. Yet all these instruments are generally said to be the inventions of Anaximander, and in fact his career was associated with Lacedaemon (Diogenes Laertios 2.1–2). It is probable that Pliny was simply in error in attributing the gnomon to him.

2.188. There were different ways of considering when the day began and ended, as continues to be the case today. Pliny, in one of the most complete passages on the topic, gave several examples, perhaps taken from Hipparchus. The last sentence, regarding the effect of the slant of the zodiac, somewhat repeats *NH* 2.81.

2.189–90. The belief that the physical and personal characteristics of people were influenced by the environment and climate in which they lived was a long-standing point of view, first set forth in the treatise of the fifth century BC, *Airs, Waters, and Places* (5–7, 13–24). The Aethiopians were seen as the epitome of the effects of climate, in their case heat, because they were believed to be closer to the sun. Those in the north had opposite characteristics because of their cold climate. But the people of Italy, in the middle latitudes of the inhabited world, had a temperate climate and thus were influenced by both the south and north, and were more balanced. These issues were discussed in greater detail by Vitruvius (6.1.3–4, 9–11; see also Strabo, *Geography* 2.3.7, 17.2.1). Thus the Italians could keep both the southerners and northerners under control, a statement with political overtones. Animals were also affected by their climate.

2.191. A section on earthquakes (through *NH* 2.200) is largely dependent on Aristotle (*Meteorologika* 2.8). The Babylonians believed that earthquakes – and indeed all things – were influenced by planetary forces. Pliny suggested that such influences were a possible reason that Anaximander was able to predict an earthquake in Lacedaemon, a seismic event so great that parts of Mt. Taÿgetus, to the west of Sparta, also collapsed. He may have been able to make his prediction for the same reason as Pherecydes of Syros, because of hints of tectonic activity in the water levels in a well. Since the two were contemporaries, the earthquakes may have been the same event. The belief that Pherecydes was the teacher of Pythagoras was a common assumption, although the evidence is vague and uncertain (Diogenes Laertios 1.119).

2.192–5. After a passing comment on the divine wisdom of early scholars, Pliny, continuing to follow Aristotle, attributed earthquakes to the winds, but this was just one of many suggested causes (Seneca, *Natural Questions* 6.5–20). There was also some understanding that birds could sense a coming quake, and a suggested connection with exhalations from caves.

The effects of earthquakes were widespread. The eastern Mediterranean, as far west as Italy, is a major earthquake zone, and thus there was extensive evidence about what happened when a quake occurred, yet relatively little understanding of the ultimate causes of seismic activity. But there was a realization that they were somehow connected with the various underground passages through which the phenomena could pass. Presumably Pliny's knowledge of the Alps – the implication is that his understanding of local quakes was due to a lengthy period of time spent in the region – was due to his upbringing in Comum and perhaps his military service in Germania, which might have required several journeys through the mountains. Gallia and Egypt are not major earthquake zones, and it was believed that the seeming lack of bedrock in Egypt meant few quakes, with the muddy soil lacking the openings in the earth that produced them (Seneca, *Natural Questions* 6.26). The association of earthquakes with specific times of day is purely anecdotal. Aristotle was somewhat diffident about making any connection between quakes and eclipses.

2.196. Sailors encountered earthquakes at sea, which might be called tsunamis today, and it was realized that they were due to the motion of the seabed (Seneca, *Natural Questions* 6.24.3). The unusual clouds that were said to be seen before quakes were attributed by Aristotle to the disappearance of the wind.

2.197. Pherecydes had connected earthquakes with unusual activity in wells (*NH* 2.191), yet it was believed that an abundance of water channels, especially in an urban environment, was a protection against them. Arches and the corners of buildings were thought to be safer than the rest of the structure, because of their elasticity, and brick was believed to be less vulnerable than stone masonry. Many of these assumptions are popular wisdom not easily proven, although Neapolis (modern Naples) was, and remains, one of the most earthquake-prone cities in the Mediterranean.

2.198. There are three types of earthquakes. A quivering one is the safest, less so is a swelling and sinking one, and a fluctuating one is the most dangerous. But there is no evidence of an earthquake ever lasting more than a few minutes: the length of 40 days up to a couple of years presumably refers to aftershocks or a series of quakes. These figures were taken by Pliny directly from Aristotle's *Meteorologika.*

2.199. A list of known earthquakes is based on material that Pliny said he found in Etruscan documents, but the reason for this is not clear, since with one exception all are from the first century BC and presumably would have been recorded elsewhere.

L. Marcius Philippus and Sextus Julius Caesar were consuls in 91 BC. Mutina (modern Modena) is in a flat plain; the quake occurred in the Apennine Mountains to the south of the city, with its epicenter around modern Fiesole (E. Boschi *et al.*, "Catalogue of Strong Italian Earthquakes From 461 B.C. to 1997," *Annals of Geophysics* 43 [2000] 845). The Via Aemilia, built by M. Aemilius Lepidus in 187 BC, extended from Ariminium (modern Rimini) on the Adriatic coast along the northern edge of the mountains to Placentia (modern Piacenza). Pliny did not explain why the sightseers were mostly equestrians, but as an equestrian himself he may have had access to information from members of that order, and throughout the *Natural History* he relied on equestrian informants (Murphy, *Pliny* 61). The earthquake, whose extent and damage were widespread, could be seen as an indirect portent for the Marsic (Social) War that began the following year.

Nero committed suicide on 9 June AD 68, so the last year of his principate would have begun the previous summer. There were two Italian quakes that year, one with an epicenter at Rome itself, and the other at Teate (modern Chieti), which is the more likely since it was in the territory of the Marrucians (Boschi, "Catalogue" 846). They lived on the Adriatic side of Italy, inland from Aternum (modern Pescara) and along the Aternus (modern Pescara) River. Running through its valley was the Via Claudia Valeria, a secondary road and probably the one mentioned by Pliny. Vettius Marcellus, also mentioned in an inscription (*CIL* 9.3019), was Nero's land agent in the region.

There is a rare reference to Pliny's historical work. It covered a period from the mid-50s AD (when the history of Aufidius Bassus [*FRH* #78] ended) to the triumph of Vespasian and Titus in AD 71 (Pliny the Younger, *Letters* 3.5.1–8; *FRH* 1, 532–3). Only a handful of fragments survive (*FRH* #80, F4–8).

2.200. Earthquakes at sea are essentially tsunamis and cause coastal surges. The earthquake of AD 17 was one of great devastation, occurring in the volcanically active region of Lydia and Caria, where a number of towns were struck, especially Sardis and Magnesia. Tiberius provided 10 million sesterces and remission of taxes as disaster relief (Strabo, *Geography* 12.8.18; Tacitus, *Annals* 2.47).

The battle at Lake Trasimenus between the Carthaginians and the Romans was in the spring of 217 BC, at the beginning of the Second Punic War (Livy 22.4–7). The lake lies to the west of modern Perugia. Earthquakes as premonitions was an Etruscan belief, still respected in the time of Cicero (Cicero, *de divinatione* 1.35).

2.201–2. The next topic (through *NH* 2.206) is the creation and destruction of land, generally islands or in coastal regions, and seen as a natural attempt to create balance. Siltation was – and largely remains – notoriously visible at the several places mentioned. Granted, Homer's report (*Odyssey* 4.354–7) that the island of Pharos was a day and a night from the mainland can hardly be sustained, since it has been within the limits of Alexandria since that city was founded. But Herodotus' statement (2.10) that the Nile valley was once a gulf of the sea is probably the most extreme example of siltation. Cercei (Greek Kirkaion, modern Monte Circeo), in southern Latium and the alleged home of the sorceress Circe, is an isolated promontory on the coast (541 m. high) that was certainly once an island (Strabo, *Geography* 5.3.6).

Extension of coastlines occurred at the places mentioned and continues today, especially on the eastern shore of the Aegean, where localities that were once coastal, such as Ilium (Troy) and Ephesus, are now well inland. The most visible example is the lower Maeander River (modern Menderes), where the coast has advanced over 40 km. since the sixth century BC.

Delos floated until Poseidon anchored it so that Leto might have a place to give birth to Apollo and Artemis (Callimachus, *Hymn* 4.28–40). Rhodes emerged from the sea as Helios' portion of the earth (Pindar, *Olympian* 7.54–70). Anaphe, in the southern Aegean, was created by Apollo for the Argonauts when they were desperate for a landing place (Apollonios 4.1717–20). All these may be early memories of seismic events. Nothing is known about the situation at Neae (near Lemnos) or Halone (in the Propontis).

Therasia, Hiera, and Automate are all part of the Thera caldera system, in the southern Aegean, and continue to be active volcanically. The eruptions listed by Pliny were (probably) in 237 BC, 197 BC, and July AD 19 (when M. Junius Silanus and L. Norbanus Balbus were consuls). The one of 197 BC was the largest, when the island of Hiera (also called

Automate, "Spontaneous," modern Palaia Kaimena) was created (P. Y. Forsyth, "After the Big Bang: Eruptive Activity in the Caldera of Greco-Roman Thera," *GRBS* 33 [1992] 191–204).

2.203. The Aeolian Islands are the modern Eolie or Lipari group north of Sicily, part of the same volcanic complex that includes Etna and Vesuvius. They remain highly active today. Pliny may be referring to the appearance of the island of Hiera (modern Volcano), an event that Posidonius noted had happened within his memory, in the late second century BC (Posidonius F227; Strabo, *Geography* 6.2.11). The island near Crete cannot be identified, nor can the one appearing in 126 BC in the "Tuscan Bay," a toponym too vague to be located. But the effects of seismic phenomena on fish were well known. Pithecussae is modern Ischia, at the north end of the Bay of Naples (Pliny's Campanian Bay), and nearby Prochyta is modern Procida, with Epopos (or Epomeus) the highest point on the former. These islands were so active volcanically as to discourage settlement (Strabo, *Geography* 5.4.9).

2.204. There is a lengthy catalogue of islands that have either become separated from the mainland, or joined to it. There is no reason to doubt the basic geological principles involved. Sicily, Cyprus, and Euboea are among the largest islands in the Mediterranean. Atalante is off the coast of mainland Greece near Euboea, and Macris, or Makris, is southeast of Euboea (*NH* 4.64). A major earthquake of 426 BC, described by Thucydides, may have made significant adjustments to the land forms in this region (Thucydides 3.89; Strabo, *Geography* 1.3.20).

Besbicus (modern İmralı Adası) is an island in the Propontis. Leucosia (modern Licosa) is off the coast of Lucania, named after one of the Sirens, whose body was cast ashore here after she committed suicide (Strabo, *Geography* 6.1.1). But the Promontory of the Sirens is some distance away, at modern Punta Campanella, the southern end of the mainland portion of the caldera that forms the Bay of Naples and a region also connected with the Sirens.

There are 12 localities listed where islands have become joined to the mainland. Most of them are on the coast of western Asia Minor, which has the highest amount of siltation in the Mediterranean. Hybanda, the most notorious example, is probably the hillock of Özbaşı in the Maeander plain southeast of Priene. If Pliny's *nunc* ("now") refers to his own era and not that of his source, by the middle of the first century BC it was 200 stadia (about 25 miles or 40 km.) from the coast.

2.205. Plato's Atlantis was seen as the best proof of extensive land that had disappeared (Plato, *Timaios* 24–6; Posidonius F49; Strabo, *Geography*

2.3.6). Each of the examples that follows is where the sea passes through a strait, such as the entrance to the Ambracian Gulf in northwest Greece, and, just to the south, the separation between Leucas (Leucadia) and the mainland. Pliny also mentioned the narrows at the end of the Corinthian Gulf (at Antirhium), the system of the Hellespont, Propontis, and Thracian Bosporus that creates the passage from the Aegean to the Black Sea, and the Cimmerian Bosporus at its north end, which is the strait to the Maeotis. All these are locations that were believed to have once been land but which were turned into straits by the action of the sea.

Cibotus (Kibotos) was one of the names of Apamea in Phrygia (Cohen, *Europe* 281–2) but no mountain of that name is otherwise known, nor any place called Carice (Karike). Mt. Sipylus (modern Spil Dağı, in Manisa) and nearby Magnesia (modern Manisa) were the site of an early earthquake, associated with the mythical king Tantalos (hence the site of Tantalis), whose punishment created the word "tantalize" (Homer, *Odyssey* 11.582–92; Pindar, *Olympian* 1.54–71; Strabo, *Geography* 1.3.17). Galene (or Galenis) cannot be specifically located. Gamale lies east of the northern Sea of Galilee. No known seismic event can be associated with either. Pliny's patron and the future emperor Titus had been at Gamale during the Jewish War, and may have been the source of his information (Suetonius, *Titus* 4). Phegium in Aethiopia is otherwise unknown.

2.206. Pyrrha and Antissa are common toponyms in the Greek world, but no places with those names are known around the Maeotis (modern Sea of Azov). Helice (near modern Aigion) and Bura (near modern Diakofto) were two small towns in Achaia, in the northern Peloponnesus, which were destroyed by an earthquake in 373 BC. This is a well-documented seismic event, since Helice was visited shortly afterward by Heraclides of Pontus (F26a) and then by Eratosthenes in the following century (*Geography* F139; see also Diodorus 15.48.3; Ovid, *Metamorphoses* 15.293–5; Strabo, *Geography* 8.7.2, 5). The remains of Helice were long visible offshore, but Bura was rebuilt (Pausanias 7.25.8–9).

Cea (Ceos) lies just southeast of the Attic mainland in the Aegean. A collapse of 30,000 paces (30 miles or 48 km.) is impossible (the island is only about 12 km. long), but this is probably an exaggerated report of a major coastal land slippage. Tyndaris (modern Tindari) is on the northeastern coast of Sicily, the closest town to the volcanically active Aeolian islands (see *NH* 2.203). Eleusis in Boeotia is at the southwest corner of the Copaic basin and was destroyed by a winter flood – lying in a region subject to such events – rather than any earthquake (Strabo, *Geography* 9.2.18; Pausanias 9.24.2). These are Pliny's final comments about earthquakes.

2.207. There follows a transitional passage regarding various phenomena exuding from the earth (the stone that blocks every kind of light except the sun is probably the Cappadocian *phengites* of *NH* 36.163, still found today). Some of the phenomena are beneficial, others less so. Particular attention was given to holes or caves that send forth deadly fumes. Mt. Soracte (modern Soratte) is an isolated mountain (691 m. high) immediately north of Rome, where there was a poisonous spring (*NH* 31.27). Sinuessa (Sinoessa) is on the border of Latium and Campania, and was a therapeutic spa (*NH* 31.8, Strabo, *Geography* 5.3.4), but there were presumably also dangerous exhalations in the region. Puteoli (modern Pozzuoli), just west of Naples, was in such a noxious location that it was said the name meant "stink" (Varro, *de lingua latina* 5.25; Strabo, *Geography* 5.4.6). Fumes are still prominent in the region today.

2.208. *Spiracula*, or vents, were also called Charonea, or shrines to Charon, because they were believed to be openings to the underworld, generally emitting dangerous fumes. Many of these existed, especially in Asia Minor (Strabo, *Geography* 12.8.17, 13.4.14, 14.1.44), although Pliny's first example is from Italy, at Ampsanctus, a crater lake east of Naples in the territory of the Hirpinians. It was described by Vergil (*Aeneid* 7.563–71), and is in a region where sulphur vapors are still apparent.

Hierapolis (modern Pamukkale) in Phrygia was an ancient sanctuary to the Great Mother. Strabo provided an eyewitness account of the effect of its fumes on all living creatures except the priest (Strabo, *Geography* 13.4.14). Delphi was well known as a place where subterranean exhalations were said to give the religious officials their prophetic ability, created by the rotting corpse of the python killed by Apollo (Homeric *Hymn to Apollo* 370–4).

2.209. Unstable earth was found in a number of places, such as at Gabii (modern Castiglione) just east of Rome. An area of 200 *iugera* would be about 200 feet or 60 m. on a side. Reate (modern Rieti) is in northern Latium near the border of Umbria. Both these locations are in regions that were historically marshland, which is probably the reason for the circumstances reported by Pliny.

Floating islands were common in the Italian lakes, easily explained as moving masses of vegetation. Lake Vadimonis (modern Laghetto di Bassano) is on the right bank of the Tiber north of Rome. Aquae Cutiliae was a spa east of Rome, located at the small lake of Cutiliensis (modern Laghetto di Paterno).

The Calaminae islands have not been located; presumably they were a refuge for those escaping the massacre of Italians in the province of Asia ordered by Mithridates VI in 88 BC (Duane W. Roller, *Empire of the Black*

Sea: the Rise and Fall of the Mithridatic World [Oxford 2020] 148–51). Pliny did not specify which of the many places named Nymphaeum was meant, but the Saliares were a group of Roman priests whose ritual included dancing, originally to the gods of war (Dionysius of Halicarnassus, *Roman Antiquities* 2.70; Varro, *de lingua latina* 5.85). The Tarquinsian (normally Volsiniensis) Lake, modern Lago di Bolsena, is the largest of the crater lakes of Etruria.

2.210. Unusual phenomena were often associated with religious establishments. Rain did not fall at a portion of the shrine of Paphian Aphrodite on Cyprus (Tacitus, *Histories* 2.3). Nea (or Nee) was a little-known sanctuary of Athena on the coast of the Troad between Alexandria Troas and the Scamander River (*NH* 5.124).

Harpasa (modern Arpaz) is a village in northern Caria. Therapeutic soil remains a feature of many religious shrines today. Parasinus is an unknown location on the peninsula of the Tauric Chersonesus (modern Crimea). Sarcophagus ("flesh-eating") stone is a type of limestone found in the region of Assos (in the southwestern Troad) and elsewhere (*NH* 36.131). Also known as Assian stone, it had medicinal uses, but could be deleterious (Dioskourides 5.124).

2.211. Magnetic stones were a feature of accounts of India, beginning with Ctesias in the early fourth century BC (*FGrHist* #688, F45.6; Karttunen, *India and the Hellenistic World* 233). These have never been fully explained, but magnetism is a property of many gems (Donald B. Hoover, "Determining Garnet Composition From Magnetic Susceptibility and Other Properties," *Gems and Gemology* 2011, 272–85). The other items in this section are anecdotal: Locri and Croton are in the southernmost part of Italy, Ilium is the site of ancient Troy, and Lycia a district in southern Asia Minor. The territory of Arpi is northwest of modern Foggia in Italy, where the toponym Arpino Nova survives (*NH* 3.105; Strabo, *Geography* 6.3.9). The Mucian Altars were near Veii (just north of Rome), and may be associated with C. Mucius Scaevola, a Roman hero of the late sixth century BC (Livy 2.12–13; Dionysius of Halicarnassus, *Roman Antiquities* 5.24–31). Tusculum is southwest of Rome at modern Tuscolo, and the Cimician woods are around the lake and mountain of the same name in northern Etruria. Crustuminium is in the Marcigliana district of northern Rome.

Hydrology

2.212–17. The next topic is the tides, something hardly understood in antiquity. The only source mentioned is Pytheas, the Arctic explorer of the late fourth century BC. He was an early theorist about the tides and was probably Pliny's primary informant (Pytheas T19). Some of his material may also be from Posidonius (F215, 217), who may himself have partially relied on Pytheas, but who added his own observations and conclusions based on his residency in Gades.

The only comment specifically attributed to Pytheas is that tides reached a height of 80 cubits (Pliny's translation of the Greek *peches*), about 120 feet or 35 m. This is impossible, and assuming that the figure is a correct report, Pytheas may have experienced a storm surge (Roller, *Through the Pillars* 76–7). Yet he was probably the first to connect tidal activity to the moon, and to be explicit about the daily tidal cycle. Pliny had to make it clear that this was measured in constant hours, not the variable ones used in everyday life. The data about the relationship of the lunar cycle to the tides are remarkably detailed, but not accurate – as Pliny knew – and is indicative of the limited knowledge of the process in antiquity (Pliny, *Histoire naturelle* 2, 253). He was the first extant source to record the nearly regular lunar-tidal intervals and the daily lagging of the tides (Georgia L. Irby, "Hydrology: Ocean, Rivers, and Other Waterways," in *A Companion to Science, Technology, and Medicine in Ancient Greece and Rome* [ed. Georgia L. Irby, Chichester 2016] 186).

2.218. Tides in the Mediterranean are limited – no more than about a meter (at the head of the Adriatic) – and this was reasonably explained by its enclosed nature. The ability to sail from Utica in Africa to Italy (a point not specified but probably the Naples area) without the use of sails was hardly due to tidal action, but due to the prevailing current across the Sicilian Strait and along the north coast of the island. Ocean currents do not seem to have been well understood.

It was known that tidal times varied in different places, and that this was due to celestial activity. It is not clear what Pliny had in mind about the Syrtes: the thought seems incomplete. They are the southernmost part of the Mediterranean, off the coast of Libya and Tunisia. This was a region of notorious shoals, and the usable sea depth could be affected by even the weak Mediterranean tides. The *Argo* was said to have had difficulty there (Apollonios 4.1232–49), and in 252 BC a Roman fleet ran aground (Polybius 1.39.3–4).

2.219. The Tauromenitanian Strait is a peculiarly localized name for the strait between Sicily and the mainland, derived from the town of Tauromenium (modern Taormina) in Sicily. The currents – rather than tides – in this region were difficult to understand, and navigational hazards had been known since prehistoric times (Homer, *Odyssey* 12.101–14; Strabo,

Geography 1.2.36). Even more problematic was the Euboean Strait, the Euripus between Euboea and the mainland of Central Greece, whose activity still baffles hydrologists (Aristotle, *Meteorologika* 2.8.366a; Strabo, *Geography* 9.2.8; Paul W. Wallace, *Strabo's Description of Boiotia* [Heidelberg 1979] 34–8).

The wells at Gades were studied by Posidonius during his time in the city (Posidonius F217; Strabo, *Geography* 3.5.7). They were said to be affected by the tides, and the phenomenon was reported elsewhere in the region, such as on the Baetis (modern Guadalquivir) River, and at Hispalis (modern Seville). Yet the explanations are problematic, because they ignore the fact that the wells are fresh water but the sea is salty. There is a regular current of 3–5 knots out of the Black Sea into the Aegean, but again it does not seem to have been understood that this would never change direction.

2.220. The sea was thought to be a living being, an idea first developed in the fourth century BC, especially by Pytheas, who saw it breathing as if a type of lung (Pytheas F5; Strabo, *Geography* 2.4.1). Messana (modern Messina) and Mylae (modern Milazzo) are towns a short distance apart in northeastern Sicily, presumably places that debris from the sea would collect. The oxen of the sun (Helios) were pastured in Sicily; Odysseus encountered them after passing Charybdis and Scylla (Homer, *Odyssey* 12.260–70). The citation of Aristotle (F287) cannot be connected with any particular work. The Gallic Ocean is usually taken to mean the modern Bay of Biscay, and a variant version of humans dying only when the tide was falling was known at Gades (Philostratos, *Life of Apollonios* 5.2).

2.221. Since the moon was the cause of the tides, it was believed to have breath (*spiritus*), or be able to create it, which affected many things, including humans, especially when it was full. Pliny discussed this topic in greater detail at *NH* 18.321–5.

2.222. After describing the sun as male, because of its heat, Pliny digressed to consider the salinity of the sea, advancing several possible theories. The first, that all sweetness has been removed and only the bitter is left, may have been the view of Anaximander (F34–6 Graham), eventually developed by Aristotle (*Meteorologika* 2.2). The second theory, which Pliny found less credible, is that the saltiness is due to the perspiration of the land; this originated with Empedokles (F100 Graham). The final suggestion, that the earth itself was the source of the saltiness, using the analogy of healing springs, was that of Xenophanes (F59 Graham). The Dionysius in question is Dionysius II of Syracuse (Pliny's use of the word *tyrannus* demonstrates this), who ruled from 367 to 357 BC. Pliny is the only source for this anecdote.

2.223. Returning to the moon, Pliny pointed out that it was believed to be female and thus a counterpart to the male sun. It exuded moisture at night and relaxed all things, a belief that may have been Egyptian (Plutarch, *On Isis and Osiris* 41). The moon was also associated with fresh water (the sun with sea water), the view of Posidonius (F10).

2.224. Papirius Fabianus lived in the early Imperial period and wrote on oceanic phenomena (Zucher, "Papirius Fabianus," *EANS* 610–11). There was a persistent interest in sea depth, something exceedingly difficult to determine, and it is not possible to know the methodology (Irby, "Hydrology" 186–7). Eratosthenes seems to have been the first to consider the issue (*Geography* F15), and Posidonius (F221) was also curious about the matter (Kidd, *Posidonius* 2, 793–5). The location of Fabianus' calculation (equivalent to about two miles or three km.) is not specified. The Coraxians ("Raven People") lived on the east coast of the Black Sea (*NH* 6.15). The deep point would be about 37 miles or 60 km. offshore. The Black Sea actually exceeds 2,000 m. in depth, but not in this region.

There were rivers that emptied into the sea but produced fresh water some distance from their mouth, due to the differences in specific gravity (Aristotle, *Meteorologika* 2.2). The failure of bodies of water to mix was noted as early as Homer (*Iliad* 2.751–4). Pliny named six lakes that have rivers flowing through them which retain a certain integrity: Lake Fucinus (modern Fucino) east of Rome, the four major lakes in northern Italy at the edge of the Alps (modern Como, Maggiore, Garda, and Iseo), and Lac Léman to the north, which is probably the most notable because the Rhone flows through it. The circumstances of the Orontes (modern Nahr el-Asi) in Syria are not known.

2.225. At this point there is a long discussion of unusual springs and rivers (through *NH* 2.233). No source is listed, but it may be C. Licinius Mucianus (see *NH* 2.231). Springs were associated with underground rivers, which were also of interest, yet many of these were questionable and resulted from the assumed linkage of hydraulic systems that were not actually connected, notably the Alpheus in the Peloponnesus with the Arethusa spring on Sicily. It is difficult to imagine how a proof of this could have been successfully tested, but its validity was a consistent tale. The story, whose origins may be in mythological genealogy, was known to Pindar (*Nemean* 1.1–2) and popularized by Timaeus of Tauromenium in the fourth or third century BC (F41 c), but was refuted by Polybius and Strabo (Polybius 12.4d; Strabo, *Geography* 6.2.4). Yet it was believed that there were many underground watercourses throughout the Mediterranean world, an idea perhaps partially sustained by evidence from mining and the karst topography common in certain areas (A. Trevor Hodge, "Purity of Water," in *Handbook of Ancient Water Technology* [ed. Örjan Wikander, Leiden 2000] 95–6).

Pliny cited six instances that supported the idea of underground watercourses. There were several rivers named the Lykos in Asia Minor, but Pliny probably meant the one near Laodikeia (modern Çürük Su) in Phrygia (Strabo, *Geography* 12.8.16). The Erasinus (perhaps the modern Kefalari) in the Argolid was believed to flow underground from sources in Arcadia, which is quite plausible (Herodotus 6.76; Eratosthenes, *Geography* F140; Strabo, *Geography* 8.8.4). The Tigris was thought to have underground portions near its source in Armenia, an idea to some extent due to a lack of understanding of the river systems in this remote and rugged region (Strabo, *Geography* 11.14.8).

The Spring of Aesculapius (Asklepios) is on the south slope of the Akropolis of Athens (John Travlos, *Pictorial Dictionary of Ancient Athens* [New York 1971] 138–42). Phalerum was the original port of Athens, about five km. to its southwest, but there is no other evidence of a believed connection between it and the Athenian spring. The Plain of Atina is in Lucania, and the river in question may be the Tanager (modern Tanagro), the largest stream in the region. The Timavus (modern Timavo) is at the head of the Adriatic near Aquileia, in an extensive karst region.

2.226. The foremost of several lakes with unusual properties was Lake Asphaltitis (the Dead Sea), whose high salinity caused effects still remarkable today. It first became known to the Greek world in the late fourth century BC (Aristotle, *Meteorologika* 2.3.359a; see also Posidonius F279; and Diodorus 2.48.6–9). Aretissa (also Aresene or Arsissa, modern Lake Erçek) is a small lake in Armenia, just east of Lake Van (ancient Thopitis or Thonitis). Pliny may have confused the two, since the latter nurtured an unusual type of fish, the *darekh*, a variety of herring (Strabo, *Geography* 11.14.8). The lake near Manduria in the Sallentinus (the heel of Italy) no longer exists, and is otherwise unknown. This is a region with little surface water, so any lake would be an unusual phenomenon, but it may have been a cistern rather than a natural lake (Gabriele Burzacchini, "Sul 'Fonte Pliniano' di Manduria (*NH* 11 103 [106], 226)," *Eikasmos* 11 [2000] 297–307).

There are several examples of waters allegedly petrifying things thrown into them. The territory of the Ciconians was at the north end of the Adriatic (the specific river is not identified; see also Ovid, *Metamorphoses* 15.313–14), and Lake Velinus (*NH* 2.153) is northeast of Rome. The Surius in Colchis is otherwise unknown, and the Silerus (modern Sele) forms the boundary between Campania and Lucania. Petrification in these watercourses was not limited to wood, and the best explanation is that it is some kind of mineral accretion: this certainly would be the case with the rocks forming at the marsh around Reate (modern Rieti), near Lake Velinus.

The final clause, about olive trees and bushes in the Red Sea, which does not appear in all the manuscripts, is either a later addition or badly out of place. Yet there is no denying the validity of the statement, which refers to the mangroves that grow on the coasts of the Arabian peninsula and which could be called "olives" (Strabo, *Geography* 16.4.7). The plant was first identified in the late fourth century BC as a result of explorations of the region commissioned by Alexander the Great (Theophrastus, *Research on Plants* 4.7.7–8).

2.227. Hot springs, especially in the sea, seemed an oddity; there were several in the volcanically active region of the Bay of Naples (Pliny's Baianian Bay), where underwater vents are numerous. Perhaps more strange were places where freshwater percolated up through the sea, such as around the Chelidoniae (modern Beşadalar) Islands off the southern coast of Asia Minor, as well as near Gades. But the most notable of these oceanic springs was in the channel between the mainland and the island of Arados in Phoenicia, which was three km. offshore, and where there was a spring at about 25 m. depth from which the water was piped up to the surface (*NH* 5.128; Strabo, *Geography* 16.2.13).

It would hardly be unexpected that flora and fauna would be found in hot springs: Pliny listed several Italian examples. Springs could vary seasonally, as with the Scatebra spring near Casinum (modern Casino) in southern Latium. The *musculus aquaticus* of the famous Lake Stymphalus in Arcadia may be a type of mussel, or even a variety of turtle (Kenneth F. Kitchell, Jr., *Animals in the Ancient World From A to Z* [Abingdon 2014] 186).

2.228–9. There are many springs that set things on fire, such as that of Jupiter (Zeus) at Dodona in northwest Greece, or one at an unspecified place in Illyria. Lucretius provided a detailed explanation of such phenomena, noting that there was heat deep in the earth (Lucretius 6.879–89). This is certainly true, but it is more likely that the springs contained a flammable material such as naphtha. Many springs did not have a constant flow, or changed temperature at various times: the several examples listed are from throughout the Greek world. At Dodona, the erratic nature of the spring of Zeus led to it being called Anapauomenon, or Intermittent, which Pliny presented in Greek characters. One of the most peculiar was the Inopus (probably the spring, not the river) at Delos, said to connect with the Nile

(Strabo, *Geography* 6.2.4; Pausanias 2.5.3). These changes in water quality or flow seemed inexplicable, but can best be explained by water sources that were far from the outlet and existed under different conditions.

2.230. There is a list of six rivers, scattered throughout Italy, Greece, and Asia Minor, that allegedly changed the color of animals drinking from them; a number of such anecdotes exist (e. g. Seneca, *Natural Questions* 3.25). These notices are generally folk wisdom, although some are onomastic, such as the Xanthus in the Troad, whose name means "yellowish-red," and which was translated by Pliny as *rufus*, the word used for redheaded people. There is certainly nothing unusual about the Neminie spring near Reate, a marshy area where the flow would vary seasonally; this is a region of unusual hydrological phenomena (*NH* 2.209, 226). The spring at Brundisium (modern Brindisi) may actually have been in the harbor itself, another example of a fresh-water spring in the sea (see also *NH* 2.227). Several springs had a bitter taste that was also said to be intoxicating (Ovid, *Metamorphoses* 15.307–36; Valerius Maximus 1.8.ext. 18; Vitruvius 8.3.20).

2.231–2. C. Licinius Mucianus wrote on geographical marvels, and was Pliny's source for much of this material (*FRH* A27, F1). He was cited over thirty times in the *Natural History*, and the two were probably personally acquainted (Rhiannon Ash, "The Wonderful World of Mucianus," *BICS Supplement* 100 [2007] 1–17). The spring that tasted like wine on the day of the Theodosia (presented in Greek characters), a festival held on 5 January, was probably a contrived part of the ritual of Father Liber (Bacchus) on the island of Andros.

There are several other examples of springs with strange behavior, scattered from the northern Black Sea region to Hispania. Comum (modern Como), on Lake Larius (modern Lago di Como), was the home of the Pliny family, and Pliny's nephew wrote in detail about the local spring, which is still visible (Pliny the Younger, *Letters* 4.30). The oracle of Apollo Clarius at Colophon, in Ionia, was described by Tacitus, because Germanicus consulted it on his fatal eastern journey in AD 18–19 (Tacitus, *Annals* 2.54). Pliny also made another elusive reference to his historical work (see *NH* 2.199).

2.233–4. There is a catalogue of items that seem to act in a manner contrary to nature when placed in water. The lack of any geographical grounding beyond the references to Syros and Mt. Aetna suggests that the list is not from Mucianus (*NH* 2.231). All the phenomena can easily be explained. Lucretius provided a detailed account of why water in wells was colder in summer, noting that the seasonal heat was due to it being sent into the atmosphere from the earth, and thus the earth as well as the water in it would become colder (Lucretius 6.840–7). The stone on the island of Syros is pumice, which until recently could be found floating in this part of the Aegean (*NH* 36.130). The juxtaposition of fire and water was always intriguing: Pytheas saw this in the volcanic portions of the Arctic (Roller, *Through the Pillars* 85).

Unusual Earthly Phenomena

2.235. L. Licinius Lucullus' assault on Samosata, the capital of Commagene, was in 69 BC and part of his general campaign against Mithridates VI of Pontus. The Roman soldiers were attacked by weaponry coated with a substance called *maltha*, which was found in a marsh and was not only highly flammable but would stick to people and their equipment. The word is latinized Greek, and the substance is a type of bitumen mixed with wax in order to give it cohesion, similar to the later Byzantine compound known as Greek fire. Natural bitumen seeps occur in many places in the eastern Mediterranean world, and the Samosatans were able to use the product as an effective weapon of war.

Naphtha is liquid asphalt and bitumen is dry asphalt. Both had been used in a variety of ways since early times. Alexander and his companions encountered naphtha in the vicinity of Babylon (Strabo, *Geography* 16.1.15; Plutarch, *Alexander* 35). There are numerous tales about its dangers: as noted, it appears in the story of Medea (Seneca, *Medea* 817–19), and it may be what created the burning fiery furnace of Daniel (Daniel 3:19–27). Eratosthenes (*Geography* F90) and Posidonius (F236) were the first to discuss the material in detail.

2.236–8. There were many places where volcanic phenomena were visible. Two sources for the catalogue that follows were named: Ctesias, the historian of around 400 BC who wrote on Persia and India, and Theopompus of Chios, active in the fourth century BC, whose treatise was on the era of Philip II of Macedonia. Ctesias was referring to the region of Mt. Chimaera, discussed below (*FGrHist* #688, F45eβ); Theopompus' note is about the Crater of Nymphaeus, a volcanic feature at modern Selenicë in southern Albania near the ancient town of Apollonia (*FGrHist* #115, F316). Not mentioned by name, but implicit, is the Carthaginian explorer Hanno, the only source for Theon Ochema, the Chariot of the Gods, a feature on the West African coast, probably modern Mt. Cameroon, which is still volcanically active today (Roller, *Through the Pillars* 40). In addition, Pliny was probably also using an unspecified catalogue of volcanic phenomena.

It was considered astounding that Mt. Aetna erupted constantly even when its upper elevations were covered with snow: the mountain (3326 m. high) remains an unusual feature today. The inability to see its eruptions in the daytime was itself considered a strange phenomenon. Mt. Chimaera was in southern Asia Minor and gave rise to the myth of the monstrous Chimaera, part lion, snake, and goat, described by Glaukos in the *Iliad* (6.178–83; see also *NH* 5.100). This region is also still volcanically active. The other examples of vulcanism are scattered throughout the known world, from Bactria to West Africa, although some were visible only at night. Many remain active today. For the Aeolian islands, see *NH* 2.203.

2.239–41. Fire is the most unusual of the four basic elements established by Empedokles (F29 Graham). It seemed to be alive, and in one sense to be everywhere, but was limited in what it consumed: Pliny ironically found it astonishing that it did not burn up the entire world. Given the vulcanism of the eastern Mediterranean, there were many places where sparks and flames appeared spontaneously and for no obvious reason. Fire was also a pervasive element in the cosmos as a whole, since celestial bodies were thought to be fiery objects: this could easily be proven by using a concave mirror or lens (*concava specula)* that appeared to focus the fire of the sun on the earth.

The Nymphaeum is probably the one in Epiros, mentioned at *NH* 2.237. The springs of Scantiae are cited only here but may be in the volcanic region of Campania (Tusculum 2, 254). Mutina is modern Modena in northern Italy. There were several festival days dedicated to Vulcan, mostly in the summer (Ovid, *Fasti* 5.725; Varro, *de lingua latina* 6.20). The heat of summer may explain some of the seemingly spontaneous appearances of flames or other phenomena, but for the most part they were probably contrived parts of the religious rituals. Activities of this nature at Gnatia (modern Egnazia) in southeastern Italy earned the ridicule of Horace (*Satires* 1.5.97–100).

The burning of Lake Trasimenus in Etruria may have been due to a conflagration of its wooded islands (Pliny, *Histoire naturelle* 2, 264). Servius Tullius was the sixth king of Rome, living in the sixth century BC; flames shooting from the head of a child are a portent of future greatness. Valerius Antias (*FRH* #25) lived in the first century BC and wrote a history of Rome from its origins to his own era. He was probably the source for some of these events and portents. L. Marcius Septimus was a military tribune in Hispania during the Second Punic War, when the brothers P. and C. Cornelius Scipio were killed (in 211 BC; Livy 25.34–6). Like the case of the young Servius Tullius, flames shooting from his head were an obvious portent of the future, as Marcius took effective control of the army.

The Measurement of the Earth

2.242. The final portion of Book 2 is a discussion of the measurements of the earth. This is material that originated with Eratosthenes in the late third century BC and was later refined by Hipparchus (Eratosthenes, *Geography* F57–66; Hipparchus, *Against the Geography of Eratosthenes* F43–60). Yet Pliny's primary sources were two later geographers: Artemidorus of Ephesus, who wrote 11 books on geography at the end of the second century BC (Stiehle, "Der Geograph Artemidoros" 193–244, F1; Daniela Dueck, "Artemidoros of Ephesos," *EANS* 165), and Isidorus of Charax, active in the Augustan period (*FGrHist* #781, F6; Daniela Dueck, "Isidoros of Kharax," *EANS* 443–4). The original measurements, whether in stadia or parasangs, were converted by Pliny into Roman miles.

He reminded his readers that the inhabited earth is surrounded by the Ocean (*NH* 2.167–73). Its east–west length was measured from India to Gades: the distance of 8,578 miles was converted from 70,000 stadia, Eratosthenes' original figure (*Geography* F30, 34, 37). But as more became known about the Atlantic coast of the Iberian peninsula during Roman times, the amount became modified, since it was realized that the coast extended far west of Gades, and thus the perceived western terminus of Europe varied from the Sacred Promontory (modern Cabo de São Vicente at the southwest corner of the peninsula) to Cape Artabrum (modern Touriñán at its northwest corner). Isidorus' 9,818 miles is equivalent to 80,000 stadia.

2.243. Pliny laid out his itinerary for measuring the east–west extent, figures obtained from Eratosthenes by means of Artemidorus. This is clearly not a measuring line across the world, but a road and sailing route from the mouth of the Ganges to Gades. It went overland from India to Myriandrus (near modern Ada Tepe at the northeast corner of the Mediterranean). By Pliny's day, and even well before, Myriandrus had ceased to be of importance due to the foundation of Alexandria (modern İskenderun) a short distance to its northeast, so some of the details of Pliny's route are pre-Hellenistic. From Myriandrus the itinerary is based on shipping lanes, which do not provide the most direct distance. It curves around the Greek peninsula and the south side of Sicily, and then heads north to Sardinia before going southwest to the Pillars of Hercules and ending at Gades. But the total, 8,578 miles, was probably constructed in a way to conform to Eratosthenes' 70,000 stadia.

2.244. There was an alternative route, more land-oriented. It also originated (in the form Pliny accessed it) from Artemidorus, and followed the ancient roads from the Ganges to northern Mesopotamia and then the evolved version of the Persian royal road to Ephesus on the Aegean. It crossed the Aegean and the Isthmos of Corinth, and headed to Patras. From there it crossed the southern end of the Adriatic to Brundisium (following the standard sailing routes along the coast to Corcyra, or modern Corfu), and then headed through Rome, around the north end of the Mediterranean, and across the Pyrenees to the Iberian Peninsula and Gades. It is difficult to believe that this highly circuitous route added only 367 miles (probably 3,000 stadia) to the direct crossing of the Mediterranean; the numbers have been adjusted to make the two distances compatible. Moreover, although attributed to Artemidorus, the itinerary is Roman in its outlook, with mention of Patras, Brundisium, and Rome itself, and is merely a report on the main route crossing the Roman Empire, with a connection to India added. Thus it has nothing to do with any measurement of the inhabited world, and may ultimately come from a source more astrological than geographical, especially given the strange tale of Dionysodorus that follows (*NH* 2.248; Dicks, *Geographical Fragments* 153, 157–8).

2.245. Next Pliny provided the north–south distance across the inhabited world. Isidorus' figure of 5,462 miles (probably 45,000 stadia) is significantly longer than Eratosthenes' 30,000 stadia (Isidorus F7; Eratosthenes, *Geography* F30, 34). Pliny (and Isidorus) sustained Eratosthenes' conviction that this north–south distance was constrained by inhospitable territory both to the north and south, and thus the figure is not the extent of the land mass but of the inhabitable portion.

The itinerary begins 625 miles (5,000 stadia) south of Meroë on the shores of the Aethiopian Sea, probably incorporating Hellenistic knowledge of the East African coast reported by Juba II of Mauretania (*NH* 6.175–7). Then it continues from Meroë down the Nile and across to Rhodes, and around the west end of Asia Minor and into the Black Sea. But rather than going straight across the sea to the mouth of the Maeotis, the route follows the coast to Cape Carambis (modern Kerembe Burnu), essentially the northernmost point in Asia Minor, where it can make the shortest crossing to the Maeotis. Like the previous itinerary, this one has little to do with measuring the earth and is merely a series of sailing routes. Nevertheless it is similar to that provided by Eratosthenes.

2.246. Pliny realized that Isidorus' 1,250 miles (F7; 10,000 stadia, or 1,850 km.) to Thule was theoretical, although presumably there were some data available from Pytheas, the only possible source. This led Pliny to speculate on the vastness of the northern regions: the north shore of the Black Sea was barely more than halfway from the equator to the pole (something easily calculated: see *NH* 6.217–19). Recognition of the immense size of this northern region was sustained by the vast populations that were constantly moving around. The large islands beyond Germania are not specified, but presumably are those discovered by the Roman expeditions into the North Sea and beyond in the last years of Augustus and the first decade of Tiberius, around the entrance to the Baltic and within it, and the southern end of Scandinavia (*NH* 4.96–7).

2.247. After a eulogistic comment about the abilities of Eratosthenes, and the respect in which he was held, Pliny reported his calculation of the circumference of the earth. He was now using stadia (252,000), but also gave the equivalent in miles, which provides his conversion factor (8:1). Eratosthenes' calculation was published in his treatise *The Measurement of the Earth*, of which a few fragments remain (Roller, *Eratosthenes* 263–7). Hipparchus' objections were because Eratosthenes favored known overland distances rather than making mathematical and astronomical calculations to locate places (Hipparchus, *Against the Geography of Eratosthenes*, F11–15), but there is no obvious reason why Pliny reported that Hipparchus added 26,000 stadia (Dicks, *Geographical Fragments* 153; Klaus Geus, "Progress in the Sciences: Astronomy and Hipparchus," in *Brill's Companion to Ancient Geography: The Inhabited World in Greek and Roman Tradition* [ed. Serena Bianchetti *et al.*, Leiden 2016] 159.

2.248. To close Book 2, Pliny related the peculiar story of Dionysodorus of Melos, who claimed to have visited the center of the earth. He presumably used Eratosthenes' circumference to calculate the radius, but the circumstances were probably staged to give him greater credibility. There were several mathematical personalities with the name, and it is impossible to add any further details about this Dionysodorus, except that he must have lived between the late third century BC and the mid-first century AD (Alexander Jones, "Dionusodoros," *EANS* 266). The additional 12,000 stadia reconciles Dionysodoros' radius with Eratosthenes' circumference, but there is no known source for the statement that the earth is 1/96 of the universe.

BOOK 3

Southern Europe

Book 3 of the *Natural History*, the first geographical book, confines itself to southern Europe, from the outlet of the Mediterranean at the west to the mouth of the Danube at the northeast, excluding the Greek peninsula. After a brief introduction about geography and Pliny's technique (1–2) and comments about Europe as a whole (3–5), the narrative moves through Hispania (6–30), Gaul (31–7), and Italy from the Alps to Sicily (38–138). The Italian portion is nearly two-thirds of the book, which concludes with the regions east of the Adriatic as far as the Danube (139–51). Discussion of mainland Greece is reserved for the following book (*NH* 4.1–39), whose beginning follows directly on the end of Book 3. The account of the lands west of Italy is limited to the Mediterranean coasts: the remaining portions of the Iberian peninsula and Europe are examined in Book 4 (94–120).

There were many precedents for beginning a geographical survey at the entrance to the Mediterranean and continuing along the coast of Europe, the process followed by Hecataeus of Miletus (F39, 41, 356) and others, including Pliny's immediate predecessor, Pomponius Mela (1.25), both of whom were known to Pliny. But there was also an ideological component to the structure of Book 3. The Iberian peninsula was the first region outside Italy to become heavily romanized: Romans had been in the region since the third century BC. And it was there that locals were first integrated into the Roman elite, notably L. Cornelius Balbus from Gades. In fact, it seems no accident that the first non-Italian to obtain the consulship (40 BC) came from the most famous and historic town in the peninsula (*NH* 7.136; Broughton 2, 378). His homonymous nephew also had a distinguished career (see further, *NH* 5.36). Thus by beginning his geographical account with Hispania, Pliny subtly positioned his narrative beyond topography, demonstrating that Hispania was linked to Italy, and together they were the essential portions of western Europe, using the topography to support ideology. In fact, Pliny moved as quickly as possible from Hispania to Italy, with only a brief account of the Gallic districts in between, connecting the two regions as the heart of the contemporary Roman world.

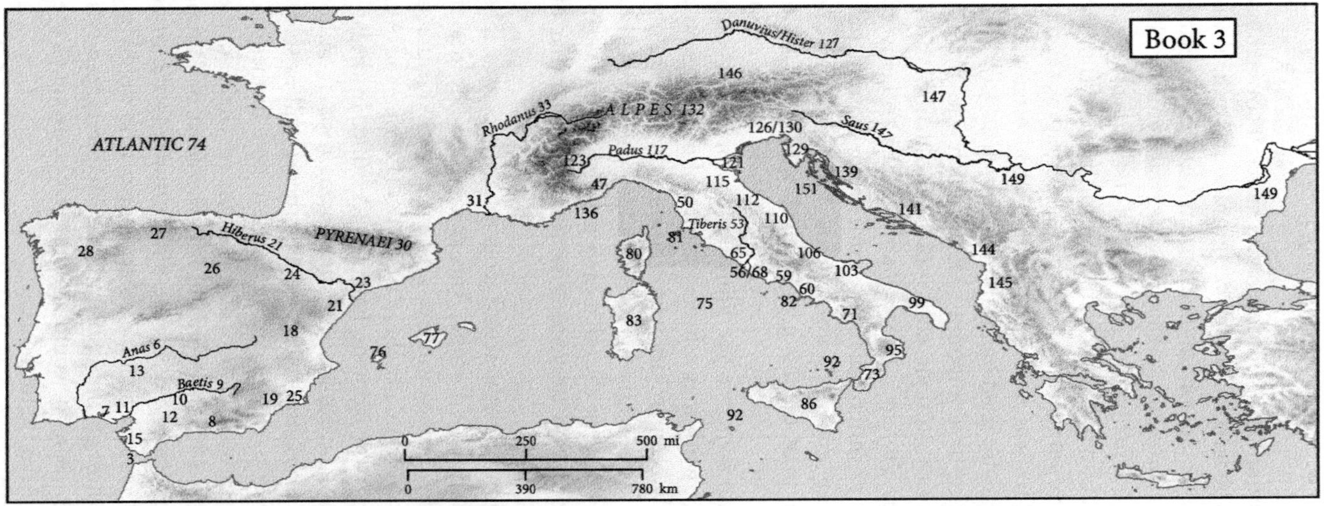

Map 3 Places cited in Book 3 of the *Natural History*, according to chapters (courtesy of Ancient World Mapping Center).

Preface

3.1–2. After reminding his readers of the material covered in Book 2, Pliny then began his geography proper, recognizing the difficulty of his task and noting that he would rely on the various authorities that he considered most reliable (*verissimus*). In this he was certainly inspired by Herodotus. The immensity of his project (in terms of the *Natural History* as a whole) was a point that he had already made in his preface (14). But rather than geography, Pliny's actual goal was more chorography, the naming of places and peoples. This was a term that perhaps had originated with Polybius (34.1.5), and was also used by Strabo (*Geography* 10.3.5). Pliny was familiar with the treatise of Pomponius Mela, which was titled *Chorographia*. Although Pliny's technique conforms to the definition of chorography (best set out by Ptolemy, *Geographical Guide* 1.1), he avoided using the word, perhaps to differentiate his work from those of his predecessors. These would include both Pomponius Mela and the *Chorographia* of Cicero, which may be the geographical work that he mentioned to Atticus in 59 BC (Cicero, *Letters to Atticus* #24; see also Priscian, *Institutiones grammaticae* 6.83). Pliny's rejection of any polemic quality to his geographical work may reflect disapproval of those who wrote in such a way, especially Hipparchus, whose *Against the Geography of Eratosthenes* was a strongly worded critique of the earlier scholar's methods. Yet at the same time Pliny made it clear that mere listing of names was an incomplete view of the world and universe, a sentiment similar that of Pomponius Mela.

3.3. The beginning of the actual geographical survey establishes the three continents of antiquity: Europe, Asia, and Africa. The idea of dividing the inhabited world into continents originated with Hecataeus of Miletus around 500 BC. Originally there were two (Europe and Asia); the third, Libya, or Africa to the Romans, had been identified by the late fifth century BC (Herodotus 4.42; Roller, *Ancient Geography* 50–1), although, perhaps for political reasons, the two-continent theory still existed into the first century BC (Varro, *de lingua latina* 5.31; see further, *NH* 3.5). The Gaditanian Strait was named after Gades, or Gadeira, the Phoenician outpost – modern Cádiz – on the Atlantic coast of Europe. Gades lies about 60 km. northwest of the western entrance to the strait, but Pliny's description reflects a navigator's point of view, who would be positioned at Gades to pass through the strait into the Mediterranean (Pliny's Internal Sea of *NH* 3.4). Seamen would also have to be aware of the current that flows from the Atlantic to the east.

Pliny laid out all three continents as if visible to someone passing through the strait. Europe and Africa defined its sides, but the closest point of Asia is 3,000 km. away. Yet he created a literary map of the continents spread out before a hypothetical viewer, all the way to the farthest extremity of the Mediterranean system at the mouth of the Tanais River (modern Don). He also noted the dividing lines between the continents, beginning with the Gaditanian Strait. The Tanais had long been generally accepted as the boundary between Europe and Asia (Herodotus 4.45), but that between Asia and Africa tended to be disputed between the Nile and the Red Sea – Herodotus was ambivalent about which to choose (Herodotus 2.16) – with Pliny preferring the former (see further, *NH* 6.177).

Pliny's distance for the length of the strait (15 miles, or about 22 km.) is difficult to interpret. As a whole, it is about 55 km. long, but Pliny must have been referring to the narrowest part, that immediately west of modern Algeciras, whose length approximates his distance. The width of five miles (about 7.5 km.) is peculiar, as it is never narrower than about 15 km., and Pliny's itinerary from Mellaria (or Menlaria), near modern Tarifa, to Cape Album (not certainly located but probably near modern Tangier) is more like 25 km. He had stated his intent to use local sources, and thus he cited Turranius Gracilis, who lived in the Augustan period and was from the region, perhaps even from Gades (*NH* 9.11).

3.4. Despite Pliny's reliance on a local source, the width of 7–10 miles (11–16 km.) provided by Livy (F59) and Cornelius Nepos (*FRH* #45) is more accurate. There are sandbanks in the strait, most notably southwest of modern Cape Trafalgar, rising to as little as 12 m. below the surface (Tusculum 3–4, 205); hence the idea that they are the threshold by which one enters the Mediterranean. The term Internal Sea is cited here for the first time in Latin, a direct translation of a Greek phrase; Pomponius Mela used Our Sea (*Mare nostrum*, 1.6), more common in Latin.

The two peaks that mark the strait, Abila (also Abile or Abilyx) at the south, modern Jebel Musa, and Calpe at the north (modern Gibraltar), are prominent today. It may seem obvious that they are the Pillars of Hercules, but such was not always the case in antiquity. The peaks were first identified with the Homeric Pillars of Atlas (Homer, *Odyssey* 1.51–4), but eventually became associated with Herakles/Hercules. Moreover, they were variably placed, all the way from Gades to the east end of the strait (Herodotus 4.8, 152). It was only by the late fourth century BC that they became associated with Calpe and Abila (Dicaearchus F125; Eratosthenes,

Geography F106; Strabo, *Geography* 3.5.5). The hydrological efforts of the hero were alluded to by Pindar (*Nemean* 3.19–29), yet the specific story seems only to be from Hellenistic times, and was described in detail by Diodorus (4.18.5). Pliny would have been familiar with the account in Seneca's *Hercules Furens* (235–8), as well as that by Pomponius Mela (1.27), a close parallel.

3.5. In his eulogistic comments about Europe, Pliny reverted to the ancient two-continent theory, which had a certain political correctness in the Roman world and which seems to have reappeared in the first century BC (Varro, *de lingua latina* 5.31). The suggestion that Europe is half the known world reflects Roman dominance – Europe was the location of Rome and Italy – with the ocean given an animistic quality as it sweeps in from the Atlantic, overwhelming everything in its path, like the spread of Roman military prowess and culture. Moreover, the ocean keeps the continents separate from one another (Lucretius 5.203), and is thus a divider and assists in governing the structure of the *Natural History*. It terrifies all the lands that may resist its progress, a military metaphor for Roman advances throughout Europe (Rhiannon Evans, "The Cruel Sea? Ocean as Boundary Marker and Transgressor in Pliny's Roman Geography," *Antichthon* 39 [2005] 105–19; Catherine Connors, "A River Runs Through It: Waterways and Narrative in Strabo," in *The Routledge Companion to Strabo* [ed. Daniela Dueck, Abingdon 2017] 210). The removal of Africa from the canonical list of continents reflects the Roman subjugation of Carthage and thus the perceived absorption of Africa into Europe (Lucan, 9.411–13; see also Sallust, *Jugurtha* 17.3).

The concept that the southern coast of Europe was a series of gulfs had been established in a reverse manner by Eratosthenes, who conceived of it as a series of peninsulas protruding into the Mediterranean (Eratosthenes, *Geography* F134). Pliny, with his emphasis on the power of the Mediterranean, saw it in the opposite way. The four bays are the Tyrrhenian Sea (that west of Italy), the Adriatic, Aegean, and Black Seas. Such an orientation continues to stress the navigational aspect of the account. At the end of the section, Pliny defined the extent of the first gulf: from Calpe in Hispania to Locri at Cape Bruttium. Locri (Lokroi) is actually on the east side of the southern tip of Italy (and thus theoretically in the Adriatic Bay), but it was probably chosen because it was the first significant town on a relatively deserted coast. Cape Bruttium is probably the Herakleion, modern Cape Spartivento (Strabo, *Geography* 6.1.7), the southeast corner of the toe of Italy, a better terminal point for the first gulf than Locri, which lies 40 km. farther along the coast.

Hispania

3.6. The first section of the regional geographical survey is devoted to Hispania, or the Iberian peninsula (through *NH* 3.30), beginning with establishing its four Roman divisions. Pliny had been a procurator in Hispania Tarraconensis (Pliny the Younger, *Letters* 3.5.17), so was well informed about the administrative structure of the region. The original Roman provinces in Hispania were created at the end of the third century BC, after the Second Punic War, at which time the peninsula was divided into two, Hispania Citerior (Nearer) and Hispania Ulterior (Farther). The concept of Nearer and Farther reflected the coastal sailing distance from Italy, so that Nearer was in the northeast (starting at the Pyrenees), and Farther in the south. The boundary between the two was not established until 197 BC (Livy 32.28.11), and was located somewhere beyond (or southwest) of Carthago Nova, or modern Cartagena (Livy 40.41.10; W. V. Harris, "Roman Expansion in the West," *CAH* 8 [second edition, 1989] 121). Pliny's boundary, in the Murgitanian region, may be somewhere in the district of modern Almería. In the reorganization of the Iberian peninsula by Augustus, the new provinces of Baetica and Lusitania were created out of Hispania Ulterior (Strabo, *Geography* 3.4.20). The division was at the Anas River, the modern Guadiana, which in its southern portion is the modern boundary between Spain and Portugal (Strabo, *Geography* 3.1.6). The source of the river is in La Mancha, southeast of Madrid near Alcázar and the ancient town of Laminium, which was noted for its whetstones (*NH* 36.165). The Anas flows for 818 km., first west and then south, to the Atlantic.

The northern part of the Iberian Peninsula became the province of Tarraconensis with the Augustan reorganization. It was no longer limited to the coastal strip south of the Pyrenees, but (in theory) extended across to what Pliny called the Gallic Ocean (the modern Bay of Biscay). East of the province was the Hiberian Sea, a localized (and somewhat anachronistic) term for the sea that lies immediately off the mouth of the Hiberus River (the modern Ebro: see *NH* 3.21).

Tarraconensis took its name from the city of Tarraco (modern Tarragona), south of the Pyrenees on the Mediterranean. Augustus was in residence there from 27 to 24 BC as he reorganized the western parts of the empire, which had long been neglected because of the attention paid to issues in the East (Dio 53.22–6). The various mountains mentioned as separating the provinces cannot be certainly identified among those of the Iberian peninsula, but Solorius is probably the Sierra Nevada, along the border between Tarraconensis and Baetica. Yet it is doubtful that there were distinct boundaries, especially in the rugged interior of Hispania.

Baetica

3.7. Baetica, which is roughly equivalent to modern Andalucía, took its name from the Baetis River (modern Guadalquivir), which flows west across southern Spain and then turns south into the Atlantic after a route of 657 km. It emptied into a large estuary (now largely reclaimed), just northwest of modern Cádiz. The agricultural richness of Baetica, especially in the river valley, was well known in antiquity and is still apparent today (Strabo, *Geography* 3.2.3; Evan W. Haley, *Baetica Felix* [Austin 2003] 21–7). The province was divided into four areas, each called a *conventus juridici*, which identified the districts that were used by the governor, who would move through his province and hold legal proceedings. Julius Caesar held sessions in Gaul and elsewhere, and Cicero recorded that in October 51 BC he was having them at various places in Asia Minor (Caesar, *Gallic War* 1.54, 6.44; Cicero, *Letters to His Friends* 3.8.6; G. P. Burton, "Proconsuls, Assizes and the Administration of Justice Under the Empire," *JRS* 65 [1975] 92–106). The four *conventus* districts were Gades, Corduba (modern Córdoba), Astigi (modern Écija), and Hispalis (Hispal, or Ispalis, within modern Seville). All are in or near the Baetis valley, which perhaps provides a hint about population distribution within the province.

Pliny was working from an official report about the organization of the province, enumerating the 175 towns and the various categories of their status, with everywhere identified in terms of its relationship to Rome (Sorcha Carey, *Pliny's Catalogue of Culture: Art and Empire in the Natural History* [Oxford 2003] 33–4). These official reports, originating in the Augustan period, were used throughout the portions of the *Natural History* describing Roman territory, but more in the west and Italy than in the east (A. H. M. Jones, *Cities of the Eastern Roman Provinces* [second edition, Oxford 1971] 502–8). First are the *coloniae* (nine in number), a word generally translated "colony" but perhaps with significant differences from the colonies of modern times. They were settlements, often agricultural, that spread Roman populations into conquered regions or areas otherwise acquired. The process began as early as the fifth century BC in Italy. In the Iberian peninsula, *coloniae* were most common in the rich agricultural south, in Baetica, but existed in varying degrees of density throughout the entire peninsula (Hartmut Galsterer, "Coloniae," *BNP* 3 [2003] 549–50).

Next are the *municipia civium Romanorum* ("municipalities of Roman citizens"), of which there were ten in Baetica. These were localities where the inhabitants had been granted Roman citizenship, again a feature largely limited to the western provinces. They could be the conversion of existing towns to the new status, or the creation of new ones in a region of villages or rural populations. By giving such people Roman citizenship, a bond was created between them and the central government in Rome (Hartmut Galsterer, "Municipium," *BNP* 9 (2006] 293–7).

There were 27 towns in Baetica given ancient Latin rights. This was a status that to some extent preceded the granting of citizenship, although such communities could continue to exist alongside of *municipia*. It was an extension of a process that had begun in Italy in the late fourth century BC, when autonomous communities retained their own (rather than Roman) legal status (Gottfried Schiemann, "Ius," *BNP* 6 [2005] 1130–1; David Espinosa Espinosa, *Plinio y los 'Oppida de antiguo Lacio': el Proceso de Difusión del Latium en Hispania Citerior* [*BAR-IS* 2686, London 2014] 1–27).

There were 6 towns which were free (presumably with no involvement with the Roman provincial system), 3 had treaties with Rome (again suggesting autonomy), and 120 paid *stipendaria*. The term actually refers to the pay given to soldiers, rather than an actual tribute, since a locality that had been hostile to Rome had to bear certain costs of pacification, such as maintaining a Roman garrison. As Cicero explained, "a fixed tax has been imposed, which is called a *stipendarium*, such as on the Hispanian and most of the Pontic people, as if the reward of victory or the penalty of war" (Cicero, *Against Verres* 2.3.12).

Next Pliny provided a list of the important localities and features of Baetica (through *NH* 3.15), a catalogue that begins at the western end of the province, at the Anas River. The first portion lists the toponyms along the coast as far as Calpe (Gibraltar). This itinerary is in the style of a *periplous*, the literary genre of a coastal sailing route, something that had existed since early times. Since most Greek sailing was coastal, having data concerning harbors, river mouths, and settlements was a high priority. Although *periploi* were originally navigational tools, they came to be important documentations of the topography and demography of coastal regions. Much of Greek literature describing coasts had its origin in *periploi*, often detected in the narrative by their offshore perspective. The earliest known is probably that of Skylax of Caryanda (*FGrHist* #709), who was commissioned by Darius I of Persia in the late sixth century BC to sail down the Indus to its mouth and then across to the Red Sea. His account is not extant but was used by Hecataeus, Herodotus, and others, perhaps even Pliny. The first surviving *periplous* is the Greek summary translation of the voyage of Hanno of Carthage down the west coast of Africa, around 500 BC, known in at least some version to Pliny (*NH* 2.38, 5.10; Roller, *Through the Pillars* 129–32).

The suggestion that Carteia was called Tartesos (usually Tartessos) by the Greeks is not totally accurate. Carteia (near modern Algeciras), noted for its maritime industries (Strabo, *Geography* 3.2.7), was a Roman base during the Second Punic War (Livy 28.30.3), and became an important town thereafter. But Tartesos was an ancient and general term for southern Iberia, perhaps equated with biblical Tarshish, where King Solomon obtained riches (2 Chronicles 9:21; Carolina López-Ruiz, "Tarshish and Tartessos Revisited: Textual Problems and Historical Implications," in *Colonial Encounters in Ancient Iberia: Phoenician, Greek, and Indigenous Relations* [ed. Michael Dietler and Carolina López-Ruiz, Chicago 2009] 255–80). It was a region of early Phoenician settlement, and known to the Greeks from at least the seventh century BC (Herodotus 4.152). Pliny's equation of a Roman settlement with this ancient pre-Greek region is difficult to sustain, and it may be a local anecdote that he learned when in the area.

3.8. Pliny, contradicting Marcus Agrippa (F37), noted that Carthaginian settlement did not extend west of the Anas River. He was only partially accurate; to be sure, the Carthaginians were primarily east of the river, but the Phoenicians, at least, had a presence on the Portuguese coast around Lisbon and somewhat to its north (Ana Margarida Arruda, "Phoenician Colonization on the Atlantic Coast of the Iberian Peninsula," in *Colonial Encounters in Ancient Iberia: Phoenician, Greek, and Indigenous Relations* [ed. Michael Dietler and Carolina López-Ruiz, Chicago 2009] 113–30). This is the first of over thirty citations of the geographical work of M. Vipsanius Agrippa, whose map was visible in Rome (see further, *NH* 3.16–17; Arnaud Pascal, "Texte et carte de Marcus Agrippa: historiographie et données textuelles," *GA* 16–17 [2007–8] 80–4).

The Baestulians and Turdulians (see also *NH* 3.14–15, 19) were closely related, and the latter, at least, also known as the Turditanians, may be the same as the Tartessians (S. Celestino Pérez and Carolina López-Ruiz, *Tartessos and the Phoenicians in Iberia* [Oxford 2016] 55–6). Regardless of their exact ethnicity, they were considered the most peaceful and civilized indigenous population in southern Iberia (Livy 34.17.2; Strabo, *Geography* 3.1.6), creators of a prosperous agricultural culture that, in time, formed the basis of Roman Baetica.

Using Varro as a source, Pliny provided a chronological ethnic history of the region, beginning with the Hibernians, believed to be the original inhabitants of the Iberian peninsula, who gave their name to it. Presumably they were originally located around the Hiber (Hiberus or Iber) River, the modern Ebro, in northern Spain. The next settlers were said to be the Persians, but this has been questioned by many commentators, and it is generally assumed that either the Tartessians or Pharusians are meant. The latter were an ethnic group in northwest Africa, notorious for their destruction of numerous Carthaginian settlements on the Atlantic coast (Strabo, *Geography* 17.3.3, 7; Pliny, *Histoire naturelle* 3, 98). But there was confusion between the ethnyms Persian and Pharusian (*NH* 5.46), and in fact the Persians did have a fleeting presence in the region, when early in the fifth century BC a certain Sataspes was sent by King Xerxes to circumnavigate Africa, thus passing through the Pillars of Hercules. His attempt failed, and he turned back only to be executed by the king upon his return to Persia (Herodotus 4.43; Roller, *Through the Pillars* 20–1). Thus it is possible that knowledge of Sataspes' expedition led to a belief in a Persian presence in southern Iberia.

The remaining three ethnic groups are not to be doubted. The Phoenicians may have arrived as early as the eighth century BC, and eventually their cities in Iberia evolved into Carthaginian ones. Celtic peoples moved at an early date into northern and western Iberia, avoiding the Phoenician, Carthaginian, and Greek settlements on the eastern and southern coasts.

As a linguist, Varro was interested in etymologies, and derived the name "Lusitania" from words associated with Father Liber (Bacchus or Dionysos), either Latin *lusus* (playfulness) or Greek *lyssa* (frenzy). As with much ethnic linguistics from antiquity, neither of these is reasonable, and both presuppose a Greco-Roman context. There is little doubt that the name is indigenous, from the Celtic Lusitanians, who vigorously resisted the Romans in the second century BC (Appian, *Iberike* 56–60). The association of Pan with the region is because of the similarity of the name "Hispania," which is actually derived from Phoenician "î-she-phannîm" (Pliny, *Histoire naturelle* 3, 98).

Although seemingly accepting the Greco-Roman role in establishing the early toponyms, Pliny chose to reject various Greek mythological tales associated with the region. Hercules (Herakles) was deeply connected with southern Iberia, as his Pillars and other place names show, as well as his last Labors. On the other hand, Pliny seems to have accepted the stories about the hero in his discussion of northwest Africa (*NH* 5.3). Yet, needless to say, these accounts were not possible until Greeks were in the region, and thus probably were no earlier than the seventh century BC, which nonetheless made them of great antiquity in Pliny's day (Herodotus 4.152). Pyrene, the eponym of the Pyrenees, was the daughter of King Bebryx, and was seduced by Herakles and came to a dismal end (Silius Italicus 3.420–41). Saturn, the Greek Kronos, was the name given by the Romans to the Phoenician god Baal.

3.9. Pliny described the route of the Baetis River (modern Guadalquivir) from its Tarragonensian source in the forested heights of the modern Sierra de Segura (probably the ancient Orospeda Mountains). Mentesa (at modern Villanueva de la Fuente) lies well north of the river and is actually the source of a tributary (the modern Guadalmena), but the main stream rises to the southeast near the divide between the Atlantic and Mediterranean watersheds. Its origin is near that of another river, the Tader (modern Segura), which flows to the southeast. The Baetis, which today is the second longest river entirely in Spain (657 km.), crosses from Tarragonensis to Baetica in Ossigetania (the district of Ossigi, a town that lies about 15 km. south of the river at modern Cerros de Alcalá). Flowing somewhat south of west, it passes Corduba (modern Córdoba), and then just north of Seville makes a sharp turn to the south, draining into the Atlantic.

The Tomb of Scipio is the grave of one of the two brothers, Cn. Cornelius Scipio (consul of 222 BC) or P. Cornelius Scipio (consul of 218 BC), who were killed 29 days apart in 211 BC when fighting the Carthaginians (Livy 25.36.13–16). Since the river seems to curve around the tomb, it was probably placed on a natural promontory. Pliny's description of the river flowing gently past towns on both sides remains true today.

3.10. The description catalogues the localities south of the Baetis, between it and the coast. Over twenty places are named, almost all without comment, in two separate lists. The first, nine towns, is in no obvious order, and all but one have both indigenous and Roman names. A second list of fourteen towns has only indigenous names, in alphabetical order. These two separate catalogues suggest that Pliny was working from official registers rather than any geographical account.

Next there is a list of ten towns in the Corduba district, ending with Corduba itself, which stands at the head of navigation on the Baetis. In Strabo's day, half a century before Pliny, the river could be navigated somewhat farther (Strabo, *Geography* 3.2.3), but this may not represent a serious difference. Today it is still navigable for small boats as far as Córdoba.

Two towns a short distance below Corduba are also mentioned, Carbula and Detumo, and the account ends at the boundary between the Corduban and Hispalensian districts, near the Singilis River (modern Genil, which preserves the ancient name), the largest southern tributary of the Baetis (358 km. long), whose origin is in the modern Sierra Nevada east of Granada.

3.11. The next area is the Hispalensian district, with a dozen towns mentioned. Unlike the ones at 3.10, this list seems to follow a definite geographical itinerary, first heading down the right (north) bank of the Baetis – where most of the settlements seem to have been – starting at the district boundary just below Detumo, and then crossing the river into the town of Hispal. Next are a number of towns on that bank (now, below Hispal, the east side). The Maenuba River (modern Guadiamar) is a short stream (82 km. in length) west of and parallel to the Baetis. According to Pliny it was a tributary, but it may have drained directly into the estuary of the larger river. Hydraulic activities from at least the first century BC changed the landscape of this region: there were many canals and it was subject to tidal flows (Strabo, *Geography* 3.2.5). This was a prosperous agricultural region which had commerce directly with Rome and Italy.

The final list of five towns, said to be within (probably meaning "on") the estuary of the Baetis, are on its southern side, although Asido Caesarina (at modern Medina Sidonia) is a few kilometers inland to the southeast. The unusual surname of Hasta (Regia) may be due to an ancient cultic center at the site, which became a shrine of Juno Regina (Daniel Jesús Martín-Arroyo Sánchez, "*Interpretatio* de la Titulatura Cívica de *Hasta Regia* a Partir del Culto a Juno Regina," *DHA* 43 [2017] 175–211).

3.12. The Singilis River (*NH* 3.10) joins the Baetis from the south downstream from Corduba near Segida. The most important town on it is Astigi, or Augusta Firma (modern Écija), about 30 km. above its mouth and at the head of navigation. The 17 towns listed – organized by political status – are mostly in the uplands between the Singilis and the Baetis, which rise to nearly a thousand meters above sea level and were the most populated part of Baetica outside the river valleys. Most of the towns (10 of the total of 16) were tribute-paying, and are listed alphabetically, indicating that Pliny was using an official register.

The most famous of these towns is Munda, perhaps near modern Montilla, about 40 km. south of Córdoba. It was the site of two important battles in Roman history. In 214 BC the Carthaginians were defeated by Cn. Cornelius Scipio (Livy 24.42.1–8), and in 45 BC Julius Caesar defeated Gnaeus Pompeius the Younger and his brother Sextus at the same location, in the last battle of that phase of the Roman civil war (*de bello hispaniensi* 28–31). The section closes with the citation of three towns on the navigable Maenuba, which lie along a stretch of 20 km. and are listed from south to north. They are directly west of modern Seville.

3.13. The final district of Baetica is the region west of the Baetis River and as far as the Anas, the southwestern corner of modern Spain. This was called Baeturia and was believed to be a region of Celtic inhabitants, who presumably had been less assimilated into the Roman system than those farther east. A portion of the region was still called Celtica (*NH* 3.14). More inland was another indigenous group, the Turduli (*NH* 3.8).

The Celtiberi were a mixture of Celtic and indigenous Iberian populations, concentrated in the west-central parts of the Iberian peninsula (Strabo, *Geography* 3.4.5). Their heartland was on the upper Hiberus (Ebro) River. The Greco-Roman impression of them was an uncivilized people with bizarre cultural practices (Silius Italicus 3.340–3). Yet their military tactics were remarkable, and the Romans, who contended with them for over a century, adopted their type of sword (Polybius F96), a reflection of their proficiency at metallurgy and their use of the local mineral resources. Pliny believed there had been a movement of Celts from Lusitania into Hispania, but this is by no means certain and the Celts may have come more directly from the north. Their religion and language are still little known, although there are epigraphic examples of the latter (Jürgen Untermann, "Hispania, Iberia: Languages" *BNP* 6 [2005] 392–5). Celtiberian remnants survived in the peninsula into late antiquity and probably far beyond (William E. Mierse, *Temples and Towns in Roman Iberia* [Berkeley 1999] 300–2).

3.14–15. Pliny drew attention to the composite toponyms of Baetica, indigenous names with a Roman surname. His first list of seven names follows this pattern, although he or his source left out Callenses Aeneanici and added it to the end. These towns are in the uplands between the coast and the Anas River, an ancient and wealthy mining district (the modern Río Tinto) which may have given rise to the stories about the wonders of Tartessos (Strabo, *Geography* 3.2.11). The second list consists of eight towns that do not have Roman surnames, and then there are four additional ones in the region of the Turdulians, more to the east and north of Corduba. The catalogue concludes with nineteen towns in the district of Gades, again divided into two lists. First there are five with Roman surnames and then the remainder with only indigenous ones. These lie to the east of Gades, in its hinterland.

3.16. It was Pliny's custom to end many geographical sections with distances, often comparing those from different sources, with or without a discussion of their relative accuracy. He was well aware of the vagaries of long-distance measurements, a problem that had existed since early Hellenistic times and the first attempts to establish distances over a large extent and to use methods other than travel times. But there were still issues regarding conversion between different units of measurement, and the exact length of those units (Diller, "Ancient Measurements" 6–9). Roman miles (equal to 1,480 m.) were the first invariable means of calculating long distances, although when they were converted from other units, or formulas based on travel times, problems could occur (Roller, *Eratosthenes* 271–3). Pliny noted another difficulty: the starting and ending points of measured distances were not always known, and different routes may have been used. He was also astutely aware of changes in the alignment of roads and boundaries of political units, as well as alterations to the physical landscape, such as the course of rivers. A particular issue was the coastline of the Mediterranean; in certain places, siltation and tectonic forces could cause changes over a short period of time. This was especially noticeable in the river estuaries of western Asia and at the head of the Persian Gulf; at the latter, siltation was as much as 80 km. over two centuries (*NH* 6.138–40). Pliny's discussion of these problems is one of the most succinct and informative extant from antiquity.

It is probable that he generally used Agrippa's commentary on his map (Agrippa F4; see further *NH* 3.17–18), rather than the actual map itself, which may not have had many, if any, distances. Yet there are certainly cases where Pliny seems to have been operating from a visual perspective and which were based on the map itself. The road distance of 475 miles (700 km.) from Carthago Nova (modern Cartagena) to the Anas River compares remarkably with the current highway distance (682 km.), which follows Roman roads much of the way. The endpoints of the north–south distance of 258 miles (380 km.) are difficult to determine (see *NH* 3.17), but it approximates that from Cádiz to the Tagus River at Mérida, a possible width of Baetica. Pliny's data demonstrate that Agrippa published at least some of his material before the Augustan reorganization of the Iberian provinces after 27 BC (see *NH* 3.6 and 3.17).

3.17. Pliny then provided contemporary figures for the extent of Baetica, presumably obtained from official records when he was procurator. From the eastern boundary at Murgi to Gades was 250 miles (370 km.), with an additional 25 miles (37 km.) from Murgi to the coast (near modern Almería); this total approximates a modern road distance of 436 km. Agrippa's width (*latitudo*) is in fact the coastal distance from Carteia (near modern Gibraltar) to the Anas River; this may be an attempt to see political units as regular geometrical forms, a concept originating with Eratosthenes (*Geography*, F66). The 234 miles (346 km.) compares with the modern road distance of 330 km., although it is far from direct today, and Agrippa's width remains inscrutable. Since the figure seems to conform to no possible direct route between Carteia and Gades, the error may be Pliny's, or reflect boundary adjustments that were not reconciled. Unfortunately there is no further explanation or correction, and the text is vague, implying that Agrippa was criticized for using pre-Augustan distances, which seems unfair. Yet Pliny was not always approving of Agrippa's data and expressed doubts about his figures for Scythia, Germania, and the total east–west length of the Mediterranean (*NH* 4.91, 98; 6.207).

The Porticus Vipsania in Rome was begun by Vipsania Polla, the sister of Agrippa (Dio 55.8.4). Its exact location is not known beyond being somewhere on the lower slopes of the Quirinal (Richardson 319–20). The structure was completed by Augustus after 7 BC. As noted, it contained Agrippa's map; its relationship to any publication of the information remains uncertain, and the project as a whole may have been based on one left unfinished by Julius Caesar. Pliny is the primary source for the map and the published commentary – although it is often difficult to determine which is being cited – and mentioned either or both over thirty times (O. A. W. Dilke, *Greek and Roman Maps* [Ithaca 1985] 39–54; Pascal Arnaud, "Marcus Vipsania Agrippa and His Geographical Work," in *Brill's Companion to Ancient Geography: The Inhabited World in Greek and Roman Tradition* [ed. Serena Bianchetti *et al.*, Leiden 2016] 205–22).

Hispania Citerior

3.18. The next province discussed is Hispania Citerior, which included much of the Mediterranean coastal regions and the northern interior of the peninsula all the way to the Atlantic (Pilar Ciprés, "*Hispania Citerior* en la Geografía de la *Naturalis Historia* de Plinio," *Veleia* 31 [2014] 15–32). In noting that the shape of the province had been altered (*forma mutata est*), Pliny was subtlely acknowledging his debt to the map of Agrippa. The monuments of Pompey the Great were built after his defeat of the Roman adventurer Q. Sertorius in 73 BC (see also *NH* 7.96; Sallust, *Histories* 3.63). Some remains are visible in the Col de Panissars on the border between France and Spain. There were several such monuments, whose inscriptions listed the 866 or 876 towns that Pompey claimed to have subjected during his proconsulship from 77 to 71 BC. This may be an exaggeration, but since Pompey organized the province after the death of Sertorius, it perhaps refers to the total number of urban centers in the territory. Pliny discussed Pompey's career, including the trophies in the Pyrenees, in greater detail at *NH* 7.95–9 (Mary Beagon, *The Elder Pliny on the Human Animal*: Natural History *Book 7* [Oxford 2005] 281–8).

The seven administrative districts reflect the Augustan reorganization and are listed roughly from the Mediterranean toward the west. Carthageniensis is on the southern coast around Carthago Nova, and Tarragonensis between it and the Pyrenees. The remaining five are in the interior and west to the Atlantic north of Lusitania. For the islands in the Mediterranean, see *NH* 3.76–9. As before, the towns in the province are catalogued according to legal status.

3.19. Pliny began his account of Hispania Citerior with a listing of the peoples living along its western boundary, from Carthago Nova to the northwest. The Bastuli were closest to the coast (Strabo, *Geography* 3.4.1), and inland from them were the little-known Mentesani. To the north, near the source of the Baetis River and the city of Castulo, were the Oretani (Strabo, *Geography* 3.3.2). Farther north, on the Tagus River, near modern Toledo, were the Carpetani, who, along with the Oretani, came to Roman notice during the Second Punic War, when Hannibal sent them to Africa for the defense of Carthage (Polybius 3.33.9). The Carpetani were said to be the strongest people in the region and were problematic Carthaginian allies (Polybius 3.14.1–3). The Vaccaei, Vettones, and Arevaci were more to the northwest. Hannibal encountered the Vaccaei when he penetrated deep into Iberia just before the Second Punic War (Livy 21.5.5), and their eventual opposition to the Romans lasted into the early Augustan period (Dio 39.54, 51.20–5). The Vettones were to their southwest (probably actually in Lusitania: Tusculum 3–4, 208–9), and were less involved in the wars of the peninsula, but were defeated by the Romans in 193 BC (Livy 35.7–8). The Arevaci, situated between the Tagus and Durius (modern Douro/Duero) Rivers, were a Celtiberian people who caused difficulty for the Romans in the mid-second century BC (Polybius 35.1–3).

After listing the indigenous peoples, the account returns to the coast and catalogues the cities and districts along it, from the border of Baetica to the Pyrenees. The first two towns, Urci and Barea, are actually in Baetica. The itinerary continues past Carthago Nova (modern Cartagena) and the mouth of the Tader River (modern Segura). Cape Saturn (189 m. above sea level, modern Cabo de Palos), about 30 km. east of Carthago Nova, was especially important because it was a major navigational point to and from Caesarea (Kaisareia) in Mauretania (see *NH* 5.20), the capital of Juba II and Cleopatra Selene, the allied monarchs of the Augustan period. After the early AD 40s it was the capital of the province of Mauretania Caesariensis. Caesarea lies to the southeast of Cape Saturn (the distance of 197 miles – 290 km. – is remarkably accurate). Inclusion of the sailing distance demonstrates that Pliny continued to be dependent on a nautical *periplous*.

3.20. The coastal itinerary continues from north of the Tader to the Udiva River, which is not certainly located but may be the modern Palancía at Saguntum. The pleasant lakes in Editania were probably south of Valentia (modern València); remnants of them survive in the modern Albufera, a nature reserve noted for its biodiversity.

Saguntum (modern Sagunto, a nineteenth-century restoration of the ancient name) was given special notice because it was implicated in the beginning of the Second Punic War. It became famous for its heroic resistance to Hannibal in 218 BC, ending when the locals burned the city and themselves (Livy 21.6–15).

3.21. Continuing to the north, the *periplous* crosses the Hiberus River (Greek Iber, modern Ebro). Pliny's length, 450 miles (666 km.), is far too short (it is actually 930 km. long), and its upper course may not have been carefully surveyed. The river forms a major barrier across the northern Iberian peninsula, most famously as the border between Roman and Carthaginian territory in the third century BC. It was always a major route into the interior, navigable, as Pliny noted, for many miles (Vareia, the head of navigation, is in modern Logroño, where the name survives in a barrio). Since there is another Hiberus River far away in southwestern Iberia (Avienus 248, probably the modern Río Tinto), Pliny's suggestion that the northern Hiberus gave its name to the peninsula is certainly the reverse of the actual situation: the name is probably an ethnym that attached itself both to the territory and the rivers.

The suggestion that Tarraco (modern Tarragona) was founded by the Scipios is somewhat inaccurate. The city was probably an indigenous settlement – it had a good anchorage (Eratosthenes, *Geography* F152) – and was established well before the Scipio brothers made it their base in the Second Punic War. Archaeological evidence suggests a town from at least the sixth century BC, probably a Phocaean or Massilian settlement (Tusculum 3–4, 218).

The Ilergaones were on the south side of the Hiberus, west of Tarraco, and may have been Carthaginian allies early in the Second Punic War (Livy 22.21.6). In the Roman civil war, they supplied Julius Caesar with grain (Caesar, *Civil War* 1.60). Vareia is at a crossing of the Hiberus, at modern Vareia in Logroño (Strabo, *Geography* 3.4.12), and Cessetania is near Tarraco (Polybius 3.76.5; Livy 21.60–1). The Ilgertes – perhaps closely related to the Ilergaones – were inland from Tarracon and on the north side of the Hiberus, around modern Lérida, which preserves their name. They were Carthaginian allies in the Second Punic War (Livy 21.61.5). The Laeetani and Indigetes were on the coast northeast of modern Barcelona (Strabo, *Geography* 3.4.8). The latter may have been the first indigenous people to encounter Greek traders, at Rhode (modern Rosas), probably the oldest Greek settlement on this coast (Duane W. Roller, *A Historical and Topographical Guide to the Geography of Strabo* [Cambridge 2018] 156).

3.22. The final section of the Iberian *periplous* describes the last portion of the coast, from Tarraco to the Pyrenees (Pliny's Pyrenaeus). Not part of the *periplous*, but also included here, are some of the populations on the southern slopes of the Pyrenees. The Ausetani, near the coast, had been subdued by Hannibal on his march to Italy, and then allied with and supplied Julius Caesar when he was in the region (Livy 21.23; Caesar, *Civil War* 1.60), but otherwise are little known. The Lacetani were to the southwest and were also defeated by Hannibal. The Cerretani ("Mountaineers"), west of the Ausetani on the mountain slopes, were probably located south of Andorra. They were famous for their ham, which is still prized today as the Serrano variety (Strabo, *Geography* 3.4.11; Martial 13.54). The Vascones, or Vaconi (perhaps the modern Basques), extended as far as the Bay of Biscay, with their center at Pompaelo (modern Pamplona). They seem to have been unknown to the Greco-Roman world until encountered by Pompey the Great in 72 BC (Sallust, *Histories* 2.80; Plutarch, *Sertorius* 21).

Pliny noted the unusual situation of Emporiae (Emporion, modern Ampurias/Empúries), a probable Phocaean settlement from around 600 BC. There were two parts to the town: the Greek trading post on an offshore island, and an indigenous settlement on the mainland. There is also evidence for a separate Roman district. The community was an unusual example of locals and Greek settlers creating a harmonious relationship (Strabo, *Geography* 3.4.8; Roger Collins, *Spain: An Oxford Archaeological Guide* [Oxford 1998] 53–9).

The Ticer River cannot be distinguished with certainty among several small streams between Emporiae and the Pyrenees. Pyrenaea Venus is a sanctuary of Aphrodite, the Aphrodision, near modern Port-Vendres, the southernmost town on the French coast of the Mediterranean, which preserves the Roman name of Portus Veneris. The sanctuary was on modern Cape Béar (see Strabo, *Geography* 4.1.3; Avienus 559–60).

3.23. Pliny's discussion of the individual districts of the province begins with Tarraconensis, at the northeastern corner of the Iberian peninsula. Rather than list towns, as in Baetica, Pliny catalogues this region by ethnic groups, beginning with the Dertosani, the people of Dertosa (modern Tortosa), where the coastal road crossed the Hiberus, about 20 km. from its mouth. The Cerretani (see also *NH* 3.22) were divided into two groups, the Iuliani and the Augustani, which presumably represents legal recognition of parts of them by Julius Caesar and then Augustus.

The Bisgargitani lived in the uplands north of modern Barcelona. The Edetani primarily lived south of the Hiberus, but there were a few groups of them on the other side, around Caesaraugusta (see *NH* 3.24; Strabo, *Geography* 3.4.1). In the winter of 219/8 BC, P. Cornelius Scipio (Africanus), whose headquarters were at Tarracon, gained the confidence of their leader, Edeko, whose name may be the eponym of his people. This significantly increased the status of the Romans in Iberia (Polybius 10.34–5).

The Gerundenses are little known, but gave their name to Gerunda, modern Girona/Gerona, northeast of Barcelona. The Iessonienes are also obscure; their name survives in modern Guisona (probably ancient Iesses), northwest of Barcelona. The Julian Teari are essentially unknown, although they must have achieved some recognition by Julius Caesar.

The Aquicaldenses are perhaps at the Hydata Therma of Ptolemy (*Geographical Guide* 2.6.70), but which of the several hot springs around Barcelona is meant cannot be determined (Pliny, *Histoire naturelle* 3, 116). The Aesonenes are reflected in modern Isona, southwest of Andorra, and the Baeculonenses are otherwise unknown unless their town was the Bailouka of Ptolemy (*Geographical Guide* 2.6.70). Generally these peoples seem to have been little involved in the romanization of the region.

3.24. Pliny listed 29 out of the 55 ethnic groups in the district of Caesaraugusta (modern Zaragoza). Many of them are little known, but some are reflected in modern toponyms around Zaragoza. As usual, they are classified according to status, beginning with those having Roman citizenship. The Bilbiltani were southwest of Caesaraugusta, around the town of Bilibis (near modern Calatayud), famous as the birthplace of the poet Martial (Martial 10.103). The Celsenses lived in the vicinity of Celsa (modern Gelsa), downstream from Caesaraugusta, where there was a bridge across the Hiberus (A. Beltrán, "Celsa," *PECS* 210–11). The Nasican Calagurritani were around Calaguris (modern Calahorra), on the Hiberus northwest of Caesaraugusta. This region was the site of several battles during the Roman civil war (Strabo, *Geography* 3.4.10). The Surdaonian Ilerdenses were a group of the Ilergaones (see *NH* 3.21), living on the Sicoris River (modern Segre), which joins the Hiberus about halfway between Caesaraugusta and the coast. The Oscenses were associated with Osca (modern Huesca), about 50 km. north-northeast of Zaragoza. It was the headquarters of Sertorius in the 70s BC (Plutarch, *Sertorius* 14). The Turiassonenses were around Turiaso (modern Tarrazona), about 65 km. northwest of Caesaraugusta, and were noted for their ironworking (*NH* 34.144).

Several ethnic groups had Latin status. The Cascantenses were at Cascantum (modern Cascante), a few kilometers north of Turiaso. Their territory was ravaged by Sertorius in the 70s BC (Livy 91.F18). The home region of the Ergavicenses, who allied with the Romans in 179 BC (Livy 40.50.1), is undetermined. The Graccurritani were the settlers in a town just north of Cascantum that was established by Ti. Sempronius Gracchus in 179 BC (Livy, *Summary* 41). The Leonicenses and Osicerdenses were probably in the same region.

The Tarracenses were the only treaty population that Pliny listed, and have not been located. Their organization may reflect the era of Sertorius in the 70s BC (Javier Andreu Pintado, "Foederatos Tarracenses: Sobre una Comunidad de la Tarraconense en los Listados de la *Naturalis Historia* de Plino el Viejo," *Gerión* 35 [2017] 461–70). Pliny catalogued 17 tributary peoples, in alphabetical order by initial letter. The name of the Arcobrigenses may survive in modern Arcos de Jalón, west of Calatayud. The Andelonenses were perhaps around modern Andión, just southwest of Pamplona, and the name of the Aracelitani may survive in modern Arakil about 30 km. to the northwest. The Bursaonenses, whose territory was plundered by Sertorius (Livy 91.F18), were northwest of Caesaraugusta. One group of the Calagurritani (see above), called the Fibularienses, had a treaty relationship with Rome.

The Conplutenses were to the southeast of modern Segovia, near modern Alcalá de Henares, but most of the rest of the populations listed in the Caesaraugusta district cannot be located, except through assumptions made about similar modern toponyms (Pliny, *Histoire naturelle* 3, 118–19). Exceptions are the Lacetani, Pompelonenses (at modern Pamplona), and the Segienses, all of which are northwest of Caesaraugusta. Few of these peoples played any prominent role in Greco-Roman history. Caesaraugusta itself, whose name survives in modern Zaragoza and which is at an ancient crossing of the Hiberus (Ebro) River, was located at a previous settlement named Salduba, whose population was probably brought together to create the Roman city (Francisco Pina Polo, "De la ciudad indígena *Salduie-Salduvia* a la colonia romana *Caesar Augusta*," *Gerión* 35 [2017] 541–50).

3.25. The district of Carthago Nova (modern Cartagena) included 65 populations, of which Pliny listed 19, both towns and ethnic groups. It extended west and north of Carthago Nova as far as Toletum (modern Toledo). The Castulonenses lived in the mountainous region near the headwaters of the Baetis, around the town of Castulo (near modern Linares; see *NH* 3.17), which figured in the Second Punic War (Polybius 10.38.7; Livy 28.20). This was a mining district with lead and silver mines under Imperial control (Strabo, *Geography* 3.2.10; Alfred Michael Hirt, *Imperial Mines and Quarries in the Roman World* [Oxford 2010] 155). The people received the surname of Caesarii Iuvenales ("young Caesars") in honor of C. Caesar, the grandson of Augustus (*CIL* 2.3267–8).

The Saetabitani (whose name is preserved in modern Xátiva) were just south of modern València, and received their rights under Augustus. The name of the Valerienses survives in modern Valera in the uplands west of València. The remainder of the ethnic groups in the district were tribute-paying, listed alphabetically. Many are unlocated or are speculatively placed in the vicinity of modern towns that seem to preserve the ancient name, such as the Bastitani around Basti (west of Cartagena), or the Consaburrenses at Consuega (south of Toledo). The Dianenses were around Dianium, an ancient sanctuary of Artemis, probably of Phocaean origin, located at the prominent promontory northeast of Cartagena which still retains the ancient name of Dénia (Strabo, *Geography* 3.4.6).

The Egelestani were probably coastal, perhaps around modern València, since they had a notable salt industry (*NH* 31.80). The Ilorcitani, Laminitani, and Mentesani (also known as the Oretani) have not been located with certainty (Pliny, *Histoire naturelle* 3, 121). The Germanian Oretanians were for some reason called "genuine" (*germanus*); the adjective has nothing to do with the Germans, and it is unlikely that any Germanic peoples had penetrated this far south by Pliny's day. The Viatienses and Virgilienses are also obscure; the name of the latter may be an indigenous ethnym that became somewhat latinized.

3.26. The next *conventus* is Clunia. The town itself lies on the upper Duero River near the modern village of Peñalba de Castro, in north-central Spain, a major archaeological site (Collins, *Spain* 114–17). The district extended to the Bay of Biscay. Pliny's catalogue of this region is somewhat different from the previous ones, since the ethnic groups and towns are not organized by status (it is probable that Roman political organization was scant in this remote region), but in terms of major groups, each of which is divided into lesser ones, most of which were not named by Pliny.

The first prominent ethnic group is the Varduli (or Bardyllians, Strabo, *Geography* 3.4.12; see also Pomponius Mela 3.15), who were south of the western end of the Pyrenees. They had 14 sub-groups, but Pliny only mentioned one, the Alabanenses. Then there were the Turmogidi, more to the west, with their sub-groups the Segisamonenses and the Segismaiulienses (modern Sasamón probably preserves their name). This is perhaps the Segasame which was the headquarters of Augustus during the Cantabrian War of the 20s BC (Florus 2.33.48), which may explain why part of the people received recognition with a Roman surname.

The Carietes and Venneses – and their five sub-groups – cannot be precisely located. The Pelondones, at the source of the Durius River (modern Douro/Duero), included the best-known peoples in the district, the Numantini, who gave their name to the Numantine Wars of 143–133 BC, when the Romans repeatedly fought the Celtiberians (Appian, *Iberike* 76–98). The site of Numantia itself lies at modern Cerro de la Muela on the river. Four of the 17 groups of the Vaccaei are named. Their territory was on the middle Durius (see *NH* 3.19), with their major settlement Palantia (modern Palancia, north of Valladolid), the region of the Palantini. This was an area of Roman wars in the second century BC; the Intercatienses were attacked by L. Licinius Lucullus in 151 BC (Appian, *Iberike* 53). The Caucenses (around modern Coca, south of Palencia) were badly treated by Lucullus, since through a subterfuge all their adult males were killed and the city sacked, an event remembered years later as a major Roman disgrace. These engagements were part of repeated Roman efforts against the Celtiberians in the second century BC (Strabo, *Geography* 3.4.13).

3.27. Next are the Cantabrians, whose name survives in modern Santander, and the Astures, farther to the west. The former generally remained outside of Roman interests until an expedition under the personal command of Augustus (his last campaign) moved into the region in 26 BC, for uncertain reasons, and initiated hostilities that lasted for seven years (Augustus, *Res gestae* 26; Strabo, *Geography* 3.3.7–8; for additional sources, see Erich S. Gruen, "The Expansion of the Empire Under Augustus," *CAH* 10 [second edition 1996] 163–6). The more western Astures were also subdued at this time, although the Cantabrians remained restive.

In these later sections of Pliny's summary of Hispania he seems to have become less interested in recording specific ethnic groups, naming one only of the seven subgroups of the Cantabrians (those at Juliobriga, probably at modern Reinosa southwest of Santander). The Autrigones lived north of the upper Hiberus; Pliny named two of their ten towns. For the Arevaci see *NH* 3.19.

3.28. The Astures were west of the Cantabrians; their name survives in modern Astorga, southwest of Leon, which (as Asturga) was their main town. Their territory extended north to the ocean. They were divided into two groups, probably those in the lowlands (the Augustani, around Asturga), and those across the mountains to the north, the Transmontani. Several of the smaller groups were listed. The Gigurri were around modern Valdeorras (Ptolemy, *Geographical Guide* 2.6.38). The Paesici were on the coast, and the Lancienses were east of Asturga; their city, Lancia, was taken by the Romans in 25 BC (Dio 53.25.8). The Zoelae were southwest of Asturga, near the modern Portuguese border. They were noted for their flax industry; the local product had begun to be imported into Italy in Pliny's day (*NH* 19.10).

To end his discussion of the district of Clunia, Pliny gave a population figure of 240,000, and also for the final two districts of Hispania, the Lucensian (166,000) and the Bracarian (285,000). There is no exact indication where these numbers came from, but presumably there had been some official census of the population. Augustus enumerated Roman citizens three times (28 BC, 8 BC, and AD 14: *Res gestae* 8), finding a total rising from around four million to nearly five million over those 42 years, which suggests that each successive census covered a wider area. Pliny's numbers may be based on one of these, but details are lacking, especially whether – or how many – people outside Italy were included. The recording of census figures for only the last three districts of Hispania, as well as the scant demographic data provided, suggests that Pliny was using a different source than previously.

The Lucensian district was in the extreme northwestern part of the Iberian peninsula (modern Lugo preserves the name). The word "lucus" refers to a sacred grove, a feature not uncommon in Celtic regions (for another example see *NH* 3.37), and presumably it was an ancient location that had been appropriated by the Imperial cult, taking the name Lucus Augusti (*CIL* 2.2638; Ptolemy, *Geographical Guide* 2.6.24). The Lemavi were probably south of the grove, near modern Lemos.

Objections to the strange sounds of foreign names was a regular feature of Greco-Roman geographical writing. Both Strabo and Pomponius Mela made similar comments about the names in northwestern Iberia. The former stated that they were unpleasant to write, and the latter that they could not be rendered in Latin (Strabo, *Geography* 3.3.7; Pomponius Mela 3.15).

The final district of Hispania was that of the Bracares, whose name is preserved in modern Braga in northern Portugal. The Biballi were to the northeast of Braga. The location of the Coelerni and Equaesi is unknown, but the Callaeci occupied the extreme northwest corner of the Iberian peninsula. They were the subject of an expedition by D. Junius Brutus (an ancestor of the tyrannicide), the consul of 138 BC, who received the title Callaicus for his efforts (Strabo, *Geography* 3.3.1–4). The Limici and the Querquerni were northwest of Braga (the Lima, or Limia, River is in northernmost Portugal).

At this point Pliny brought his analysis of the provinces of Hispania to a temporary end. The discussion of the third Iberian province, Lusitania, was reserved for later (*NH* 4.110–20), since due to its Celtic population it was seen to some extent as an extension of Gallia.

3.29. There are some statistics and comments about Hispania as a whole; the use of the term Hispania Ulterior (which was obsolete in Pliny's day) indicates a pre-Augustan source. The distance of 607 miles (898 km.) from the Pyrenees to Castulo (on the upper Baetis) is remarkably close to the modern road distance of 899 km., as is the other figure, 307 miles (454 km.) from Tarraco to Olarson (at modern Irún on the Atlantic coast), which is comparable to the modern road distance of 485 km.

The shape of the Iberian peninsula was problematic to ancient geographers. Pliny saw it as a wedge, and Strabo (*Geography* 3.1.3) as an oxhide (a rectangle with the corners drawn in). Both were reflecting Eratosthenes' tendency to see geographical units as descriptive forms. Pliny was the first extant author to understand the orientation of the Pyrenees, and previous difficulties regarding this may have contributed to issues concerning the shape of the peninsula as a whole. As late as Strabo, and thus even after the intensive Roman expeditions in the region, the mountains were said to run north–south (they actually extend from east-southeast to west-northwest). Pomponius Mela (2.85), perhaps attempting to reconcile Roman knowledge with the data from geographical writers, wrote that the mountains changed direction. It was not until Pliny (*NH* 4.110) that the proper orientation was established, but difficulties still remained.

3.30. The Iberian peninsula was noted for its mineral resources, and mining of them began long before Greeks and Romans reached the area. Portions of southwestern Spain – the modern Río Tinto – were still being exploited until recently, and in antiquity produced silver and lead (Hirt,

Imperial Mines 38–41). Inland from Carthago Nova was an extensive region of silver mines, which employed 40,000 workmen in the second century BC (Polybius 34.9.8–11). Gold was found in the northwestern part of the peninsula (Hirt, *Imperial Mines* 77), and there was copper north of Corduba. It was said that all of Iberia produced metals of the highest quality (Strabo, *Geography* 3.2.3, 8–10).

Another Iberian product was *lapis specularis*, a form of mica or *selenite* (not modern moonstone) that could be used for windows, and, when, crushed into powder, medicinally (Strabo, *Geography* 12.2.10; Dioskourides 5.141). *Minium*, or cinnabar (mercury sulphide), was a red pigment used in painting and for limited medicinal purposes (*NH* 33.111–22). Iberian cinnabar had been known to the Greek world as early as the fifth century BC (it is found in the same context as silver), and in Roman times it came from the region north of Corduba.

Iberian marble was not commonly used outside the peninsula, but the variety known as *broccatello* came from Dertosa (modern Tortosa) and was shipped down the Hiberus to Italy and elsewhere. It was used in Roman structures in Hispania, such as the pavement of the theater at Caesaraugusta (Marc Mayer and Isabel Roda, "El Broccatello de Tortosa: Testimonios Arqueológicos," *Pallas* 50 [1999] 43–52).

The Iberian peninsula was particularly affected during the Roman civil war of AD 68–9. Two of the brief emperors of that period had been posted there, Ser. Sulpicius Galba to Tarraconensis and M. Salvius Otho to Lusitania. Galba was proclaimed emperor at Carthago Nova in the summer of AD 68, with Otho's support, who eventually revolted and engineered the death of Galba and his own succession the following January. He was replaced by A. Vitellius three months later, who was followed by Vespasian shortly thereafter (T. E. J. Wiedemann, "From Nero to Vespasian," *CAH* 10 [second edition 1996] 256–82). Given the involvement of Roman officials in Hispania in these events – and others besides Galba and Otho were implicated – Vespasian, when his rule was secure, felt that it was wise to ensure the loyalty of the entire peninsula (which can be the only meaning of "universae Hispaniae," not merely the single province of Hispania) by raising the status of its inhabitants (Javier Andreu Pintado, "Entre la Literatura y la Historia: Plin. *Nat.* III 30 y la Latinización de *Hispania*," in *"Urbs aeterna," Actas y Collaboraciones del Coloquio Internacional "Roma entre la Literatura y la Historia"* [ed. Alonso del Real, Pamplona 2003] 187–210). This probably happened while the emperor was censor along with his son Titus in AD 73–4 (Géza Alföldy, "Spain," *CAH* 11 [second edition, 1996] 450–1).

As a final note, Pliny literarily crossed the Pyrenees from Hispania to Gallia Narbonensis. The mountains extend across the entire northern edge of the Iberian peninsula, thus separating it, physically and culturally, from the rest of Europe. They rise to a height of 3,404 m. near the middle of the range. The toponym (as Pyrene) was known to the Greek world as early as the fifth century BC (Herodotus 2.33), but only vaguely, and the exact nature of the mountains was not fully understood until the first century AD (see *NH* 3.29).

Gallia Narbonensis

3.31. After going past the Pyrenees, Pliny's account continues in the form of a *periplous* along the coast of Gallia Narbonensis, the Mediterranean litoral of modern France (through *NH* 3.37). Interior Gallia (the rest of France) is reserved for *NH* 4.105–9. Gallia Narbonensis extended from the Italian boundary at the Varus River (modern Var) to Hispania at the Pyrenees. The region was conquered by the Romans during a series of campaigns between 125 and 118 BC; Greeks had founded towns on the coast as early as around 600 BC. The name Bracata refers to the trousers that the locals wore (*bracae*, an indigenous word used in Latin by the first century BC: Cicero, *pro Fonteio* 33; Cicero, *Letters to His Friends* 9.15.2). It was unlikely to be an official term, and was probably no older in Latin than the time of the creation of the province of Narbonensis.

The eastern end of the province was the Varus River, just west of modern Nice, which remained the boundary between France and Italy until 1860. The northern edge was defined by the mountain ranges of the Alps, the Iures (modern Jura), and the Cebenna (modern Cévennes), which were seen as creating a barrier that kept Narbonensis separate from the rest of Gallia. This concept is only partially true, since the Cebenna is isolated from the Alps by the broad valley of the Rhône (ancient Rhodanus), which creates a major route into the interior. Pliny's comment that Gallia Narbonensis was virtually an extension of Italy is well taken, and can still be understood today in the striking archaeological remains and the survival of the name Provence.

3.32. The account begins just north of the Pyrenees, in the territory of the Sordones (Pomponius Mela 2.84). Inland from them, on the north slope of the mountains, were the Consuarani (perhaps the Consoranni of *NH* 4.108), whose name survives in modern Saint-Lizier-en-Conserans, about 70 km. south of Toulouse. The narrative continues north along the marshy coast (the Rubrensis Lake is probably the modern Bages and Sigean lakes), past Narbo (modern Narbonne), the capital of the province. Its official name was Narbo Martius (Velleius 1.15.5). Here Julius Caesar settled veterans after his campaigns in Iberia (Michel Gayraud, *Narbonne antique des origines à la fin du IIIe siècle* [Paris 1981] 175–81).

3.33. The *periplous* continues from Agatha (modern Agde), at the western end of the Rhone delta. The Volcae were the primary indigenous group of this region, with the Tectosages in the more western portion, closer to the Pyrenees.

Pliny seems to have confused Rhoda, a little-known town in the Rhone delta (abandoned in his day), with Rhode (modern Rosas in Spain). There is some evidence of settlement on the Iberian coast from the Greek island of Rhodes on the Iberian coast (Rosa-Araceli Santiago, "Enigmas en torno a Saguntum y Rhoda," *Faventia* 16 [1994] 56–64), but none for any in the Rhone delta. Rhoda did not give its name to the Rhodanus River (modern Rhone); the reverse is true. "Rhodanus" is an indigenous name, known in various forms (such as "Rhine") for rivers in western Europe. The mouth of the Rhodanus is the most important topographical feature of this coast, providing access to much of the interior and, through its tributaries, an easy passage to the Rhine system and thus northern Europe. Although the Greek traders around the lower Rhodanus knew about its upper course from early times, it came into more general knowledge at the time of Hannibal (Polybius 3.47). Yet the passage of the river through Lake Lemannus (modern Lac Léman or Lake Geneva) does not seem to have been known until the first century BC (Strabo, *Geography* 4.6.6).

The number of mouths of the Rhone varied, and presumably reflected siltation and water levels. Pliny knew of three; Polybius had only two (34.10.5), and Apollonios reported seven (4.634). Today there are two, the Petit Rhône to the west and the Grand Rhone to the east.

3.34. In 102 BC the consul C. Marius restructured the outlets of the Rhodanus so that ships were forced to go past a place where the Massilians charged tolls. His works were called the Fossae Marianae (Pomponius Mela 2.78; Strabo, *Geography* 4.1.8). The Campi Lapidei ("Stony Plains") were southeast of modern Arles on the eastern edge of the delta; Strabo described them in great detail (*Geography* 4.1.7). There were a number of geological explanations, but Pliny only referred to the role of Hercules (Herakles), who fought the indigenous Ligyans there by throwing stones at them. The story was recorded in Aeschylus' *Prometheus Unbound* (Roller, *Historical and Topographical Guide* 180–1).

The Anatili, Dexivates, and Cavares were ethnic groups east of the lower Rhone along the Druentia River (modern Durance). The Cavares (in the lowlands east of modern Avignon) were the most prominent (Strabo, *Geography* 4.1.11). The Tricores and Tritolli were on the upper Druentia, with the Vocontii to their north (see *NH* 3.37). The Segovellauni lived around modern Valence. The Allobroges covered an extensive area east of the Rhodanus and into the Alps, roughly from modern Lyon to Geneva. Hannibal encountered them, and Q. Fabius Maximus Aemilianus, the consul of 121 BC, defeated them near modern Valence, in what was considered one of the great victories of the Roman Republic, receiving the surname Allobrogicus (Cicero, *pro Fonteio* 36; Strabo, *Geography* 4.1.11). Hannibal also had contact with the Tricores and Vocontii (Livy 21.31).

Massilia (Greek Massalia, modern Marseille), was the primary Greek city on this coast. It was founded around 600 BC by Phocaeans who had emigrated from their city in Ionia (Herodotus 1.163; Strabo, *Geography* 4.1.4–5). At its peak, Massilia had settlements all along the coast as far as the northern Iberian Peninsula, and exploited the trade routes up the Rhone into the interior of Europe. Although it had been incorporated into the Roman provincial system by Pliny's day, it was still the major city of the region. Despite the modern urbanization, its impressive location – with a hidden harbor and waterfront springs – is still apparent, and there are many visible remains (James Bromwich, *The Roman Remains of Southern France* [London 1993] 168–77). The Camactulici, Suelteri, and Verucini were all minor ethnic groups to the east of Massilia, with the Verucini inland of the other two.

3.35. The *periplous* of the coast of Narbonensis concludes with the territory east of Massilia, as far as the Varus River. Forum Iuli (modern Fréjus) was at the head of a small bay, and this made it suitable for a naval base, established by Julius Caesar (Cicero, *Letters to His Friends* 10.15, 17). Ships captured from the Battle of Actium were stationed here (Tacitus, *Annals* 4.5), and some of the harbor installations are still visible (Bromwich, *Roman Remains* 252–67).

The Oxubii and Ligauni were in the hinterland of Forum Iuli, and the Suebri, Quariates, and Adunicates, all of whom are hardly known, were more inland. The Deciates were farther east along the coast. The Quariates may have given their name to modern Queyras, an isolated mountain valley about 50 km. southeast of Briançon, a nature reserve where traditional ways still exist.

The Varus River marked the boundary of Italy. It is not a major stream (114 km. long) but is a definite break in the topography of the coast, flowing out of the Maritime Alps east then south, through a picturesque gorge.

3.36. The coastal *periplous* essentially ends at this point, and does not resume until *NH* 3.47. Pliny devoted the rest of his description of Gallia Narbonensis to the peoples and towns of the interior, reverting somewhat to the organization by status that he used for parts of the Iberian peninsula. But the literary structure is somewhat different than previously, not only providing town names but often the indigenous ethnic groups in whose territory they were located. The first list – alphabetical – names the five towns that were *coloniae*. For the Cavares and Allobroges, see *NH* 3.34. Then there are 28 towns with Latin status, also listed alphabetically, spread across the southern part of modern France, as far as the Spanish border and into Aquitania.

3.37. The list of towns with Latin status includes the two capitals of the Vocontii (for whom, see *NH* 3.34), and then 43 unnamed towns, over half of which are in the hinterland of Nemausus (modern Nîmes).

The emperor Galba, during his brief reign (summer AD 68 to the following January) managed some reorganization of the province, adding two ethnic groups, the Avantici, who lived on the upper Druentia River, south of modern Gap, and the Bodiontici, more to the south around Dinia (modern Digne-les-Bagnes). This does not seem to have enlarged the limits of the province, but may have brought some isolated mountain peoples more under Roman control. In Pliny's mind, Galba's achievement was not so much outright conquest as adding to the register (*adiecit formulae*). Thus the register and the conquest became merged (Pascal Arnaud, "Pline, *Nat. Hist.* III, 37, et la cité de Digne: Problèmes de toponymie historique," in *Παλαιὰ Φιλία: Studi di topografia antica in onore di Giovanni Uggeri* [ed. Cesare Marangio and Giovanni Laudizi, Galatina 2009] 593–600).

The 370 miles (548 km.) recorded by Agrippa (F9) cannot be equated precisely with any particular extent of the province. It is not the coastal sailing distance from the Pyrenees to the border of Italy, which is more like 250 km. Presumably it is from the border to an undetermined point in or near Aquitania. The north–south distance (248 miles or 367 km.) is again problematic, but the modern Marseille–Geneva distance is roughly the same.

Italy

3.38. Having completed his description of Gallia Narbonensis, Pliny's account next begins a lengthy discussion of Italy (through *NH* 3.138, except for 75–94, devoted to the islands of the western Mediterranean), nearly two-thirds of Book 3. To some extent he abandoned the detailed *periplous* that he had followed since the Pyrenees (and before, on the coasts of the

Iberian Peninsula), and organized his section on Italy in terms of the 11 regions that Augustus established, an effort that originated at the beginning of his principate (Dio 52.22). Pliny's thorough account of the regions is the primary source for their territorial extent (see further, *NH* 3.46).

Yet Pliny was still somewhat beholden to the *periplous* format, since his summary of the indigenous populations of Italy followed the coast, beginning at the Varus River with the Ligurians. He then moved down the west coast to the Bruttians at the extreme south, and after a digression on the islands, continued around to the north and the head of the Adriatic, and then proceeded to the Istrians and Liburnians, areas that today are part of modern Croatia but were largely part of Italy into modern times and remain in some ways culturally Italian.

3.39. After completing his brief coastal survey, Pliny eulogistically described the benefits that Italy (actually the Roman state) provided the rest of the world (through *NH* 3.42). This is an important exposition of the civilizing power of the Roman Empire: Rome gave *humanitas* to the human race (Greg Woolf, *Becoming Roman: The Origins of Provincial Civilization in Gaul* [Cambridge 1998] 57). Such a role had divine authority and was effected by the absorbing of other states into the Roman system, perhaps primarily a reference to the Hellenistic kingdoms of the East. The use of a common language was also part of the dominance of Italy (ignoring the fact that Greek, not Latin, remained the primary language of much of the empire). As at *NH* 3.28, there is the suggestion of the unpleasant sounds of foreign names (*discordes ferasque*).

Such views about Italy had been common since at least the first century BC and were especially popular at the beginning of the Augustan period. Vergil was probably among the first to express them, writing that no place could be praised more than Italy (Vergil, *Georgics* 2.136–76). Pliny's debt to Vergil was strong, and reminders of the poet occur throughout the *Natural History* (Richard T. Bruère, "Pliny the Elder and Virgil," *CP* 51 [1956] 228–46). Eulogistic views about Italy were not limited to Romans: Dionysius of Halicarnassus (*Roman Antiquities* 1.37) and Strabo (*Geography* 6.4.1) expressed similar ideas.

3.40–2. Pliny, unable to mention all the glories of Italy, singled out two places worthy of note: the city of Rome itself and Campania. Even today few could deny the beauty of Campania, despite rampant urbanization and a change in its physical aspect since the eruptions of Vesuvius. The name Campania (first documented as Kampania, in the fourth century BC: Pseudo-Skylax 10), is probably descriptive, related to Latin *campus*, and suggests prosperity and fertility (Diodorus 12.31.1). Presumably the exuberant tone of *NH* 3.41 also refers to Campania, at least at the beginning, but then evolves to a general eulogy on Italy as a whole. In fact, there is slippage in these three sections (40–2), because by the final one the topic is Magna Graecia, the southern part of the Italian peninsula. Pliny's enthusiasm replaces any geographical precision.

The term Magna Graecia (in the Greek form Megale Hellas) is extant from the early third century BC (Timaeus F13), but it consistently had Pythagorean connections, and thus may have been in use as early as the sixth century BC (Polybius 2.39.1; Valerius Maximus 8.7, ext. 2). It originally referred both to the southern part of Italy itself and Sicily (Strabo, *Geography* 6.1.2). Pliny appears to have suggested that because the Greeks called part of Italy "Greece" (even giving it the additional name of "great"), this was an endorsement on their part of the superb qualities of the region, a rather obscure image. Much more striking is the comparison of the innumerable glories of Italy with the innumerable stars. He also made use of the long-standing rhetorical device that he could not possibly list all the virtues of the region.

3.43. The attempt to understand geographical regions by simple shapes seems to have originated with Eratosthenes in the latter third century BC, who was perhaps affected by Euclidean geometry (Eratosthenes, *Geography* F66 = Strabo, *Geography* 2.1.22). By the second century BC, Italy was seen as a triangle, although in time this was believed to be too simple a concept (Polybius 2.14.4–6; Strabo, *Geography* 5.1.2). The oak leaf simile offers little beyond the concept of an extended rectangle with protrusions.

The image of an Amazon's shield is notable but also problematic. Pliny used the word *parma*, a common but foreign word for shield, but not necessarily connected to the Amazons. Their shields could take many forms, including the traditional Greek round shape, as seen on the frieze of the Temple of Apollo at Bassai, now in the British Museum. The Amazon Hippolyte was buried in Attika in a tomb that was rhomboid and said to have been shaped like her shield (Pausanias 1.41.7; Plutarch, *Theseus* 27). Neither of these seems particularly appropriate to the shape of Italy, and Pliny probably had in mind a shield that was curved on one side and had two or more cavities on the other separated by sharp projections. This type appears regularly in Greek art.

How this can be fitted to the shape of Italy is difficult to understand, but it was most likely inspired by the bays at the southern end of the peninsula. In fact, the outline of the Amazons' shield conforms surprisingly to the Bay of Naples and the adjoining Gulf of Salerno, with the sharply projecting Sorrento peninsula between them. Yet Pliny explicitly wrote otherwise: to him the shape was represented by the shallow bays along the outer side of the heel of Italy, the ancient Bruttius and Scyllaeus Gulfs, separated by Cocynthos (modern Punta Stilo). But these bays are insignificant, the Bruttius so much so that it does not even seem to have a modern name (the Scyllaeus is the modern Golfo di Squillace). It is possible that there has been some confusion in applying the description and the exact locale that was meant.

The source for this image of Italy is not known. The only author cited by Pliny in these passages is the polymath of the first century BC, M. Terentius Varro (*NH* 3.45), and it may be that Pliny had in mind the map of Italy that Varro saw in the Temple of Tellus in Rome (Varro, *de re rustica* 1.2.1; Pliny, *Histoire naturelle* 3, 137). It stood on the Carinae, north of the Flavian Amphitheater (the Colosseum, which was under construction as Pliny wrote). No remains of either temple or map are known (Richardson 378–9).

The measurements of Italy may also come from this map, or Varro himself. The length from Augusta Praetoria (modern Aosta, at the northwest corner of Italy) to Regium (modern Reggio di Calabria, opposite Sicily) is a road distance, recorded at 1,020 miles (1,510 km.); the modern driving distance, whose route approximates the ancient one, is 1,428 km. Lacinium (modern Cape Colonna), the easternmost point of the heel, is an additional 257 km. around the southernmost point of Italy and up the east side. It is near ancient Croton (modern Crotone), the home of the Pythagoreans, and thus may again suggest a Pythagorean origin for some of the general material Pliny presented about Italy.

3.44. The width of Italy was variable enough that Pliny provided several measurements. The maximum, in the north between the Varus and the Arsia Rivers, was 410 miles (607 km.). The Arsia is the modern Raša in Croatia, a short stream in eastern Istria, and a peculiar point to consider as the eastern boundary of Italy since Pliny included the Liburnians, beyond the Arsia, in the territory (*NH* 3.38). The choice of the Arsia may be because it had a broad estuary with a fertile interior basin, and thus the river might be better known than the inconsequential streams in Liburnia. Demonstrating his Italian perspective, Pliny referred to the Adriatic and Mediterranean by their relative location to Italy, or above and below, perhaps a reference to Agrippa's map.

The middle distance, from the mouth of the Aternus River (modern Pescara) on the Adriatic to that of the Tiber, is 136 miles (201 km.). The peninsula is about 150 km. across between these two points, and it is essentially the same between Castrum Novum (at modern Giulianova on the coast about 30 km. northwest of the Pescara River) and Alsium (modern Palo on the coast west of Rome). Pliny's distances are rather long, but may reflect travel routes. The entire circuit of Italy (2,049 miles or 3,033 km.), from the Varus to the Arsia, may be a measured sailing itinerary.

3.45. The various distances from points in Italy to places in surrounding areas are all sailing itineraries, crossing the seas that surround the peninsula. Many do not have any points of origin or termination specified. Elsewhere (*NH* 3.129) Pliny recorded that it was 120 miles (178 km.) from Pola (modern Pula in Istria) to Ancona. The distance between Epiros and Illyricum is probably between Hydruntum (modern Otranto) and Apollonia (modern Pojan in Albania), 50 miles or 74 km., reported at *NH* 3.100.

The mention of Varro is enigmatic, since it cannot be determined whether he was responsible only for the African distance or all the ones in section 45. The route to Africa is not specified; 200 miles (269 km) would perhaps be the sailing itinerary from ports on the south side of Sicily to Carthage. The 120 miles (178 km.) to Sardinia is the distance from Ostia to the northern tip of the island. The width of the Strait of Messana from the mainland to Sicily is anomalous (the channel is about five km. wide today), but it may have been narrower in antiquity, since there have been tectonic changes in this highly active region. That from Italy to Corcyra (modern Corfu), less than 80 miles (118 km.), is probably calculated from Hydruntum. Issa (modern Vis in Croatia) is the outermost of the islands on the Croatian coast. The Italian destination is not provided, but 50 miles (74 km.) would place one in the middle of the Adriatic. It is implausible that there would be such an error on this well-travelled route, and either the distance was improperly recorded or there is a textual difficulty. The orientation of Italy between the first and sixth hour in midwinter (the morning) would be south-southeast, but it is actually closer to due southeast.

3.46. The 11 Augustan regions were numbered from Latium (Region 1) toward the south, and then along the Italian peninsula to the north, ending with Region 11, Transpadana, in the northwest (essentially modern Piedmont and Valle d'Aosta). Pliny's account is one of the most definitive, although he chose not to describe the regions in numerical order but to follow the coastline south from the Varus River to the end of Italy, and then around into the Adriatic and the north. Augustus evidently had created a list of the settlements in each region, alphabetically arranged, that Pliny followed and which is similar to the pattern used in Iberia and

Gallia Narbonensis (Rudi Thomsen, *The Italic Regions from Augustus to the Lombard Invasion* [tr. Anne Monrad and Paul Monrad, Copenhagen 1947] 17–46, 74–144). Yet he was well aware that boundaries were ever changing, singling out the Ingauni as an example, who occupied the hinterland of Albuminganum (modern Albenga in Liguria), and whose territory changed every few years. But all the same Pliny chose to rely on the documents purely for demography, and continued his *periplous* format for geographical data (Edward Bispham, "Pliny the Elder's Italy," *BICS Supplement* 100 [2007] 41–67).

Liguria

3.47–9. The first region Pliny examined was the ninth, or Liguria. It was more extensive than the modern Italian province, extending from the Varus River to the Macra River (modern Magra, still the eastern boundary of Liguria), a distance of 211 miles (312 km.). The northern boundary was the Padus River (modern Po, now in Piedmont).

This was historically the land of the Ligurians (Greek Ligystikians or Ligyans), a large ethnic group whose populations were not only along the coast but north into the Alpine regions on the modern French–Italian border. Greeks had encountered them from the time of the establishment of Massilia around 600 BC; the Romans arrived in the second half of the third century BC (Livy, *Summary* 20). There were many Ligurian subgroups, and Pliny enumerated both those on the coast and inland, in a rough geographical order from west to east. In 14 BC a new province, Alpes Maritimae, was created in the mountainous region between Liguria and Gallia Narbonensis (Dio 54.24.3), but Pliny's source – perhaps Varro or Agrippa – is earlier than this.

The beginning of Pliny's account of western Liguria repeats some of the material from eastern Gallia Narbonensis (*NH* 3.35). This is not that the two territories overlapped, but represents a new source, or suggests that the indigenous people were wide-ranging. Many of the locals in western Liguria were known as the Capillati, a Latin descriptive word meaning "Long-Haired," clearly a generic term not attached to any specific group. The Vediantii were in the uplands above Nicaea (modern Nice). The Sallvi were also in the hinterland, with a village at modern Entremont, where there are extensive Iron Age remains (Barry Cunliffe, *The Ancient Celts* [Oxford 1997] 213–14). There were also the Deciates and Oxubi (*NH* 3.35). These were seen as "beyond the Alps," meaning not the ranges in modern Switzerland but the Maritime Alps, which to some extent created a barrier between Italy and Gallia. On the Italian, or near, side of the mountains was Liguria proper. Pliny listed 10 peoples that were inland, extending east from the Veneni and Soti on the eastern slopes of the mountains. Other Ligurians were in the valley of the Padus River, on its south side. Pliny's order of names is not strictly geographical but nevertheless roughly from east to west.

Interspersed with the catalogue of the inland ethnic groups is a brief coastal account, which implies that the actual Ligurian coast was only considered to begin east of Portus Herculis, the home of the Minoecians, at modern Monaco. The initial point of Ligurian territory was perhaps at the Rutuba (or Rotuba) River, the modern Roya. The coastal account continues east through Genua (modern Genoa) to the Macra River at the eastern end of Liguria. It is in *periplous* fashion, including towns, ports, and rivers.

The Appenninus Mountains (English Apennines) stretch the length of Italy. Geomorphically they are divided into three portions, the Northern, Central, and Southern, with their highest point Corno Grande (2,912 m.) northeast of Rome. Their northern end is connected with various Alpine ranges in Liguria and Piedmont, around Genoa (Strabo, *Geography* 4.6.1, 5.1.3). After 1,200 km. they terminate at the end of the Italian peninsula in Calabria. Because they include the highest territory in Italy outside the Alps, their ecology was considered unusual, with high-quality forests, as Vergil noted (*Aeneid* 12.701–3). Despite their conspicuousness as a major geographical feature of the Mediterranean world, the earliest extant detailed material about them is what Strabo recorded in Books 4–7 of his *Geography*.

The ancient Ligurian region included territory as far as the south (right) bank of the Padus. Most of the towns named by Pliny survive today as regional centers and are spread along the river and its drainage up to about 75 km. east of modern Turin. The region continued downstream to a point just east of Dertona (modern Tortona), with its terminus perhaps at the modern Piedmont–Lombardy boundary.

Etruria

3.50. The next region (through *NH* 3.55) is the seventh, Etruria, which includes most of modern Tuscany but also extends as far as the right bank of the Tiber, territory that today is in Umbria or Lazio (Cecilia Gábli, "Pliny the Elder on the Etruscans," *AAntHung* 52 [2012] 137–45). The northwestern boundary is at the Macra River, today still that between Liguria and Tuscany.

Pliny's account of the seventh region begins with a brief historical digression. The Umbrians were assumed to be the original population of Etruria. They had come into the area from the north at the beginning of the Iron Age, after 1000 BC, and settled across a broad swath of central Italy, from coast to coast (Guy Bradley, *Ancient Umbria* [Oxford 2000] 19–23). The next ethnic movement was represented by the Pelasgians, who were said to have expelled the Umbrians from west of the Tiber. They were one of the most enigmatic populations of antiquity, located in many places but with a probable homeland in Thessaly, where the ethnym Pelasgikos was still known into Roman times (Strabo, *Geography* 9.5.5). Yet their extent, both chronologically and geographically, was widespread: in Etruria they were said to have founded the city of Caere, at modern Cerveteri (*NH* 3.51; Strabo, *Geography* 5.2.2–5). This is probably the tradition that was known to Pliny, although there is no connection between the Pelasgians and Umbrians in other extant sources.

In time the Tyrrhenians (Etruscans) arrived, according to Pliny; his account roughly parallels the version known to Herodotus (1.94) and Dionysius of Halicarnassus (*Roman Antiquities* 1.27), which attributed their origin to Lydia and their name to the Lydian prince Tyrrhenus. It was believed that the Greek word Touskoi (latinized into Tusci) was from Greek *thyoskoos*, meaning a sacrificing priest (Homer, *Odyssey* 21.145). Pliny's source for this information may also have been Dionysius of Halicarnassus (*Roman Antiquities* 1.30).

Following the digression on ethnic movements, there is a catalogue of the coastal cities of Etruria, from Luna (modern Luni) as far as Populonium (modern Piombino). The harbor at Luna was notable because it was the shipping point for the famous quarries a few kilometers inland. The Luna quarries (whose name is a translation of Greek Selene, due to the glow of the marble in the moonlight, still apparent today) were opened in the first century BC (*NH* 36.48; Hirt, *Imperial Mines* 29–30). The marble became a primary building material in Rome. Since medieval times the quarries have been known as Carrara, and are still active.

The story that Pisae (modern Pisa) was founded by the descendants of Pelops, who gave his name to the Peloponnesus, does not seem to have been current before the first century BC. It is probably nothing more than an onomastic association between Italian Pisae and the Pisa near Olympia in the northwest Peloponnesus (Dionysius of Halicarnassus, *Roman Antiquities* 1.20.5; Vergil, *Aeneid* 10.179; Strabo, *Geography* 5.2.5). Populonium was unusual because it was the seaport of the Etruscans; all their other cities were inland (Diodorus 5.13.1; Strabo, *Geography* 5.2.6).

3.51–2. The itinerary continues along the coast to the end of Etruria at the Tiber River. The Umbro River is the modern Ombrone, which flows into the Mediterranean southeast of Piombino. It is unlikely that the name Umbro, as well as Pliny's Tractus Umbriae, are vestiges of Umbrian settlement, but this is far from certain. Any assumed connection between the Umbrians and these toponyms probably reflects the similarity of names and the strength of the ancient traditions (Bradley, *Ancient Umbria* 23; Giacomo Devoto, "Umbri e Etruschi," *SE* 28 [1960] 263–76).

After completing the Etrurian coast, the account considers the interior. As usual, the toponymic catalogues are a mixture of the alphabetical and geographical. Most of the towns are defined by their ethnyms (such as the Florentini or Arretini), but there are occasional intrusions of the actual town names, such as Perusia. None of the peoples listed is an actual ethnic group. Many of the localities are quite familiar and have remained important into modern times – even if Pliny did not provide the actual town names – such as Siena, Florence, Arezzo, or Perugia. In addition, notice was taken of the Arnus River (modern Arno), the largest completely in Etruria. Some of this material came from Cato's *Origines* (*FRH* #5, F53), written in the second century BC, probably the earliest history written in Latin.

The Tiber River

3.53–5. Before continuing on to the following region, the first (Latium), Pliny digressed for a lengthy discussion of the Tiberis River (modern Tiber). It is the third longest in Italy (406 km.), but by far the most important culturally and historically. The river originates on the western slopes of the Apennines, northwest of modern Arezzo, and flows essentially due south, emptying into the Mediterranean at Ostia, the port of Rome. In the Augustan system it separated Region 7, Etruria, on its right bank, from (north to south) Regions 6, 4, and 1 (Umbria, Samnium, and Latium) on its left.

The names Tiberis and Thybris are not historically sequential, as far as can be determined, but different forms of the same toponym, probably reflecting indigenous usage at different points along the river. Tiberis is documented by the second century BC in both Latin and Greek (Ennius, *Annals* F145; Polybius 6.55.1); Thybris is a more poetic form that became popular in the Augustan period (Vergil, *Aeneid* 3.500; Ovid, *Fasti* 5.635–7). Albula is also not recorded before the first century BC (Varro, *de lingua latina* 5.30; Livy 1.3.5; Vergil, *Aeneid* 8.332), but was always believed to be the earlier name. It survived with the Aquae Albulae, east of Rome near Tibur (modern Tivoli), which were springs used as therapy for wounded Romans (*NH* 31.10; Vitruvius 8.3.2). The locality remains a spa today.

The Tinia (modern Timia) and Glanis (modern Chiana) are tributaries that flow into the Tiber from the east near Perugia; the latter was the effective head of navigation on the Tiber, since its volume was much greater downstream. Below Rome, the river could take any size vessel. In Pliny's time Rome probably still had extensive international shipping traffic, despite the large harbor built by Claudius at Ostia, which, although relatively new, was rapidly becoming obsolete and would be expanded by Trajan (Casson, *Ships* 368–70).

Nine aqueducts served Rome when Pliny wrote, from the Appia of 312 BC to the Anio Novus of AD 38–52 (see the chart, Tusculum 3–4, 240); their remains are conspicuous around the city. Presumably whatever water was not drawn off ended up in the river. Most of the springs are hidden today – or dry – yet there were many in antiquity (Richardson 152–3).

Flooding has always been a problem on the lower Tiber, and flood markers remain conspicuous in modern Rome. Floods have been documented at regular intervals (Tusculum 3–4, 241). The earliest recorded was in 241 BC, followed by a devastating fire (Orosius 4.11.5–9). More relevant to Pliny was the flood of spring AD 69 (Tacitus, *Histories* 1.86; Plutarch, *Otho* 4.5; Suetonius, *Otho* 8.3). It was particularly violent, with buildings undermined by the waters and supplies flooded in the markets, causing scarcities for many days. It occurred when the high waters broke down the Pons Sublicius, the oldest bridge in Rome, whose remains then blocked the river (Richardson 299). Despite their devastation, Tiber floods were considered a religious portent: that of AD 69 was an omen for the coming death of the emperor Otho.

Latium

3.56. The first Augustan region was known as Latium et Campania (discussed through *NH* 3.70). Pliny examined the two sections separately, beginning with Latium, which extended from the Tiber to Cercei (modern Monte Circeo), a high isolated peak (541 m.) that was once an island (*NH* 3.57). It is located 50 miles (74 km.) south of Rome, where the Italian coast turns sharply to the east. Pliny implied that at one time this was the limit of Roman territory. From Rome to Circei was Latium Antiquum, or Ancient Latium, and the parts immediately to the east of the region of Circei, as far as the Liris River (modern Garigliano), were Adiectum Latium, or Extended Latium (*NH* 3.59).

Six ethnic groups are listed, all primitive inhabitants of the region south of Rome. The Aborigines were believed to have been the earliest inhabitants of Latium, although it is not certain whether their name is descriptive (*ab origine*) or an actual ethnym (Roller, *Historical and Topographical Guide* 245). For the Pelasgi, see *NH* 3.50. The Arcades were believed to have been settlers from Arcadia in the Peloponnesus; it was commonly thought that there were connections between that region of Greece and central Italy. The Siculi should be the indigenous people of Sicily, but it is not explained why they would be in central Italy (both in Latium and Campania: see *NH* 3.71). The Aurunci lived in the southeastern corner of Latium along the Liris River around Minturnae. The Romans had encountered them at the end of the sixth century BC and had assimilated them by the fourth century BC (Livy 2.16–17, 8.15). The Rutuli were immediately south of Rome around Ardea, and were one of the major opponents of Aeneas, mentioned frequently in the *Aeneid*; they were assimilated at an early date (Livy 4.11.4; Strabo, *Geography* 5.3.2)

The distinction that Pliny made between Ancient and Extended Latium, or the territory on either side of Circei, is somewhat artificial, since he has already mentioned the Aurunci, who were about 60 km. beyond Circei. He may have changed sources here, since the Ausones are probably the Greek version of the Aurunci (Pliny, *Histoire naturelle* 3, 148). This was almost a generic term for the ancient people of southern Italy, who were said to be descended from Auson, the son of Odysseus and Circe (Dionysius of Halicarnassus, *Roman Antiquities* 1.35.3; Strabo, *Geography* 2.5.20, 5.3.6). The Volsci lived east of Circei around modern Terracina, and were vigorous opponents of Rome in the fifth and fourth centuries BC, famous for their leaders Coriolanus and Camilla (Livy 2.33–40; Vergil, *Aeneid* 11.532–835; Strabo, *Geography* 5.3.4). The Osci (Oscans) were on the borders of Campania, but had died out long before Pliny wrote (Strabo, *Geography* 5.3.6). These peoples, nevertheless, came to be considered among the Latins, and thus the boundary of Latium was eventually extended about 75 km. to the east in order to include them.

The first town mentioned in Latium is Ostia (Rome is reserved for *NH* 3.66–7), traditionally founded by Ancus Marcius, the fourth king of Rome, in order to establish a presence at the mouth of the Tiber and to control the local salt deposits (Livy 1.33.9). Other locations immediately south of Rome are also listed, all within 30 km. of the city and toward the coast. The story that Ardea (modern Casalazzora, renamed) was founded by Danaë, the mother of Perseus, was popularized by Vergil (*Aeneid* 7.410–13), but any earlier version is unknown.

3.57–8. The additional localities listed are farther south of Rome. Circei was assumed originally to have been an island, not so much for geomorphic reasons as because it was identified with Aiaia, the home of Circe as described by Homer (*Odyssey* 10.133–574). Whether association of sorceress and island predated the Hellenistic period is not known; the earliest extant citation is by Apollonios (4.659–63; see also Vergil, *Aeneid* 3.384–7, 7.10; Strabo, *Geography* 5.3.6). According to Pliny, Circei was still an island in 314/13 BC, when Theophrastus published his *Research on Plants*, but he actually misquoted the earlier scholar, who in fact wrote that it had recently been added to the mainland (Theophrastus, *Research on Plants* 5.8.3). Theophrastus' interest in the region was due to its unusual flora, seen to be a vestige of Circe's presence; Pliny's information provides the date of publication of Theophrastus' treatise.

Following is a digression on the earliest Greek authors to mention Rome. Pliny qualified his discussion by noting that Theophrastus was the first to write especially carefully (*diligentius*) about the subject, presumably because he knew enough about Italian topography both to discuss the local flora around Circei and to know the circumference of the mountain (80 stadia or about 16 km.; it had increased in size by the time Pliny wrote). By contrast Theopompus, slightly earlier, only referred in passing to the Gallic invasion of the early fourth century BC (*FGrHist* #115, F317). Clitarchus (Kleitarchos), who published his account of Alexander's expedition late in the fourth century BC (at the earliest), was responsible for the much-discussed comment that a Roman embassy had visited Alexander (*FGrHist* #137, F31; Lionel Pearson, *The Lost Historians of Alexander the Great* [New York 1960) 232–3). Although such an embassy was highly improbable, and may have been a construct from as late as the first century BC (W. W. Tarn, *Alexander the Great 2: Sources and Studies* [Cambridge 1948] 23–6), it nevertheless seems certain that Greek awareness of Rome was a feature of the late fourth century BC, even though the various citations of this admit of uncertain interpretations. Pliny seems to have been unaware of another early Greek source on Rome, Aristotle, who knew about the Gallic invasion (Plutarch, *Camillus* 22), and whose comment may be the earliest known Greek mention of Roman history.

3.59. The last section on Latium is about the coastal strip from Circei to the Liris River. The Pomptine Marsh (or Pomentine Plain) was a swamp and malarial region noted for its brigandage (Juvenal 3.305–8; Strabo, *Geography* 5.3.6). There were various attempts to drain it in the late Republic, and whatever reclamation was possible turned parts of it into a fertile agricultural district (Vitruvius 1.4.12; Suetonius, *Divine Julius* 44.2;

Plutarch, *Caesar* 58). C. Licinius Mucianus, consul in AD 57, 70, and 72, seems to have been particularly interested in hydraulic issues, but it is difficult to credit the 24 cities in the marsh, whose maximum extent is no more than about 50 km. (Keyser, "Licinius Mucianus" 507–8).

Formiae (or Hormiae, modern Formia) is, like Cercei and the Sirens, one of several local places associated with Odysseus. Sinope, the Greek antecedent of Sinuessa, was known to Livy (10.21.8), although he was dubious about its actual existence. It remains problematic and is probably a conflation of similar names, since Strabo, whose homeland was Pontus in Asia Minor, the location of the original Sinope, did not mention any such Latin town. In placing Sinuessa in Latium, Pliny contradicted what he wrote at *NH* 3.56 about the boundary of Extended Latium at the Liris River; Sinuessa is about 12 km. to its southeast (the river is at Minturnae). Presumably he failed to reconcile two different sources, one for the Augustan first region (which probably placed the boundary at the Liris), and an older one, which placed it at the next river, the Savo (modern Savone; see *NH* 3.61).

Campania

3.60. Campania (through 3.65) is the second part of Region 1. It was treated eulogistically by Pliny, as was common (Polybius 3.91; Vergil, *Georgics* 2.217–25; Strabo, *Geography* 5.4.3). Its wine was legendary, including Setinian (the favorite of Augustus), Falernian, and Caecuban (*NH* 14.61–3).

The competition between Father Liber (Bacchus or Dionysos) and Ceres (Demeter) is to some extent metaphorical, with fertile Campania noted for its wine and cereal grains, products of interest to the two divinities. They were seen as complementary (Euripides, *Bacchants* 274–83), and, more importantly, they allowed Venus (Aphrodite) to perform her duties, without whom the human experience would be incomplete. These were often depicted together in post-antique art, most notably the version by Rubens in Kassel.

To prove the importance of Bacchus, Pliny enumerated the most famous wine-producing districts of the region (Caecubus, Falernus, Setinus, and Calenus). Actually, the first two are in southeastern Latium, but the entire coast from Circei to Naples remains one of the best wine-growing areas of Italy, although the exact locations of the ancient vineyards are difficult to determine, in part because of intensive tectonic changes, beginning with the eruption of Vesuvius in AD 79. Elsewhere Pliny noted that Caecuban wine was no longer available because of the destruction of the vineyards due to a construction project during the reign of Nero (*NH* 14.61).

In terms of cereal grains, the Leborinian fields were noted for their wheat crops – several each year – which produced a variety with particularly thick stalks. The district was located between modern Naples and Capua – a region now heavily affected by urban sprawl – and produced *alica* (emmer meal), a product made from the grains of zea wheat (Dalby, *Food* 131–2). The industry was important enough that it was subsidized by Augustus (*NH* 18.112–14).

The various hot springs are conspicuous today, and nourished the agricultural bounty. The fishing industry was also significant. Oysters were imported from the Adriatic and raised in ponds near Baiae. A certain M. Sergius Orata developed the concept of fish farming in the early first century BC (*NH* 9.168–9; Strabo, *Geography* 5.4.6), and the *scarus* (perhaps a wrasse) was brought from the Troad during the principate of Claudius and seeded all along the coasts of Latium and Campania (*NH* 9.62–3). Campanian olive oil was also renowned, especially that from Venafrum (modern Venafro), at the northern edge of the region, where a gravelly soil was particularly suitable for cultivation of the olive tree (*NH* 15.8, 17.31).

The five ethnic groups listed are presented in chronological order, giving a summary of the demographic history of Campania. For the Oscans, see *NH* 3.56. Greek settlement around the Bay of Naples began as early as the eighth century BC, when Euboeans came to the island of Pithecussae, or modern Ischia (John Boardman, *The Greeks Overseas: Their Early Colonies and Trade* [fourth edition, New York 1999] 165–9), and the region retained its Greek character well into Roman times. The Umbrians were not far north of Campania, although there is no evidence for any specific local presence. The Tusci (Etruscans) arrived in the seventh century BC (Dionysius of Halicarnassus, *Roman Antiquities* 7.3), but then faded under local pressures by the fifth century BC (Diodorus 12.31). Any distinct Campanian ethnicity is difficult to isolate, and oddly the Campanians are the last on this chronological list (which would put them after the sixth century BC), but Pliny may have been referring to the multi-ethnic population that developed through so many settlers from elsewhere.

3.61–2. The first catalogue of localities in Campania covers the coast from the Savo River (modern Savone), just southeast of Sinuessa. At some time this was probably the boundary between Latium and Campania. The account passes through the familiar towns of the Bay of Naples, including

Pompeii (ironically noting that it is close to Mt. Vesuvius), and continuing as far as Surrentum (modern Sorrento), near the end of Cape Minerva (Greek Athenaion Promontory; modern Punta Campanella).

Conspicuous in the history of this region were the Sirens. Neapolis (modern Naples), settled from Chalcis on Euboea (Boardman, *Greeks Overseas* 192), was originally named Parthenope, after a Siren who was buried at the site (Vergil, *Georgics* 4.563–6; Strabo, *Geography* 5.4.7). Pliny placed the home of the Sirens on Cape Minerva (also called the Seirenoussai Promontory). Another tradition located them on the three islets to the south, called the Seirenes in antiquity (modern Li Galli), where they positioned themselves to harass shipping (Strabo, *Geography* 5.4.8). Homer only recorded that the Sirens were the first peril Odysseus encountered after he left Circe (Homer, *Odyssey* 12.37–54, 165–200), but arguments about Homeric topography become dangerously circular. Parthenope was said to have committed suicide, with her body washing up at the future site of Neapolis, but this is post-Homeric (the fate of the Sirens is not stated in the epic), and first appears in the *Alexandra* attributed to Lykophron, of uncertain Hellenistic date (*Alexandra* 712–37). By this time the Sirens had evolved from two to three and had acquired names, reflecting the Hellenistic interest in Homeric exegesis.

Pliny also specified that the distance reported from Circei to Cape Minerva was by sea. This was probably because the sea itinerary was quite different from any on land, and shipping would take a straight line southeast from Circei directly to the cape, passing between Misenum and Pithecussae, a distance of 78 miles (115 km.).

3.63–4. The first seven towns in Campania are listed in no obvious order. They are the most interior of the region, lying against the foothills of the Apennines, from Sora in the north to Nola in the south. The name Capua may be related to *campus*, and it is possible that Campania took its name from the town. Strabo (*Geography* 5.4.3) suggested "Capua" meant that it was the Etruscan capital (*caput*) of the region, but this sounds more like a Latin witticism (Roller, *Historical and Topographical Guide* 270–1).

Next there is a lengthy alphabetical list of nearly sixty Campanian localities, mixing local ethnyms and toponyms. The list is wide-ranging: two of the places, Auximum and Cingulum (actually appearing as ethnyms in the list) are in Picenum (Region 5, on the Adriatic; see *NH* 3.111), and another, Cora, is well into Latium at the eastern edge of the Pomentine Plain. The list is from some official document, and unlike previous ones, it merely includes towns or their ethynms, with no physical features mentioned. Only one parenthetical comment appears, that the people of Cora (the Corani) were descendants of Dardanus, the ancestor of the Trojan royal line. This fact is not documented elsewhere, but Cora was an early Roman outpost (Livy 2.16.8), and was close to the portion of Latium that figures in the story of Aeneas.

Rome

3.65. The idea that Rome had secret names is by necessity obscure, but has been the source of much speculation by modern scholars, with no conclusive results (Macrobius, *Saturnalia* 3.9.3; Pliny, *Histoire naturelle* 3, 160–1). There are few details about the career of Q. Valerius Soranus, but he may be the erudite associate of Cicero (Cicero, *de oratore* 3.43), who was plebian tribune in probably 82 BC (Broughton 2, 68). He evidently published the secret name and either was killed or committed suicide (Plutarch, *Roman Questions* 61).

The goddess Angerona was a primitive Italian or Roman deity, perhaps connected with the winter solstice (Varro, *de lingua latina* 6.23; Macrobius, *Saturnalia* 1.10.7). It was necessary that she keep a secret, and this may have something to do with ensuring an auspicious beginning to the new year. Her cult statue at Rome was in the shrine of Volupia, near the Velabrum (Richardson 433).

3.66–7. The description of Rome emphasizes its physical layout, with its gates and dimensions, in order to demonstrate its size (*magnitudo*). In a way, this was seen as a microcosm of the empire as a whole. Later in the *Natural History* (36.101), Pliny created the astonishing image of piling all the buildings of Rome on top of each other, a striking vision of Roman power.

Eulogistic reports on the city were a common literary feature of the first century BC, each with its own agenda (Cicero, *Republic* 2.11; Varro, *de lingua latina* 5.41–54, 143–65; Strabo, *Geography* 5.3.7–8; Hubert Zehnacker, "La description de Rome dans le livre 3 de la *NH*," in *Pline l'Ancien, témoin de son temps: Conventus Pliniani internationalis Namneti 22–26 Oct. 1985 habiti acta* [ed. Jackie Pigeaud and José Oroz Reta, Salamanca 1987] 307–20). The three gates of Romulus are probably the Mugonia (near the future site of the Arch of Titus), the Romana or Romanula (at the west), and presumably one somewhere around the Aventine. A fourth gate might have been on the southeast, an ancestor of the Porta Appia. No remains are known of any of these (Richardson 304, 308).

By AD 73, when Vespasian and his son Titus held a joint censorship, the city had expanded to its traditional seven hills and had a perimeter of slightly over 13 miles (19 km.), having far outgrown the so-called Servian Wall of the Republic, which extended for 11 km. and whose remains are visible at various points in the modern city (Aldo Borlenghi, "Mura Serviane e Mura Aureliane: l'Occupazione di Spazi Pubblici e di Aree Private in Occasione dell'Abbandono e della Costruzione delle due Cinte Urbane," in *Les confiscations, le pouvoir et Rome, de la fin de la République à la mort de Néron* [ed. Clément Chillet *et al.*, Bordeaux 2016] 301–21). In 7 BC, Augustus divided Rome into 14 regions (Suetonius, *Augustus* 30.1; Dio 55.8.7; Richardson 331–2). Details of their boundaries only come from the fourth century AD, so the exact Augustan limits remain speculative. The milestone in the Forum, probably made of gilded bronze and near the Temple of Saturn, was where all the roads into Rome came together; it was put in place by Augustus in 20 BC (Plutarch, *Galba* 24.4; Tacitus, *Histories* 1.27; Dio 54.8.4; Suetonius, *Otho* 6.2; Richardson 254).

It is not possible to identify all the 37 gates. Nearly 60 are known today from literature and physical remains (Richardson 299–310), but many of these are later than Pliny's time, especially from the third century BC when the Aurelian Wall was built. Double gates were common: the Porta Praenestina (modern Porta Maggiore) is the best surviving example. Pliny's measurements of over 20 miles (31 km.) of roads to the gates and 60 miles (89 km.) of streets are impossible to reconstruct.

Significantly, the only building in Rome mentioned by Pliny, the career military man, was the Castra Praetoria, the army barracks, whose remains are conspicuous today at the northern edge of the ancient city. It was built by Tiberius in AD 21–3, and situated just outside the urbanized area (Suetonius, *Tiberius* 37.1; Tacitus, *Annals* 4.2; Dio 57.19.6; Richardson 78–9).

The earthern rampart (*agger*) of Tarquinius Superbus was part of the defensive construction of the Republican wall of the city, now generally called that of King Servius Tullius but also attributed to various other kings and actually built in the fourth century BC (Livy 1.38.6, 1.44.4; Dionysius of Halicarnassus, *Roman Antiquities* 4.54; Strabo, *Geography* 5.3.7). The earthworks are probably earlier than the walls and were about 10 m. high. Their purpose was to defend against any incursion from the flat land to the northeast of the city, where the natural topography was less to Rome's advantage (Axel Boëthius, *Etruscan and Early Roman Architecture* [second edition, Harmondsworth 1978] 122). This would be roughly in the vicinity of the modern Stazione Termini, where remains of the wall are conspicuously visible. After it and the earthworks ceased to serve any defensive purpose, the latter became a place for walking, noted for its fine view of the city (Horace, *Satires* 1.8.14–16). There were early villages within the future limits of Rome, and close by, which were assimilated as the city expanded: see further, *NH* 3.68.

3.68–9. Pliny listed 53 towns and local ethnyms of the area immediately around Rome that were no longer in existence, and which had been incorporated into the growing city. Some of them had joined in an alliance against Rome in the first year of the Republic (Dionysius of Halicarnassus, *Roman Antiquities* 5.61.3). Others are villages that were actually on the site of Rome itself (*NH* 3.67). Saturnia was an ancient name for the Tarpeian Rock, the southern escarpment of the Capitol, and allegedly an early name for Rome itself (Dionysius of Halicarnassus, *Roman Antiquities* 1.34.1; Varro, *de lingua latina* 5.42; Ovid, *Fasti* 6.31). Antipolis, on the Janiculum across the Tiber, probably was incorporated into Rome during the regal period (Livy 1.33.6). Its Greek name, "Opposite the City," while topographically valid, is a curiosity. Of particular interest are the towns around Mt. Albanus that "received meat" (*carnem accipere*), a memory of some primitive ritual. Nevertheless the date and accuracy of the list remains disputed (A. Momigliano, "The Origins of Rome," *CAH* 7.2 [second edition 1989] 65).

3.70. Pliny also cited four towns south of Rome, as far as the Bay of Naples, that no longer existed. They are spread through a chronological range of several hundred years, and seem to have been chosen at random to demonstrate the changing demographics of Italy from earliest times to Pliny's own day. Stabiae, near modern Castellammare di Stabia on the Bay of Naples, was destroyed in 89 BC during the Marsic (Social) War, although it later prospered as a resort until the eruption of Vesuvius, and gained notoriety as the place where Pliny died. Taurania may be in the district still called Taurano, near Nola, and was probably incorporated as that city expanded. Casilinum was effectively a suburb of Capua. Apiolae was probably in northern Latium, near Mt. Albanus (Valerius Antias [*FRH* #25] F19; Dionysius of Halicarnassus, *Roman Antiquities* 3.49; Livy 1.35.7; Strabo, *Geography* 5.3.4). Its destruction during the regal period produced a large amount of booty, which Rome used to its own advantage, either to begin the Capitoline Temple or to lay out the Circus Maximus.

After the digression on Rome and its environs, Pliny returned to Campania (from *NH* 3.64) for some comments about its southern portion. The Picentines were the indigenous people on the Adriatic coast northeast of Rome (see *NH* 3.110–11). They supported Hannibal, and thus were later dispersed, with some of them being settled in southern Campania in what came to be called the Picentinus Ager (modern Piana del Sele), where the toponym Picenzia survives, near modern Salerno (Strabo, *Geography* 5.4.13). This region had been Etruscan territory from the seventh century BC until around 400 BC (Roller, *Historical and Topographical Guide* 271). The Silerus River (Silaris, modern Sele) marked the southern end of Campania. At its mouth was a famous sanctuary of Argive Hera (Juno to Pliny). Crediting its foundation to Jason is probably a report taken from Timaeus, who had a proclivity for locating Argonautic tales in Italy (Lionel Pearson, *The Greek Historians of the West: Timaeus and His Predecessors* [Atlanta 1987] 63–5). Yet archaeological evidence suggests that it was only established in the seventh century BC (Strabo, *Geography* 6.1.1; Boardman, *Greeks Overseas* 180–1). There are extensive remains at the site today called Foce del Sele.

Bruttium and Lucania

3.71–3. The third Augustan region, Bruttium et Lucania, extended from the Silerus River to the end of the toe of Italy, essentially the modern Italian provinces of Basilicata and Calabria, along with some of the southern parts of Campania. The Bruttians (Greek Brettians) were the southernmost indigenous peoples of Italy, who were perhaps escaped slaves from Lucania (*brettioi* was said to mean "rebels" in Lucanian: Strabo, *Geography* 6.1.4). The Lucani (Greek Leukanoi) moved into the region from the north and the area of their relatives the Samnites (Greek Saunitians; see *NH* 3.107) at the beginning of the fourth century BC (Diodorus 14.101; Strabo, *Geography* 6.1.2–3). The name of Lucius, the Lucanian leader, is probably nothing more than a created ethnym: the Roman praenomen is generally thought to be from *lux* ("light," signifying a child born before dawn: Varro, *de lingua latina* 9.60).

As before, Pliny emphasized the ethnic changes. For the Pelasgians, see *NH* 3.50. The Oenotri were believed to have been the earliest settlers in the area, led by Oinotros, who arrived 17 generations before the Trojan War (Aristotle, *Politics* 7.9; Dionysius of Halicarnassus, *Roman Antiquities* 1.11.2; Strabo, *Geography* 6.1.4). They were the ancestors of the Italians, originally a localized population in extreme southern Italy (Strabo, *Geography* 5.1.1, 6.1.4; Dionysius of Halicarnassus, *Roman Antiquities* 1.35.1–2; 2.1.2). The toponym Italia began to spread north at some uncertain date after the fifth century BC, and eventually included the entire peninsula. The Morgetes were in the southernmost part of the region and also in Sicily, where their name survived with the city of Morgantion (modern Morgantina, Strabo, *Geography* 6.1.6, 6.2.4). The Siculi were the original inhabitants of Sicily, who presumably also spread to the mainland. This ethnic history may be from Cato's *Origines* of the second century BC, also used by Pliny at *NH* 3.51.

Greek settlement in far southern Italy was to protect the sailing route through the Straits of Messina to the Bay of Naples, where Euboeans had established themselves in the eighth century BC. In the south, Regium (Rhegion) was probably the earliest Greek outpost. Within a few generations there were Greek towns along both coasts of Italy (Boardman, *Greeks Overseas* 165–89).

The towns along the western coast extended from Paestum (Greek Poseidonia), now with magnificant remains, to the Regian Column (Columna Regia) on the Straits of Messina (at modern Reggio di Calabria). It marked the crossing point to Sicily and survived until at least the second century AD (*Antonine Itinerary* 98.4–5).

Pliny's first itinerary is the sailing route from Paestum past the Palinurus Promontory to Regium, cutting straight across the gulf. The 100 miles (148 km.) is remarkably short, as the distance is more like 275 km. Then there is a road itinerary, which adheres to the coast except for a brief transit across the Taurianum Promontory (modern Capo Vaticano). Consentia (modern Cosenza) is the most important inland town, about 20 km. from the coast. The distance reported for the crossing of the Straits of Messana is shorter than today, perhaps due to tectonic changes. From Caenus (Greek Kainos, modern Punta del Pezzo) to Regium is 93 stadia, about 19 km., an accurate distance. The use of stadia indicates a Greek source, which is not identified. Strabo, whose account of these regions in the most complete extant, based his material on the author of the fifth century BC, Antiochos of Syracuse (*FGrHist* #555), who was not mentioned by Pliny.

The Western Mediterranean Seas and Islands

3.74–5. The itinerary summarized the route around the end of Italy to Locri on the east coast, which Pliny had already established as the end of his first gulf of Europe (see *NH* 3.5). He reverted to a Roman source, using miles. The distances are difficult to understand because there is one missing from Regium to the Sila forest, but 51 miles (75 km.) from Leucopetra (see *NH* 3.43) to Locri is reasonable. Yet the 303 miles (448 km.) from the Silerus River to Locri is unusually long (the modern driving distance is 392 km.).

Having thus completed the account of the coastal territories of the first gulf of Europe (as well as some of the hinterland), Pliny then provided an overview of the gulf itself, describing the seas within it from its entrance at the Atlantic. It was common practice to name seas from the peoples on their coasts; hence the use of alternative names for essentially the same sea (Hispanian, Hiberic, or Baliaric), reflecting different eras and locations along the coasts. Some of Pliny's names originated in different languages (e. g. Tuscan and Tyrrhenian), and others demonstrate the geographical perspective of those on the coast, such as the Notian (Southern) Sea (how the Greek cities in Liguria viewed it), or Inferum (Lower), a Roman term for the western Mediterranean based on the orientation of Italy.

At the end of the passage Pliny contrasted two of the early geographical authors regarding the nomenclature of the seas of the western Mediterranean. Polybius (34.15.4) used the poetic "Ausonian" for the sea south of Italy, named after Auson, the son of Odysseus and Circe. Eratosthenes (*Geography*, F127) had the term "Sardoan" (Sardinian) for the entire western Mediterranean, reflecting that in his era (third century BC) it was not well known, especially to those in the East. As such the term is quite anachronistic to Pliny's time (Pliny, *Histoire naturelle* 3, 175–7).

3.76. A lengthy section on the islands in the western Mediterranean (through *NH* 3.96) may reflect Pliny's navigational perspective. Pityussae is a descriptive name provided by the first Greek settlers, and was somewhat obsolete in the Roman period. The more common name was Ebusus (Greek Ebysos, modern Ibiza). Pliny was not quite correct in reporting that both islands had this name, since the southern was called Ophiusa ("Snaky," see *NH* 3.78; modern Formentera), again a descriptive Greek name (Strabo, *Geography* 3.5.1). The two islands lie about 10 km. apart with several islets in between. Dianium (modern Dénia, *NH* 3.25) is the closest point on the mainland (700 stadia or about 140 km. to the west-southwest); again, Pliny has used stadia, demonstrating a Greek source, such as that of Timosthenes of Rhodes from the early third century BC. The distance seems a little long (it is actually 90 km.), but the length of the stadion is always uncertain (Diller, "Ancient Measurements" 6–9).

Pliny's description is a little confused, because after mentioning the Baliares – east of the Pityussae – he then cited Colubraria, which is in fact Ophiusa, the southern of the two Pityussae, having perhaps left it out of his previous description. It is odd that it is oriented opposite the Sucro River, which is north of Dianium, the sailing point for the more northern of the Pityussae, so that the two routes from the islands to the mainland would cross each other in the open sea, but this may reflect constraints not known today.

3.77. The ability of the locals in the Baliares with slingshots was well known, and gave the islands their name (from Greek *ballein*, to throw).

The original Greek name, the Gymnasiae, reflected the belief that the inhabitants went naked in the summer (Diodorus 5.17.1). No indigenous name has survived, and the modern names are Mallorca (the larger, to the southwest), and Minorca, derived from Latin Maior and Minor, which evolved from purely descriptive to toponymic. The lengths of both Mallorca and Minorca are too large. The distance of 30 miles or 44 km. between them is accurate.

3.78. After naming the several small islands around the Baliares, Pliny considered the matter of their unusual fauna. The connection between the Baliares and snakes was derived from the name of one of them, Ophiusa, applied by Greek settlers. Presumably it was descriptive, and reflected actual conditions at an early date, but the incompatibility of the snakes with imported soil seems more of a folk tale, since specific properties of earth from particular regions was a common belief (*NH* 35.202; Vitruvius 8.3.24). The issue of the rabbits from Ebusus was so serious a problem that the locals petitioned Augustus for military assistance in their removal, since they were causing extensive crop damage (*NH* 8.217–18; Strabo, *Geography* 3.2.6).

3.79. The islands mentioned are all on the coast of Gallia, from the mouth of the Rhodanus east to the vicinity of Antipolis (modern Antibes). None is more than a few kilometers from the mainland.

Metina cannot be specifically located and may have been absorbed by siltation. Blascon is modern Brescou, and the three Stoechades, the Row Islands (otherwise First Island, Middle Island or Pomponiana, and Hypaea) are still conspicuous as the modern Îles de Hyères. Sturium, Phoenice, and Phila may be part of the same group, or perhaps some of the smaller islands near modern Marseille. The name Phoenice shows that it was once a Phoenician or Carthaginian outpost. Lero and Lerina are the modern Îles de Lérins, southwest of Antibes.

3.80. The dimensions of Corsica (Kyrnos in Greek) are not accurate. Pliny's 150 by 50 miles (222 by 74 km.) elongate the island at the expense of its width (the actual figures should be 183 by 83 km.). Vada Volaterrana (modern Vada) is northeast of the island, not the closest point on the mainland but probably the end of a sailing route. Pliny's 62 miles (92 km.) is an accurate distance to the north end of Corsica. Greeks had settled the island in the sixth century BC (Herodotus 1.165–7), and it eventually came under Roman control in the third century BC, but intensive Roman occupation did not begin until the early first century BC, through the efforts of C. Marius and P. Cornelius Sulla, as noted.

Oglasa (probably modern Montecristo) is a small island (only three km. across) that is 65 km. east of central Corsica. Planasia (modern Pianosa) lies 30 km. northwest of Oglasa and, unusually for Mediterraean islands, is flat, with a high point of only 29 m. above sea level; thus it was a particular hazard to shipping. It is most famous as the place of banishment of Agrippa Postumus (Tacitus, *Annals* 1.3; Dio 55.32.2).

3.81. There are 10 islands listed that generally lie between Corsica and the mainland, with a small group south of Antium (modern Anzio). Except for Ilva, all are obscure and some cannot be identified. Pliny's interest in these minor islands demonstrates his continued reliance on navigational data. Ilva (Greek Aithalia, modern Elba), lying southwest of Populonium, was noted for its iron deposits, exploited from early times (Strabo, *Geography* 5.2.6; Oliver Davies, *Roman Mines in Europe* [Oxford 1935] 67–8).

3.82. The islands listed are generally in the Bay of Naples (Puteolanus Sinus), loosely applied. Pandateria (modern Ventotene) is well west of the bay, and was famous as a place of exile for Julio-Claudian women. Prochyta (modern Procida) is just southwest of Misenum, and is the northernmost of the islands that make the outer circuit of the volcanic caldera represented by the Bay of Naples. It was said to be named after a relative of Aeneas, a tradition from at least the third century BC (Naevius, *Bellum Punicum*, F17; Dionysius of Halicarnassus, *Roman Antiquities* 1.53.3), but the toponym is probably descriptive, "that which is poured out," referring to its violent volcanic activity (*NH* 2.203; Strabo, *Geography* 5.4.9).

Aenaria and Pithecusa (Pithecussae) are the same island, modern Ischia, just southwest of Prochyta. The Homeric citation (*Iliad* 2.783) is only marginally relevant, not only based on a variant reading not generally accepted today, but assuming that the battle between Zeus and Typhon took place in this region. To be sure, Typhon was said to have been buried under Mt. Etna (Strabo, *Geography* 5.4.9), but the connection with the island is an example of the extreme Homeric exegesis of Hellenistic times. Equating Aenaria and Aeneas is equally dubious.

The name "Pithecusa" was often said to refer to indigenous simians, but Pliny offered an alternative explanation, from *pithos*, which is itself quite possible, since the island was the earliest Greek settlement and may have had a ceramic industry (Boardman, *Greeks Overseas* 163–8).

Capreae (modern Capri), to the southeast of Pithecussae, was a private possession of Augustus and became famous for the Villa Jovis of Tiberius, in a stunning location at the south end of the island. It was the first major Imperial residence and an important factor in the development of Roman domestic architecture (J. B. Ward-Perkins, *Roman Imperial Architecture* [second edition, Harmondsworth 1981] 198–201).

3.83. Leucothea is probably modern Licosa, a small islet at the south end of the Gulf of Salerno, the next large caldera south of the Bay of Naples (Strabo, *Geography* 6.1.1). Pliny may have another description of it, from another source, at *NH* 3.85.

The account then examines the island of Sardinia (through *NH* 3.85), which is only a few kilometers south of Corsica, with many small islets in between, including those called the Cuniculariae, or the Rabbit Burrows. To Pliny, Sardinia was not visible from the mainland, but his vantage point was probably around the Bay of Naples, 400 km. away. Strabo (*Geography* 5.2.8) noted that one could see it from southern Etruria, about 200 km. away; today it is occasionally faintly visible from the mainland.

Fossa/Taphios may be one of the many small islands between Sardinia and Corsica, but the name, as Pliny noted, is more appropriate to the straits themselves (with Fossa meaning "Ditch"), the modern Straits of Bonifacio.

3.84. The measurements for the extent and perimeter of Sardinia are probably coastal sailing routes of uncertain itineraries. The circumference of 565 miles (836 km.) seems to approximate Strabo's 4,000 stadia (perhaps 800 km.; Strabo, *Geography* 5.2.7), close to the official perimeter today of 790 km. (Pliny, *Histoire naturelle* 3, 182–3).

Cape Caralis (modern Capo Carbonara), the southeastern corner of the island, was said to be 200 miles (296 km.) from Africa. The African coast is in fact only about 200 km. away, but no terminal point of Pliny's measurement was specified, and the distance suggests the sailing route from Caralis (modern Cagliari, the largest town on the island) to Carthage or Utica. A distance of 1,250 miles (1,850 km.) to Gades is quite long, but again the actual sailing route is not provided. There is also a list of islands around Sardinia (into *NH* 3.85).

3.85. After completing the catalogue of islands around Sardinia, the account lists nine ethnic groups that inhabited the island. The Ilienses were in the northeastern part and were said to be the most ancient (Pomponius Mela 2.123). The Balari were to their north and were Carthaginian deserters who hid out in the mountains: their name was said to mean "fugitive" in the local language (Pausanias 10.17.9). The Corsi are otherwise unknown but their name suggests that they were Corsican immigrants, and as such probably lived in the extreme north of Sardinia. The Sulcitani were at Sulchi, a Carthaginian foundation around modern San Antioco on the east coast. The Valentini, who seem to have a Roman name, lived in the south (Ptolemy, *Geographical Guide* 3.3.6); the

Neapolitani, whose Greek name is unexplained, were in the southwest near modern Santa Maria di Napoli. The Bitienses were perhaps at the locality of Bithia at the southern end of the island (Ptolemy, *Geographical Guide* 3.3.3). The Caralitani were the inhabitants of the major city of Sardinia, Caralis (modern Cagliari), where extensive remains are visible. It was originally an indigenous settlement that was successively Phoenician, Carthaginian, and Roman. The Norenses lived at Nora, the oldest town on Sardinia, at modern Pula on the southern coast, founded by Iberians or Phoenicians from Cyprus in the ninth century BC (Pausanias 10.17.5; D. Manconi, "Nora," *PECS* 628). Nora may be the origin of the world *nuraghi*, the distinctive prehistoric constructions on the island.

The names applied by Timaeus ("sandal-like," *FGrHist* #566, F63) and Myrsilus ("footprint," *FGrHist* #477, F11) descriptively refer to the shape of the island. The accounts date from the late fourth or third centuries BC. Obviously neither name could have been applied before Greeks knew about the island, probably no earlier than the sixth century BC (Boardman, *Greeks Overseas* 214–16).

To conclude the section, Pliny returned to the Italian coastal islands. Leucosia seems to be a duplicate of Leucothea (*NH* 3.83). There is no doubt that this entry refers to modern Licosa, because of its association with the Siren Leukosia (Strabo, *Geography* 6.1.1). The islands mentioned probably extended along the coast south from the area of Paestum, but cannot be easily identified today. Some may have been joined to the mainland, and others may have disappeared due to tectonic activity. For the Oenotri, see *NH* 3.71.

3.86. Sicilia (Sicily), the largest island in the Mediterranean, was examined in detail (through *NH* 3.94). It had several names: Thrinakia, used by Homer (*Odyssey* 11.107, etc.), was said to be the original one and Iberian in origin (*King Nikomedes Periplous* 268). By the fifth century BC it had evolved into Trinakia (Thucydides 6.2), or Pliny's Trinacria, a name assumed (probably erroneously) to refer to the three capes of the island. Sicania was also said to be Iberian, and became the common name (as Sikelia in Greek), although in recent times there has been a resurgence of Trinacria.

There is little doubt that Sicily was once attached to the mainland. It is suggested that the name Regium (Greek Rhegion) reflects the separation, from Greek *rhegnymi*, "to break apart." Needless to say, this would mean that the founders of the city, in the eighth century BC, had a conception of the geologic history of the region (Strabo, *Geography* 6.1.6; Boardman, *Greeks Overseas* 171–2). This seems questionable, but the derivation may have been documented as early as the fifth century BC (Aeschylus F402). The strait is wider today than Pliny's 1½ miles. The end points of the reported length, 15 miles or 22 km., cannot be determined. The circumference of Sicily, 618 miles (915 km.), from Agrippa (F13), is reasonable.

3.87. The Homeric perils of Scylla and Charybdis have long been located on opposite sides of the straits. Scylla, on the mainland, was at the northern entrance. The Skyllaion Promontory (modern Rocca di Scillo) preserved the name (Strabo, *Geography* 6.1.5). Across from it was the traditional location of the great whirlpool, Charybdis (Thucydides 4.24.4), of which only a vestige, modern Garofalo, remains, just north of Messina. Both were major navigational hazards.

The three promontories of Sicily defined the island, regardless of the meaning of the name Trinacria. The 440 miles (651 km.) to the Peloponnesus would be a sailing route and not to the closest point, but to a major port, probably Corinth. From Lilybaeum (modern Capo Lilibeo), at the southwest corner of the island, it is far less than the stated 180 miles (266 km.) to the African mainland; it is only about 150 km. to Cape Mercurii (modern Cape Bon in Tunisia), so again the distance reported must be a sailing route between two unmentioned ports. The 190 miles (281 km.) to the end of Sardinia is reasonable. Generally the distances between the three capes of Sicily are too long.

3.88–90. Pliny's circuit of the coastal cities of Sicily – interspersed with other topographical data – begins at Pelorum (modern Capo Peloro at the northeast corner) and makes a clockwise route. Most of the locations remain well known. The Mamertini were Oscan mercenaries, named after Mars (in his Oscan form, Mamors), who had been brought to Sicily in the early third century BC and became brigands, settling around Messina (Strabo, *Geography* 6.2.3).

The crater of Aetna (Etna) is constantly changing, and eruptions have been recorded since the fifth century BC (Pindar, *Pythian* 1.19–28; Aeschylus, *Prometheus Bound* 365–74). Before the eruption of long-dormant Vesuvius in AD 79, Etna was the most visible volcano in the Mediterranean.

3.91. Nearly fifty towns in the interior of Sicily are listed (as ethnyms), most of which were tribute-paying. This catalogue was somewhat anachronistic in Pliny's day, since it does not reflect the changes to the status of Sicilian cities that occurred shortly after the death of Julius Caesar (Diodorus 13.35.3; Cicero, *Letters to Atticus* #366; Pliny, *Histoire naturelle* 3, 191–6).

3.92. The islands near Sicily are described in a roughly clockwise fashion, beginning with Gaulos and Melita (modern Gozo and Malta). The list includes all the islands that can loosely be said to be associated with Sicily. Some cannot be located. The most anomalous is Galata (see *NH* 5.42), modern Galite, about 110 km. off the Tunisian coast, far west of Sicily and actually closer to Sardinia. Osteodes and Ustica are probably the same island, modern Ustica, in an isolated location north of Palermo. For the Aeoliae or Liparaei, see *NH* 3.93–4.

3.93–4. The seven Liparaean or Aeolian islands (modern Eolie or Lipari) are conspicuous from the north shore of Sicily, and remain visibly active volcanically. From at least the fifth century BC they were believed to be the home of Aeolus and the winds (hence the dual name, which survives today); the earliest notice of the connection is by Antiochos of Syracuse (*FGrHist* #555, F1; see also Thucydides 3.88; Pausanias 10.11). There are seven major islands, with Lipara (modern Lipari) the largest and most central. Strongyle (modern Stromboli) exuded unusual flames that were more liquid (*liquidior*) than normal, with a transparent quality (Constantino Moro, "*Liquidiore flamma*: un problema lessicale in Plinio il Vecchio (*Nat. Hist.* III 93 s.)," *PP* 57 [2002] 115–20). Elsewhere Pliny reported on a volcanic event from around 90 BC, as well as a previous one in 126 BC, which created the island of Hiera, or modern Volcano (*NH* 2.203, 238; see also Posidonius F227; Strabo, *Geography* 6.2.10–11). The winds produced by the volcanic activity supported the belief that these islands were the home of Aeolus.

With these islands, Pliny completed what he called the first gulf of Europe, extending on the north side of the Mediterranean from its western end to the south end of Italy, with the peculiarity that it was considered to go around the toe of Italy to Locri (*NH* 3.5, 3.74).

Italy from Locri to Ravenna

3.95–6. Pliny realized the odd ending to the first gulf, and thus he began the second gulf at the Lacinian Promontory (modern Capo Colonna), the easternmost part of the toe of Italy. Between it and Locri was an interval of 75 or 86 miles (111 or 127 km.) that he defined (narrowly) as Magna Graecia. Why Cape Cocynthum (modern Punta Stilo) should be regarded as the longest promontory in Italy is not obvious, but in the second century BC it was believed (erroneously) to be the south end of Italy (Polybius 2.14.5). Pliny's statement may reflect this misapprehension, and his "longest promontory" is probably the entire toe.

The narrowest part of Italy – from Scolacium (near modern Squillace) across to the head of the gulf on the west side north of Vibo Valentina – is not 40 miles (59 km.) across, but much less than that. Pliny's figure is probably the road distance through Teura (modern Tiriolo). Castra Hannibalis is not otherwise known, but the memory of the Carthaginian invader was strong in southern Italy. Dionysius I (432–367 BC), tyrant of Syracuse, planned a wall across this isthmus in the early fourth century BC, as part of his unsuccessful plan to create a physical connection between the lands to its south and Sicily. There is no evidence that the project was ever completed, or even started, but it would not have been impossible (Diodorus 14.100; Strabo, *Geography* 6.1.10).

Five islands are named off this coast, including Ogygia (the island of Kalypso), which has been located in many places throughout and beyond the Mediterranean. None of the five can be located, and in fact there is no known history of islands in this region; even the exhaustive *Geographical Guide* of Ptolemy has no mention of them. It is improbable that all could have vanished due to tectonic activity, and it is more likely that they are part of the mythical environment of Kalypso. Pliny's source for this information was Greek, as shown by the Greek genitive, Dioscoron. The 70 miles (104 km.) reported by Agrippa (F42) from Lacinium to Caulon (near modern Punta Stilo) is reasonable for a land route; the sea route is much shorter.

3.97. Pliny's second gulf of Europe is essentially the Adriatic, beginning at Cape Lacinium (and thus including the Gulf of Taranto), and ending at Acroceranium (Cape Gjuhëzës in Albania), the westernmost point of the Greek peninsula and at the mouth of the Adriatic. The distance between the two points in the manuscripts is 75 miles (111 km.), but this is absurdly short, and it is possible that the text should be emended to 175 (259 km.), which is essentially accurate.

The towns and rivers along the coast north from Cape Lacinium are presented in *periplous* fashion. Metapontum (modern Metaponto), which lies at the cultural boundary between Lucania and Apulia, was the end of the third Augustan region (Bruttium et Lucania) and the beginning of the second (Apulia et Calabria).

3.98. Abandoning the *periplous* for a while, Pliny listed the inland communities. Bruttium was rugged and lightly populated, and thus had only one town, that of the Asprutani, which cannot be located. Lucania, with more benign topography and fertile valleys, had more settlements.

Pliny continued to show his interest in changing demographics and the depopulation of certain regions. Lucanian Thebes, cited by Cato (*Origines* F68), had disappeared by the second century BC. Pandosia, whose name implies a location in a fertile agricultural district and perhaps Greek settlement, may have been southwest of Metapontum at modern Santa Maria d'Anglona. This was where Alexander (I) of Epiros (or Molossia), the uncle of Alexander the Great, who was in the service of the Tarantines, was killed around 330 BC under circumstances that have never been satisfactorily explained (Theopompus F318; Strabo, *Geography* 6.1.5; Livy 8.24).

3.99. The second Augustan region, Apulia et Calabria (discussed through *NH* 3.105), included all of the heel of Italy and north to a point beyond the Garganus Peninsula (modern Gargano), as well as much of the upland interior and even across the Apennines to their western foothills above Capua. The Hirpini were a group of Samnites (Greek Saunitians) who lived around Beneventum (Strabo, *Geography* 5.4.12). Calabria was the heel of Italy, with the Sallentini a small ethnic group at its southern extremity. Apulia was along the coast north of Calabria (Strabo, *Geography* 6.3.1). Messapia, east of Tarentum in the northern part of the heel, may have been settled by immigrants from Central Greece, although the eponym Messapus seems to be a late construct (Thucydides 7.33.4; Strabo, *Geography* 9.2.13; Pausanias 10.10.6). Peucetia was in southern Apulia, and its inhabitants may also have come from Greece, in this case from Arcadia (Pherecydes [*FGrHist* #3] F156; Dionysius of Halicarnassus, *Roman Antiquities* 1.11.3; Strabo, *Geography* 6.3.8; Pausanias 8.3.5). The territories and ethnic groups in this part of Italy are often not easily distinguished from one another (Francesco D'Andria, "Messapi e Peuceti," in *Italia: omnium terrarum alumna* [Milan 1988] 653–715).

The Tarantinus Bay (modern Gulf of Taranto) separates the toe and heel of Italy, named after the most prominent city on it, Tarentum (Greek Taras, modern Taranto). It was founded by people expelled from Sparta in Laconica at the end of the eighth century BC (Boardman, *Greeks Overseas* 184). A distance of 250 miles (370 km.) for the bay is the full coastal circuit from Lacinium to the end of the heel, and 136 miles (201 km.) from Lacinium to Tarentum is also a coastal distance. The route straight across the bay is closer to 125 km. than to Pliny's 100 miles (148 km.). That between Tarentum and Brundisium (45 miles or 67 km.) conforms to the modern road across the northern end of the heel, which still uses the extension of the Via Appia, built in the early third century BC. Sasine has not been located but may be either just to the east of Tarentum, where a small bay extends inland, or south on the coast, where the peninsula is only about 45 km. across.

3.100. After mentioning the two inland towns, Uria (modern Oria, halfway between Taranto and Brindisi) and Sarmadium (perhaps a variant for Scamnum, at Muro Tenente, just east of Oria), Pliny made a circuit of the coast of the heel of Italy. Callipolis (modern Gallipoli) is less than 75 miles (111 km.) from Tarentum, but the 33 miles (49 km.) to the Acra Iapygia is accurate. "Acran Iapygiam" is one of Pliny's more unusual linguistically hybrid forms, with a Greek accusative followed by a Latin one. The implicit Greek source was not named. The promontory (modern Capo Santa Maria di Leuca) is actually not the southernmost point of either Italy or its heel (Punta Ristola, slightly to the west, is a short distance farther south), but it is the most prominent feature on this coast and was a major sailing marker on the route between Greece and Italy (Strabo, *Geography* 6.3.5).

The 19 miles (28 km.) from Acra Iapygia is the distance to Basta (modern Vaste); Hydruntum (Greek Hydrous, modern Otranto) is farther north. It is at the easternmost point of Italy, and the region of Apollonia (modern Pojan in Albania) is visible on clear days 50 miles (74 km.) across the narrows at the mouth of the Adriatic.

3.101. Pyrrhus' plan to bridge the strait is not documented elsewhere, but presumably would have been part of his preparations for invading Italy in 280 BC. Varro, the polymath often quoted by Pliny, who owned land in Apulia, was in the region while a naval commander under Pompey the Great in the campaign against piracy of 67 BC (*NH* 7.115–16; 16.7; Varro, *de re rustica* 2, *Preface* 6; Appian, *Mithridateios* 95; Florus 1.41.10). This would have been the context for any Roman project to bridge the straits.

The itinerary continues to Brundisium (Greek Brentesion, modern Brindisi), 50 miles (74 km.) north of Hydruntum, which remains the primary port for the crossing from Italy to Greece. The distance of 225 miles (333 km.) to Dyrrachium (modern Durrës in Albania) is the sailing route, which went straight across to the island of Salon (modern Sazani in Albania), just off the Illyrian coast, and then headed north to Dyrrachium (Strabo, *Geography* 6.3.5). The winds tended to be more favorable for the Brundisium crossing than the Hydruntum route.

3.102. The Poediculi were essentially the same as the Peucetians (Strabo, *Geography* 6.3.1). They lived around Barium (modern Bari) and in the uplands to the west; the story of the origins of their ethnic subdivisions is not documented elsewhere. The onomastic connections of Iapyx may have been known as early as the fifth century BC (Antiochos of Syracuse F13); if so, this would give him more credibility as an eponym than the numerous ones documented only from the first century BC (Strabo, *Geography* 6.3.2).

3.103. The southern border of Apulia was at Canusium (modern Canosa), although, as noted, ethnic boundaries were variable in this region. The Daunii were another of the vague local ethnic groups (Strabo, *Geography* 6.3.8–9). Like so many of the local eponyms, their leader is not documented before Roman times (Ovid, *Fasti* 4.75–6; Ovid, *Metamorphoses* 14.458–9). They were associated with the town of Luceria (Greek Loucheria, modern Luceria), southwest of Garganus (*NH* 3.104). Another late tradition told how Diomedes was not welcome at home after the Trojan War and came to Italy; his ability to survive a confrontation with his unfaithful wife was contrasted with the experience of Agamemnon (Lykophron, *Alexandra* 594–632; Juba F5; Vergil, *Aeneid* 11.225–90).

Hannibal was at Salapia (modern Salpi) in 214 BC, but the primary source (Livy 24.20.15) does not record any personal relationship there. The Garganus (modern Gargano) is the distinctly visible protrusion on this coast, a part of the Apennines and a unique ecosystem. Pliny's 234 miles (346 km.) is the sailing route from the Iapygian Promontory to the Garganus and around it, perhaps to the harbor on the Fertor River (modern Fortore). The Frentanians were a Samnite people located west of the Garganus and north of the river, who were assimilated by the Romans in the late fourth century BC (Livy 9.16; Strabo, *Geography* 5.4.2).

3.104. The Teani lived in the valley of the Fertor, around the town of Teanum Apulium (near modern San Paolo di Civitate). Their leader is otherwise unknown but presumably was yet another immigrant from the Greek world. Pliny used the poetic word Graius (instead of Graecus), which may give a hint as to the nature of his source. The Lucani (*NH* 3.71) were generally farther south, but presumably one of their groups ventured north to the region inland from Garganus; the Atinates were one of their populations (*NH* 3.98). Calchas, the famous prophet of the *Iliad*, travelled throughout Asia Minor (Strabo, *Geography* 6.3.9; Roller, *Historical and Topographical Guide* 331), but was also said to have had a shrine at Drion (perhaps a peak of the Garganus), if this is the same person.

The ethnic groups that Diomedes destroyed, the Monadi and Dardi, are otherwise unknown, but presumably were small and perhaps even mythical, as seems the case with Apina and Trica. Both of these toponyms are similar to their Latin homonyms meaning "nonsense" (Plautus, *Curculio* 613; Martial 14.1.7), and Pliny may have been subtly referring to the mythical nature of the names.

3.105. The Hirpini were a Samnite population, named from their word for "wolf," an example of the lupine foundation tales common in central Italy (Strabo, *Geography* 5.4.12; Rafael Scopacasa, *Ancient Samnium: Settlement, Culture, and Identity Between History and Archaeology* [Oxford

2015] 22–4). They lived around Maleventum (or Malventum), whose indigenous name was misinterpreted by the Romans as an evil omen. After the defeat of Pyrrhus of Epiros there in 275 BC, it was changed to the more auspicious Beneventum (Livy 9.27.14). The town is at modern Benevento on the western side of the Apennines.

There are three alphabetical lists of ethnyms. Most of the names represent town populations rather than rural peoples spread over a district. Some cannot be identified. The catalogues cover a wide region, from the southernmost part of Calabria (the Veretini – from Veretum, modern Vereto – just north of the Iapygian Promontory) to the Larinates, probably from Larinum (modern Larino), west of the Garganus. The distinction between the first two lists is not always apparent; the five local ethnyms of the third are all from a small region in the heel of Italy.

The two groups of Ligures were transplanted from their homeland into this area of the Apennines because of their persistent resistance to the Romans, and named after those effecting the movement, the consuls of 181 BC, P. Cornelius Cethegus and M. Baebius Tamphilus (Livy 40.38). The Cannenses lived around Cannae, near the coast southeast of Garganus, an unimportant village that became notorious when Hannibal defeated the Romans there in 216 BC (Polybius 3.113–17; Livy 22.45–50).

3.106. The fourth Augustan region (discussed through *NH* 3.109), Samnium, began at the Tifernus River (modern Biferno) and extended northeast of Latium to the edge of Etruria and Umbria. The country away from the coast was wild and remote, including the highest and most rugged portions of the Apennines. This was the territory in which the Marsic or Social War erupted in 91 BC, lasting for four years and causing Rome much difficulty (Strabo, *Geography* 5.4.2). It may have been what Pliny had in mind when he commented on the resilience of the locals.

For the Frentani, see *NH* 3.103. The demographic lists for the fourth region are structured in terms of both ethnic groups and toponyms. The Carecini were divided in two, a lower (Infernates) and upper (Supernates), based on their position on the Sagrus River (modern Sagro). The Marrucini were to their northwest, along the Aternus River (modern Pescara), with the Paeligni more upstream, and the Marsi – the instigators of the war that bears their name – even farther upstream in the heart of the Apennines. The Albenses were around the Fucinus Lake (for which, see *NH* 3.108) on the western drainage of the mountains, and the Aequiculani farther west and downslope toward Rome. They were famous for having been defeated by L. Quinctius Cincinnatus in the fifth century BC (Livy 3.26–9; Strabo, *Geography* 5.3.2), and their name may survive in the toponym Cicolano, northeast of modern Tivoli. These four ethnic groups formed a strip across Italy, from east of Rome to the Adriatic.

3.107. The Vestini were in the eastern Apennine foothills north of the Aternus River. The Samnites were one of the major ethnic groups of central Italy; their variant names indicate their spread and diversity. They had an extensive series of wars with the Roman Republic in the fourth and third centuries BC, but were steadily defeated despite an alliance with Pyrrhus of Epiros. Nevertheless, they retained their identity until the first century BC. At the Samnite town of Bovianum (modern Boiano), Vespasian settled members of the Legio XI Claudia, which had supported his efforts to obtain the principate (Tacitus, *Histories* 4.68). The Sabines lived northeast of Rome. Famous for their early war with the Republic and the role of women in bringing it to an end, they were soon assimilated.

3.108. The Aequicoli were presumably related to the Aequiculani (*NH* 3.106), if not the same group of people, since they lived near each other on the western slopes of the Apennines. Marsyas, the Lydian eponym of the Marsi, is another of those founders of Italian groups who does not seem to have an early history and, in this case, is otherwise unknown. The Fucinus Lake is now drained (its basin is known as the Piano del Fucino); it was once one of the largest lakes in Italy, located in the midst of the Apennines. It had no natural outlet and its level changed dramatically, causing repeated problems for the locals. There were various Roman attempts to drain it, beginning in the first century BC, but none was successful (Strabo, *Geography* 5.3.13; Suetonius, *Divine Julius* 44).

The etymology of "Sabini" was believed to be from Greek *sebo* ("to worship"), but there is no reason to believe this. The Velini Lakes were probably in the basin of the upper Avens River (modern Velino), northeast of Rome, where today there are several small lakes. Reate (*NH* 3.109, modern Rieti), at the south end of this basin, was one of the ancient Sabine cities (Strabo, *Geography* 5.3.1).

3.109. The two main tributaries entering the Tiber from the east are the Nar (modern Nera), which is the larger and more northern (with its own tributary, the Avens), and the Anio (modern Aniene). Both drain the west slopes of the Apennines. The country of the upper Anio – southeast of Tibur (modern Tivoli) – was considered especially beautiful, and at Sublaqueum (modern Subiaco) Nero built a villa, damming the river to form lakes (Tacitus, *Annals* 14.22; Frontinus, *de aquis* 93). Extensive remains of both villa and dams remain (M. Torelli, "Sublaqueum," *PECS* 864).

Lake Cutilia (at modern Terme di Cotilia), near the source of the Avens, was considered to be the center of Italy, where, in theory, one could see

Latium, Picenum, and Umbria. The local springs were medicinal (*NH* 31.10) and are still an active spa. The floating island was probably a vegetal mass (Dionysius of Halicarnassus, *Roman Antiquities* 1.15; Varro, *de lingua latina* 5.71; Seneca, *Natural Questions* 3.25.8–10).

3.110. Picenum, the fifth Augustan region, extended from the Adriatic into the eastern foothills of the Apennines, north of the Samnite territory. Despite the heavy local population, this was the smallest of the 13 regions. The Picentines were consistently believed to be an offshoot of the Sabines. A Sacred Spring (*Ver Sacrum*) was a ritual connected with the reduction of a population by encouraging emigration. Several instances are known in Italian history (Dionysius of Halicarnassus, *Roman Antiquities* 1.16, 24). The Samnites had lengthy wars with the Romans, and may have been economically pressed and thus persuaded some of their people to move north (Scopacasa, *Ancient Samnium* 35–6). Allegedly the migrants followed a woodpecker (*picus*) to their new home, becoming the Picentines. This was one of a number of stories in Greco-Roman culture of animal-led population movements (Strabo, *Geography* 5.4.12). The Picentine territory extended from the Aternus River along the coast to Ancona, and inland to the foothills of the Apennines. The Liburni were on the east side of the Adriatic (*NH* 3.139–40), but occasionally gained footholds in Italy (see also *NH* 3.112); their survival in Italy may be the remnant of an Iron Age population distribution (Edward Togo Salmon, "The Iron Age: the Peoples of Italy," *CAH* 4 [second edition 1988] 696–7).

3.111. There are two catalogues of the towns of Picenum, one of the coastal populations (with town names listed) and the second of those in the interior, ethnically structured. The 183 miles (271 km.) from Garganus to Ancona is accurate.

3.112. The sixth region, Umbria, has a small coastal frontage (from Ancona to near Ariminium, modern Rimini), but also extends inland as far as the Tiber, and south to the borders of the Samnite territory. At Ariminium the Apennines begin to curve to the west, and the broad Padus (modern Po) valley is beyond to the north. This was considered to be the boundary of traditional Roman Italy and the beginning of the Gallic territory, occupied historically by Celtic peoples. Although the Umbrians (as Ombrikoi) had been known to the Greek world since at least the fifth century BC (Herodotus 1.94), the ethnym only became topographical with the establishment of the Augustan regional system (Bradley, *Ancient Umbria* 19–20). Nevertheless, the Umbrians retained an ethnic identity even into Pliny's time (*NH* 3.115; see also Strabo, *Geography* 5.1.10). Provincial organization of the region began at the end of the third century

BC (Bradley, *Ancient Umbria* 193–4), and the term Togata Gallia came into use when the locals had become romanized enough to receive citizenship. The province was dissolved when the Augustan system was created, but the name was still in use a century later.

The Siculi were a little-known indigenous group, not associated with Sicily. For Liburnian settlement in this area, see *NH* 3.110. Pliny's account of the ethnic evolution of this region is simplified. It is possible that the Umbrians were the first settlers, but other groups such as the Etruscans and Gallians were more contemporary than sequential, with the former coming from the southwest and the latter from the northwest. The Umbrians and Etruscans had a long history of hostility (Strabo, *Geography* 5.1.10).

3.113–14. Pliny's catalogue of towns in the sixth region is largely based on local ethnyms, with four Roman *coloniae* listed first, then those living along the coast, and finally those inland, through and beyond the Apennines. Several towns had disappeared; the source for this information is probably Cato's *Origines*, which also recorded the date for the founding of Ameria (modern Amelia in the uplands just north of where the Tiber and Nar join). This was the equivalent of 1134 BC, 963 years before the war with Perseus, the last king of Macedonia (Cato, *Origines* F49). The town, in a well-fortified location, may be unusually old, and is noted today for its impressive standing walls.

The Padus River

3.115. The eighth region, Aemilia, is the territory on the south side of the Padus, south to the Apennines and west to the boundary of the ninth region, Liguria, probably at the modern Piedmont–Lombardy boundary near Tortona. The Rubico River (modern Rubicone) is an insignificant stream geographically, less than 30 km. long, which was designated the northern boundary of Italy late in the second century BC (F. W. Walbank, *A Historical Commentary on Polybius* [Oxford 1957–79], vol. 1, 396–7). It is not mentioned in extant literature before Caesar's illegal crossing in 49 BC (Cicero, *Philippic* 6.5; Velleius 2.49.4). It is unlikely that the Sabines were as far north as Ravenna; Pliny may have meant to write "Umbrians" here, which was the accepted ethnic origin of the city (Strabo, *Geography* 5.1.11). Yet nearby Butrium (perhaps around modern Commachio) was actually said to be Umbrian. The 105 miles (155 km.) from Ancona to the Bedesis River (modern Ronco, at the south edge of Ravenna) is accurate. After the list of towns and rivers on the coast there is an alphabetical series along the

Via Aemilia, towns that generally still remain important today, such as Bononia, Mutina, and Parma (modern Bologna, Modena, and Parma).

3.116. The towns and local ethnyms of the eighth region are catalogued in a single alphabetical list. Pliny continued to use Cato's *Origines* as a primary source, noting that the Boii had disappeared. They were a large group, who had come from central Europe (their name is the origin of "Bohemia") at some early date, and had situated themselves around modern Bologna. They were defeated by the Romans in the early second century BC, shortly before Cato wrote (Livy 32.29–31, 36.38; Strabo, *Geography* 5.1.6, 7.1.5, 7.3.11). Whether or not they had 112 tribes cannot be determined, but they certainly were widespread, and Boii even survived in central Europe into the first century BC.

The Senones were Celtic peoples who entered Italy at the beginning of the fourth century BC and eventually took part in the attack on Rome (Livy 5.34–5). In time they settled in the region of the Sena River (modern Misa), northwest of Ancona on the coast. This district came to be called the Ager Gallicus, a name that persisted long after they were assimilated or dispersed. The name Senigallia survives in this region.

3.117. The Padus River (modern Po) is the longest and largest river in Italy, and Pliny discussed it in detail. He was perhaps inspired by the fact that his hometown of Comum (modern Como) was not far from the river, although, to be sure, other writers also wrote about it (Polybius 2.17; Strabo, *Geography* 4.6.5). Its headwaters are at modern Pian del Re, a village at 2,020 m. above sea level, just east of the modern Italian–French border. Mt. Vesulus (modern Viso) is just to the south, and rises to 3,841 m., the highest peak in the Cottian Alps and an isolated summit that is prominent from far away.

Underground rivers were believed in antiquity to be common, with the most famous example the Nile, whose upper course, thought to be in the western Sahara, allegedly ran underground for long distances (*NH* 5.51–4). In high mountainous regions rivers might seem to be below the surface because of glacial debris. Vibi Forum (see *NH* 3.123) is at modern Cavour, southwest of Turin, where the river emerges from the mountains into the floodplain.

The Eridanus was first mentioned by Hesiod in his catalogue of the great rivers at the fringes of the world (Hesiod, *Theogony* 338). Herodotus, although uncertain that it even existed, had reports that it emptied into the northern sea, which suggests that the Eridanus was either the Rhine or the Vistula (Herodotus 3.115). But the river also came to be localized in the west and by the fifth century BC it was thought to be the Padus (Pherecydes

[*FGrHist* #3] F74), although as late as the first century BC even its very existence was still doubted (Strabo, *Geography* 5.1.9). Association with Phaethon is in the context of the theft of the chariot of his father Helios and his subsequent self-destruction. His sisters wept for him on the banks of the Eridanus and their tears were turned into amber, which again suggests that the river was in the north. The tale seems to have developed, especially in drama, by the fifth century BC (Euripides, *Hippolytos* 735–41; see also *NH* 37.32–5).

The rising of the Dog Star (Sirius) is now in August, but was earlier in antiquity. It coincided with the melting of the Alpine snows, and thus the flooding of the river and the creation of the fertile floodplain that is still prominent today.

3.118. Pliny's unusual statement about the length of the Padus (a total of 388 miles, or 574 km.) may reflect a road distance through the plain, with the additional length of the river. The official length today is 652 km. There were 30 known tributaries, of which Pliny named eight from the south and 10 from the north. These are all well known today, and most preserve modern derivations of their ancient names.

3.119–21. As with any river of large volume, erosion along the banks of the Padus remains a perennial problem, although its size is much reduced today because of modern usage. Vergil, who was born near the river, described its bucolic atmosphere (Vergil, *Aeneid* 9.680). The reported width of the delta, 120 miles (178 km.) from Ravenna to Altinum (modern Altino), is excessive, and it is more like 125 km. The seven mouths (there are eight today), implicit in the toponym Septem Maria, invite comparison with the multiple mouths of the Nile and Rhodanus (*NH* 3.33, 5.64; Tusculum 3–4, 302).

The canal known as the Fossa Augusti extended from the southernmost mouth of the Padus for a distance of about 20 km. through the marshes and lagoons to Ravenna. It provided a way for ships to go from the naval base at Classis, just south of Ravenna, into the Padus without going out to sea. Claudius' British triumph was in AD 44, and his interest in hydraulics was famous, most notably with his construction of the harbor at Ostia.

Some of Pliny's information on the lower Padus came from Greek writers. Timaeus may have been the source for the Spinetian treasury at Delphi, although whether it actually existed is questionable (Pearson, *Greek Historians* 84). Another was perhaps Metrodorus of Scepsis (*NH* 3.122), who provided the distance for the circumference of the delta (2,000 stadia or perhaps 400 km.). This is far too large, and he may have meant the entire coastal plain from the Apennines to the Alps.

3.122. Metrodorus of Scepsis (*FGrHist* #184) spent time at the court of Mithridates VI of Pontus. The king was regularly interested in invading Italy from the north, as if a second Hannibal, and Metrodorus may have done some of the topographical research for such an expedition (Roller, *Empire* 195–6). There was a little-known *pados* (or *pedos*) tree (Theophrastus, *Research on Plants* 5.7.6), but whether the word is associated with the river is not known. In one instance Padus is cited as a Celtic personal name (Silius Italicus 4.232), but this may be a literary formation from the name of the river.

Italy Across the Padus

3.123. The eleventh Augustan region, Transpadana ("Across the Padus"), discussed through *NH* 3.125, encompassed the extreme northwest of modern Italy, from the north bank of the river (beginning upstream of Cremona) north and west to the perceived crest of the Alps. This is largely at the modern border with France and Switzerland, but the region also included the modern Swiss canton of Ticino, which remained under Italian control until 1803. Transpadana is the only Augustan region that had no seacoast, but was linked economically to the Padus. Augusta Taurinorum, the head of navigation on the river, is modern Turin. Augusta Praetoria is modern Aosta, a town with extraordinary Roman remains.

The two passes through the Alps are generally accepted as the Petit (Graiae) and Grand (Poeninae) St. Bernard. It is highly questionable that Hannibal used the Poeninae, despite endless and continuing debate concerning his route (Walbank, *Historical Commentary* 1, 382–7). There was no strategic reason for him to have been that far north (and every reason for him not to be); there is little doubt that he used one of the passes to the south in the Mt. Cenis or Mt. Genèvre region. But Hannibal's association with the St. Bernard passes has taken on a mythical quality, especially since the time of Napoleon and the paintings by David. For Hercules' crossing, see *NH* 3.134.

Eporedia (modern Ivrea) is located where the Dora Baltea leaves the Alpine valley today known as the Val d'Aosta. There is no other evidence that the Sibyl told the Romans to found the city, but details of the Roman campaign of 143 BC in the region were directed by the oracle (Julius Obsequens 21). The assumed indigenous name indicates a previous settlement.

3.124. For the Sallvi, who were an Alpine highland people, see *NH* 3.47; for the Vocontii, who lived inland east of the Rhone, see *NH* 3.34. The

Laevi and Marici were small groups on the north side of the Padus; their city, Ticinum, is modern Pavia. In the early fourth century BC the Boii (*NH* 3.116) moved south from central Europe, crossing the Padus on rafts. The Insubres were also in the region at the same time (Mediolanum is modern Milan).

The Oromobii, who were in the Alpine foothills, lived around Comum (modern Como) and Bergomum (modern Bergamo). Pliny was continuing to use Cato's *Origines* (F57) as his primary source, but also added the comment by Cornelius Alexander (better known as Alexander Polyhistor, *FGrHist* #273, F104), who was later than Cato. Yet the suggestion that the name Oromobii referred to their mountainous origin (from Greek *oros* and *bios*, "mountain dwellers") is not believable.

3.125. For the Oromobii, see *NH* 3.124. The connection between the otherwise unknown town of Parra and Bergomum was only mentioned by Cato, yet the additional comment (whether by Cato or Pliny) suggests that Parra was north of Bergomum in the foothills of the Alps.

The Caturiges, if the same as those mentioned at *NH* 3.137, came from much farther west (in the Cottian Alps), and presumably were among those Celtic peoples on the move in the early fourth century BC. For the Insubres, see *NH* 3.124, and for the Boii and Senones, *NH* 3.116. The capture of Veii (Cornelius Nepos, F10) was in 396 BC; Melpum is otherwise unknown, but given the context and its destruction by what seems to have been a coalition of three Gallic peoples, it was probably a wealthy city in the Padus valley.

3.126. The tenth region of Italy, Venetia et Histria (discussed through *NH* 3.131), was the largest, extending from the north bank of the lower Padus and the boundary of the eleventh region above Cremona north into the Alps and east essentially to the affluents of the Danube, as well as including Histria (Istria). The account begins at Altinum (modern Altino) and describes the settlements and topographical features around the head of the Adriatic as far as Aquileia.

3.127. The Carni were mountain people whose territory extended south to the head of the Adriatic (Strabo, *Geography* 4.6.9, 7.5.3). The Romans encountered them in the early second century BC (Livy 43.1.8). Their name survives in the Karnian Alps on the Italian–Austrian border. The Iapudes (or Iapodians) were more to the east, in modern Slovenia and northern Croatia. Both peoples were the object of a campaign by Octavian in 35 BC (Strabo, *Geography* 4.6.10; Appian, *Illyrike* 16–21; Dio 49.35). Castellum Pucinum is just northwest of modern Trieste at Prosecco, whose name demonstrates the survival of its reputation for viniculture. Livia believed that her longevity was due to her taste for the local wine (*NH* 14.60–1).

The Formio River (modern Rižana in Slovenia) lies just south of Trieste (six miles or nine km.). It became the boundary of Italy in 42 BC when Gallia Comata was annexed (Dio 46.55), and also marked the northern limit of Histria, the triangular peninsula, now mostly in Croatia, that protrudes into the northern Adriatic (*NH* 3.129). From Aquileia to Tergeste (Trieste) was 33 miles (49 km.) and from Ravenna to the Formio 189 miles (280 km.); both these distances are accurate.

The matter of a connection between the Danube basin and the Adriatic was a persistent problem. Comprehension of the issues was complicated by the onomastic resemblance between Histria, the territory, and the Greek name for the Danube, the Istros (or Histros, latinized as Hister). Whether or not the names were related, the fact that the Danube and Adriatic drainages are only a few kilometers apart north and east of Trieste, as well as the karst topography of the region with its many sinkholes, made it easy to believe that the two watersheds were connected (Strabo, *Geography* 5.1.8). Moreover, Pliny seemed to know of a Hister River that was an affluent of the Danube, but recognized that there was confusion by noting that the Danube was also the Hister. The tributary Hister is probably nothing more than a misapplication of the Greek river name first cited, as Istros, by Hesiod (*Theogony* 339). Greeks originally encountered the river at or near its mouth, and hundreds of years later Romans, perhaps Julius Caesar, approached the upper river near its source in what is now southern Germany, and applied a local name, Danubius, or Danuvius (Caesar, *Gallic War* 6.25; Diodorus 5.25.4). As is common with long rivers, it had more than one name, but by the time of Strabo it was realized that the Danubius and Istros were the same stream (Strabo, *Geography* 7.3.13).

3.128. Continuing his discussion of the Danube and the Adriatic, Pliny rejected any physical connection between the two, rightly pointing out that the issue had become confused because of the presumed local role of the Argonauts. By Hellenistic times it was believed that they had returned from Colchis by going up the Istros/Danube and crossing over to the Adriatic, a theory based on geographical knowledge not available previously. Nevertheless there was early information about the trade connections between the rivers of central Europe and the Adriatic, something that Herodotus seems to have vaguely comprehended (Herodotus 4.48). In fact, Pliny outlined such a route, from the Hister (actually the Danube, but to Pliny it was a tributary) to the Savus (modern Savo), which joins the Danube just below modern Belgrade, and then up the Savus to its tributary, the Nauportus (modern Ljubljana). From the upper Nauportus it was a portage of about 30 km. to the Frigidus (modern Vipava), which is in the

Adriatic drainage. Yet ancient understanding of the river systems in this region was always confused, and can defy modern explanation (Karol Mikula *et al.*, "What Was the River Ister in the Time of Strabo? A Mathematical Approach," *Tatra Mountains Mathematical Publications* 78 [2021] 1–47).

The idea that the Argonauts essentially followed ancient trade routes (although not specified as such) was given more credence with the Hellenistic idea that sea levels were not static, and that at one time the Black Sea (and thus the Ister/Danube drainage) and the Adriatic were contiguous (Eratosthenes, *Geography* F16; Strabo, *Geography* 1.3.15). Thus it only remained for Apollonios of Rhodes, in the third century BC, to popularize in his *Argonautika* (4.289, 325–6) that the Argonauts reached the head of the Adriatic from the Istros.

The Head of the Adriatic

3.129. Histria (modern Istria) is now divided between Slovenia and Croatia, but has often been Italian territory and retains an Italian cultural outlook. The dimensions reported (40 by 125 miles, or 59 by 185 km.) are accurate. Since Liburnia is the coastal region southeast of Histria, it is difficult to apply similar dimensions to that region, and the greatly varying numbers probably indicate uncertainty about the boundaries of both Liburnia and Iapudia (Jehan Desanges, "Pline l'Ancien et l'Istrie: Anomalies et hypothèses," *CRAI* [2004], 1181–1203). Generally Liburnia is the coastal strip and Iapudia the interior. The Flanatic Bay is the large bay just east of Histria, modern Kvarner or Carnaro Gulf.

C. Sempronius Tuditanus, the consul of 129 BC, led an expedition against the Iapudians (not the Histrians), for which he celebrated a triumph, despite mixed results (Livy, *Summary* 59; Appian, *Illyrike* 10). He seems to have had a penchant for erecting public inscriptions commemorating his activities (*ILS* 8885). The Titius River (modern Krka) is presumably his farthest point of advance, 2,000 stadia (perhaps 400 km.) from Aquileia, his base. A point of interest is his use of stadia, indicating that he relied on local Greek sources for his distances.

Pola (modern Pula) was another supposed remnant of the Argonauts' passage, allegedly founded by the pursuing Colchians (Callimachus, *Aitia* F11). The town is ancient, and the tale may reflect this. It is situated in a prominent location at the southern end of Istria at the end of a sailing route to Ancona, and has notable remains (M. Zaninović, "Pola," *PECS*

720–1). Both the distance to Tergeste (modern Trieste) and that to Ancona are short.

3.130. In the interior of the tenth region were a number of *coloniae* but most of the towns listed were indigenous settlements, with varying degrees of a Roman presence. The Cenomani were one of the Celtic peoples who moved into Italy in the early fourth century BC, and settled between Brixia (modern Brescia) and Cremona. Their first contact with the Romans was in the third century BC (Livy 21.55). The Veneti, living at the head of the Adriatic in the modern Veneto, were believed to be the Homeric Enetians (*Iliad* 2.852), who, under their leader Antenor, had come to this region after the Trojan War. As was often the case, this was a tale with no early history; Cato (*Origines* F58) may have been the first to suggest the connection between the Enetians and the Venetians, and the story became common in the first century BC (Livy 1.1.3; Vergil, *Aeneid* 1.242–53; Strabo, *Geography* 1.3.2, 5.1.4). It is probably nothing more than similar-sounding names and the desire to find Homeric remnants in northern Italy.

Mantua was said to be an Etruscan town, a tale popularized by Vergil, who was a local (*Aeneid* 10.198–203). The Raetians were the peoples of eastern Switzerland, where the name survives, and were believed to have been Etruscans who had been driven north when the Celts arrived in their homeland (*NH* 3.133). For the Euganians, see *NH* 3.133; for the Carnians, *NH* 3.127.

3.131. As usual, perhaps following Cato's *Origines*, Pliny listed towns that had disappeared. For the Veneti, see *NH* 3.126; for the Carni, *NH* 3.127. According to L. Calpurnius Piso (*FRH* #9, F37), a certain M. Claudius Marcellus destroyed a town 12 miles (18 km.) from Aquileia. This was probably the consul of 183 BC, who was at odds with the Senate because he wanted to make a war against the Histrians (Livy 39.55). The account is somewhat confused because Aquileia was not founded until the following year; the Senate was debating whether to establish it when Claudius Marcellus sought permission for his campaign. He may have advanced 12 miles toward Histria and made a show of destroying a single town before being recalled. The lakes enumerated are those in the southern Alps, extending from the Benacus (Lago di Garda) to the Verbannus (Lago Maggiore). The passage provides a transition to the next topic, the Alps (through 3.138).

The Alps

3.132. Caelius Antipater, who lived in the second century BC, wrote about the Second Punic War; hence his interest in the Alps (*FRH* #15, F51). By

"Superum" and "Inferum" (Upper and Lower) Seas, Pliny was referring to the Adriatic and Mediterranean, as seen from an Italian perspective. The distance of 1,000 miles (1,480 km.) is reasonable – even though the terminal points are not specified – but the width is problematic. Timagenes' slight difference may be a matter of conversion (*FGrHist* #88, F7). The 100 miles (148 km.) of Cornelius Nepos would only be partway through the mountains, and the 3,000 stadia (about 600 km.) reported by Livy is excessive. In the first century BC, when these two wrote, little was known about the interior of the mountains, or their extent. Moreover, it is often difficult to determine what parts of the complex Alpine system are meant; some of Pliny's figures may refer to the Ligurian Alps. The 745 miles (1,103 km) is probably the road distance south of the mountains, from the Varus River (*NH* 3.31), the western boundary of Italy, to the Arsia River in Histria, the eastern limit, yet this is a much larger figure than that reported at *NH* 3.44.

3.133. Pliny's catalogue of Alpine peoples begins at the east, north of Histria and Tergeste. Many of them are only vaguely known. The Fecusses may have been the indigenous people of Histria. The Subocrini have a hybrid Latin/indigenous name, demonstrating that their homeland was below Mt. Ocra, the modern Nanos plateau north of Trieste (Strabo, *Geography* 4.6.10). The Catali were in the same region (Pliny, *Histoire naturelle* 3, 256). The Menocaleni were perhaps in northern Histria. For the Carni, see *NH* 3.127.

Taurisci seems a broad ethnic term, used primarily for peoples in the Alps south of the Danube – a name that survives today in the Tauern Mountains of Austria (Strabo, *Geography* 4.6.9) – but who lived all along the Alps (*NH* 3.134). In the second century BC, a local chieftain, Cincibilus, became friends with the Romans and promoted the major trade center now known as the Magdalensburg, northeast of Klagenfurt in Austria (Livy 43.5). This is a site with prominent remains, located on the route between the Adriatic and Baltic, which received goods from a wide area of Europe. For the Raeti, see *NH* 3.130. The Vindelici lived north of the Alps, and were also a large ethnic group. Their main center was eventually called Augusta Vindelicum, at modern Augsburg. In 15 BC the brothers Tiberius and Drusus led the first Roman expedition to cross the Alps, reaching the upper Rhine and the source of the Danube (Horace, *Odes* 4.4.14; Suetonius, *Augustus* 21). The Euganians were in the plains of the Veneto, and had been there since early times (Livy 1.1.3).

3.134. The Trumplini (or Trumpilini) were north of modern Brescia, where the toponym Val Trompia survives, and the Camunni were to their north around what is now the Val Camonica (Strabo, *Geography* 4.6.8; Dio 54.20). The Lepontii were in the modern Val Leventina, in the Ticino canton of Switzerland, although Julius Caesar placed them at the source of the Rhine (Caesar, *Gallic War* 4.10), which is unlikely and probably a mixing of information. The Salassi lived in the modern Val d'Aosta, in extreme northwest Italy and along the access to the St. Bernard passes. They controlled a gold-mining industry and were notorious for brigandage, exorbitant tolls, and misuse of water rights, a problem that was only solved in 25 BC by a Roman expedition (Strabo, *Geography* 4.6.7). For the widespread Tauriscians, see *NH* 3.133.

The story of Hercules (Herakles) in the Alps evolved from his tenth labor, the cattle of Geryon, and how the hero returned from the far west to the Greek world. The earliest account has no mention of the mountains (Hesiod, *Theogony* 287–94), which were probably unknown to the poet. The extant fragments of Stesichorus' *Geryoneis* also show no knowledge of a return through the mountains. An Alpine journey seems to have become a popular element by the third century BC, when Hannibal's crossing of them became compared to that of the hero (Livy 21.41.7; Silius Italicus 4.1–5, 11.134–8). Livy was dubious about the story, and if it has any early origin, it probably referred merely to the Ligurian Alps, near the coast. Nevertheless the hero's journey probably became enhanced in order to make that of Hannibal more impressive.

There is no reason to believe in the Greek linguistic origins of the various ethnyms, a point of view perhaps from Cato's *Origines* (F59–60), although the stories demonstrate that Greek mythology penetrated far beyond the regions of Greek settlement. The Stoeni were perhaps the Stonians of Strabo (*Geography* 4.6.6), whose name may survive in the village of Stenico west of Trento.

3.135. The Vennonenses and the Sarunetes lived near the source of the Rhenus (Rhine), which probably means in the valley of the Vorderrhein, west of Chur. The sources of the Rhine and the Rhône (Rhodanus) are only about 20 km. apart in the heights of the Alps; there was some perception of this in the first century AD, but probably no detailed understanding of Alpine topography. A group of Lepontii (*NH* 3.134) also lived in the region.

The Octodurenses and Ceutrones were in less remote regions to the west, on the upper Rhone (Octodurus is at modern Martigny, where the Rhone makes its sharp turn to the north). These regions would have been reached from Gallia Narbonensis by travelling up the Rhone, and thus the locals who were more accessible had been given Latin rights. The Ceutrones were known to Caesar (*Gallic War* 1.10).

The Cottian communities were along the modern French–Italian border in the region still called the Cottian Alps, southeast of Briançon. M. Julius Cottius was the local ruler during the Augustan period, and after initial opposition became an allied monarch of Rome, the only such person in western Europe. His capital was at Segusium (modern Susa), where he erected an arch to Augustus in 9/8 BC, which is still standing and one of the earliest extant examples of the genre (Ward-Perkins, *Roman Imperial Architecture* 171–2). Its inscription reveals the extent of his kingdom and various subject peoples. Other remains are visible in the town (S. Finocchi, "Segusium" *PECS* 819–20). Cottius also built and maintained the road over the Montgenèvre pass, a project that the Romans were happy to contract out to a local (*PIR* I188; Strabo, *Geography* 4.1.3), and which is still partially in use. He was buried in Segusium and remembered with respect over 300 years later (Ammianus Marcellinus 15.10.2–7). His son, who had the same name, succeeded him. Claudius enlarged his territory, but when the king died in AD 63, Nero provincialized it (Suetonius, *Nero* 18; Dio 60.24.4). For the Turri, Bagienni, and Capillati, see *NH* 3.47; they were between the Cottian Alps and the Mediterranean.

3.136–7. One of the most important archaeological and demographic monuments of the Maritime Alps is the Tropaeum Alpium, located on the ancient Via Julia Augusta at the summit of a ridge that marked the boundary between Italy and Gallia Narbonensis (Bromwich, *Roman Remains* 270–5). Its remains are in the village of La Turbie (a survival of the ancient name) just northwest of Monaco. There is a rectangular podium, surmounted by another, and then a cylindrical colonnade, constructed of local limestone with details in Luna (Carrara) marble. There are various military decorations sculpted on the monument, and an inscription on the west side that can be restored by Pliny's text. The structure rose to about 50 m. in height. Its style is derived from the late Republican monumental mausolea still visible outside of Rome on the Via Appia (C. Goudineau, "Tropaeum Alpium," *PECS* 936–7; Ward-Perkins, *Roman Imperial Architecture* 171).

The inscription gives a date of construction – or, at least, the commissioning – of 7–6 BC and lists 45 Alpine peoples who had come under Roman control due to the efforts of Augustus. Many of these had already been mentioned by Pliny: for a complete list, see Pliny, *Histoire naturelle* 3, 258–64. The ethnic groups cover a wide geographical area, from the Ambisontes in modern western Austria to the Seduni and Uberi on the upper Rhone, and the Brigiani in Alpine France. The monument is a physical reminder of Augustus' statement regarding the pacification of the Alps from the Adriatic to the Tuscan (Mediterranean) Seas (Augustus, *Res gestae* 26; Suetonius, *Augustus* 21).

3.138. The Cottian ethnic groups were under the control of King Cottius, and thus it was not appropriate to list them on the Tropaeum, although some appear on the arch at Segusium. Cn. Pompeius Strabo (the consul of 89 BC) passed legislation that gave Latin rights to the towns north of the Padus and perhaps in some of the Alpine regions, thus beginning the process of romanization (E. Gabba, "Rome and Italy: the Social War," *CAH* 9 [second edition 1994] 126–7). Pliny's catalogue organized the communities and peoples of the Alps by legal status, as he had done more explicitly in Hispania: the categories are conquered groups (on the Tropaeum), those controlled by Cottius (on his arch), and those with Latin privileges.

The consulship of L. Aemilius Papus (not Paulus) and C. Atilius Regulus was in 225 BC. At that time various Celtic peoples were moving into the Padus valley, and continued farther south. They were eventually and decisively defeated at Telamon (modern Talamonaccio) on the Etrurian coast (Polybius 2.23–31). The number of troops in the battle seems to originate from the contemporary account of Fabius Pictor (*FRH* #1, F10), but like all such enumerations it is disputable (Walbank, *Historical Commentary* 1, 196–203). Pliny's point is political: that even before the incorporation of the Transpadane populations the Romans were able to mobilize massive forces in order to repel invaders.

In early times there was legislation restricting the exploitation of the mineral resources of Italy (*NH* 33.78, 37.202). This included limiting the amount of manpower that could be used in mining operations. The Salassians (*NH* 3.134) were notorious for misuse of their mining industry – in this case gold – and were expelled from their homeland (the modern Val d'Aosta) when the Romans arrived (Strabo, *Geography* 4.6.7).

Illyricum and Liburnia

3.139. The remainder of Book 3 is an account of the eastern shore of the Adriatic, from the boundary of Italy east of Histria to the northern edge of Macedonia. There is also mention of some of the interior populations of the west side of the Balkan Peninsula and a few of the Adriatic islands. The region was generally called Illyricum, an indigenous term that was known to the Greek world (as Illyria) from the fifth century BC (Hecataeus of Miletus [*FGrHist* #1] F93–101; Herodotus 1.196, 4.49). The Romans began to move into the area in the late third century BC (R. M. Errington, "Rome and Greece to 205 BC," *CAH* 8 [second edition 1989] 85–94). After various campaigns Illyria was provincialized in the first century BC.

There were three major ethnicities in the region, and a vast number of smaller groups. The Iapudes were the more northern, occupying the mainland from Histria to the Telavium River (modern Zrmanja). On the islands and part of the mainland south of the Iapudes were the Liburnians (*NH* 3.38, 45), extending to the Titius River (*NH* 3.129). South of these were the Delmatians, as far as Macedonia (see *NH 3*.141).

The Mentores and Himani cannot be located with precision. The Encheleae, or Eel People, were known to Greeks as early as the fifth century BC and were associated with Cadmus and Harmonia, the founders of Thebes in Boeotia, who ended up in this region (Strabo, *Geography* 7.7.8). Their connection to Greek myth implies a southern coastal location, probably in northern modern Albania. The Bulini (perhaps the Byllonians of Strabo) were probably around Byllis (modern Hekal in Albania). For the Peucetii, who spread across the Adriatic into Italy, see *NH* 3.99; Callimachus F107. The Scardonians were at Scardona (*NH* 3.141), modern Skradin. It was an administrative center, with several towns and islands under its jurisdiction.

3.140. This section is from a different source than the previous one and is a *periplous*, with the Iapudian towns listed in geographical order from north to south, from Nesactium (modern Visače), just north of Pola/Pula to Aenona (modern Nina). The Pasini were probably in the same region but cannot be located. Aenona is actually in Liburnia, and the *periplous* continues to the town of Iadar (modern Zadar). It probably represents a shipping route from Histria to that point (160 miles or 237 km.), perhaps even continuing 30 miles (44 km.) farther to Colentum (the modern island of Murter) and another 43 miles (64 km.) to the Titius River. Given the orientation of the coast and the vagaries of shipping lanes it is impossible to determine the accuracy of these figures but they seem reasonable.

Dalmatia

3.141. Scardona was on the Titius River and marked the dividing line between Liburnia and Delmatia (more commonly Dalmatia: Strabo,

Geography 7.5.3). Greeks had reached the southern portion of Dalmatia as early as the late seventh century BC (Boardman, *Greeks Overseas* 225–9). The Romans arrived in the third century BC, primarily interested in restraining the local piracy (Polybius 2.10–11), and over the next two centuries increased their involvement, although the region was not pacified until the early first century AD (Gruen, "Expansion" 177–8).

The Tariotae are not otherwise known but presumably were in the northern part of Dalmatia, since the account continues the *periplous* from Scardona to Salona, and the other toponyms can easily be identified along this coastal stretch of 112 miles (166 km.).

3.142. Salona (or Salon), inland from modern Split, was an important regional center and is a major archaeological site (M. Zaninović, "Salona," *PECS* 799). The *decuriae* were the smallest administrative unit of a Roman town, or, as in this case, a provincial jurisdiction. The Deuri cannot be located. The Ditiones lived in the uplands north of Scardona, and the Maezei farther north (Strabo, *Geography* 7.5.3). The Sardeates were between the two previous groups. All these people would have come down to Salona if they had business with the Roman government. The jurisdiction of Salona also included the population of four islands: Issa, Solentia, and Epetium are immediately south of Salona, and Separi would be in the same vicinity but is otherwise unknown.

Narona, the seat of the third jurisdiction, is at modern Vid, located in an unusually large floodplain on the Naro River (modern Neretva). It was probably the northernmost Greek settlement on this coast, established by the middle of the fourth century BC (Pseudo-Skylax 24). Although it was 20 miles (30 km.) inland, ocean-going vessels would come up the river to the trading post.

3.143. Pliny, writing a century later than Varro, noted the decline in rural population in the jurisdiction of Narona. The Cerauni were probably on the upper Naro. The Daversi (perhaps the Daorizians of Strabo, *Geography* 7.5.5) were downstream from Narona. The Desitiates were to the northeast, remembered for the two brothers named Baton, who put together a coalition and instigated the local revolt against Rome in AD 6 (Strabo, *Geography* 7.5.3; Velleius 2.110–16; Suetonius, *Tiberius* 20; Dio 55.29–34). The name of the Docleatae survives at Duklja in Montenegro. The Dindari, Glinditiones, Melcumani, Naresi, and Scirtari are little known but presumably were on the upper Naro or in its hinterland.

The Vardaeans, or Ardaeans, lived on the coast southeast of Narona, and were infamous for their piracy. The consul Ser. Fulvius Flaccus defeated them in 135 BC and forced them to move inland and become farmers, something almost impossible because of the local karst topography. By the late first century BC they had essentially died out (Broughton 1, 488–9; Strabo, *Geography* 7.5.6).

The Ozuaei are hardly known. The Parthini, living in what is now Albania, were early allies of Rome, but became involved in the Roman civil war and were not subdued until 39 BC (Polybius 2.11.11; Caesar, *Civil War* 3.41; Appian, *Civil War* 5.75). The Caui, Haemasi, Masthitae, and Arnistae were little-known populations southeast of Narona.

3.144. The account continues along the coast from the Naro River, cataloguing the towns in *periplous* style, beginning with Epidaurum (modern Cavat). It is only about 60 km. from the Naro, but the distance of 100 miles (148 km.) is the inter-island sailing route. The towns are listed in geographical order from north to south.

There is no reason to see a connection between Colchinium (modern Ulcinj) and the Colchians other than the similarity of names and the prevalent stories about Argonautic activity in the Adriatic. In fact, the name may have been altered from an indigenous (or Greek) form, Oulkinion (Ptolemy, *Geographical Guide* 2.16.5), in order to create the correspondence.

A list of seven ethnic groups that no longer existed follows, perhaps having been forced to change their lifestyles in ways similar to the Vardaei/Ardaeans (*NH* 3.143). The Labeateae were at Scodra (modern Shkodra), which they had fortified by the early second century BC (Livy 44.31). The Senedi, Rudini, Sasaei, and Grabaei are otherwise unknown. The Taulanti were around Epidamnum (Dyrrachium, modern Durrës) and were known to Greeks as early as the fifth century BC, managing to remain free of foreign control until conquered by Octavian in 35 BC (Thucydides 1.24; Strabo, *Geography* 7.7.8; Appian, *Illyrike* 16). The Pyraei, if they are the Pleraians of Strabo (*Geography* 7.5.5), seem to have been near the Naro River. Lissum (modern Lezha in northern Albania) was the beginning of Macedonia.

3.145. A full discussion of Macedonia was reserved for *NH* 4.33–9. At this point Pliny was only interested in its short coastal segment along the Adriatic, from Lissum to the Acroceraunian Mountains (*NH* 3.97). For the Parthini, see *NH* 3.143. The Dassaretians were in the interior around modern Lake Ohrid (Polybius 5.108.2; Livy 43.9.7; Strabo, *Geography* 7.5.12). Mt. Candavia is south of the lake, in a region notoriously difficult of access, crossed by Pompey the Great on his way to Pharsalus (Lucan 6.331; Seneca, *Letters* 31.9).

Denda, Epidamnum, Apollonia, and Oricum were the major towns of the Adriatic coast of Macedonia, listed north to south. Epidamnum

(Epidamnus or Epidamnos, modern Durrës) is the original name of the Greek settlement of the seventh century BC (Boardman, *Greeks Overseas* 226); the Roman town was a short distance away and was called Dyrrachium – probably a local toponym – and this gradually became used for the entire urban area (Strabo, *Geography* 7.5.8; Pausanias 6.10.8). "Epidamnum" allegedly sounded inauspicious in Latin ("damno" means "to condemn").

The Nymphaeum at Apollonia was an area of bituminous springs (perhaps located around modern Selenizza) and was famous as a phenomenon; the region was also notable for the local mineral resources (Posidonius F235; Strabo, *Geography* 7.5.8). In addition, there was an oracle, where the properties of the fires could provide information, and which was still functioning in the third century AD (Dio 41.45).

The Amantes were at Amantis, south of Apollonia, and the Buliones were probably from Bullis (or Byllis), a short distance away across the Aous River. The pervasiveness of Argonautic connections is shown by the presumed Colchian foundation of Oricum (Greek Orikon, at modern Orikumi) on the Ceraunian Promontory. It was one of the earliest Greek settlements on this coast, already established by the sixth century BC (Hecataeus of Miletus [*FGrHist* #1] F106; Herodotus 9.93; Strabo, *Geography* 7.5.8), which may have assisted in the belief that it was a Colchian foundation, or one made by people returning from Troy (Lucan 3.187). The distance from the Salentine Promontory (Iapygian Cape) in Italy, 80 miles (118 km.), is not a direct route, but reflects coastal sailing south from Oricum before making the crossing at the narrowest point. Epiros (discussed in detail at *NH* 4.1–4), the first district of Greece proper, began just south of Oricum in the Ceraunian Mountains.

Northeast of the Adriatic

3.146. The next topic (through *NH* 3.149) is the interior of Illyricum, as far as the Hister (Danube). The Carni (*NH* 3.38, 127–33) were a wide-ranging population located all across south-central Europe. For the Iapudes, see *NH* 3.139. Mention of them along with the Raeti and Boii – other widespread peoples (*NH* 3.124–5, 130–5) – demonstrates that Pliny's discussion covers much of modern Austria, at the east end of the high Alps.

3.147. Pannonia is the region south of the Danube as far as the divide between the Black Sea and the Adriatic drainages, consisting of much of modern Austria, interior Slovenia and Croatia, and western Hungary. It was largely beyond any areas known to the Greeks, but the Romans penetrated into the region in the late second century BC, and eventually conquered the indigenous Pannonians, especially through an expedition led by Tiberius in 9 BC (Augustus, *Res gestae* 30; Strabo, *Geography* 7.5.2–3; Velleius 2.96). In the following century it became a Roman province. As Pliny noted, Pannonia was at the eastern end of the Alps, with no high mountains except in its western portions. In fact it was a region of great rivers – not only the Danube but the Draus (modern Drava) and Saus (modern Sava) – with extensive plains in its northern portions. The Draus and Saus run roughly parallel – the 120 miles (178 km.) between them is excessive – from sources in the eastern Alps into the Danube.

The Serretes and Serapilli remain vague and are hardly known. The Iasi were along the Draus, with the Andizetes probably on the lower river near its junction with the Danube, in northern Croatia (Strabo, *Geography* 7.5.3). The Colapiani would have been on the Colapis River (the modern Kupa) in western Croatia south of modern Zagreb, and the Breuci on the lower Saus northwest of modern Beograd.

3.148. An alphabetical list of the lesser populations of Pannonia may come from the report of Tiberius' campaign. None of these ethnic groups is well known; the Azali may have lived west of modern Budapest, and the Cornacates would have been the inhabitants of Coracum (modern Sutin on the Danube in Croatia). The Eravisci, or Araviscians (Tacitus, *Germania* 28), were south of the Azali, along the Danube, and the Hercuniates were to their south. The Latovici, Oseriates, and Varciani were on the upper Saus.

Mt. Claudius was a promontory or ridge that figured in the local campaigns of Tiberius and was given his family name (Velleius 2.112.3). It has not been located. Its position between the Scordisci and Taurisci is of little help, since these are wide-ranging ethnic groups: the former were along the Danube above modern Budapest, and the latter (*NH* 3.131) were farther to the west along the Draus, both too vaguely located to assist in finding the mountain. It is described from the perspective of someone approaching from the south, as the Roman army might have done. It is difficult to imagine that the Saus contains the largest island known to be in a river (those in the deltas of major streams are certainly larger), but Metubarbis may be the extensive marshy regions upstream from modern Sremska Mitrovica in Croatia (which, as ancient Sirmium, Pliny cited in the same sentence).

3.149. Moesia was the territory on the south side of the Danube, extending from the eastern limit of Pannonia (at Singidunum, modern Belgrade, where the Saus joins the Danube) to the Black Sea. Except for the

Black Sea coastal areas, it was hardly known to the Greek world, and the Romans did not enter the region until the early first century BC, conquering it in 29–28 BC and eventually provincializing the territory (Dio 51.23). It was a developing region in Pliny's day, and his list of ethnyms is somewhat anachronistic (J. J. Wilkes, "The Danubian and Balkan Provinces," *CAH* 10 [second edition 1996] 579–80). Significantly, he mentioned no settlements, only the local ethnic groups and the rivers.

The Dardani were extinct by the early first century AD (Strabo, *Geography* 7.4.6), or, more probably, broken up into local ethnic groups. The Celegeri are otherwise unknown. The Triballi had had contacts with the Greek world since the fifth century BC (Thucydides 2.96) and had caught the interest of Alexander the Great, who set forth to subdue them in 335 BC, making a seminal overland journey to the Danube (Strabo, *Geography* 7.3.8; Arrian, *Anabasis* 1.1–7). Almost constantly at war with their neighbors, and occasionally Greek and Macedonian invaders, they were eliminated by the second century BC (Appian, *Illyrike* 2–3).

The Timachi are not otherwise known, and the Moesi gave their name to the territory and were connected ethnically to the Thraces, or Thracians (Strabo, *Geography* 7.3.2), which itself was a general term for the populations between the head of the Aegean and the Danube, just as the Scythae (Scythians) were the peoples north of the river (Strabo, *Geography* 1.2.28).

3.150. As was often his custom, Pliny ended a regional survey with various distances. He used two sources, one unnamed, and the other Marcus Agrippa (F16, 47). The width of Illyricum (325 miles or 481 km.) has no locations specified, but would allow one to reach various points on the Danube from Aquileia, and may be a military report. The length of 530 miles (784 km.) from the Arsia to the Drinium (modern Drina, an affluent of the Sava above Belgrade and which is the modern Bosnian–Croatian border) is excessive, but this is almost the distance from the Arsia to the Black Sea, and may be an attempt to calculate the width of the Balkan Peninsula.

From the Drinium to the Acroceraunian Promontory (175 miles or 260 km.) is also too long, but presumably reflects the twists and turns of routes in rugged country. These are all attempts to provide distances across little-known territory, and might be expected to be erroneous. The distance for the Italian and Illyrian bays – terms not used previously but necessarily meaning the Adriatic – of 1,700 miles (2,516 km.) is again far too large (the extent of the Adriatic from Brindisi to Trieste is about 600 km.), but if Pliny's distance reflects the actual coastline, as followed by shipping, it might be reasonable. For the distinction between the Adriatic and Ionian Seas (north and south of the Hydruntum–Epidamnus narrows), see *NH* 3.100.

Islands in the Ionian and Adriatic Seas

3.151–2. Another common feature at the end of a geographical section of the *Natural History* is a catalogue of the local islands. Those in the Ionian Sea (south of the Adriatic narrows) are all on the Greek side (*NH* 4.52–5). Diomedia (modern Isole Tremiti, north of Gargano; there are actually several islands) was the most conspicuous token of the hero in Italy (*NH* 10.127, 12.6). Teutria may be part of the same group and the origin of the modern name.

The Illyrian coast of the Adriatic, or Hadriatic, is noted for its vast number of islands and narrow shallow passages between them. Pliny's account is north to south. For the springs at the mouth of the Timavus – no longer extant – see *NH* 2.229. The Absyrtides (or Apsyrtides, modern Cresa and Lošinj) are the northernmost major islands off the eastern shore of the Adriatic, and were another alleged vestige of the tale of the Argonauts. Apsyrtus, the brother of Medea, was said to have been killed there (Apollonios 4.480–1). As usual, the name of the islands was probably the assimilation of a local toponym into Greek myth.

The Electrides (Amber Islands) were a mythic locale connected with the story of Phaethon and his sisters, whose tears were turned into amber after his death (*NH* 3.117). As early as the fifth century BC it was realized that the story had topographical difficulties (Herodotus 3.115): the islands were eventually located at the mouth of the Padus, but as knowledge of the upper Adriatic increased it was realized that there were no islands in this region (Strabo, *Geography* 5.1.9). Pliny used this account to criticize the Greeks, but a further discussion at *NH* 37.32 outlined the topographical issues. Yet the story is not to be totally dismissed, because it represents early yet inaccurate information about the source of amber, which was in the Baltic. It came to the northern Adriatic from there, by means of a route perfected by the Middle Bronze Age if not earlier (Barry Cunliffe, *The Extraordinary Voyage of Pytheas the Greek* [New York 2001] 136–52). Thus it becomes an example of the familiar failure to comprehend trade routes in their entirety but only as far as the last point known to those on the receiving end of the trade. And, perhaps more significantly, there were amber-working industries in the northern Adriatic in the Late Bronze Age, most notably at the site of Frattesina in the Po delta, which could have given rise to the belief that this was the source of the material (Carolynn E. Roncaglia, *Northern Italy in the Roman World* [Baltimore 2018] 9–10). Nevertheless most amber objects found in northern Italy, generally from the Bronze Age, are Baltic in origin (I. Angelini and P. Bellintani, "Archaeological Ambers From Northern Italy," *Archaeometry* 47 [2005] 441–54).

Other islands along the east coast of the Adriatic are listed, including several already mentioned. Some cannot be precisely identified, but Issa is modern Vis and Pharia modern Hvar. Corcyra Melaena is not the famous Greek island of Corfu, but Black Corcyra (Kerkyra Melaina in Greek – Pliny used a latinized Greek adjective – at modern Korčula), so named because of its unusual dark forests (Apollonios 4.569–71). Melite (modern Mljet) was often confused with Melita (modern Malta) south of Sicily, and there was uncertainty as to which locality produced the distinctive dogs (Callimachus F579; Strabo, Geography 6.2.11). Sasonis, or Sanos (modern Sazani), lies off the coast at the end of the Acroceraunian Promontory, and was a sailing point on the crossing to Italy. The island was not only a pirate refuge but also had dangerous shoals (Lucan 5.650). Book 3 ends abruptly, as does Book 5, with no transition to the following book, unlike Books 4 and 6.

BOOK 4

The Greek Peninsula and Interior Europe

Book 4, the shortest of the geographical books of the *Natural History*, begins at the Acroceraunian Promontory on the west coast of the Balkan Peninsula, at the northern edge of Epiros. It thus connects with *NH* 3.145, where Pliny completed his discussion of the Adriatic coast. The book includes the entire Greek peninsula, described in a counter-clockwise fashion from Epiros through the Peloponnesus and north to Macedonia (4.1–39), Thrace and the Aegean (4.40–51), the Greek islands (4.52–74), and the European side of the Black Sea (4.75–93). Rather than cross into Asia at this point, and continue along the eastern coast of the sea, the narrative heads back west into northern Europe (3.94–101), and then passes through the British Isles into the interior parts of Gallia and Hispania (4.102–120), regions that were not discussed in Book 3. The circular itinerary, unusual in Greco-Roman geographical studies, allowed Pliny to complete Europe before moving to the other continents, yet retain the coastal orientation that is an essential part of the geographical portions of the *Natural History*. But it meant that the central nature of Europe – the location of Rome – was emphasized before moving to the other continents.

Northwest Greece

4.1. Having completed what he called the first gulf or bay of Europe (*NH* 3.6–96) – basically the western Mediterranean between the Iberian Peninsula and Italy – and the second (*NH* 3.97–145), the Adriatic, Pliny began his discussion of the third gulf, from Acroceraunia to the Hellespont. To modern observers this pattern may seem geographically anomalous, since the Ionian Sea (west of the Greek peninsula) is thus connected with the Aegean (on its east). In part this is because Pliny was providing a reverse image of Eratosthenes' third promontory of Europe, which was basically the Greek peninsula (Eratosthenes, *Geography* F135 = Strabo, *Geography* 2.4.8).

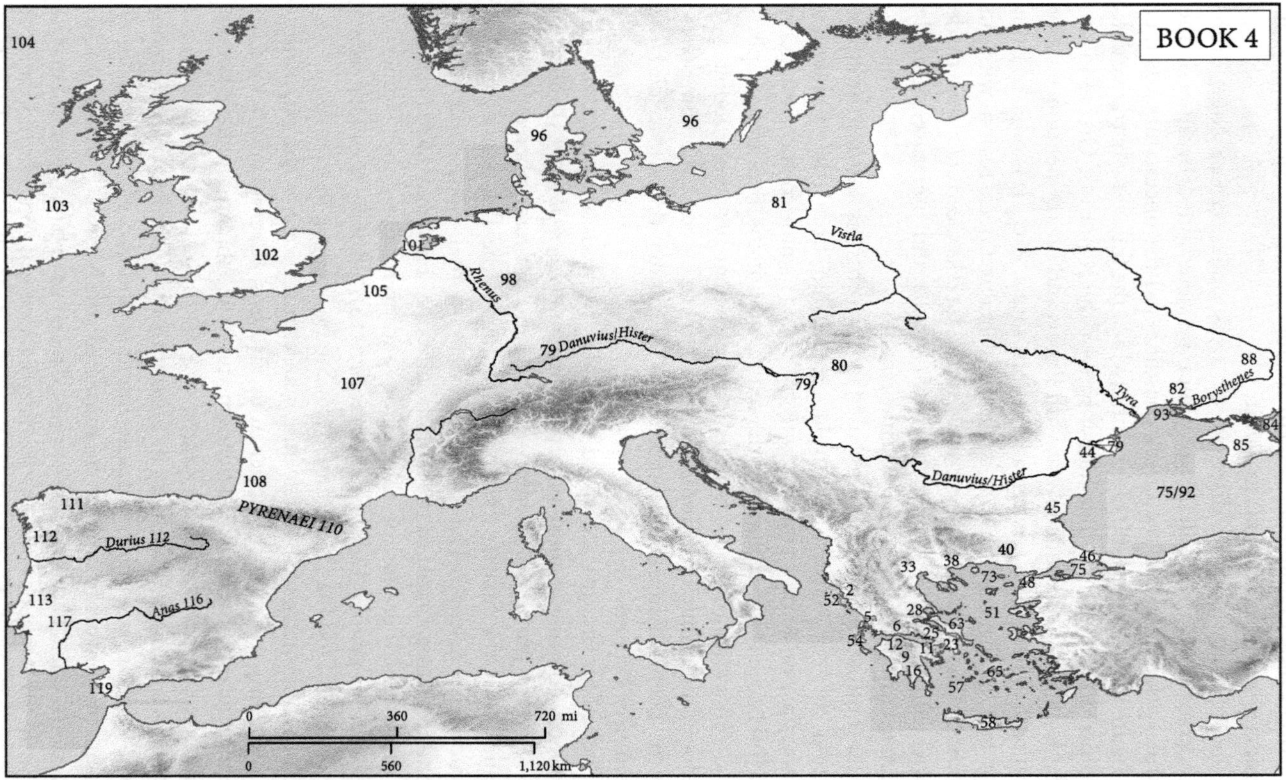

Map 4 Places cited in Book 4 of the *Natural History*, according to chapters (courtesy of Ancient World Mapping Center).

Moreover, with Eratosthenes' system in mind, Pliny probably wanted to consider the Greek peninsula as a unit.

The accuracy of the distance of 1,925 miles (2,849 km.) from Acroceraunia to the Hellespont cannot be determined because of the complex indentation of the Greek coastline and the uncertainty whether the figure includes the Gulf of Corinth (which the following narrative does consider). The two lists that follow present the regions of Greece in the order used in the subsequent text, with Phocis and Locris mentioned twice because they have both western and eastern coasts. Arcadia is omitted from both lists, perhaps because it is the only major Greek region with no seacoast. Pliny, who could be critical of the Greeks (see *NH* 3.152), made a brief eulogistic statement about their myths and literature, a highly selective comment about Greek achievement.

4.2. Epiros (or Epirus; Pliny used both forms) is the northwestern portion of the Greek peninsula, extending from the Ceraunian Mountains, or Acroceraunian Cape (*NH* 3.97) south to the Ambracius Bay (NH 4.4). It is a strip west of the central mountains, the Pindus. Epiros was a diverse region, most famous for the ancient oracle of Dodona (or Dodone), mentioned by Homer (*Iliad* 2.750 etc.), and as the home territory of both Olympias, the mother of Alexander the Great, and the adventurer Pyrrhus.

The ethnic groups are listed generally from north to south. The Chaones, one of the most powerful (Strabo, *Geography* 7.7.5), lived in the southern portion of modern Albania, and are remembered for their alliance with the Spartans at the beginning of the Peloponnesian War (Thucydides 2.80–1). The Thesproti were to their south and slightly inland, reaching almost to the Ambracian Gulf. They may have been one of the original groups of the region and were noted by Homer as a prosperous people (*Odyssey* 14.314–20, 17.526); allegedly they established the oracle at Dodona (Herodotus 2.56; Strabo, *Geography* 7.7.11).

The Antigonenses are not so much an ethnic group as the population around Antigoneia, on the upper Aous, or Aoos, River (modern Vjosë in Albania) near modern Lekli. The town and its inhabitants are little known (Ptolemy, *Geographical Guide* 3.14.7).

Aornos is an oracle of the dead in Thesprotia that Pausanias called Aornon, where Orpheus attempted to rescue Eurydice and also where Periander of Corinth sent envoys to question his murdered wife Melissa (Pausanias 9.30.6; Herodotus 5.92). The effusions that were dangerous to birds suggested that its name meant "without birds"; such an explanation was applied to a locality of the same name in the volcanic region around the Bay of Naples, where disagreeable odors were prevalent (Strabo, *Geography* 5.4.5). The site of Aornos may be that discovered at ancient Ephyra, near the village of Mesopotamos, where there is physical evidence of an oracle of the dead (S. Dakaris, "Ephyra," *PECS* 310–11).

The Cestrini were on the coastal plain along the modern Greek–Albanian border, north of the Thyamis River. The region figured in the early stages of the Peloponnesian War (Thucydides 1.46). By all accounts the Perrhaebi were the people of northern Thessaly (Homer, *Iliad* 2.749; Strabo, *Geography* 9.5.19), and thus seem out of place in Pliny's catalogue. Equally inscrutable is the association of them with the Pindus, the central mountain chain of the northern Greek peninsula, which is well to their west. This suggests that there might have been more than one group of Perrhaebi, a view supported by the indication at *NH* 4.6 that they were in northwestern Greece.

The Cassiopaei lived on the coast north of the Ambracian Gulf. The Dryopes were to their east, north of the gulf, and were one of the primitive populations of Greece. Their name ("Oak People") implies forest dwellers, and peoples with this name were recorded in several places in central and northern Greece, demonstrative of population movements (Strabo, *Geography* 8.6.13, 9.5.10). Herodotus (8.46) placed Dryopes on the Aegean islands, and seems to have used the ethnym generically.

The Selloe and Hellopes were in the interior, originally in the vicinity of Dodona (Strabo, *Geography* 7.7.10); the latter ("Marsh People") were another early population that may have moved around Greece, perhaps originating in the marshy regions northeast of Dodona. The Selloe are more obscure, and were possibly connected to the oracular interpreters of the same name at Dodona (Homer, *Iliad* 16.234–5).

The Molossi were, along with the Chaonians, one of the powerful groups in northwest Greece (Strabo, *Geography* 7.7.5). They controlled the oracle at Dodona and became increasingly prominent. As a distinct ethnic group they were eliminated when the Romans took over northern Greece in 167 BC (Polybius 30.15).

Remains of the Temple of Zeus (Pliny's Iovis) at Dodona are visible, dating from the fourth century BC and later (Y. Béquignon, "Dodona," *PECS* 279–80). Mt. Talarus (also known as Tomaros or Tmaros) rises dramatically just to the southwest of the site; some of the oracular interpreters were called Tomouroi (Strabo, *Geography* 7.7.11). Theopompus' *Philippika*, a biographical history of the world of his contemporary Philip II of Macedonia, would have discussed this region in detail, but the hundred springs are no longer visible (*FGrHist* #115, F319).

4.3. Section 3 covers a broad region, citing ethnic groups and toponyms over much of the northern Greek peninsula and Balkans as far as the Aegean, Black Sea, and Danube. It may be that the material was gathered from the map of Agrippa or a similar source. Magnesia (*NH* 4.32) is the coastal strip along the Aegean side of Thessaly, so it is inaccurate to suggest that it and Epiros are contiguous, although Macedonia (*NH* 4.33) does extend into the central mountains and thus can be said to adjoin Epiros in the highlands of northern Greece. The Dassaretae (*NH* 3.145) probably lived south of modern Lake Ohrid and were relatively obscure, but opposed Macedonian expansion and were early supporters of the Romans, thus retaining a certain level of independence (Polybius 5.108.2; Livy 45.26.14; Strabo, *Geography* 7.5.7, 12). The Dardani inhabited the remote territories to the north and were known for their primitive quality, living in houses said to be made of dung hills. This was perhaps merely a type of roofing that was a way of providing warmth, a constructional device known from colder regions (Strabo, *Geography* 7.5.7; Tacitus, *Germania* 16). The Dardani who participated in the Trojan War were an ethnic group in the northern Troad who were mentioned frequently by Homer, but had no connection to those in the Balkans (Strabo, *Geography* 12.4.6).

The Triballi were vaguely known to the Greek world as early as the fifth century BC and lived somewhere on the lower Danube (Herodotus 4.49; Thucydides 2.96; Strabo, *Geography* 7.3.8). The Moesians were also on the lower Danube, located on both sides of the river (*NH* 3.149; Strabo, *Geography* 7.3.2–3), and were often confused with the Mysians in northwest Asia Minor. The Moesians were said to have practiced an abstemious lifestyle, a common Mediterranean view of many northern peoples (Posidonius F277A). They were subdued by M. Licinius Crassus (the grandson of the triumvir) in 28 BC (Dio 51.23–7). The Maedi lived along the Strymon River, and also had been known to the Greek world since the fifth century BC, coming under Roman rule at the time of Crassus' expedition (Thucydides 2.98; Strabo, *Geography* 7.5.7). The Denseletae (Dantheletae or Dentheletiae) were to their north, also along the Strymon, and attacked the province of Macedonia in 16 BC (Polybius 23.8.4; Strabo, *Geography* 7.5.12; Dio 54.20.3).

The Thracians were specifically the peoples directly north of the Aegean, but, as Pliny noted, the ethnym could be used to describe populations as far as the Danube. He used a latinized Ionic form for them (Threces), based on that used by Homer and Herodotus (Threike), indicative of his source at this point, which cannot be specifically identified (Homer, *Iliad* 9.5 etc.; Herodotus 3.90, etc.). Normally Pliny used the toponym Thracia.

In the final sentence of section 3 he seems again to have been looking at a map, seeing the various populations of the region from northern Greece to the Danube encircling the two great mountains of the district. Rhodope extends across southern modern Bulgaria just north of the Aegean. Haemus (Haimos) is the modern Balkan range and is parallel to and north of Rhodope (Strabo, *Geography* 7.5.1).

4.4. After the general discussion of the demographics of the territory north and east of Epiros, Pliny returned to its coast, beginning at the fortress of Chimera. This may be the Cheimerion mentioned by Thucydides (1.46) that figured in the opening of the Peloponnesian War. But if this is the case, Pliny misplaced it, putting it on the Acroceraunian Mountains. Yet Thucydides clearly located Cheimerion opposite Kerkyra (Corcyra, modern Corfu), which is about 90 km. farther down the coast. Needless to say, Pliny may have meant another site, perhaps around Himara (which seems to preserve the name Chimera) in Albania. Since Buthrotum (Bouthroton, modern Butrint) is definitely located in southern Albania opposite Corfu (Strabo, *Geography* 7.7.5), and is the third town listed after Chimera, it is more probable that Chimera and Cheimerion are not the same.

After Buthrotum, no other toponyms are mentioned for an extent of 100 km. until the Ambracius Bay. This is the large and conspicuous gulf that opens out after a short inlet from the sea (somewhat longer today that Pliny's half a mile), a region best known because the promontory of Actium is on the south side of the entrance (*NH* 4.5). Pliny's length (35 miles or 52 km.) is probably a circumference and, if so, an accurate figure, as is the width of 15 miles or 22 km. The Acheron River does not flow into the gulf, but directly into the sea about 35 km. to the north, passing through the Acherusia Lake and the probable location of the Aornos oracle (*NH* 4.2); it is possible that Pliny inserted the Acheron after having left it out previously. The thousand-foot bridge is otherwise unknown, but it may have been a flood control project (Tusculum 3–4, 336). The Aphas (modern Louros) and Arrathus (or Arachthos) flow into the Ambracius Bay from the north. Anactorica (Greek Anaktorion, near modern Nea Kamarina) is on the south side of the gulf, but Pandosia may be on the Acheron River some distance away (Strabo, *Geography* 7.7.5).

4.5. Acarnania extends from the south side of the Ambracius Bay to the western end of the Gulf of Corinth. The toponym Curetis (Greek Kouretis) seems originally to have been associated with the coastal plain in southwestern Acarnania (Pausanias 10.31.3) rather than the district as a whole; Strabo also knew of a Mt. Kourion in this region. The names were connected to the ethnic group known as the Kouretians, or Curetians, mentioned by Homer (*Iliad* 9.529, 589). They often became tangled with the mythical beings of the same name who were involved with the birth of Zeus on Crete. These issues, both cultic and toponymic, were discussed in detail by Strabo (*Geography* 10.3.1–18).

Actium (Greek Aktion) is at the end of the long peninsula that marks the south side of the entrance to the Ambracius Bay. It was the location of a cult center to Apollo that had probably existed since Corinthian settlers moved into the region in the seventh century BC (Thucydides 1.29.3; Nikolaos of Damascus [*FGrHist* #90] F57). Nevertheless the locality became famous because of the battle in September of 31 BC between the forces of Cleopatra VII and Antonius and those of Octavian. On the north side of the entrance to the gulf Octavian founded his victory city of Nicopolis, a major archaeological site today (Strabo, *Geography* 7.7.6; James Wiseman and Konstantinos Zachos, "The Nikopolis Project: Concept, Aims, and Organization," in *Landscape Archaeology in Southern Epirus, Greece 1* [ed. James Wiseman and Konstantinos Zachos, *Hesperia Supplement* 32, 2003] 1–22).

The Leucadian shore is that opposite the island of Leucadia (Greek Leucas), which lies only a few hundred meters off the coast. Neritis (Homeric Nerikos, *Odyssey* 24.377–8) was on the mainland. Leucadia has been connected to the mainland at various times (Strabo, *Geography* 10.2.8), and there was a canal separating the two, called the Dioryctos ("Cutting Through," Polybius 5.5.12). It kept silting up, and the channel was eventually abandoned at some time after the third century BC. By the late first century BC a bridge had been constructed over the sand, whose remains are still visible.

Siltation in this region has always been extensive, and the topography today is quite different from that in antiquity (*NH* 2.204). That produced by the Achelous River was noted as early as the fifth century BC (Herodotus 2.10), and Artemita was originally one of the Echinades Islands, which were joined to the mainland in the Classical period. For Aetolia see *NH* 4.6.

4.6. Aetolia is the district east of Acarnania, along the north side of the Gulf of Corinth as far as the western edge of Locris (*NH* 4.7). Pliny's seven ethnic groups allegedly in Aetolia are actually spread across a wide area of Central Greece, including peoples to the north and east of the traditional boundaries of the district, which again suggests information gathered from viewing a map.

The Athamanes were mountaineers, the most inland of Aetolian peoples, and were surprisingly prominent for a brief period in the third and early second centuries BC, eventually being assimilated with the Roman takeover (Diodorus 14.82.7; Strabo, *Geography* 9.4.11; Stewart Irvin Oost, "Amynander, Athamania, and Rome," *CP* 52 [1957] 1–15). The Tymphaeans were well north of the traditional boundaries of Aetolia, in the highlands of Mt. Pindus (Strabo considered them Epirote: *Geography* 7.7.8), and the Perrhaebi (*NH* 4.2) were to their east in northern Thessaly. The Ephyri were presumably the inhabitants around a town named Ephyra, but the toponym is common in central and southern Greece: Strabo knew of seven places so named, including one in Aetolia, which has not been located (Strabo, *Geography* 8.3.5).

The Aenienses (Ainianians) were at the northeastern limits of Aetolia, near Mt. Oeta, and the Dolopes were in the mountains of western Thessaly (Strabo, *Geography* 9.5.5, 11; 10.2.1). The Maraces were probably neighbors of the Dolopes, but are hardly known except for their involvement in the brief empire of Jason of Pherai in the early fourth century BC (Xenophon, *Hellenika* 6.1.7). The Atraces were around Atrax. The known locality with this name is in Thessaly (*NH* 4.29; Strabo, *Geography* 9.5.17), although the Atrax River that drains into the Ionian Sea, otherwise unknown, suggests a place west of the Pindus Mountains. Yet the fact that so many of these ethnyms seem to have connections with Thessaly indicates Pliny was probably using a source from the environment of Jason of Pherai, which Pliny attached to his discussion of Aetolia, since it adjoined Jason's territory on the southwest.

After the excursus on Thessaly, the remainder of Section 6, except for the final sentence, is unquestionably about Aetolia, moving roughly from west to east. The account begins at Calydon (near modern Evenochori), the most important Bronze Age site in Aetolia – as demonstrated by the tale of the Calydonian Boar Hunt – and moves along the coast to Eupalimna (the Eupalion of Strabo, *Geography* 9.4.8), at the Locrian border. Pliny also noted the narrows of the Corinthian Bay (Gulf of Corinth) at Antirhium (on the Aetolian side) and Rhium (on the Peloponnesian side), which were not towns but localities marking the point where the gulf is at its narrowest (today 2.3 km. across).

Pleuron – generally called New Pleuron, as it was a Hellenistic foundation replacing an older town on the coast – is northwest of Calydon at modern Kyria Irini. Halicarna (Greek Halikyrna) lies closer to Calydon (Strabo, *Geography* 10.2.21). Both these towns were only a few kilometers from the coast, and while there were inland settlements in Aetolia, especially around Thermon (or Thermos), about 20 km. northeast of Calydon, most of the interior was lightly populated.

The section closes with a catalogue of mountains in northwest Greece. Tomarus, Crania, and Aracynthus were prominent isolated summits. Achaton and Macynium cannot be located. Panaetolium would seem to be a sanctuary or the seat of a regional assembly, perhaps around Thermon, where the Aetolian Federation met and which was the locale of a large festival and market (Polybius 5.8; Pliny, *Histoire naturelle* 4 [ed. Hubert Zehnacker and Alain Silberman, Paris 2015] 115).

4.7. Ozolian Locris was a small territory on the Gulf of Corinth just east of Aetolia; the inhabitants were called Ozolae to distinguish them from other groups of Locrians, including the Epicnemidi (*NH* 4.27), who were not contiguous. The derivation of the name "Ozolae" is uncertain, and it was said improbably to mean "Strong-Smelling," because of an odoriferous spring in the territory, or the personal habits of the locals; in fact there were many explanations (Strabo, *Geography* 9.4.8; Plutarch, *Greek Questions* 15; Pausanias 10.38.1–3).

The eastern end of Locris is at the Crisaeus Bay, which led to the Cirrhaean Plain, unusually large for this coast, and the beginning of Phocis, which extended inland to the western end of Boeotia. Phocis, also a small territory, was famous for Mt. Parnassus and the Delphic oracle, whose primacy had been a part of Greek culture ever since it was chosen as the most authoritative by Croesus of Lydia (Herodotus 1.47–56).

4.8. After a brief mention of some of the topographical features of Delphi and a catalogue of the towns of Phocis (including one ethnym, the Bulensi, or the people of Bulis), the account returns to the coast of the Gulf of Corinth, mentioning three towns in Boeotia, but in a confused fashion. First is Siphae, unquestionably identified at modern Aliki (John M. Fossey, *Topography and Population of Ancient Boiotia* [Chicago 1988] 168–73). Next is the peculiar "Thebae also called Corsiae," said to be next to Mt. Helicon. No such place is known, and the famous Boeotian city of Thebae (Thebes) is not only well away from the mountain but is not documented with such an alias. It seems that Pliny probably wrote "Thebae" in error for "Thespiae," which is about six km. north-northwest of Siphae on the southern slope of Mt. Helicon, and conflated it with a reference to Corsiae, at modern Kastro Khostion on the coast just north of Thespiae (Fossey, *Topography* 187–94). Then Pliny's account continued southeast along the coast to Pagae (modern Alepochori), which is not in Boeotia but Megaris (*NH* 4.23–4), on what he called the neck of the Peloponnesus, the narrow strip of land connecting it to Central Greece.

The Peloponnesus

4.9. After completing Phocis, Pliny moved across the Gulf of Corinth to the Peloponnesus (through 4.22). Beginning at the Isthmos, the account circles the Peloponnesus in a counter-clockwise fashion, concluding with Arcadia in its center.

The name Apia was favored by the dramatists as an ancient name for the region, perhaps originally meaning only the Argolid and Corinthia (Aeschylus, *Suppliant Women* 260–7, *Agamemnon* 256–7; Sophocles, *Oedipus at Colonus* 1303; Pausanias 2.5.7). Pelasgia is associated with the assumed first inhabitants of the Greek peninsula, the Pelasgians, who were probably originally from Thessaly but came to be connected with many locations (Herodotus 7.95; Thucydides 1.3.2; Ephorus F113; Strabo, *Geography* 5.2.4; Roller, *Historical and Topographical Guide* 231).

The comparison of the Peloponnesus to a plane tree (*Platanus orientalis*), a species still common in Greece, was a frequent image but of uncertain origin. The simile was first documented by Strabo (*Geography* 8.2.1), but was probably older, and reflects Eratosthenes' tendency to see the inhabited earth in terms of recognizable forms (Eratosthenes, *Geography* F66). The leaf is roughly square with many protrusions, an apt image for the region.

The circuit of 563 miles (833 km.) is one of several extant estimates (cited by Strabo, *Geography* 8.2.1), whose accuracy is difficult to determine because of the numerous bays and gulfs, as well as the uncertain length of the Greek stadion. Nevertheless it seems reasonable. Isidorus of Charax, who wrote a geographical treatise during the Augustan period (*FGrHist* #781, F12), may have been one of Pliny's major sources for the Peloponnesus.

The actual description of the Peloponnesus begins at its northeast corner with the Isthmos, with it seen as spreading out from that locality. Pliny vacillated between the Greek form, Isthmos, and the Latin one, Isthmus. Its five-mile (7.4 km.) width is probably a conversion of the 40 stadia used by Strabo (*Geography* 8.1.1), but obtained by Pliny from an earlier source.

4.10. Sailing around the Peloponnesus was always difficult, as Jason, Menelaus, and Odysseus all learned. The first roadway across the Isthmos was probably built at the time of the tyranny of Periander around 600 BC (Thucydides 3.15; Strabo, *Geography* 8.2.1; Diogenes Laertios 1.99). It was called the Diolkos ("the Pulling Through"), and extensive remains of it are visible (David K. Pettegrew, "The *Diolkos* of Corinth," *AJA* 115 [2011] 549–74).

Attempts at constructing a canal are documented from Hellenistic times, beginning with Demetrius Poliorcetes at the end of the fourth century BC, whose engineers halted the project because it was believed that the flow (1–3 knots today) from the Gulf of Corinth to the Saronic Gulf would cause flooding (Eratosthenes, *Geography* F16). Julius Caesar, Gaius Caligula, and Nero all had similar ideas (for a summary of the projects, see James G. Frazer, *Pausanias's Description of Greece* [New York 1965], vol. 3, 6–8). How far any of these progressed is uncertain, but Nero's was the most advanced, and extensive remnants of his excavations could be seen until the construction of the modern canal at the end of the nineteenth century. The three Romans were all assassinated. Demetrius died of natural causes, but at the age of 55 due to a dissolute lifestyle and the collapse of his health (Plutarch, *Demetrios* 52): all four demonstrated to Pliny that attempts to modify the natural world were sacrilegious.

4.11. Corinth is actually somewhat west of the Isthmos proper, but away from either coast. After its destruction by the Romans in 146 BC, the city languished, and was reestablished by Julius Caesar as Colonia Laus Julia Corinthiensis. It was populated by a large number of Romans (Strabo, *Geography* 8.6.23; Plutarch, *Caesar* 57.8; Suetonius, *Divine Julius* 42; Dio 43.50.3–5; John Harvey Kent, *Corinth 8.3: The Inscriptions 1926–1950* [Princeton 1966] no. 130). Ephyra was connected with the story of Bellerophon (Homer, *Iliad* 6.152–205). Its location is unknown, but eventually it was believed to be at or near Corinth (Apollodorus, *Bibliotheke* 1.9.3). Pliny's use of a Greek accusative ("Isthmon") and stadia demonstrate a Greek source, perhaps still Isidorus, but his reversion to miles for the remainder of the section shows a change to another one.

Pliny recorded the distance from Leucas to Patrae (modern Patras) as 88 miles (130 km.), and from Patrae to the Isthmos at 85 miles (126 km.); these are both reasonable. Yet to place Patrae less than a mile from the Evenus River is a major understatement since the two localities are about 30 km. apart; Pliny seems to have confused matters with the Rhium–Antirhium crossing east of Patrae. Patrae was the residence of Cleopatra and Antonius just before the Battle of Actium (Plutarch, *Antonius* 60.2–3), and thus Octavian paid careful attention to it thereafter, giving it special privileges as a *colonia* (Pausanias 7.18.7). It long has been a major port of embarkation for Italy.

Pliny's statement that Patrae is situated on the longest promontory of the Peloponnesus is peculiar. It is probable that he was referring to the extensive coastal district southwest of the town, as far as Araxus and Chelonates (*NH* 4.13), the broadest protrusion of the Peloponnesus (about 50 km. across).

4.12. The name Achaia is as old as Homer, and originally seems to have had a broad scope, since the Achaians were the forces that attacked Troy. Eventually the term became localized as the northern coastal regions of the Peloponnesus, often called Aegialos (Greek Aigileia, Strabo, *Geography* 8.7.1), or "the Coast," also a toponym known to Homer (*Iliad* 2.575).

As a political term Achaia became prominent with the establishment of the Achaian Federation, a major political force in Hellenistic times, especially in the first half of the second century BC (Polybius 2.41; Pausanias 7.7). It was dissolved by the Romans in 146 BC, but in time the name Achaia was used for the Roman province established in the early Augustan period, which included the Peloponnesus and much of Central Greece (Strabo, *Geography* 17.3.25).

Beginning at Lecheae (Greek Lechaion), near the east end of the Gulf of Corinth, Pliny listed 11 additional towns in Achaia. The single ethnym, the Pellenaei, refers merely to the inhabitants of Pellene, which is several kilometers inland near a modern town of the same name, and in a precipitous location, alluded to by Pliny (Strabo, *Geography* 8.7.5).

Pliny's description is focused on the earthquake in the winter of 373 BC, which destroyed Helice and Bura, towns about seven km. apart in western Achaia. Eratosthenes, on one of his few field trips, visited the region in the following century and provided a report (Eratosthenes, *Geography* F139), which was preserved by Strabo (*Geography* 8.7.2; see also Diodorus 15.48–9); thus the earthquake is one of the best documented from antiquity. Refugees went to several towns in Achaia, as Pliny noted. Helice seems to have been abandoned but Bura was rebuilt and regained its prosperity (Pausanias 7.25.8–9).

Cleonae (Kleonai) is southwest of Corinth, actually in the Corinthia near the border with Achaia, but Hysiae (at modern Achladokampos) was in the western Argolid on the border with Arcadia (Strabo, *Geography* 8.6.17, 19), nowhere close to Achaia (see further, *NH* 4.13). Assuming that this well-known town was the one meant by Pliny, it is difficult to understand why it was placed in Achaia.

4.13. A list of toponyms in western Achaia follows. Even if Scioessa was the best-known mountain in the region, it is otherwise undocumented. The closest possibility, both orthographically and topographically, is Skollis or Skollion, a ridge (rising to 1,016 m.) about 10 km. south of Dyme (Strabo, *Geography* 8.3.7, 10). Because it stands alone, it might be Pliny's Scioessa.

Dyme, about 12 km. southwest of Patrae, presumably became a Roman *colonia* as part of the reorganization of the region after the Battle of Actium (Strabo, *Geography* 8.7.5; Pausanias 7.17.5–6). Most of the remaining toponyms are in the coastal plain of the extreme northwestern Peloponnesus. The exception is Phlius (Phleious), which is in the southwestern part of the Corinthia, near Cleonae (*NH* 4.12), and is badly misplaced. It was indeed believed to be Homeric Araethyrea (Homer, *Iliad* 2.571; Strabo, *Geography* 8.6.19). Phlius is on the Asopis (Asopos) River, the longest in the Corinthia, which flows north into the Gulf of Corinth. That three toponyms, only 12 km. apart (see also *NH* 4.12), should be erroneously placed in Achaia suggests the intrusion of another source into Pliny's narrative.

4.14. South of western Arcadia is the territory of the Elii, or Eleia, around the town of Elis. Whether the Eleans were the same as the Epians was an ongoing dispute in antiquity, since Homer mentioned the district of Elis and the Epian peoples several times, placing them in the same vicinity (Strabo, *Geography* 8.3.9). The town of Elis was only established in 471 BC (Diodorus 11.54.1), so could not figure in any arguments about Homeric topography, but it added to the confusion of similar toponyms. Olympia, with its famous Temple of Zeus (Pliny's Iovis) is just west of the city of the Pisaeans, or Pisa.

There were several places named Pylos in the western Peloponnesus: Pliny mentioned two (see also *NH* 4.15). Another long-standing dispute concerned which Pylos was the home of Nestor (Strabo, *Geography* 8.3.7, 27–9), an issue only resolved with the discovery of the Mycenaean palace at Ano Englianos in Messenia in 1939. The establishment of Olympiads as a calendrical device had its beginnings in the fifth century BC, when Hippias of Elis created a list of Olympic victors (*FGrHist* #6, F2). Although it was said to be unreliable (Plutarch, *Numa* 1), it was the basis of later works by Timaeus, Eratosthenes, and others (Katherine Clarke, *Making Time for the Past* [Oxford 2008] 64–70).

The itinerary crosses the Alpheus River and continues south along the west coast of the Peloponnesus as far as Cape Platanodes (Greek Platamodes) in northern Messenia. This coast runs north–south, and thus the localities are turned to the west, as Pliny phrased it.

4.15. The itinerary runs along the Peloponnesian coast through Messenia, into the Asinaeus or Coronaeus Bay (two variant names for the same body of water), to Cape Taenarum, the southernmost point of the Peloponnesus and the dividing line between Messenia and Laconica. Cyparissius Bay is normally the indentation from Elis to Messenia, but 75 miles (111 km.) would be the length of the entire west coast of the Peloponnesus from the Chelonates Bay to Methone. With the exceptions of Messene, Ithome, and Dorion, the interior towns are only speculatively located. It is difficult to determine what the unnamed gulf in the final sentence is: 80 miles (114 km.) is virtually the entire north–south extent of the Peloponnesus, but the distance may reflect the vagaries of shipping routes in the gulf. The maximum width of the Asinaeus or Coronaeus Bay is, as noted, about 30 miles (43 km.).

4.16. After Taenarum the account enters Laconica. The form "Laconia," widely used today, appears only occasionally in Latin literature and is probably an error (D. Graham J. Shipley, *The Early Hellenistic Peloponnese* [Cambridge 2018] 25). The unnamed bay is presumably the Gytheates, mentioned below, south of the outlet of the Eurotas River. Again the circuit distance (106 miles or 157 km.) seems excessive, but that across it (38 miles or 56 km.) is reasonable. The town of Gythium is on the west side of the homonymous gulf. It was the port of Sparta and was long an embarkation point to Crete. Cape Malea marks the east side of the gulf. Although it is not the southernmost point of the Peloponnesus (Taenarum is about 12 km. farther south), it was, and remains, the sailing point from the Aegean into the central Mediterranean and for rounding the Peloponnesus, a place of danger since earliest times.

4.17. From Cape Malea the *periplous* continues up the east side of the Peloponnesus, into and around the Gulf of Argos, and along the coast of the eastern Argolid to Cape Scyllaeum (Skyllaion), modern Cape Spathi, the easternmost point of the Peloponnesus. The meaning of the 50 miles (74 km.) cannot be determined among the many possible distances in this region, but 162 miles (240 km.) may be a reasonable figure for the entire eastern coastal route.

The list of towns actually begins on the southern part of the eastern shore of the Gulf of Gythium at Boea and continues around Malea, and north along the eastern Laconican coast into the Argive plain. An odd exception in the list is Mantinea, which is well to the west of Argos in Arcadia, and may be an error on Pliny's part, since the locality was mentioned again at *NH* 4.20. Only the spring of Amymone can be identified among those mentioned: it was near Lerna and famous for its association with the Hydra (Strabo, *Geography* 8.6.8; Pausanias 2.37.4). The other two springs are probably also in the Argolid.

4.18. Pliny completed the *periplous* of the Peloponnesus with the west side of the Saronic Gulf, from Scyllaeum to the Isthmos; the Greek accusative "Isthmon" shows his continued use of a Greek source, probably Isidorus. A distance of 80 miles (118 km.) seems excessive, but the east coast of the Argolid is highly indented. The first towns listed are actually west of Scyllaeum, and the only known place named Coryphasium is on the coast of Messenia, although there is a ridge just south of the Asklepeion near Epidaurum (more commonly Epidauros) called Koryphon (Pausanias 2.28.2). The etymology for the name "Saronic" is not the common one (for the sources see Pliny, *Histoire naturelle* 4, 140), but there is a rare Greek word, *saronis*, which seems to mean a type of oak (Callimachus, *Hymn* 1.21).

The Aesculapium (Asklepieion), located nine km. southwest of the port of Epidaurus, is one of the major cultural and archaeological sites of the Peloponnesus (N. Yalouris, "Epidauros," *PECS* 311–14). The *periplous* continues to Cenchreae (*NH* 4.10), the eastern port of Corinth, at the Isthmos. It is actually on the near side of the Isthmos, according to the route of the *periplous*, but Pliny's reference to the "other side" may mean relative to the starting point of the circuit at *NH* 4.10.

Cenchreae was famous for its proximity to the Temple of Poseidon (Neptune) at Isthmia, four km. to the north, where the Isthmian games were held, one of the four major festivals of the Greek world (the others were the Olympian, Nemean, and Pythian). Both temple and games had their origins in the Archaic period during the predominance of Corinth (Donald G. Kyle, *Sport and Spectacle in the Ancient World* [Malden, Mass. 2007] 140–3). Remains of the Classical temple and features associated with the games are visible. Originally they seem to have been celebrated every other year (Pindar, *Nemean* 6.40), but this may have changed in later times, especially after the disruptions caused by the Roman destruction of Corinth in 146 BC.

4.19. Since the Bronze Age the coasts of the Peloponnesus have been considered particularly rough to sailors, as the various tales about rounding Cape Malea show. Pliny's catalogue of the seas circles around counter-clockwise, like his *periplous*, from the northwest (the Ionian Sea) to the Myrtoan Sea at the east. In one sense the Aegean should not be on the list, since it lies to the east beyond the Myrtoan (which is defined as the sea off the eastern Peloponnesus, Megaris, and Attica). "Winter east," reflecting the position of the sun at sunrise, is southeast; "solstician east" is northeast.

4.20. Pliny next considered Arcadia, the interior highlands and the only territory in the region with no seacoast. The locals were so unacquainted with seamanship that Agamemnon had to provide them with ships so that they could join the Trojan expedition (Homer, *Iliad* 2.612–14). The ancient name, Drymodes (perhaps "Oak People"), is not cited elsewhere, but may refer to the woodlands common in the region. Pelasgis, a variant of Pelasgia (*NH* 4.9), is an early name for various regions and peoples of the Greek peninsula (Strabo, *Geography* 5.2.4). More specifically, the name

Pelasgia or Pelasgiotis was associated with the Peloponnesus from at least the fifth century BC (Euripides, *Archelaos* F228; Ephorus, F113). The first known king of Arcadia, Lycaon, was said to have been a son of Pelasgos (Hesiod, *Catalogue of Women* F111; Akousilaos [*FGrHist* #2] F25a).

The list of towns for Arcadia (over 30) is much more thorough than for any other region in the Peloponnesus, but there are no distances, and Mantinea and Cleonae are repeated from previous sections (*NH* 4.12, 17). All this indicates that Pliny was using a different source than the coastal account of *NH* 4.9–19. To write that Nemea is between Clitorium (Greek Kleitor) and Cleonae is literally true, but peculiar. Clitorium is in north central Arcadia, and the two towns are over 60 km. apart, but Nemea is only eight km. from Cleonae. This is another indication that Pliny had difficulty with the region around Cleonae.

The connection between the town of Pallantium (Pallanteion, at modern Berbati in southeastern Arcadia) and the Palatine Hill in Rome is based on the story of Euandros, or Evander, who was from the Arcadian town and who came to Italy before the time of the Trojan War. He settled on the Palatine and named it after his hometown. This was a tale current in Rome from at least the second century BC (Acilius [*FGrHist* #813] F1), and may have been promulgated or reinforced by Polybius, who came from the vicinity of Arcadian Pallanteion (Strabo, *Geography* 5.3.3; Roller, *Historical and Topographical Guide* 247).

4.21. The toponymic thoroughness of Pliny's account of Arcadia continues with a list of seven mountains and two rivers. The mountains seem to be in no particular order. The sanctuary of Zeus (Iovis) on Mt. Lycaon (Lykaion), in west central Arcadia, was the most important mountain shrine of the region. Extensive remains are visible in a saddle near the summit at the 1,200 m. elevation.

4.22. To complete the survey of the Peloponnesus, Pliny returned to Achaia, using a different source than the *periplous* of *NH* 4.12–13, and one that provided an alphabetical list of ethnyms rather than toponyms. Yet there are only three initial letters – A, P, and T – suggesting that the list has been summarized. Most of the towns were decayed or abandoned by Pliny's day, perhaps indicating that he was not using the map of Agrippa, but a historical source (Ulrich Kahrstedt, "Zwei Probleme im kaiserzeitlichen Griechenland," *SO* 28 [1950] 66–75).

During his visit in AD 66–67, Nero announced that the Greek provinces would be free from taxes. This was not merely the action of an emperor noted for his hellenophilic qualities, but also part of a reorganization of the Imperial tax structure. It lasted for only two years, since Vespasian cancelled the edict early in his reign (Pausanias 7.17.3–4; Philostratos, *Life of Apollonios* 5.41; Susan E. Alcock, *Graecia Capta: The Landscapes of Roman Greece* [Cambridge 1993] 16).

To conclude the Peloponnesus, Pliny provided a number of distances. The 190 miles (281 km.) from Cape Malea to Aegium is long (it is actually about 210 km. in a direct line). From Elis to Epidaurum is close to the 125 miles (185 km.) reported, as is the 68 miles (101 km.) between Olympia and Argos.

Although expressed essentially in philosophical terms about the animistic conflict between the sea and land, the ever-changing nature of the surface of the earth had been realized since the fifth century BC. Scholars such as Xanthos of Lydia and Straton of Lampsacus, whom Pliny could access through Eratosthenes' *Geography* (F15), were the first to consider this issue (Strabo, *Geography* 1.3.4). The Peloponnesus, while not overtly volcanic, had especially rugged terrain and would have seemed an ideal example of this mutability.

Central Greece

4.23. The account moves into Central Greece, which Pliny called Hellas in order to distinguish it from the Peloponnesus (see also *NH* 4.9). The name is Thessalian in origin (Strabo, *Geography* 9.5.6), but eventually came to mean all of the central part of the Greek peninsula. The Latin term Graecia is probably derived from Graikoi, an ethnym from the region of Dodona in Epiros (Aristotle, *Meterologika* 1.14), whose inhabitants were perhaps some of the first Greek peoples that the Romans encountered.

Acte, the ancient name for Attice (more commonly Attica), was from Greek *akte* ("promontory"), referring to the peninsular shape of the territory. The name Attice is not documented until the fifth century BC (Aeschylus, *Eumenides* 681; see also Euripides, *Helen* 1673). The peninsula that connects the Peloponnesus with Central Greece, the Megaris, was seen as a shoulder, another attempt to explain geographical forms by recognizable shapes. Pagae and Aegosthena are actually on the Gulf of Corinth side of the Megaris. The section concludes with a *periplous* along the coast of the Saronic Gulf from Schoenos (at the narrowest point of the Isthmos) to Eleusis, just northwest of Athens.

4.24. Various topographical features in Attica were noted. First there are two towns in the north, Oenoe and Probalinthos, which had been depopulated. Then the *periplous* continues from where it had left off at the end of *NH* 4.23, with a digression to mention Athens, which is five miles (7.4 km.)

inland. There were actually two walls leading to Athens, built in the fifth century BC and which created a secure corridor. But they had fallen into decay and were finally demolished by Sulla in the 80s BC during the war against Mithridates VI of Pontus (Appian, *Mithridateios* 30). After its recovery from the war, Athens retained its independence within the Roman system (Strabo, *Geography* 9.1.20).

There follows a summary of the springs and mountains of Attica, and then Pliny returned again to the *periplous*, omitting any toponyms on the west coast of southern Attica. Beginning at Cape Sunium, the southern end of the peninsula, the account continues up the east side to Oropus, in the borderlands between Attica and Boeotia.

4.25. The narrative next considers Boeotia in Central Greece (through *NH* 4.26), following a similar pattern as before of listing towns, springs, and mountains. The catalogue of towns (22 in all) is particularly long, in part because of the account in the Homeric Catalogue of Ships (*Iliad* 2.494–510), which allowed a more detailed record of local demography than was available for most other places. There is also a blending of three separate lists (John M. Fossey, "Notes on the Texts of Pliny's *Natural History*," *Euphrosyne* 3 [1969] 215–23). Much of the section is oriented on Thebes (Thebae), which is perhaps reasonable, given that it was the primary Boeotian city. Yet it also suggests that Pliny was using a source with a focus on Theban topography.

There were several other towns named Thebes, especially the one in Egypt and that in Phthiotis (Strabo, *Geography* 9.5.6); hence Pliny's precision in noting that Boeotian Thebes was meant. Dionysos (mentioned in his Roman version as Liber) and Herakles (here Hercules) were both associated with the city (Homeric *Hymn to Herakles* 1–3; Euripides, *Bacchants* 1–3).

The Sanctuary of the Muses is about 12 km. west of Thebes in a valley on the east side of Helicon, the highest mountain in Boeotia (Strabo, *Geography* 9.2.25; Pausanias 9.29–31). Mt. Cithaeron (Kithairon) is just to the south of the city, and its wooded northern slopes are visible from there. It was famous because of its role in the story of Oedipus. The Ismenus is a small stream to the north of the city (Strabo, *Geography* 9.2.24). Pliny noted an unusual number of springs in this section, a total of eight. Not all can be located, but those around Thebes were cited in the dramatic accounts of the Theban dynasty (Aeschylus, *Seven Against Thebes* 273–4; Sophocles, *Antigone* 104–5).

4.26. With an orientation still on Thebes, there is a long list of Boeotian towns. It is somewhat scattered: Eleutherae, the first cited, is actually south of Cithaeron in Attica, and Copae is northwest of Thebes, not south toward Megara. Generally, however, the towns are roughly ordered in a clockwise circuit around Thebes, ending at Tanagra, the southeasternmost major city in Boeotia. In Pliny's day it was a free city, perhaps due to the restoration of M. Cocceius Nerva, who was probably the great-grandfather of the emperor and active in the 30s BC (*IG* 7.570). As an indication of its status, the city continued to strike its own coinage to the end of the second century AD (Barclay V. Head, *Historia Numorum* [Oxford 1911] 348–9).

The Euripus is the strait between the mainland and the island of Euboea. Its strange currents have intrigued hydrologists since antiquity. In Roman times the strait was two *plethra* across (about 60 m.); it is narrower today (Aristotle, *Meteorologika* 2.8.366a; Strabo, *Geography* 1.3.11–12, 9.2.8). There may be a certain irony in noting that Aulis (just south of the Euripus) had a spacious harbor: the opposite is true, and it was a problem for scholars in antiquity to consider how the Achaian fleet could have gathered there.

The Hyantes were an early Boeotian population. The similarity of the ethnym to *hys*, "pig" suggested that they were originally called the Pig People, a point of view first expressed by Pindar (*Olympian* 6.87–90). This was the origin of the well-known slander that the Boeotians had swinish qualities (Duane W. Roller, "The Boiotian Pig," in *Essays in the Topography, History, and Culture of Boiotia* [*Teiresias Supplement* 3, ed. A. Schachter, Montreal 1990] 139–44). The name may have survived in the Boeotian town of Hyampolis (*NH* 4.27).

4.27. Epicnemidian Locris was on the coast of the Euboean Strait, northwest of Boeotia. The name, which distinguishes it from the other portions of Locris (*NH* 4.7), refers to its location around the town of Cnemis, or Knemides (Strabo, *Geography* 9.4.4). The Leleges were another of the early populations of the Greek peninsula (Strabo, *Geography* 9.2.3), located in many places, but often associated with western Asia Minor (Homer, *Iliad* 10.429; Strabo, *Geography* 7.7.2). By the first century BC they no longer survived (Roller, *Historical and Topographical Guide* 384–5). Pliny's source for this section is probably different from that for the account of Boeotia, mentioning only towns and rivers, with no mountains or springs.

Although Pliny had already discussed southern Phocis (*NH* 4.7), the territory also included the Cephisus valley west of Boeotia, and a small strip extending north to the coast of the Euboean Strait, which split Locris in two. The eastern portion of the Locrians was generally called Opuntian, from the town of Opus (Strabo, *Geography* 9.4.2), but Pliny did not make the distinction.

Doris and Thessaly

4.28. Doris was a small region near the sources of the Cephisus River, named after the Dorians, one of the primeval ethnic groups of the Greek peninsula (Herodotus 1.56; Strabo, *Geography* 9.5.17). Although associated with Herakles (Hercules) – he died on Mt. Oeta – in historical times the region played hardly any role in events.

Haemonia is an ancient name for Thessaly, after a certain Haemon (which of several personalities with that name is not known), whose son was Thessalos (Strabo, *Geography* 9.5.23). For Pelasgis, see *NH* 4.20. Here Hellas is a town, but normally it was a region of Thessaly (*NH* 4.23; Strabo, *Geography* 9.5.6). Dryopis was a small and obscure area around Mt. Oeta (Strabo, *Geography* 9.5.10). These are all ancient names which were interpreted as representing the primitive populations of Central Greece, all of which had ethnyms of unfamiliar primitive kings. King Graecus was a son of Zeus, and Hellen a son of Prometheus (Hesiod, *Catalogue of Women* F2–3).

The Myrmidones, Hellenes, and Achaei were mentioned frequently by Homer. The Myrmidones were from Thessaly (the ethnic group of Achilles), although they were also associated with the island of Aegina (Homer, *Iliad* 2.683–5; Strabo, *Geography* 8.6.16). The nearby Hellenes were originally from southern Thessaly, where there was a town of Hellas. The Achaei were from the northern Peloponnesus, but Homer also used the ethnym as a generic term for those attacking Troy. Pliny's grouping of these three groups is a direct translation of *Iliad* 2.684, where Homer seemed to use the term "Achaioi" in a more localized sense. Interestingly, Pliny generally used the more latinized spelling, Achaeae, for the ethnic group, but the hellenized form, Achaia, for the Roman province. The Phthiotans occupied Phthiotae, the southern part of Thessaly and the homeland of Achilles (*NH* 4.29; Strabo, *Geography* 9.5.3).

4.29. Although there were several places called Orchomenus (Strabo, *Geography* 8.3.6), including one in Thessaly whose location has not been determined (Diodorus 20.110.3), Pliny's mention of the Minyans indicates that, at the very least, there has been a confusion with Boeotian Orchomenus, at the west end of the Copaic basin. The Minyans were another early population of Central Greece, found in several places (Strabo, *Geography* 8.3.19, 9.2.40), although Homer only associated them with Orchomenus (*Iliad* 2.511; *Odyssey* 11.284). The site has long been famous for its impressive tholos tomb, one of the outstanding examples of its genre, which Pausanias (9.36.6, 9.38.2) called the Treasury of Minyas.

Pliny may have been confused because there is a toponym Minya in northern Thessaly near Halmon, or Halmonia (Stephanos of Byzantium, "Minya"; *IG* 9.2.521), as well as the obscure Thessalian Orchomenus.

The remaining toponyms of the section, insofar as they can be located, are all in Thessaly, with an emphasis on the southern portions. Mt. Nymphaeum is not otherwise cited, so it cannot be determined why it was notable for its landscapes.

4.30. This section is entirely devoted to the mountains and rivers of Thessaly, with the only demographic comments referring to the Lapithae and the existence of 75 cities. Carrying over from *NH* 4.29, a total of 13 mountains are named, suggesting that Pliny may have used a work on the topic, such as that by Dicaearchus of Messana, an author familiar to him (*NH* 1.4; Dicaearchus, F118–20). There is also a list of hydraulic features, which may come from the same work.

The Lapithae were a mythical people of Thessaly (Homer, *Iliad* 12.128, 181), who have a rich mythology and were frequently represented in art, most notably on the Parthenon metopes. For the wooded glen of the Penius River, see *NH* 4.31. A length of 500 stadia (about 100 km.) for the river is reasonable.

4.31. In an unusual digression from his strict topographical account, Pliny devoted much of section 31 to a description of the beauties of the gorge of the lower Penius (Peneios) River, known familiarly as the Vale of Tempe. Despite modern intrusions, it remains one of the most pleasant localities in Greece. As he noted, the predominant color is green. Since antiquity it has been a major route into Central Greece, used well before it was made famous by Xerxes (Herodotus 7.129–30). It was a favorite of bucolic poets and others (e. g. Theokritos 1.67; Vergil, *Georgics* 4.317).

There is no known Horcos River flowing into the Penius: Pliny misinterpreted a passage in the *Iliad* (2.751–5) to assume such a stream. It describes the Titaresios River (modern Xerias), the major northern affluent of the Penius, whose waters do not mix with the Penius, and tells how the Titaresios is a branch of the Styx, "the terrible river of oaths" (*horkoi*), which Pliny read as a Horcos River. The Dirae were the Roman version of the Erinyes, or Furies (Vergil, *Aeneid* 4.473, etc.), which further emphasized the connection with the Styx.

4.32. Magnesia (or Magnetis) is the coastal strip of eastern Thessaly, on the Aegean (Strabo, *Geography* 9.5.15). Although it was known to Homer as a toponym (*Iliad* 2.756), and the region was the home of Philoktetes, as a geographical unit it was obscure until the fourth century BC and the rise of Jason of Pherai. Pliny's account is purely topographic, with a catalogue of towns and a few geographical features. The distances at the end of the section are probably from the map of Agrippa. Both are long (297 miles or 440 km. by 490 miles or 725 km.) and may reflect road routes.

Macedonia

4.33. This section initiates a lengthy discussion of Macedonia (through *NH* 4.39). Pliny emphasized its historic prominence under Philip II and Alexander the Great, without naming them. The reported 150 ethnic groups is probably a statistic from an official record (Pliny, *Histoire naturelle* 4, 178). To Homer, Emathia was a mountain summit (Homer, *Iliad* 14.226), but it evolved into a general toponym describing the central part of Macedonia (Strabo, *Geography* 7F6).

For Epiros, the Dardani, and Triballi, see *NH* 4.1–4. Aegae (near modern Vergina) is famous for the Macedonian royal tombs, excavated since the 1970s. Pieria is the southern part of Macedonia along the coast (Homer, *Iliad* 14.226; Strabo, *Geography* 7F7). The word is certainly an indigenous toponym but as a common noun means "rich" or "abundant," essentially an agricultural term (Homer, *Iliad* 18.541, etc.), and this may have been what Pliny was alluding to.

4.34. The actual account of Macedonia begins with a *periplous*, starting just north of the outlet of the Penius and heading north to the Haliacmon, the major river of southern Macedonia. Then there is a change of source to a catalogue of inland towns, using both toponyms and ethnyms. This list seems to have a focus on the Axius River, as far north as Stobi, which is actually in Paeonia.

4.35. There is a summary of the various ethnic groups around the perimeter of Macedonia, especially in Paeonia, repeating the Dardani from *NH* 4.3. The Treres were a Thracian people, around modern Sofia (Thucydides 2.96.4). For the Pieres, see *NH* 4.29.

Paeonia was a small region along the upper Axius. The locals had been one of the more remote allies of the Trojans (Homer, *Iliad* 2.848–50), and eventually became part of the kingdom of Philip II of Macedonia (Diodorus 16.3.4). The surprisingly long list of Paeonian ethnic groups suggests a source that discussed the region in detail, perhaps Theopompus' *Philippika*. They are little known except for the Pelagones (*NH* 4.33) and the Mygdones, whom the Greek world encountered by the fifth century BC and who were Thracians living east of the Axius (Herodotus 7.123; Thucydides 2.99.4).

There follows a further list of 17 ethnyms and two towns. Most of those which can be identified are in eastern Macedonia. Yet the Lynchestae are around southern Lake Ohrid; their king, Arrhabaios, was the great-grandfather of Philip II of Macedonia (Strabo, *Geography* 7.7.8). The Othryonei, mentioned only by Pliny, were probably in the same region.

Other peoples were farther afield. The Bullidenses were from Bullis, in coastal Epiros, and the Dienses were from Dium, or Dion, in extreme southern Macedonia. The Tymphaei were on the upper Penius (Strabo, *Geography* 7.7.9) and the Toronaei were in southeastern Macedonia on the Chalcidean peninsula (Herodotus 7.22).

4.36. The Thermaic Bay is the northwest corner of the Aegean, reaching as far as Thessalonica. The distance to Dyrrhacium, on the Adriatic, is a measured route on the Via Egnatia, built in the 140s BC (Polybius 34.12.2–8; Strabo, *Geography* 7.7.4; Erich S. Gruen, *The Hellenistic World and the Coming of Rome* [Berkeley 1984] 435–6). Strabo's distance, obtained from Polybius but with uncertainties as to the method of conversion of stadia to miles, is 267 miles (395 km.), slightly more than Pliny's 245 miles (363 km.). The four mountains are unique to the *Natural History* and may be ridges or summits southeast of Thessalonica.

4.37. The next area is around the Chalcidean Peninsula, marked by the three long promontories that extend into the northwestern Aegean. Most of the description is devoted to the easternmost, Athos. It was noted for the canal built by the Persian king Xerxes (Herodotus 7.22–4; Strabo, *Geography* 7F15). Evidence for it survives, and excavations in the 1990s clarified its details (B. S. J. Isserlin *et al.*, "The Canal of Xerxes: Summary of Investigations 1991–2001," *BSA* 98 [2003] 369–85). Pliny's length of 1,500 paces (about 1.2 km.) is approximately correct. A length of 75 miles (111 km.) for the Athos peninsula is far too long; the circumference, 150 miles (222 km.), is probably merely a doubling of the previous figure.

4.38. The final section on Macedonia concerns its easternmost portion, as far as the Strymon River, which was the original border of the territory (Diodorus 31.8.8). The towns listed are in the area between the Athos peninsula and the river, including some in the uplands of the immediate interior. Amphipolis is actually on the east (left) bank of the Strymon, but came under Macedonian control in 358 BC (Diodorus 16.8.2–3). The Bisaltae were a wealthy population on the floodplain of the lower Strymon, above Amphipolis; their prosperity was reflected in their high-quality coinage (Herodotus 7.115; Strabo, *Geography* 7F16; Head, *Historia Numorum* 199–200).

Until the time of Philip II the Strymon was the eastern boundary of Macedonia, but thereafter, under both the Macedonians and the Romans, the political limits of the territory were well to the east. The Strymon is

415 km. long and actually rises in uplands at the 2,180 m. level just west of modern Sofia and south of Mt. Haemus. There were several lakes on the lower river, most of which are now agricultural plains.

4.39. To complete the discussion of Macedonia, Pliny provided an extravagant summary of its widest geographical extent in the years after the conquests of Alexander the Great and before the Seleucid contraction in the third century BC. In this context Pliny used "Macedonia" in its broadest sense of any territory conquered by Macedonians, in other words, those with and immediately after Alexander the Great. Seleucid control of Hiberia (Iberia) and Albania (territories on the west side of the Caspian Sea) was minimal to non-existent, and "the entire East" is somewhat exaggerated.

Alexander was conscious of his relationship to Dionysos (Father Liber) and Herakles (Hercules), and this connection and the constant search for physical proof of it was a feature of the eastern expedition (A. B. Bosworth, *Alexander and the East: The Tragedy of Triumph* [Oxford 1996] 119–23). By contrast, L. Aemilius Paullus, the victor over King Perseus of Macedonia at Pydna in 168 BC in the battle that brought the Macedonian kingdom to an end, became notorious for his widespread destruction throughout Macedonia, and as far west as Epiros (Polybius 30.15; Livy 45.34; Strabo, *Geography* 7.7.3; Plutarch, *Aemilius Paullus* 29). It was said that 150,000 were sold into slavery, and 70 (or 72) cities destroyed, with the local valuables confiscated. The Molossians were effectively eliminated. The circumstances were reported not long after the events by Polybius, and even allowing for exaggeration, it was a major disruption in the social fabric and an unfortunate inauguration of Roman rule in the Greek peninsula. Yet it was a Roman attempt to demonstrate that, insofar as their expansion into the Greek peninsula was concerned, resistance was futile (Gruen, *Hellenistic World* 516–17). Pliny placed the events into a moral context.

Thracia (Thrace)

4.40. After Macedonia, the next section is Thracia (Thrace, through *NH* 4.49), broadly interpreted as the territory north of the Aegean as far as the Black Sea and the Danube. One of Pliny's sources may have been the account of the expedition into the region by Alexander the Great in 335 BC (Arrian, *Anabasis* 1.1–6). The Thracians had been known to the Greek world since the time of Homer – they were Trojan allies (*Iliad* 2.844–5) – and their power, wealth, and extent became part of Greek lore (Herodotus 5.3–8). By the second century BC they had made contact with the Romans (Polybius 23.8), but they remained largely independent and under the rule of allied monarchs, with provincialization not beginning until late in the following century.

For the Denseletae and Maedi, see *NH* 4.3, and for the Bisaltae, 4.38. The Diggeri have not been located. The Bessi were on the upper Hebrus River (the modern Evros, Meriç, or Matisa) in central Bulgaria, and could be problematic for the Romans (Dio 51.25.5, 54.20). The Diobessi were probably one of their sub-groups. The Haleti and Carbilesi are otherwise unknown. The Brigae can refer to a wide-ranging group of people (as demonstrated by the frequency of European toponyms with variants of the form "brig"), but these Brigae were believed to be the ancestors of the Phrygians (Herodotus 7.73; Strabo, *Geography* 7.3.2). The Sapaei were in the uplands immediately north of the Aegean, and were best known for their encounter with Archilochos (F5) in the seventh century BC. The Odomantians were to their west on the lower Strymon (Herodotus 7.112; Strabo, *Geography* 7F16).

The Odysians were a population which extended from the upper Hebrus to the Black Sea. The Greek world first saw them at the time of the Scythian expedition of Darius I of Persia (Herodotus 4.92). There is a further list of nine additional peoples: the first six are obscure and probably lived along the upper and middle Hebrus. The Corpili were presumably in the uplands west of the lower river (Strabo, *Geography* 7F20), and the Bottiaei farther west, in the Chalcidean peninsula of Macedonia (Herodotus 8.127). They are rather anomalous on the list, and may have migrated from elsewhere in early times. The Edoni were on the lower Strymon and to its east, but seem to have been eliminated by the fourth century BC (Herodotus 5.11, 124; Thucydides 4.102). They are most remembered for their king Rhesos, the subject of the extant drama bearing his name.

4.41. The next group of Thracian peoples includes several on the Aegean coast, although the Thyni were on the Thracian coast of the Black Sea (Strabo, *Geography* 7.6.1). The Coelaletae lived in two separate groups south of Mt. Haemus; the six-mile (9 km.) ascent to the summit is probably the length of the trail. It was said – erroneously – that one could see both the Aegean and the Black Sea from its heights (Polybius 24.4; Livy 40.21.2; Strabo, *Geography* 7.5.1).

On the north side of Mt. Haemus and beyond were increasingly remote populations, only vaguely known in antiquity. The Moesi occupied the region between the mountain and the Danube. The Getae were also along the lower river, and were constant opponents of Rome: their king Diokomes promised support to Antonius before Actium (Plutarch,

Antonius 63). The Aedi, Scaugdae, and Clariae are otherwise unknown but were probably also on the lower Danube.

The Sarmatae (or Sauromatians) were a powerful group north and west of the Black Sea, who since at least the fifth century BC had been trade intermediaries between the Mediterranean world and northern Europe. The Scythae (Scythians) had long been considered generically as the people of the barbarian north (Ephorus F30a), but here Pliny may have been referring to the inhabitants of the region known as Little Skythia, the territory between the northern curve of the lower Danube and the Black Sea, in modern Romania (Strabo, *Geography* 7.5.12). The other ethnic groups listed by Pliny are unknown elsewhere but were probably north of the Danube.

4.42–3. The account returns to coastal Thrace, combining a *periplous* with an itinerary along the Via Egnatia, which is often a few kilometers inland. The distance from Philippi (famously where Brutus and Cassius were defeated in 42 BC) to Dyrrhachium, 325 miles (481 km.), is a measured route along the road. There are several anomalies in the list of towns: Scotusa, previously mentioned (*NH* 4.35), is well up the Strymon, and Olynthos and Heraclea (the latter cited at *NH* 4.38) are in the Chalcidice, to the west of Thrace. Eastern Thrace was the territory of Diomedes, king of the Bistonians and the victim of the eighth labor of Herakles (Hercules).

The *periplous* continues east along the north coast of the Aegean. Xerxes counted his forces at Doriscum (Doriskos), in the western edge of the Hebrus estuary, which had been a Persian outpost since the sixth century BC. Presumably it was possible to do this because of the extensive plain of the Hebrus valley. It resulted in the famous description of the Persian contingents preserved by Herodotus (7.59–83).

Reaching the northeast corner of the Aegean, Pliny's account continues to the Thracian Cherronesus (Chersonesus), the long peninsula on the European side of the Hellespont. At its interior end, where the peninsula was only 10 km. across, there was a wall that had been built by the elder Miltiades in the sixth century BC, and which was kept repaired and continued to serve a strategic purpose into at least the fourth century BC (Herodotus 6.36; Plutarch, *Perikles* 19.1; Xenophon, *Hellenika* 3.2.10; Strabo, *Geography* 7F21).

4.44. Pliny then considered the Thracian shore of the Pontus (the Black Sea), from the Hister River south (through 4.45). The earthquake that destroyed Bizone (near Kavarna in Bulgaria) was probably in the first century BC (Strabo, *Geography* 7.6.1). "Aroteres" may be a Greek descriptive term ("Plowmen") rather than an indigenous ethnym, describing the population of Little Skythia, in the Danube curve. The first four Scythian towns cannot be precisely located. Parthenopolis is probably the town taken by C. Licinius Lucullus in 72 BC, which is near Tomi (Eutropius 6.10).

Gerania is near Dionysopolis (*NH* 4.45). There was a localized version of the Homeric story of the Pygmaeans and the cranes (*Iliad* 3.1–6), based on the name of the town, which means "cranes" in Greek. But this was not a region known to Homer, and by all accounts the story, if localized at all, is generally placed in sub-Saharan Africa. The name Gerania may indeed be Greek, but it may also be an indigenous toponym, and "Pygmaei" was probably a local ethnym (Stephanos of Byzantium, "Kattouza").

4.45. The *periplous* continues south along the coast of the Black Sea as far as the outlet of the Thracian Bosporus, listing towns and rivers, the most important data for coastal sailing. Mt. Haemus (*NH* 4.3, 38, 41) is indeed a long east–west ridge that would have been visible from out to sea. It is nearly 300 km. long, with its highest part at the west, sloping down east toward the sea. The one exception to the *periplous* is the insertion of Develton (modern Develt in Bulgaria), which is not on the coast but four km. inland at the head of Bourgas Bay. It was established as a veterans' settlement by Vespasian (*CIL* 6.3828), and thus was probably inserted by Pliny into the coastal narrative, but in the wrong place (it is between Anchialum and Thynias).

The unspecified "others" who provided a distance of 500 miles (740 km.) from the Hister to the Thracian Bosporus probably include Varro, whose Black Sea coastal distances north of the Hister are at *NH* 4.78. Both this distance and the 560 miles (829 km.) of Agrippa (F50) may be based on a Greek source, perhaps the one providing the figure of 3,830 stadia quoted by Strabo (Geography 7.6.1), which may have been converted to miles in different ways by the Latin writers. The 150 miles (222 km.) to the Chersonesus Wall is reasonable, but the last figure, 126 miles (186 km.), said to be to Chersonesus, is difficult to interpret since the peninsula is only 85 km. long. It may be the sailing distance completely around it, from one end of the wall to the other.

4.46. Continuing the *periplous*, the itinerary moves along the European side of the Bosporus, as far as Cape Chryseon Ceras (the Golden Horn) and Byzantium (Istanbul). This was 711 miles (1,052 km.) from Dyrrhachium on the Adriatic, and is presumably a road distance along the Via Egnatia, which had been extended east from the Hebrus River by the early 50s BC (Cicero, *de provinciis consularibus* 4). The easternmost 10 km. of the road followed the coast of the Propontis, the sea between the Bosporus and the Hellespont.

4.47. The *periplous* ends at Byzantium, and from there the itinerary largely follows the Via Egnatia, but with reference to localities both north and south of the road. It continues on the road as far as Apros, or Aproi (modern Germeyan in Turkey), shortly after it leaves the coast of the Propontis.

Bizye (modern Vize in Turkey) lies 70 km. north of Perinthus and was a royal residence of the Thracian kings. The familiar story of how Tereus, a Thracian king, assaulted his sister-in-law Philomela and cut out her tongue, who then wove the details into a tapestry, did not take place here, but rather when Tereus was in Central Greece (Ovid, *Metamorphoses* 6.424–674; Strabo, *Geography* 9.3.13). Yet later, after further gory events, Philomela was changed into a swallow.

From Bizye to Apros is 50 miles (74 km.), and the 189 miles (280 km.) to Philippi is along the Via Egnatia. The section closes with a diversion to the Thracian Chersonesus (*NH* 4.43, 45) and the town of Lysimachea (probably at modern Ortaköy, in Turkey), founded by Lysimachos – who had been a companion of Alexander the Great – in 309 BC and eventually destroyed by the Thracian king Diegylis in the mid-second century BC (Diodorus 33.14; Cohen, *Europe* 82–7).

4.48. The survey of the Thracian Chersonesus continues, comparing its narrows, where the Long Wall was located, with the Isthmos of Corinth. Here Pliny translated into Latin the name of the wall and used the plural (Longi Muri), but at *NH* 4.43 it was presented as a singular in latinized Greek (Macron Tichos). The origin of the name Cardia (Kardia), a Milesian settlement, is because it was at the "heart" of the Chersonesus. Pliny provided a concise summary of the nature of Hellenistic city foundation, which usually meant the incorporation of existing cities and villages into the new settlement. In the process of founding Lysimachea, the older town, Cardia, which may have been within the limits of the new one, was eliminated (Pausanias 1.9.8).

Resisthos is not cited before its mention in the *Natural History*, and may be the Rhaidestos of Prokopios (*On Buildings* 4.9.17). A location 22 miles (33 km.) from Apros would not be in the Chersonesus, but near Bisanthe (*NH* 4.43), and it may be an insertion Pliny made into his text, for he suggested that the name was recent.

4.49. The Hellespont is the strait between the Aegean and the Propontis. Its width of seven stadia (about 1.3 km.) only applies to its narrowest part, near the Aegean at Sestos and Abydos. Seven-Stadia Strait was probably an ancient descriptive name used as a toponym (Strabo, *Geography* 7F22). Unlike his account of the Thracian Bosporus, Pliny saw the Hellespont and its shores as a unit, pairing the towns on its coasts. The Cynossema (Dog's Grave Monument) was probably at modern Kilitbahir on the European side, where the narrows widen toward the Aegean. It commemorated the location where Hecuba, the wife of Priam of Troy, was turned into a dog and leapt from the ship transporting her to Greece (Euripides, *Hecuba* 1270–3; Ovid, *Metamorphoses* 13.565–75; Strabo, *Geography* 7F21).

The anchorage and tower of the Achaeans, where their forces landed to besiege Troy, is probably not the locality with the same name on the western side of the Troad (outside the Hellespont) near the site of Troy, at modern Hacı Ahmet Ağa (Pseudo-Skylax 98.2; Strabo, *Geography* 13.1.31), but another landing point on the Hellespont. Protesilaus was the first to die in the Trojan War (Homer, *Iliad* 2.698–9), and his shrine was in existence by the early fifth century BC (Herodotus 7.33). Since he was killed while disembarking from his ship, the Achaian forces may have had a base near the future site of the Cynossema.

4.50. With the Melas Gulf (*NH* 4.43, 48–9), the northeasternmost part of the Aegean, Pliny has completed the coastal survey of his third gulf of Europe, which began at the Acroceraunian Mountains on the west coast of the Greek peninsula (*NH* 4.1). The description has been primarily in *periplous* form, moving around Greece and along the European shore of the Aegean, but with many digressions into the interior. He also added a description of the Black Sea coast south of the Hister and through the Thracian Bosporus, Propontis, and Hellespont, which should have been part of his fourth gulf (*NH* 4.75).

The distance from southern Macedonia through Thrace to the mouth of the Hellespont, calculated at 720 miles (1,066 km.), is probably from the Penius River. The width (384 miles or 568 km.) may be from the head of the Aegean to the Hister/Danube, but this cannot be determined. The figures are probably from the map of Agrippa.

The Islands of the Third Gulf

4.51. There follows a lengthy discussion of the islands off the coast of Pliny's third gulf (through *NH* 4.74). again showing the maritime orientation of much of his geography. The source may be an unknown work on islands, or from a navigational guide such as that written by Timosthenes of Rhodes in the early third century BC. Such an emphasis for the narrative is shown by the matter of the rock named Aex (or Aix). Tenos (here Tenus) and Chios (here Chius) are Aegean islands about 90 km. apart, and the main north–south route through the Aegean would pass between them. But Achaia (not on the Aegean) and Antandros (on the southern coast of the Troad) are two points far distant from one another, and there would be no direct route between the two. This probably reflects a sea journey that Pliny or his source took, which passed by the hazardous Aex ("Goat") Rock. It cannot be located today, and may have vanished due to erosion or changing sea levels.

Pliny's source for the etymology of the name "Aegean" was Varro (*de lingua latina* 7.22). As one interested in nautical matters, Pliny found his explanation more credible than the familiar tale involving the fate of King Aegeus of Athens. The remainder of the section catalogues the various divisions of the Aegean, distinguishing between Greek and Roman terminology. In antiquity it was common for seas to be named after the peoples on their coasts, and thus a single body of water might have many overlapping names.

4.52. Pliny's catalogue of islands begins off the Epirote coast with Corcyra (Greek Kerkyra, modern Corfu) and those in its immediate vicinity. Corcyra was precisely placed, and the distances from Buthrotum (12 miles or 18 km.) and Acroceraunia (50 miles or 74 km.) are accurate. Yet the island is nowhere near 97 miles (144 km.) long, and the number is probably the distance of the sailing route around it. The Temple of Zeus Kassios (Iovis Cassius) was the first place in Greece that Nero visited during his tour of AD 66–7 (Suetonius, *Nero* 22). Equating Corcyra with Homeric Scheria and Phaeacia or calling it Drepane ("Sickle") after its shape are assumptions that date from at least the fifth century BC (Hellanicus [*FGrHist* #4] F77).

4.53. The catalogue continues south along the west coast of the Greek peninsula as far as the mouth of the Gulf of Corinth. Phalacrum (Phalakron, at modern Agios Taxiarchos) is a promontory on the west side of Corcyra, where ships coming from the south would turn west to Brundisium in Italy. There are numerous rocks in the vicinity, some of which even look like ships, and one of these was identified as the Phaeacian ship that was returning from having deposited Odysseus (Ulixes) on Ithaca (Homer, *Odyssey* 13.161–4).

The Echinades are a group off the southwest coast of Aetolia, but Pliny's list of the ten is questionable. Some belong in the Aegean (especially Thia and Thera), and others may be in the Strophades (*NH* 4.55). Pomponius Mela (2.110) seems also to have entangled the Echinades and the Strophades, and Pliny, in part, continued his error.

4.54. Next are six major Ionian islands, located north and south of the outlet of the Gulf of Corinth. Cephallania (Kephallenia) is the largest; the 10 miles (15 km.) recorded from Paxos (to its north toward Corcyra) is so much in error (it is actually 80 km.) that it is certainly a transmission issue. The circumference of 93 miles (138 km.) is reasonable.

Same is a Homeric toponym (*Odyssey* 1.246, etc.) that refers to an island in the vicinity, but the name did not survive as such into later times. There was a town on Cephallania with the name (Strabo, *Geography* 10.2.13), which was destroyed by M. Fulvius Nobilior in 189 BC (Livy 38.28–30). Yet Pliny seemed uncertain whether the toponym referred to an island or town, unless his "three towns" were merely surviving neighborhoods of the destroyed city.

This topographical confusion persisted with the location of Zacynthus, the southernmost of the Ionian islands. It is southeast of Cephallania and hardly between it and the mainland, but its assumed position probably reflects the fact that ships from the latter stopped at the former on their way to the mainland. Moreover, unless Pliny was using Achaia in the Homeric sense of all of the southern Greek peninsula, the positioning of Zacynthus becomes even more enigmatic, as historic Achaia only bordered the Gulf of Corinth, far from Zacynthus. Moreover, there is no other evidence that the island was once called Hyrie, although that was a common toponym in the Greek world.

Zacynthus and Cephallania are only about 15 km. apart, but the 25 miles (37 km.) is probably the sailing distance between major ports. The circumference (36 miles or 53 km.) is far too low and may merely be a doubling of the maximum length of Zacynthus. Mt. Elatus is presumably the high point of the island (modern Vrachionas, 756 m.), but there is no particular reason why it should be especially famous, although Elatos was one of the suitors of Penelope (Homer, *Odyssey* 22.267).

4.55. Ithaca lies to the east of Cephallania across a narrow strait (four km. across). Pliny's 15 miles (22 km.) may be, as usual, the distance between island ports. The circumference of 25 miles (37 km.) is reasonable. Mt. Neritus (or Neriton) was mentioned by Homer (*Iliad* 2.632; *Odyssey* 9.22), and is presumably the high point on the island. It is far more than 15 miles (22 km.) from Araxus, one of the northwestern promontories of the Peloponnesus, which is about 60 km. away.

The remainder of the section catalogues various small islands in the region, as far south as Messenia (where Plotae is actually located: Strabo, *Geography* 8.3.23). The Eurus was one of the original cardinal winds (Homer, *Odyssey* 5.295; Strabo, *Geography* 1.2.21), from the east, but the

Strophades are actually south of Zacynthus, somewhat closer than the 35 miles (52 km.) reported by Pliny.

4.56. The next group of islands are those south of the Peloponnesus, in the Asinaeus and Laconic Bays (*NH* 4.15–16). Also included is Cythera, the large island in the strait between the Peloponnesus and Crete. The narrow channel between it and the mainland was the preferred sailing route in early times, because after rounding Malea upon leaving the Aegean (itself a difficult passage) ships could head straight west along the southern edge of the Peloponnesus rather than having to go around Cythera.

The next region of interest is the islands on the east coast of the Peloponnesus as far as those near Troezen in the southeastern Argolid, and also including Belbina (modern Agios Georgios), an isolated island in the Myrtoan Sea that is actually south of Attica. The 50 miles (74 km.) that Calauria is said to be off the coast is an error, since it practically touches the Peloponnesus. It is odd that Calauria is the only island for which a location is provided, and that it is so incorrect; the best emendation is 50 paces, or about 75 m. (Pliny, *Histoire naturelle* 4, 237).

4.57. Further islands in the Saronic Gulf are catalogued, from those near Epidaurus around to Megara, and including Aegina at the outer edge of the gulf. There are a large number of distances in this section, which may reflect a change of source.

The four Methurides are presumably some of the islets off the Megarian coast (Stephanos of Byzantium, "Methouriades"). The association of this region with King Minos of Crete – one of the local islands is named Minoa (Thucydides 3.51.1; Strabo, *Geography* 9.1.4; Pausanias 1.44.3) – may have led Pliny to insert his Cretan section at this point, and the middle of a sentence he shifted from the Saronic Gulf to the south end of the Peloponnesus. Aegila is not near Megara but rather the island known today as Antikythera, between Cythera (which is 15 miles or 22 km. to its northwest) and Phalasarna, the nearest town on Crete (25 miles or 37 km. to its southeast).

4.58. The Cretan portion (through *NH* 4.61) begins with a topographical and historical survey of the island, orienting it and noting its reputation for having a hundred cities (a Homeric reference, *Iliad* 2.649); Pliny listed forty. There were many explanations for the name of the island: in a highly corrupt passage Pliny cited four sources, beginning with the early Hellenistic historian Dosiades (*FGrHist* #458, F4). Although it seems unlikely that a nymph named Crete was one of the Hesperides (the daughters of the evening), this is not impossible. Anaximander (*FGrHist*

#9, F2) is not the famous natural philosopher from Miletus, but a little-known antiquarian scholar of the fourth century BC. The eponym of Crete, Cres, may be associated with the Curetes (Kouretes), the mythical craftsmen who were also involved with the birth of Zeus on the island (Diodorus 5.64.1). Philistides (*FGrHist* #11, F2) was another little-known Hellenistic historian, who may have been from Mallos; Crates was certainly from Mallos, dating to the first half of the second century BC. Pliny was presumably using a work on Crete from no earlier than the middle of the second century BC, which included further unnamed sources that may have been cited in the same treatise.

Crete, as the southernmost part of the traditional Greek world, was noted for its fine climate. The widest part of the island (about 50 miles or 74 km.) is west of Knossos. There were many calculations of its dimensions (a list is in Pliny, *Histoire naturelle* 4, 240), most of which are in stadia and thus difficult to convert. Pliny's length of 270 miles (400 km.) is much longer than the actual 250 km. The two locations, Cape Samonium (modern Cape Sidera) at the northeast and Criu Metopon (the Ram's Forehead, modern Cape Krios; Pliny used latinized Greek endings) at the southwest, were important navigational points for routes past Crete toward the western Mediterranean and North Africa.

4.59. Pliny's unusually long list of Cretan toponyms is indirect confirmation of the belief held since the time of Homer that the island had a high degree of urbanization. The "sixty other towns" not named brings the number up to the Homeric hundred. The list of four mountains is peculiar: two of the three highest (Leuka and Aigaion) are not named (only Idaeus, or Ida, is cited), and Cadistus and Corycus (Greek Corycos) are rugged promontories rather than mountains. The list may be heavily summarized.

4.60. Pliny located Crete in terms of sailing routes to other places in the Mediterranean, using the map of Agrippa. Criu Metopon, at the southwest, was the navigational point for the Cyrenaica in Libya. Phycus, or Phykous (modern Ras Aamer, the northernmost point in Libya), was reported to be 125 miles (185 km.) to the southwest of Crete (it is actually somewhat farther). From Cadistus (*NH* 4.59) to Malea in the Peloponnesus was said to be 80 miles (118 km.), but since Cadistus is in eastern Crete, 250 km. from Malea, this is probably an error for some point in the northwest of the island. From Carpathos, one of the southernmost Aegean islands, it was reported to be 60 miles (89 km) toward the west wind to the Samonium Promontory at the northeast corner of Crete, but it is quite a bit farther.

4.61. Pliny cited 20 islands around Crete, listed in roughly a clockwise fashion from Corycus, presumably at the northwest corner of the island (probably modern Gramvousa and Agria Gramvousa). The source in this section is probably the one about islands regularly used in the *Natural History*, rather than any about Crete, since the last six are not near Crete, but around Casos, the southernmost of the Aegean islands, 50 km. east-northeast of Crete.

4.62. Pliny returned to cataloguing the islands off Attica (*NH* 4.57). The first four are not otherwise known but must be some of the minor islands between Megara and Salamis. The only known Lichades are off the north end of Euboea (Strabo, *Geography* 9.4.4), and Pliny may have interpolated them from his Euboean discussion, beginning later in this section.

The islands off the west coast of Attica, between Piraeus and Sunium, are not listed. The catalogue continues with Helene and Ceos (Cea). The former is indeed 5 miles (7.4 km.) from Sunium, to its east, but Ceos is 18 km. farther to the east-southeast. Hydrusa is a name occasionally applied to Greek islands that have a water supply, which can be non-existent on many of them.

The distance of 500 stadia (about 100 km.) could not refer to Ceos, and the mention of Boeotia may mean that the figure is for Euboea, but the amount is not applicable to that island either. The passage is confused, but in some way reflects the tectonic changes of the earthquake of autumn 426 BC, one of the largest in Greek history (Francisco Javier Fernández Nieto, "Incidentes de una Correción Geográfica de Solino a Plinio: la Isla de Cos," in *Solinus. New Studies* [ed. Kai Brodersen, Heidelberg 2014] 90–5). Its epicenter was around the Lichades Islands at the north end of Euboea (Thucydides 3.89; Strabo, *Geography* 1.3.19). Yet the four towns mentioned are all on Ceos, perhaps indicating the spread of the quake.

Ceos had long been the center of a textile industry. It was in existence before the fourth century BC, when Aristotle (*Research on Animals* 5.19.551b) reported that it had been established by a certain Pamphile. The cloth was spun from an insect product that was probably a European silkworm (either *Saturnia pyri* or *Pachypasa otus*: William T. M. Forbes, "The Silkworm of Aristotle," *CP* 25 [1930] 22–6). Although the Coan insect is often thought to be the Chinese silkworm, this is impossible, since that insect was not known in the Mediterranean world at the time of Aristotle. Chinese (actually that from the Seres) silk was first mentioned, perhaps anachronistically, in connection with the wardrobe of Cleopatra VII (Lucan 10.141–3), yet does not seem to have become a major import until the Julio-Claudian period (*Periplous of the*

Erythraian Sea 49). There is no reason to believe that the indigenous Coan industry would have ceased soon after the arrival of silk from the Far East, and the Coan product was praised by Augustan poets (Horace, *Odes* 4.13.13; Ovid, *Ars Amatoria* 2.298; Propertius 1.2.2, 4.5.57; Tibullus 2.4.29–30).

4.63. Euboea, the long narrow island off the eastern Attic coast, was described in a certain amount of detail. For the Euripus, see *NH* 4.26. The widest part of the island is actually only about 30 km., near its center (Pliny's 40 miles or 59 km. is difficult to understand), but its narrowest part, near the south end (two miles or three km.) is about equal to Pliny's figure. The length opposite Boeotia (150 miles or 222 km.) is excessive, but the circumference (365 miles or 540 km.) is reasonable, given the indentation of the coast.

4.64. Caphereus is the easternmost point on Euboea – no doubt a navigational marker – and it would have been a straight sail northeast to the outlet of the Hellespont. But Pliny's 225 miles (333 km.) is half again too long, and the number may reflect a sailing route with some intermediate stops. There are two lists of toponyms: the first of eight cities that no longer existed in Pliny's day, and the second of a mixture of cities, springs, one sanctuary, and the major river of the island, all presumably representing its current (Roman period) status.

The most famous product of Euboea was the marble from Carystus, at the south end of the island. The quarries – which are still visible – produced a green veined stone, today called *cipollino verde*, which became popular in Rome in the first century BC (*NH* 36.48). Eventually the quarries came under Imperial control (Doris Vanhove, *Roman Marble Quarries in Southern Euboea* [Leiden 1995]; Hirt, *Imperial Mines* 157–9). The marble can be seen today in the Temple of Antoninus and Faustina in the Roman Forum.

The ancient name Chalcodontis was only cited by Pliny, but may be a variant of Chalcis, the town at the narrows of the Euripus. The three Greek sources that can be dated, Ephorus (*FGrHist* #70, F151), Aristides (*FGrHist* #444, F5), and Menaechmus (*FGrHist* #131, F7), are all from the fourth century BC, suggesting that Pliny used an early Hellenistic antiquarian source, although which of the many authors named Dionysius is meant cannot be determined. One possibility is the obscure Dionysius of Chalcis, who was thus from Euboea (Dionysius of Halicarnassus, *Roman Antiquities* 1.72.6). Copper was mined in the Lelanton Plain near Chalcis, and this may be the origin of the city name, but by the first century BC the industry had failed (Strabo, *Geography* 10.1.9).

4.65. The first islands mentioned in this section are obscurities around Euboea. The name Cyclades (Kyklades, "Encircling") is documented from the fifth century BC (Herodotus 5.30); Delos was the cultural center of the region and especially important as the birthplace of Apollo and Artemis, so it is reasonable that the other islands would be oriented around it. Pliny's account approaches the Cyclades from Euboea, with Andros first, 10 miles (15 km.) to its southeast, and continuing to Tenos, a mile (1.5 km.) farther, and then on to Delos, another 15 miles. This is a sailing route from the mainland to Delos. Again, Pliny's sources reflect the early Hellenistic period. For Hydrusa as an island name, see *NH* 4.62; Ophiusa ("Snaky") was also a revealing island toponym throughout the ancient world. This is one of Pliny's rare citations of Aristotle by name in the geographical books (F595; see also *NH* 4.66, 4.70, 5.135).

4.66. Having reached Delos, the account mentions a few other Cycladic islands, with some of their dimensions, before discussing Delos itself. The island was notable not only for its divine connections but also for its role as the mercantile center of the Aegean in Hellenistic times, until the forces of Mithridates VI of Pontus sacked it in 88 BC (Appian, *Mithridateios* 28; Strabo, *Geography* 10.5.2–4; Pausanias 3.23). It was also famous for its extensive slave market and its architectural innovation.

Delos as a floating island may reflect a memory of seismic activity, perhaps hinted at by Aristotle. His etymology is from Greek *delos*, meaning "visible" (Aristotle F488). The island experienced an earthquake in 490 BC (Herodotus 6.98) and another during the Peloponnesian War (Thucydides 2.8); these may be the two that Pliny's contemporary C. Licinius Mucianus cited. Varro also would have known about these quakes, yet if properly quoted, his comments remain enigmatic. There were many names for Delos; the tradition that fire was first discovered there is at odds with the standard mythological view connecting the event with Prometheus at an unknown location. Yet the early name Pyrpyle may be connected to volcanic activity and thus a local belief that fire originated on the island.

4.67. The discussion of the Cyclades concludes with a number of other islands, beginning with some islets around Delos. Syros, immediately to the west, has a circuit of 20 miles (30 km.) as reported; the figure of 160 from Mucianus is probably stadia, which would correspond to the other measurement. Naxus (more commonly Naxos), south-southeast of Delos, is the largest island in the Cyclades, but its wine seems otherwise to have been hardly known; Pliny did not mention it in his catalogue of Greek wines (*NH* 14.73–82).

4.68. From the Cyclades Pliny moved to the Sporades ("Scattered"), lying to the east off the Asian coast (Strabo, *Geography* 10.5.14). The list is corrupt (see also *NH* 4.53): Helene (*NH* 4.62) is off Attica, and Phacusa, Nicasia, and Schinusa are small islands in the Cyclades near Naxus. Sailing routes have been provided from the Cyclades (Naxus and Delos) to Icaros, the closest of the Sporades, which is also positioned 35 miles (52 km.) from Samos, to its east just off the Asian coast, as well as 112½ miles (166 km.) from the southern end of Euboea. This effective triangulation of Icaros is a good demonstration of the use of sailing routes to position maritime locations.

4.69. Pliny's assertion that the islands in this section are in no order is quite accurate, as the following are scattered all across the central Aegean, from Scyros in the north to Telos in the far southwest. Several Sporades are also included. The distances are highly inaccurate: the maximum length of Ios is more like 10 miles (15 km.) and that of Patmus (usually Patmos) even less.

The tomb of Homer was located in several places, but the inhabitants of Ios (one of the Cyclades) had a strong claim, and the feature was still visible in the second century AD (Pausanias 10.24.3). Telos was one of the Sporades, and the local ointment may be the variety called *telinum*, whose recipe Pliny recorded (*NH* 13.13; see also Callimachus F581), noting that it was most popular in early Hellenistic times.

4.70. Many more islands in the central and southern Aegean are listed, from the Cyclades to Casos, the farthest south Aegean island. A number of these are otherwise unknown. Particular attention was paid to Thera (25 miles or 37 km. south of Ios) and its associated islands, Therasia, Automate ("Spontaneous," or Hiera), and Thia, all part of the same volcanic complex (see *NH* 2.202). Pliny recorded three eruptions at Thera: 197 BC, 67 BC (which produced Automate) and AD 19 (which produced Thia); the major one seems to have been the first, although modern geological studies do not fully agree with Pliny's account (Strabo, *Geography* 1.3.16; Forsyth, "After the Big Bang" 191–204).

4.71. Most of the islands that can be identified in this section are in the southern Aegean. Rather than simply a catalogue, it is a sailing itinerary from Thera east through various minor islands and the Sporades, ending at Rhodes. From Carpathos there is another route back to Crete. The sailing distances are generally accurate, although the circuit of Astypalaea (88 miles or 130 km.) is several times too large. The source for this section is different from that used previously, as demonstrated by variant spellings (Casus instead of Casos). The African wind is the southwest one.

Another sailing route passes through the Euboean Euripus, from southeast to northwest, as shown by the terms "entrance" and "outlet." The alleged area of the Cyclades and Sporades, 700 miles by 200 miles (1,036 by 296 km.), would be larger than the entire Aegean, and the figure 700 may actually be stadia, or simply an irrecoverable error.

4.72. The next topic is the islands in the northern Aegean, beginning with the ones off the outer coast of Euboea and including those in the Thermaic Gulf of Macedonia and around the Chalcidice and Athos. This catalogue does not seem to reflect sailing routes but is probably from the same work on islands used previously. Imbrus, the most northeastern island in the Aegean, is out of place, lying (as noted) 22 miles (33 km.) from the Thracian Chersonesus and half again as far from Sciathos as the 88 miles (130 km.) recorded; there may be a confused vestige of a sailing route across the northern Aegean.

4.73. Despite the problems with Imbrus at *NH* 4.72, in this section it is properly located in reference to the three other large islands (Lemnos, Thasos, and Samothrace) in the northern Aegean. Mt. Athos lies slightly north of west from Lemnos (and about 60 km. away, not the 87 miles or 129 km. recorded). Even though the mountain is steep and precipitous, rising to over 2,000 m., it is hardly possible that its shadow could project into the forum of Myrina. Yet the peculiarities of its shadows were noted as early as the fifth century BC (Sophocles F776; Apollonios 1.601–4; Strabo, *Geography* 7F15). Athos was clearly an important navigational marker for the northern Aegean – there are many distances cited from it – but the persistent tales about its shadow must be local lore. The distances are generally accurate, although the 62½ miles (92 km.) between Thasos and Athos seems actually to be that between Abdera and Athos. The lack of harbors on these islands is evident and is a contrast to the highly indented shorelines of the more southern islands.

4.74. Five islands were between Samothrace and the Thracian Chersonesus, but the list is confused and cannot be related to islands in that location. Halonesos is an isolated island south of Lemnos (Strabo, *Geography* 9.5.16). Gethone is perhaps the Getone of *NH* 5.138, somewhere near the Troad. Lamponia, insofar as it is known, is a town on the west slope of Mt. Ida in the Troad (Strabo, *Geography* 13.1.58). Alopeconnesus was a town at the west end of the Thracian Chersonesus (Strabo, *Geography* 7F21).

In addition there is a second list of 21 islands, none of which can be identified. In theory they should be in the Melas Gulf, the northeastern corner of the Aegean, but this cannot be taken for granted. The toponyms cited do occur elsewhere (such as Arconesus, in the Propontis), but there is no certainty of the location of those on Pliny's list. This completes the survey of islands that began at *NH* 4.52.

The Hellespont to the Pontus and Maeotis

4.75. Pliny's fourth bay or gulf of Europe (discussed through *NH* 4.93) is the most anomalous topographically, since it includes only the western side of the Pontus (Black Sea), whose coasts were divided between Europe and Asia. Thus, following his own definition of European gulfs, the fourth ends at the strait known as the Cimmerian Bosporus (modern Kerch Strait), the eastern boundary of Europe. Examination of the remaining coast of the sea, from the Cimmerian Bosporus to the Thracian Bosporus on the Asian side, is reserved for *NH* 6.1–22, and presented in counter-clockwise order. Pliny has fallen victim to a problem inherent in ancient geographical studies, first encountered by Eratosthenes in the third century BC: attempting to define the surface of the earth in terms of broad shapes, a theoretical structure that could be useful but which might become unsatisfactory in practice (Roller, *Eratosthenes* 26–7). In addition, for ideological reasons, Pliny wanted to complete Europe before moving on to the other continents.

Because the Black Sea was called in antiquity merely the Pontus ("Sea"), there can be confusion between the proper and common noun. Pliny's orientation is, as usual, from a seaman's perspective, and his itinerary sails from the Aegean to the Black Sea and observes the sea in virtually animistic terms as it forces itself through the various narrows into the Aegean. The Hellespont (see also *NH* 4.49–50) was long known for its current into the Aegean: Homer called it "strong flowing" (*Iliad* 2.845, 12.30). Today it is two to three knots.

Xerxes' bridge was at the narrowest part of the Hellespont. There were actually two of them, about 600 m. apart and about two km. across, near the Cynossema (*NH* 4.49). Nothing remains of either (N. G. L. Hammond and L. J. Roseman, "The Construction of Xerxes' Bridge over the Hellespont," *JHS* 116 [1996] 88–107). Priapus is where the Hellespont ends and the Propontis (*NH* 4.76) begins; the unusually precise distance (86 miles or 128 km.) is not from the site of Xerxes' bridge but the entire length of the Hellespont. Alexander the Great actually crossed into Asia (spring 334 BC) farther to the southwest at the narrows of the Hellespont (Arrian, *Anabasis* 1.11–12), symbolically replicating Xerxes' crossing in reverse, and only reached Priapus somewhat later.

4.76. For someone sailing toward the Black Sea, the Propontis is after the Hellespont, and followed by the narrows of the Thracian Bosporus (*NH* 4.45–6). The bridge of Darius I of Persia, constructed in 513 BC, is only documented in passing by Herodotus (4.83–7) and Ctesias (F13.21, who wrote of "bridges"). It was said to be halfway between Byzantium and the southern entrance to the Bosporus, and inspired the first known historical epic, by Choirilos of Samos, titled *The Crossing of the Pontoon Bridge*, which may only have been a section of a longer work (Strabo, *Geography* 7.3.9; George Huxley, "Choirilos of Samos," *GRBS* 10 [1969] 12–29).

Again there is a precise measurement (239 miles or 354 km.), which is probably that of the entire passage from the Aegean to the Black Sea. The mariner's perspective is apparent in the description of how the coasts of the Black Sea recede on either side, similar to that of the opening of the Mediterranean at its west end (*NH* 3.3).

Axenus is the original indigenous name (latinized) of the Black Sea, passing through Greek from Persian *aexsaena*, meaning "dark" or "sombre." The name was turned into a euphemism, "Euxeinos," or "Friendly," but the sea was often merely called the Pontus, or "the Sea" (Stephanie West, "'The Most Marvellous of All Seas': the Greek Encounter with the Euxine," *G&R* 50 [2003] 151–67). The analogy with a Scythian bow not only has ethnic grounding but reflects the double curvature of the coasts of the sea. The comparison is of uncertain origin, but was documented by the first century BC (Strabo, *Geography* 2.5.22; see also Ammianus Marcellius 22.8.10). On the north side of the sea is the channel known as the Cimmerian Bosporus, which varies in width from six to 30 km. It was the boundary between Europe and Asia, and after a length of 50 km. opens into the Maeotis (the modern Sea of Azov).

4.77. The distance of 500 miles (740 km.) between the two Bosporus straits is accurate. There were many calculations for the circumference of the sea, demonstrating that a figure was not easily obtained. The unnamed other sources would include Eratosthenes, whose dimensions Pliny cited elsewhere (Eratosthenes, *Geography* F115–16 = *NH* 5.47, 6.3), and who had a figure of 23,000 stadia (Eratosthenes, *Geography* F114). A certain Hecataeus was allegedly the first to make the calculation, but it is not known whether this was the scholar from Miletus (*FGrHist* #1) or Abdera (*FGrHist* #264).

4.78. Varro is the primary source that Pliny used for distances along the Black Sea coast (see also *NH* 4.77), providing a sailing route from the Thracian Bosporus around the west side of the sea to Panticapaeum on the Cimmerian Bosporus. All of the mileages produce a round number when multiplied by eight, a conventional conversion factor for stadia. Varro's original source was probably Eratosthenes, since his total (recorded by Pliny as 1,338½ miles: Eratosthenes, *Geography* F115 = *NH* 6.3) is essentially the same. Agrippa's distance is significantly shorter; it may have been independently measured or have used a different conversion factor from stadia.

The Tanais (modern Don), which empties into the Maeotis at its northeastern corner, was said to originate in the Ripaei ("Tossing") Mountains, which in fact are mythical, a result of the early theory that there were mountains at the northern edge of the world (Aristotle, *Meteorologika* 1.13.350b; Strabo, *Geography* 7.3.1; Jacques Desautels, "Les monts Riphées et les Hyperboréens dans le traité hippocratique *Des airs, des eaux et des lieux*," *RÉG* 74 [1971] 289–96). The source of the Tanais was actually unknown to the Greco-Roman world: it is about 275 km. southeast of Moscow in a region devoid of any mountains. The border between Europe and Asia was disputed (the Phasis River, or modern Rioni in Georgia, was also considered), but by the fifth century BC it was being established at the Tanais (Herodotus 4.45). For Histropolis, see *NH* 4.44.

4.79. Pliny digressed from the coasts of the Black Sea to discuss the Hister (Danube) River. Greeks had known about its lower part since early times, calling it the Istros (Hesiod, *Theogony* 339), but the upper portion, the Danuvius (first noted by Julius Caesar, *Gallic War* 6.25), was not reached until the Romans moved into central Europe. The brothers Tiberius and Drusus, in 15 BC, were probably the first to discover its source (Strabo, *Geography* 7.1.5). The division between the Danuvius and Hister was at the modern Iron Gates on the Romanian–Serbian border.

The number of mouths of the Hister varied in the sources, from two (Eratosthenes, *Geography* F148) to seven (Strabo, *Geography* 7.3.15), due to changing topography and different ways of determining what constituted a mouth. Greeks had known about the outlet since the expedition of Darius I of Persia in the late sixth century BC (Herodotus 4.93–143); two centuries later Alexander the Great also reached the region (Strabo, *Geography* 7.3.8; Arrian, *Anabasis* 1.1–6).

4.80. "Scythian" had long been a generic term for peoples of the north (*NH* 4.81), although as the Romans penetrated into the Danube region, the ethnic names became more localized (Timothy C. Hart, "Evolving Ethnographies in Pliny the Elder's Transdanubian Exegesis (*HN* 4.80–1)," *CQ* 71 [2021] 1–8). The first ethnyms are theoretically those of the lower Hister, but their range is extremely broad, with populations extending all across the northern Black Sea regions from the river to the Tanais basin. These were peoples only vaguely known, and moreover were nomadic, so they might be encountered in different places at different times.

For the Getae, see *NH* 4.41. The Daci were either the same or a closely related sub-group. The Sarmatae (or Sauromatae) were often associated with the Scythians (Herodotus 4.117; Strabo, *Geography* 7.3.2) and chariot nomads, such as the Hamaxobi ("Wagon Dwellers"; Strabo, *Geography* 2.5.26, as Hamaxoikoi). They were generally located around the Tanais, but would have covered a wide region. The Aorsi were also said to be along the Tanais (Strabo, *Geography* 11.5.8). The Degenerate Scythians may be those who were ruled by women, an evolution of Herodotus' account of Scythians who mingled with the Amazons (Herodotus 4.110–17; Pomponius Mela 1.116; *NH* 6.19). The Trogodytae seem to have been within the curve of the lower Hister (Strabo, *Geography* 7.5.12). The Alani were also around the Tanais, but had moved into Armenia and Media in Pliny's day (Josephus, *Jewish War* 7.244). The Rhoxolani were originally thought to be the most northern of peoples, but were probably not far beyond the Danube (Strabo, *Geography* 2.5.7). The Sarmatian Iazygae had been encountered by Ovid in his exile at Tomis, just south of the mouths of the Hister (Ovid, *Epistulae ex Ponto* 1.2.77), but may have moved inland toward the Hungarian Plain.

4.81. The regions and peoples mentioned are on the north side of the Hister; the Marus is the modern Morava, which joins the Danube just below Vienna. The Suebi (Greek Soebians; the name survives in modern Swabia) were considered the most powerful group east of the Rhine (Caesar, *Gallic War* 1.37; Strabo, *Geography* 7.1.3; Tacitus, *Germania* 38). Vannius' kingdom may be the Vendelians (Vandals) to their northeast. The Basternae (or Bastarnae), another extensive group, lived on the upper Vistula (Strabo, *Geography* 7.2.4). The Germani were also a broad ethnic group in the north of Europe, living east of the Rhine, and were first encountered by Julius Caesar (*Gallic War* 1.36).

The 1,200 miles (1,775 km.) of Agrippa (F21) cannot be exactly placed but would approximate the distance from the lower Hister to the English Channel. The width, from the Sarmatian Wilderness, is probably a trade route from the Hister to the upper Vistla (modern Vistula). The latter river may be part of a route used by Pytheas of Massilia in the fourth century BC, but was still little known in Pliny's time (Pytheas F5 = Strabo, *Geography* 2.4.1; Pomponius Mela 3.33). For the Scythians as a generic term for northern peoples, see Ephorus, F30a. As more was learned about the populations along the Danube, the name "Scythian" tended to become more limited, and to move beyond current knowledge.

4.82. The account returns to the Black Sea and a *periplous* along its coast north from the mouth of the Hister to the Tyra (or Tyras, modern

Dniester) River, 130 miles (192 km.) away. The distance is rather long. The Tyragetae – the "Getians on the Tyra" – with their island home were probably near its mouth. The Axiacae are otherwise unknown, but their river is probably the Tiligul, about 60 km. beyond the Tyra. The Crobyzi were south of the Hister in the fifth century BC (Hecataeus of Miletus F170; Herodotus 4.49) but may have migrated north by Pliny's day.

The next major river beyond the Tyra is the Borysthenes (modern Dnieper), the second largest of those emptying into the Black Sea (Herodotus 4.53). Its lower reaches were a region of extensive Greek settlement (largely Milesian in origin), located on a large bay that is the common estuary of the Borysthenes and the Hypanis (modern Bug). These people had a certain notoriety as the northernmost Greek populations, where the climate was unfamiliarly rugged for people from the Mediterranean, and fauna and flora were alien (Strabo, *Geography* 7.3.18).

4.83. The *periplous* continues along the coast east of the Borysthenes. The Race Course of Achilles is at the south side of the Hypanis estuary, a sandy spit 65 km. long today. The cult of Achilles was common in this region, documented as early as the fifth century BC (Herodotus 4.55; Strabo, *Geography* 7.3.19).

The Sardi are otherwise unknown; the Siraci were generally located east of the Maeotis (Strabo, *Geography* 11.5.8), but seem to have moved west during the reign of Claudius, since at that time they threatened the Bosporanian kingdom, located around the Cimmerian Bosporus (Tacitus, *Annals* 12.15–16). The Enoecadoe were probably on the lower Borysthenes. The Panticapes River is not certainly located but may be the modern Ingulets or Vissum, a northern tributary of the Borysthenes and about 45 km. above its mouth (thus not below Olbia). The river had long been believed to be the limit of agriculturalism, with nothing but nomads beyond. In the fifth century BC they were known to extend north for a journey of 14 days, probably the limit of a trading network (Herodotus 4.19).

There were two rivers named the Hypanis, the modern Bug at Olbia, and the modern Kuban, which drains into the Maeotis east of the Cimmerian Bosporus (Strabo, *Geography* 11.2.9). Since one was in Europe and the other in Asia, there could be confusion.

4.84. Carcinites Bay (modern Karkinyts'ka) is the large gulf on the northwest edge of the Chersonesus (Strabo, *Geography* 7.3.18). Today it is about 90 km. from the Maeotis, but this is a region of many lakes and sloughs, and the five miles (7.4 km.) reported may reflect the location where a portage was necessary, at the west end between Lake Buces and the

Carcinites. The Coretus Bay is the northwest part of the Maeotis, and there was clearly a well-established route across this region, using bridges and canals as necessary. The Basilidae (Basileians or Royal Scythians) were a dominant indigenous group, presumably named for the nature of their political organization (Herodotus 4.20; Strabo, *Geography* 7.3.17). The Hylaei (perhaps "Forest People") lived on the (European) Hypanis.

4.85. Taurica, or the Taurian Chersonesus (the modern Crimea), was famous for its association with Iphigeneia. There was a Greek population on the southern coast and there were indigenous villages in the interior. The six village populations are otherwise unknown. The Scythotaurians – a generic term – presumably included these villagers, and perhaps others on the high ridge (the Tauric Range) of the southwest Chersonesus, which reaches 1,545 m. a few kilometers from the coast, providing a refuge from which they could harass the coastal towns (Strabo, *Geography* 7.4.3). The Stactares were probably in the lowlands to the northeast.

4.86. Criu Metopon (Ram's Head) Cape is the southernmost point of the Tauric Chersonesus (modern Cape Sarych). It marked the point where the perceived halves of the Black Sea – the two arms of the Scythian bow (*NH* 4.76) – came together. The distance across to the Carambis Promontory, one of the northernmost points of the Asia Minor shore of the Black Sea, is slightly more than the 170 miles (252 km.) quoted, but it was possible to make the crossing without losing sight of land. Theodosia, 125 miles (185 km.) east of Criu Metopon by coastal sailing, is at the western end of the isthmus that separates the Black Sea from the Maeotis.

4.87. Panticapaeum (Pantikapaion, modern Kerch), situated on the west side of the Cimmerian Bosporus, was the capital city of the Bosporanian kingdom, which had emerged in the early fifth century BC. It was integrated into the Pontic kingdom of Mithridates VI in 109 BC, and after his death became an allied kingdom of Rome, which was its status in Pliny's day (Iris von Bredow, "Regnum Bosporanum," *BNP* 12 [2008] 443–50). Panticapaeum itself had been founded by Milesians in the seventh century BC as a trading outpost within an indigenous settlement (Boardman, *Greeks Overseas* 253). Cimmerium, on the Asian shore, has not been mentioned previously in the extant text of the *Natural History*.

The cold weather of the north was always intriguing to those from the Mediterranean world, especially the idea that the sea could be frozen. Neoptolemos, a commander of Mithridates VI, fought a battle on the ice in the early first century BC, and Greeks could hardly understand a world where it was impossible to produce the grapevine and olive tree (Strabo, *Geography* 2.1.16, 7.3.18).

The Cimmerian Bosporus is about 50 km. long, and the 12.5 miles (18.5 km.) is probably the distance from Panticapaeum to one of the Asian towns. Taphrae (modern Perikop) is actually at the northwest corner of the Tauric Chersonesus, not on the Maeotis, but the 260 miles (385 km.) would conform to a sailing route between it and the Bosporus.

North of the Black Sea

4.88. From Taphrae the account enters the semi-mythical world north of the Black Sea, collecting together notices of many of the remote and little-known peoples of that region. The Auchetae (or Auchatiai, Herodotus 4.6) would have been in western Ukraine (the Hypanis or Bug is 774 km. long). The Neuroe were north of the Borysthenes region, and in the fifth century BC there was no known population to their north (Herodotus 4.17). The Budini and Geloni were segments of the same peoples, who lived in a wooden city named Gelonos, which had Greek architecture but whose location is unknown (Herodotus 4.108–9). It was said that the Geloni were originally Greeks, and it is quite possible that a settlement of Greek traders was assimilated into a local population, probably somewhere in northern Ukraine. Their name is orthographically Greek. The Thyssagetae were in Asia, according to *NH* 6.19, but were presumably also from northern Ukraine. For the Basilidae, see *NH* 4.84.

The Agathyrsi were only vaguely located somewhere in the north (Herodotus 4.100), and the Anthropophagians ("People Eaters") were placed nearby. Their name is not an ethnym, but a Greek descriptive term, and may be a generalization about certain peoples of the far north who practiced cannibalism, which did exist in this region (Strabo, *Geography* 7.3.9). For the Sauromatae (or Sarmatae), see 4.80. The Essedones, or Issidones, were another remote population, one of the farthest known, although they were enough connected with the Greek world for Herodotus to provide a detailed ethnography. He placed them in central Asia, in the context of the expedition of Cyrus the Great against the Massagetae in the sixth century BC (Herodotus 1.201).

Beyond them were the Arimaspi, who were known through the obscure work of Aristeas of Proconnesus, the *Arimaspeia*, written sometime before the mid-fifth century BC (Herodotus 4.13–14). The work is a mixture of fantasy, mythology, and ethnography, but there seems little doubt that the Arismaspi represent a remote ethnic group somewhere in central Asia.

Pterophoros ("Feather-Bearing") is a toponym not cited previously and was probably a created name, based on Herodotus' description of the feathery snow of the far north (Herodotus 4.7, 31; J. D. P. Bolton, *Aristeas of Proconnesus* [Oxford 1962] 44–5). As noted, Greeks were incredulous at the strange climate and phenomena of the north and, extrapolating from data in temperate latitudes, were early aware that matters would change the farther north one went. When Pytheas of Massilia went into the Arctic in the fourth century BC, he became the first Greek to experience the midnight sun and the long nights (Strabo, *Geography* 2.1.18; 2.5.8, 43). The Aquilo is the north wind.

4.89. Hyperborei ("Those Beyond the North") is another created Greek ethnym, describing the semi-mythical people of the distant north, even beyond the Ripaean Mountains. They were first mentioned in the Homeric *Hymn to Dionysos* (7.29; see also Hesiod, *Catalogue of Women* F98). Although Pliny's data has some actual information about the far north, the Hyperboreans belong in the context of fantasy geography, idealized societies that were free of contention and disease, and which lived in a perfect climate beyond the known world (Timothy P. Bridgman, *Hyperboreans: Myth and History in Celtic–Hellenic Contexts* [New York 2005]). Fantasy geography was a common literary phenomenon in Hellenistic times, when such cultures could be created for allegorical reasons (Roller, *Eratosthenes* 121–2). Pliny's source is probably *On the Hyperboreans* by Hecataeus of Abdera (*FGrHist* #264, F7–14), from the fourth century BC. He connected them with the Cimmerians, another remote population, but the ethnym fits into Pliny's trajectory at this point.

4.90. Further comments on the Hyperboreans placed them around the Attacians, who were well to the east, beyond India (*NH* 6.55), or even across the External Ocean, suggesting a location on an unknown continent. Yet at the same time it was believed that they were in the nearest part of coastal Asia, which would mean somewhere east of the Cimmerian Bosporus. These contradictory locations demonstrate their mythical nature. The tale about their six-month agricultural practices is an interesting way of attempting to reconcile the realities of their climate with the necessities of an agrarian people.

4.91. The Hyperboreans had an unusual relationship with the island of Delos. The story, reported in detail by Herodotus (4.32–6), is not merely a myth but an important account about early trade contacts. The connection with Delos may be because Apollo, born on the island, was associated with the Hyperboreans. Pliny greatly summarized Herodotus' account, which tells how two young women (Arge and Opis) came to Delos. What is of interest is that he provided details of their route: from their home to

Scythia to the head of the Adriatic, and then down to Dodona, across the Greek peninsula to the Malian Gulf, and to Euboea and past various islands to Delos. This is a replication of the amber route from the Baltic, which entered the Greek world at the upper Adriatic (*NH* 3.152), and had been known since prehistoric times. Herodotus' account, as Pliny noted, also recorded an essential fact about long-distance trade, that hardly anyone travelled the entire route. The termination of the trade may reflect the collapse of Delos as a mercantile center due to the depredations of Mithridates VI in 88 BC.

The distances recorded by Agrippa (F22), 980 by 716 miles (1,450 by 1,060 km.), seem to approximate those across Europe: it is about 1,200 km. from the western Black Sea to the Baltic. Despite Pliny's realization that there was a certain amount of uncertainty, and with no modern comprehension of the terminal points, the numbers are a surprisingly accurate measurement of the extent of northern Europe.

Islands in the Black Sea

4.92. It was Pliny's custom to end broad geographical sections with a discussion of the local islands. The Symplegades ("Clashers," modern Örektaşi), on the north side of the Black Sea mouth of the Thracian Bosporus, were notorious maritime perils because they appeared to move: Pliny's explanation, from a sailor's point of view, is quite reasonable. They were a problem faced by the Argonauts (Euripides, *Andromache* 795).

The Apollonatae were small islands in and around the city of Apollonia (modern Sozopul in Bulgaria). It had a famous Temple of Apollo, which M. Licinius Lucullus plundered in 72 BC, removing its cult statue, one of the largest known (30 cubits high, or about 14 m.). He placed it on the Capitol in Rome (*NH* 34.39; Strabo, *Geography* 7.6.1). For the islands at the mouth of the Hister, see *NH* 4.79.

4.93. The island of Achilles, normally called Leuke (modern Zmeinij), is in the western Black Sea, about 40 km. offshore. It was where the hero was said to be buried (*Aithiopis, Argument* 4; Strabo, *Geography* 7.3.16). The island rises sharply from the sea and is isolated, so it was a significant navigational hazard, and Pliny triangulated its location.

Because of the current from the Black Sea to the Aegean, and the encircling nature of the External Ocean, it was believed that there was a constant flow from the great rivers north of the Black Sea through it and into the Aegean, and eventually across the Mediterranean and out into the

Atlantic. This was probably the theory of Straton of Lampsacus, in the fifth century BC, who considered the nature of the surface of the earth and its hydrology (Straton F54; Hipparchus, *Against the Geography of Eratosthenes* F6; Strabo, *Geography* 1.3.4–5).

Northern Europe

4.94. Leaving the Black Sea, Pliny's narrative moves across Europe, past the hypothetical Ripaean Mountains (*NH* 4.78) to the Northern Ocean (Septentrionalis, the modern North Sea [*NH* 2.167]). Although he was attempting to make a coastal circuit of Europe, this portion of the coasts of the Ocean was totally unknown, and a land route was necessary. The name Northern Ocean was probably applied at the time of the expedition of Drusus in 12 BC. There is no evidence that Pliny was following any particular route through Europe, but was merely recording data from the map of Agrippa. An itinerary from the North Sea to Gades would cover all of the known Atlantic coast of Europe, and may reflect the voyage of Pytheas of Massilia in the fourth century BC, who was said to have covered the entire coast from the Tanais to Gades (Strabo, *Geography* 2.4.1). Yet northern Europe was so poorly known that few toponyms were recorded. Many of them remain highly disputed by modern commentators, and can be located in widely separated places in northern Europe and Scandinavia. Archaeological evidence is equally uncertain (Thomas Grane, "Pliny and the Wandering Mountain. A New Interpretation of Pliny's Account of the Northern Barbaricum With an Archaeological Context," *ARID* 37 [2012] 7–28).

Pytheas (T21) is in fact the source for much of this material, presumably handed down to Pliny by other writers, especially Timaeus of Tauromenium (F75a), and the little-known Philemon, probably of the first century AD (*NH* 4.95; 37.33, 36; Keyser, "Philemon," *EANS* 645). Xenophon of Lampsacus (Dueck, "Xenophon," *EANS* 840–1), another obscurity, wrote on the islands of the External Ocean, perhaps in the second century BC; it is not certain whether his information on the Baltic was new or derivative from Pytheas.

The island of Baunonia, where amber came from, is probably in the Baltic and may be the same as Basilia or Balcia (*NH* 4.95; Hecataeus of Abdera [*FGrHist* #264] F14; Roller, *Through the Pillars* 88–9). Amalcius presumably is a far north locale, perhaps in the upper Baltic.

4.95. All the toponyms seem to be around the Baltic, a region of little Greco-Roman exploration after that of Pytheas. The diet of the inhabitants of Oeonae (an island in the Baltic: Pomponius Mela 3.56) is reasonable enough for northern peoples but would have been peculiar to Greeks and Romans. But as was often the case in discussing populations at the fringes of the known world, the accounts can often degenerate into fantasy, such as the people with horses' hooves (obviously called the Hippopodes), or those with unusual ears. Such anatomically improbable people were also said to be common in India (Strabo, *Geography* 15.1.57). Whether the toponym "Balcia" is the earliest citation of "Baltic" is possible but highly debatable. If such is the case, the name was forgotten (except locally) until medieval times (J. Svennung, *Belt und Baltisch. Ostseeische Namenstudien mit besonderer Rücksicht auf Adam von Bremen* [Uppsala 1953] 92–3).

4.96. Pliny's account has, in part, moved away from the interior Baltic to its west end, where he could rely on more contemporary reports, such as that of Drusus and other Roman commanders over the following thirty years (Roller, *Ancient Geography* 163–4). Pytheas also remained a source, as demonstrated by the Saevo Mountains, which seem to be in Norway (although this is far from certain: see Grane, "Pliny" 11–12), where there was the requisite topography in a region that the Romans did not reach. The Inguaeones were on the North Sea coast (Tacitus, *Germania* 2).

The orientation of the account is from the Atlantic into the Baltic (the Codanus) and east as far as the Vistla (Vistula, *NH* 4.97), which probably replicates Pytheas' route. Scatinavia (Scadinavia, Scandinavia) was first cited by Pomponius Mela (3.54) and was said to be a vast land of five hundred communities that was hardly known; Greco-Roman sources were probably only aware of the south end, where variants of the name are still used locally. The Hilleviones are otherwise unknown, but the assertation that they lived in "another world" demonstrates that this was the end of Roman knowledge, and thus the effective end of the world (M. Cary and E. H. Warmington, *The Ancient Explorers* [revised edition, London 1963] 60–1). Aeningia may be the land of the Fenni (Finns), mentioned by Tacitus (*Germania* 46). These comments are all reports by the locals at the mouth of the Baltic about the extensive lands to their north.

4.97. The account mixes information from Pytheas with that of the expeditions of the Roman period, which probably did not go far into the Baltic if at all. For the Sarmatae, see *NH* 4.80. They hardly belong in the Baltic region but may have been seen generically as spreading in that direction. In fact, the list of ethnyms reflects people scattered over a wide area, the usual problem of locating little-known people in a little-known region. The Venedi may be the Venethi of Tacitus (*Germania* 46), located on the Vistula around the site of Warsaw. The Sciri and Hirri (perhaps the same people) are inhabitants of unspecified areas in eastern Europe: the Sciri, at least, had trade contacts with the people of Olbia, at the mouth of the Borysthenes on the Black Sea, as early as the third century BC (*SIG* 495.104).

The Cimbrian Cape is modern Jutland, which a Roman fleet rounded in the Augustan period, seeing the Baltic opening up before them (*NH* 2.167; Augustus, *Res gestae* 26). The expedition also became aware of the numerous islands in the region, although Pliny's list includes some that are actually in the North Sea, such as Burcana (modern Borkum, in Germany). In AD 16 the fleet of Germanicus was wrecked in this region during a bad storm (Seneca the Elder, *Suasoriae* 1.15; Tacitus, *Annals* 2.23–5). But Glaesaria, the amber island, must be in the Baltic, and Germanicus actually went that far (*NH* 37.42). The toponym may come from the indigenous word "glesum," probably related to modern Swedish "glas" (Tacitus, *Germania* 45), although Austeravia could be as far as Lithuania (Audroné Bliujiené, *Northern Gold: Amber in Lithuania (c. 100 to c. 1200* [Leiden 2011]).

4.98. The Scaldis River is the modern Escaut or Schelde in the Netherlands and Belgium. Presumably Pliny was refering to the region between it and the mouth of the Baltic; despite the Roman expeditions of the Augustan period, the interior was hardly known. There was scant information about distances in this part of the world: Caesar noted that the people of northern Europe had no means of measuring their travels other than in days (Caesar, *Gallic War* 6.25). The distance of 2,500 miles (3,700 km.), perhaps converted from 20,000 stadia, would be essentially the entire north–south extent of Europe, which may be what Pliny had in mind. That from the Baltic to Raetia and Noricum (in the eastern Alps) approximates the 636 miles (941 km.) reported by Agrippa (F17); the width (248 miles or 367 km.) is far too short for any measurement across Europe, and may refer only to that between Raetia and Noricum. Agrippa died in 12 BC, shortly after the expedition of Tiberius and Drusus into Raetia and the upper Danube, and did not have the opportunity to incorporate their material into his writings. Germania was to some extent explored by Drusus in 12 and 9 BC, as well as his son Germanicus two decades later.

4.99. The Vandili were at the southern edge of the Polish plain, around the middle Vistula. The Burgodiones, who later moved into southern France and gave their name to Burgundy, had been in central Germany east of the Rhine. The Varinnae and Charini are little known. The Gutones, perhaps the ancestors of the Goths, were east of the lower

Vistula (Strabo, *Geography* 7.1.3). The scattered nature of these ethnyms – from the middle Rhine to beyond the Vistula – shows the lack of information even in Pliny's day about the ethnography of Germany. It was only with Tacitus' *Germania*, published 20 years after the *Natural History*, that there was a detailed study of these regions.

For the Inguaeones, see *NH* 4.96. The Cimbri were originally in Jutland, with the Chaucians to their southwest (*NH* 16.1–2; Tacitus, *Germania* 35–6). The Teutoni were nearby, perhaps even extending into southern Scandinavia (Pomponius Mela 3.54).

4.100. Pliny's account has moved south, into the territory of the Rhenus (Rhine), where the Istuaeones and their sub-group the Sicambri were located. The Hermiones were on the Elbe (Tacitus, *Germania* 2). For the Suebi, see *NH* 4.81. The Hermunduri were east of the Elbe (Strabo, *Geography* 7.1.3), and the Chatti lived around modern Hesse (Tacitus, *Annals* 1.55–6). The Cherusci were between the Weser and the Elbe: these were the people of Arminius, whose coalition was responsible for the destruction of three Roman legions in AD 9 (Tacitus, *Annals* 1.60–2; Suetonius, *Augustus* 23; Dio 56.18–24). The Peucini and Bastarnae (or Basternae; see *NH* 4.81) seem to have been the same ethnic group and were to the east of the Cherusci; Tacitus (*Germania* 46) was not even certain they should be considered Germanic.

There is a list of seven notable regional rivers, extending across northern Europe from the Vistula to the Rhine (Hubert Zehnacker, "L'Europe du Nord dans l'*Histoire Naturelle* de Pline l'Ancien (N. H. IV, 88–104)," *RÉL* 82 [2004] 167–86). Then Pliny mentioned the mysterious Hercynian Mountains, perhaps the first toponym in Europe north of the Alps known to the Greco-Roman world. The earliest citation is by Aristotle (*Meteorologika* 1.13.350b), as Arkynian, a great mountain range from which all the northern European rivers originated, probably the Alps. Two centuries later, Eratosthenes believed that the toponym (which he recorded as Orkynian) referred to a vast forest, nine days across, presumably a Greek conception of the extensive woodlands of northern Europe (Eratosthenes, *Geography* F150; Caesar, *Gallic War* 6.24–5). Pliny reverted to a mountain range and placed the toponym in a region conspicuous for its lack of mountains, perhaps using Aristotle as his source.

4.101. The mouth of the Rhenus (Rhine) has various channels, connected with the Mosa (modern Meuse), and which produce a large number of islands. The distance of 100 miles (148 km.) would be about that from where the Rhine divides into channels which flow downstream to the ocean. The Batavi were on one of these islands, with the Cannenefates as their neighbors, toward the ocean.

The Frisi and the Frisiauones were probably the same (perhaps Tacitus' Greater and Lesser Frisians, *Germania* 34). They were along the coast north of the Rhine, and the Chauci (*NH* 4.99) to the north. The Sturi are otherwise unknown, and the Marsaci were south of the Mosa: many of these peoples took advantage of the Roman civil war of AD 68–9 to revolt (Tacitus, *Histories* 4.56).

Islands in the North Atlantic

4.102–3. Britannia was first explored by Pytheas in the fourth century BC (Strabo, *Geography* 4.5.1–5), who was one of Pliny's major sources for the region, supplemented by Isidorus of Charax (*FGrHist* #781, F11), writing in the Augustan period, as well as data from the expeditions of Julius Caesar in 55–54 BC (Caesar, *Gallic War* 4.20–36, 5.6–23) and Claudius a century later (Suetonius, *Claudius* 17; Dio 60.19–23). Pliny also knew about the cruise of the Carthaginian Himilco from around 500 BC, who may have reached Britannia (*NH* 2.169). The name Albion is usually presented as an ethymn, Albioni, and may refer to a specific group of people rather than a toponym; it was little used in the Roman period (Avienus, *Ora maritima* 112). The Morini were on the coast of France at the narrows of the English Channel (the modern Pas de Calais), which is only about 37 km. wide. Gerovacium was at modern Boulogne. The stated circumference of Britannia (4,875 miles or 7,215 km.) is difficult to understand and does not seem to conform to the 12,900 stadia (about 2,580 km.) recorded by Pytheas (T22; Strabo, *Geography* 4.5.1) or the 2,000 miles (2,960 km.) of Caesar (*Gallic War* 5.13), both of which are more reasonable, as are Agrippa's measurements (F8) of 800 by 300 miles (1,184 by 444 km.). Pliny's figure is so excessive that a transmission error is likely.

Calidonia (usually Caledonia), referring vaguely to Scotland, was not well known to the Roman world until the expedition of C. Julius Agricola in 83 BC, but Pytheas may have learned about the toponym (Pliny, *Histoire naturelle* 4, 332). Hibernia (Ireland) may also have been reached by Himilco, but was still only slightly comprehended by Hellenistic geographers. The comments are sparse (Diodorus 5.32; Caesar, *Gallic War* 5.13) and its size was overestimated. Pliny's material is topographic and no ethnyms were provided. The Silures were in southern Wales, hardly the closest point to Ireland but perhaps the only ethnic group near the Irish Sea known to Pliny. A crossing of 30 miles (44 km.) is about half the actual distance.

The islands listed are not near Hibernia, but around Britannia and elsewhere. The Orcades are the Orkneys, north of Scotland; the Haemodae seem to be in the Baltic (Pomponius Mela 3.54). The Hebudes are the Hebrides, west of Scotland. Their name may actually be Phoenician (due to Carthaginian activity in the area) and mean "Sheep Islands" (George Broderick, "Some Island Names in the Former 'Kingdom of the Isles': A Reappraisal," *Journal of Scottish Name Studies* 7 [2013] 4–5). The remaining islands are assumed to be west and south of Britannia, except for the Glaesariae, for which see *NH* 4.97.

4.104. Tyle is the famous Thule, or Thoule, discovered by Pytheas (T23), and probably but not certainly Iceland (Roller, *Through the Pillars* 78–87). The comments about its peculiar astronomical phenomena are from Pytheas, perhaps obtained by Pliny from Timaeus of Tauromenium (F74) or, more likely, Diodorus (5.22). The abrupt change in the middle of the section to the tin industry, and from the Arctic to Ictis, generally located in or near Cornwall, follows the pattern of Diodorus. The tin could be sent across the English Channel and then follow the Loire–Rhône route across France. It is quite possible that Pytheas, who probably followed this route himself, was aware of the industry. Boats of hide and wicker (*corium* and *vitilis*) were a common means of transport in this region; a gold model now in the National Museum of Ireland may represent one (Barry Cunliffe, *Facing the Ocean: the Atlantic and Its Peoples* [Oxford 2001] 66–8).

The final part of the section returns to the north of Europe, and has ethnyms such as Scandiae and Bergi (perhaps Bergen in Norway). The mention of Tyle and Berrice (perhaps the Faeroes or Shetlands) suggests that this is a confused report of Pytheas' route from Tyle back to the mainland. For Cronium, see *NH* 4.95. The "frozen sea" (*mare concretum*) is a phrase directly translated from Pytheas' *pepegyria thalatta* (Strabo, *Geography* 1.4.2). It is probably not the pack ice of the Arctic Ocean, but more likely glaciers or frozen inlets.

Coastal and Central Gallia

4.105. Pliny returned to the mainland of Europe and Gallia, paraphrasing the opening of Caesar's *Gallic War* and outlining the main features of the region, although Caesar was not his only source. He defined the district by its rivers, from the Scaldis to the Garunna (modern Garonne), and beyond to the Pyrenees. Since he had already discussed the Mediterranean coast and the southeast portions of Gallia (*NH* 3.31–7), the current description (through *NH* 4.109) is limited to the Atlantic litoral and the interior portions not previously considered.

Agrippa's distance (F7) of 1,750 miles (2,590 km.) is perhaps the equivalent of 14,000 stadia and thus may reflect a Greek source, perhaps Posidonius' *On the Ocean*. The distance seems long, but the coast is highly indented. The other two figures (420 and 318 miles, or 622 and 471 km.) cannot easily be positioned.

4.106. After the catalogue of rivers, Pliny's discussion of Gallia is almost totally a list of ethnyms, arranged geographically and oriented from north to south. The first series outlines those living immediately beyond the northern and eastern boundaries of Gallia, with a group of seven along the coast south of the Scaldis. This is followed by nineteen ethnyms and two toponyms naming inland populations, from the Catoslugi in the northwest to the Raurici and Helvetii in the western Alps. This third group consists of a few ethnyms east of the upper Rhine that had not been mentioned previously.

4.107. Lugdunensian Gallia is essentially central modern France, centered on Lyon. The first ethnyms are those peoples north of the Liger (modern Loire) River. Pliny also took note of the peninsula of extreme northwestern France, the home of the Ossismians (Strabo, *Geography* 4.4.1). In antiquity this was a long narrow strip of land, although the recorded circuit of 625 miles (925 km.) is excessive, and perhaps refers to the distance from the mouth of the Scaldis (or even Rhenus) to that of the Liger. Today the peninsula has broken up into a number of islands, of which the westernmost and best known is Ushant (derived from Ossismians), long important as a navigational point from the English Channel to the Atlantic. This information about the extremities of Gallia may have come from Pytheas (T3). Not all of the twenty ethnic groups in the following list are south of the Liger (e. g. the Parisii), but the catalogue makes a broad swath from the mouth of the river (the home of the Namnetes, at modern Nantes) to Lugudunum (normally Lugdunum, modern Lyon).

4.108. In the southwest of Gallia were the Aquitani, with thirteen ethnic groups, from the Ambilatri south of the mouth of the Liger to the Tarbelli at the west end of the Pyrenees. The final list of twenty peoples seem to be mostly on the north side of the mountains, although many are obscure.

4.109. The final section about Gallic ethnic groups lists those in the central part of the territory, inland from the Pictones, who were south of the Liger (*NH* 4.108), as far as the Tarnis River (modern Tarn), which is a northern affluent of the Garunna (*NH* 4.105). The Tolosani (*NH* 3.37, the inhabitants around modern Toulouse) were about 50 km. south of the Tarnis.

To complete his material on Gallia, Pliny catalogued the seas, from the Northern (the North Sea) south along the coast to the Gallic (the modern Bay of Biscay) and as far as the Pyrenees. The Venesti were on the southern coast of modern Brittany, north of the mouth of the Liger. They were notable mariners, who caused difficulty for Caesar (*Gallic War* 3.8–10). There were attempts to connect them with the Venetians at the head of the Adriatic, but this was considered problematic even in antiquity (Strabo, *Geography* 4.4.1). There were also a number of islands off the Venestian coast.

Western Hispania

4.110. Most of the remainder of Book 4 (through *NH* 4.120) is about the portions of Hispania that were not covered in Book 3 (6–30). Pliny's pattern of following the coasts of Europe has led to a strange splitting of the Iberian Peninsula into Mediterranean and Atlantic portions, with some duplication between the two books. The material in Book 4 is essentially the Atlantic coasts and some of the immediate interior.

After some general statements about the peninsula and the Pyrenees, referring back to Book 3, the account follows the ocean around the outer side of the peninsula. Pliny generally continued the pattern of Book 3, describing regions in terms of the Roman organization. For the Varduli, see *NH* 3.26–7. The Morogi are otherwise unknown.

4.111. For the Cantabri, see *NH* 3.21, 27. The account continues along the south coast of the Bay of Biscay. The distance from Victoria Iuliobrigensium (near modern Santander) to the source of the Hiberus (modern Ebro), calculated at 40 miles (59 km.), is reasonable and reflects a trade route across the Iberian Peninsula. The Orgenomesci were west of modern Santander. For the Astures, see *NH* 3.28. At the Navia River the jurisdiction of Lucensis began, and the Albiones, Cibarci, Egi, Namarinian Varri, Adoni, Arroni, and Arrotrebae were all small populations west of the river, extending as far as where the coast of Hispania begins to turn to the south (at modern Punta dos Aguillóns).

Cape Celticum (Celtic Promontory, modern Cabo Touriñán), essentially the westernmost point of this part of the Iberian peninsula, is where the coast makes its final turn to the south. The Neri and Supertamarci lived in this region. The altars of Sestius were erected by L. Sestius Quirinalis (the consul of 23 BC), who campaigned against the Cantabrians in the following year (Pomponius Mela 3.13). The Copori were in the lower Ulla River valley, east of the Celtic Promontory, with the Praestamarci to their west on the coast. The Cileni were to their south, and Corticata and Aunios were two of the small islands in this region.

4.112. The description continues south into the Bracarian jurisdiction. The Helleni, Grovi, and Tyde were all on the lower Minius River (the Minho or Miño, the modern boundary between Spain and Portugal). Because of the ethnym Helleni, it was believed that this was an area of Greek settlement, but this is based on hardly more than the Greek-sounding names (Tyde was a town in Aetolia; Silius Italicus 16.368–9).

The Leuni and Seurbi were south of the Minius. The source of the Durius River (modern Duero or Douro) is northeast of modern Madrid, among the Pendelonians, and at 824 km. it is the third longest river in the Iberian Peninsula. For the Arevaci, see *NH* 3.19; the Vaccaei, 3.19, 26; the Vettones, 3.19; the Gallaeci (or Callaecians), 3.28; the Turduli, 4.116; and the Bracari, 3.28. The mineral riches of northwest Iberia were discussed at 3.30.

4.113–14. Lusitania, essentially modern Portugal and some of the territory to its east, had been known to the Roman world since the second century BC and became a Roman province through the Augustan reorganization of 27 BC. The Veteres (Old) Turduli were on the south side of the lower Durius, and the Paesuri were around the Vagia River (modern Vaga).

Pliny conflated the names of the two western promontories of the Iberian Peninsula. Cape Artabrum is the Celtic Promontory of *NH* 4.111. The Great (Magnum) or Olisiponensian Promontory is modern Cabo de Roca, just west of Lisbon (ancient Olisipo), the westernmost point of Europe. It was a major navigational marker and had been known to Greek mariners probably since the sixth century BC (Avienus, *Ora maritima* 171–2). It was also believed to divide the coast into two parts, as well as separating the Gallic Ocean (to the north) from the Atlantic. The latter term was originally limited to the more southern region, orginally accessed through the Pillars of Hercules. They were assumed to be the ancient Pillars of Atlas, and the Atlantic derived its name from him (Homer, *Odyssey* 1.51–4; Aeschylus, *Prometheus Bound* 349–52; Herodotus 1.203). When the Greeks ventured into what would today be called the North Atlantic, they considered it to be a different ocean.

Nevertheless Pliny confused the Great Promontory with the Artabrian farther north, and was uncertain about the location of the Artabres and

Arrotrebae (probably two groups of the same people). The 60 or 90 mile (89 or 133 km.) ridge that defines the Great Promontory is probably the uplands to the northeast that are isolated from the surrounding territory, although the distance is excessive.

4.115. Using Varro as his primary source, Pliny noted some variances in distances and river names along the Lusitanian coast. The toponym Aeminius is not cited elsewhere, but Pliny recorded that the river had several names, and, moreover, he confused two streams. The Minius had been mentioned previously (*NH* 4.112); the Limaea (the modern Lima or Limia), whose mouth is about 20 km. south of the Minius, was also called the Belion. This may be the source of the name Oblivio ("Forgetfulness"), turning an indigenous name into a Latin descriptive term. Oblivio was first cited by Livy in his report on the local campaign of D. Junius Brutus in the 130s BC (Livy, *Summary* 54–5). Greek writers carried the onomastic evolution one step farther and called the river the Lethe, a translation of Oblivio (Strabo, *Geography* 3.3.4).

The Tagus River (modern Tejo or Tajo) is the longest in the Iberian Peninsula (1,007 km.) and empties into the Atlantic at the Great Promontory west of Lisbon. The distance of 200 miles (296 km.) between the Minius and Tagus is reasonable. The gold-bearing sands of the Tagus were noted by the poets of the late Roman Republic and the first century AD, beginning with Catullus (29.19–20; for a complete list see Pliny, *Histoire Naturelle* 4, 364). Elsewhere Pliny commented that the Tagus was one of several rivers in the Greco-Roman world that produced gold (*NH* 33.66).

The Sacred (Sacrum) Promontory, modern Cabo de São Vicente, is the southwestern corner of the Iberian Peninsula, about 160 miles (237 km.) from the Tagus, as noted. The idea that it projects from the middle of Hispania is probably taken from the map of Agrippa, which may have emphasized its position. A distance of 1,400 miles (2,072 km.) to the middle of the Pyrenees is difficult to critique and may reflect a road route.

4.116. Pliny's account reaches the Anas River (modern Guadiana) and the boundary between Lusitania and Baetica, 126 miles (186 km.) from the Sacred Promontory. This is where his survey of Europe began at *NH* 3.6, and thus he has made a complete circuit of the continent. He also emphasized the connection with Gades (*NH* 3.7), 102 miles (151 km.) farther.

For the Celtic Turduli, see *NH* 3.13, and the Vettones, 3.19. The Lusitani originally extended from the Durius to the Tagus, but later moved south toward the Anas (Appian, *Iberike* 50–7). The Romans first came into contact with them in 194 BC, and there were difficult military engagements for over a century, until Caesar's campaign of 60 BC (Livy 35.1.1–12; Strabo, *Geography* 3.3.3).

The horses of Olisipo would stand facing west into the Favonius wind and conceive. They would give birth quickly but the colt would only live for three years (*NH* 8.166). The association of equine conception and the winds was an old folk tale, alluded to by Homer (*Iliad* 16.148–51, 20.219–22), and located by Aristotle on Crete (*Research on Animals* 6.18.572a). Vergil also made use of the story (*Georgics* 3.271–9).

4.117–18. Referring to official Roman documents, Pliny provided a demographic survey of Lusitania – following the pattern previously used for the Iberian Peninsula – distinguishing towns and populations by their Roman status. The province was organized into three broad groups: those attached to Augusta Emerita (modern Mérida), Pacensis (modern Beja), and Scalabis (modern Santarém). Olisipo was unique in the province and was one of the oldest settlements in the region, since the estuary of the Tagus had attracted Greek traders by the sixth century BC (Avienus, *Ora maritima* 174–7).

There is an alphabetical list of 18 tribute-paying towns in Lusitania. In contrast to the similar one for Baetica (*NH* 3.15), here ethnyms rather than toponyms are used, suggesting that the official document was not of the same type as for Baetica.

The length of Lusitania, Asturia, and Gallaecia, essentially the western half of the Iberian Peninsula, calculated at 540 miles (799 km.), is reasonable. The northern end is probably the Celtic Promontory (*NH* 4.111, 114) or another of the prominent capes in the region. The southern is somewhat speculative, but the most probable is the Sacred Promontory (*NH* 4.115–16). The reported total circuit of Hispania, from one end of the Pyrenees around to the other, varies from 2,600 to 2,924 miles (3,848 to 4,328 km.). The lesser measurement conforms to 21,000 stadia, and it may be from Posidonius' *On the Ocean*.

The Atlantic Islands

4.119. The Cassiterides (Kassiterides, "Tin Islands") are one of the more mysterious toponyms of northwest Europe. The word is not indigenous, but Elamite, which had come into Greek at an early date as the word for tin, reflecting its original sources for the Mediterranean world (Homer, *Iliad* 11.25, etc.). A certain Midacritus, probably a Phocaean or Massilian, was the first to visit them and to bring tin ("white lead") to the Mediterranean (*NH* 7.197), at some time before the fifth century BC (*NH* 34.156; Herodotus 3.115; Diodorus 5.38; Strabo, *Geography* 2.5.15).

The earliest sources cite only a single island, Cassiteris, but eventually they evolved into a group. Yet by the time of Herodotus their location was not known – the traders who went there kept their position secret – and they had become a vague semi-mythical place at the fringes of the world. The Romans made a number of attempts to find them, and improbably claimed to have done so. In modern times they have been never been located with certainty (and they may not have been islands, but promontories in little-known regions); they were probably somewhere along the western end of the English Channel (Roller, *Through the Pillars* 12–14).

Pliny's account borders on the mythical, naming a number of islands of uncertain location. The Six Gods are otherwise only cited by Ptolemy (although two in number, *Geographical Guide* 2.6.76), and the Fortunate Islands are the Islands of the Blessed, the home of the heroes, somewhere in the Atlantic (Hesiod, *Works and Days* 166–73), a mythical locale that was long sought. As an actual toponym they were found in many places and believed by some to be the Canaries (*NH* 6.202–5; Roller, *Through the Pillars* 44–9).

Julia Gaditana, or Gades (modern Cádiz) was the ancient Phoenician settlement of Gadir, on an island off the southwest coast of the Iberian Peninsula (*NH* 3.7, 17), located 25 miles (37 km.) northwest of the western end of the "Strait," the modern Strait of Gibraltar. Pliny's primary source was Polybius (34.15.3), perhaps accessed through Varro. The distances of 12 by 3 miles (18 by 4.5 km.) are probably converted from 100 and 25 stadia, and refer to the island on which the town was located, which is now joined to the mainland except for a canal at the southeastern end. The distance of 15 miles is 120 stadia. The exact date that its inhabitants were given Roman citizenship is not known, but Augustus granted the city further privileges (Pliny, *Histoire naturelle* 4, 372–3).

4.120. The second island (probably at modern San Sebastián) has also become part of the mainland. It lies inland of the first at its southern end. Pliny here has used the original Phoenician name for the settlement, Gadir, although he did not reconcile his sources, contradicting himself by placing it on both islands. Erythea is an ancient toponym, where Hercules (Herakles) killed Geryon (Hesiod, *Theogony* 287–94), perhaps a descriptive name referring to the red sunsets of the region. It was a location associated with Gades by the fifth century BC (Herodotus 4.8), a view followed by the Hellenistic historian Philistides (*FGrHist* #11, F3). Timaeus of Tauromenium (F67) and Silenos (*FGrHist* #175, F7) proposed a cultic name, Aphrodisias, after Phoenician Astarte. Juno, presumably used during the Roman period, may be a version of Phoenician Tanit. This catalogue of alternate names for the city, from sources obscure today, may come from Posidonius' *On the Ocean*; he lived there for a period in the early first century BC.

The hydrology of Gades was unusual. There were wells near sea level in the town, and their properties were affected by the tides. Polybius (34.9.5–7) was the first to consider the issues, but the most detailed examinations were by Posidonius (F217) and his contemporary Artemidorus of Ephesus (F14; Strabo, *Geography* 3.5.7). Potinusa ("Drinkable") is actually a Latin descriptive term and may be from Varro.

For Tartesos, see *NH* 3.7, where it was equated with Carteia, at the Pillars of Hercules. There is much that is mysterious about Tartesos, including the Greco-Roman understanding of it, and the toponym was located in several places in southern Iberia (Strabo, *Geography* 3.2.11; Celestino Pérez and López-Ruiz, *Tartessos* 120–1). "Fence," "enclosure," or "wall" (Pliny's *saepes*) are reasonable translations of the name Gadir, perhaps referring to a fortified trading settlement. But Erythea cannot be a local toponym, since it is Greek, and could only be applied after the first Greeks arrived in the seventh century BC. The Erythrean Sea is usually located around the Arabian Peninsula, certainly an area of Phoenician interest (Strabo, *Geography* 16.3.1).

The theft of the cattle of Geryon was the tenth labor of Hercules (Herakles). Pliny had already discussed the route of the hero back to Greece through the Alps (*NH* 3.134). The Geryones are probably nothing more than a created ethnym based on the myth, and most accounts speak of a single individual, perhaps originally a local deity. The story was known to Hesiod (*Theogony* 287–94) and was the subject of an epic by Stesichorus, the *Geryoneis*, from the sixth century BC; a handful of fragments survives. By the following century the tale had become localized in the vicinity of Gades (Herodotus 4.8). But even in Roman times Erythea continued to be located in many places, including Lusitania (Pomponius Mela 3.47).

The Dimensions of Europe

4.121–2. To end his circuit of Europe, Pliny provided a number of distances, obtained from various sources of the second and first centuries BC. All that are named are Greek. That from Tanais to Gades (8,714 miles or 12,897 km., probably 70,000 stadia) is most likely from Pytheas – who explored between those two points (Strabo, *Geography* 2.4.1) – as reported by Artemidorus of Ephesus (F8) of around 100 BC and Isidorus of Charax a century later (F8).

Polybius (34.15.1) had a distance from Italy to the Ocean, but this is too vague to understand. Nevertheless he was more knowledgeable about the width of Europe than Pliny admitted, since he had either actually made the overland journey himself from Massilia to Korbilon at the mouth of the Liger (Loire), or talked to people who had (Polybius 34.10.6).

Pliny provided an analysis of Polybius' figures and a route from an unspecified point in Italy across the Alps to Lugdunum and the Morini. He was certainly familiar with the routes across the western Alps toward the Rhone, and then north to the Morini, who were probably at modern Boulogne (*NH* 4.102, 106). Pliny adjusted his account somewhat since Lugdunum (modern Lyon) did not exist in Polybius' day, but its future site at the junction of major rivers would have been an obvious place for an earlier trading rendezvous. The route provided from Italy to the English Channel would be a circuitous way to cross Europe, but the itinerary was dependent on the known trade connections.

There was also a totally separate route north from Italy to the mouth of the Rhine (1,243 miles or 1,840 km.), only possible in Roman times and probably based on information gained from Pliny's service in Germania in the AD 40s and 50s. This would have followed Roman roads, and need not have been a straight line, perhaps even connecting the Roman military camps north of the Alps (for a list, see Pliny, *Histoire naturelle* 4, 375). As a point of comparison, the official modern distance from Rome to Amsterdam is 1,646 km.

BOOK 5

Africa and Western Asia

Book 5 is oriented on the southern and eastern coasts of the Mediterranean. The discussion includes the known portions of the continent of Africa (essentially as far south as the tropics, 1–46), and then continues to Egypt, which is considered a region separate from Africa (47–64). After an account of Egypt that is limited to its more northern portions, the narrative moves through coastal Asia, from western Arabia north into Judaea and Phoenicia, and to Syria, with a digression on the Euphrates River (65–90). Then there is consideration of the coastal territories of Asia Minor, from Syria around as far as the Troad (Troas) and the Thracian Bosporus (91–127). The book concludes with the Aegean islands not previously discussed (those close to the Asian shore), some of the inland territories of western Asia, and the islands in the Propontis (128–151). The end of the book leads directly into the beginning of Book 6, continuing the examination of northern Asia Minor.

It can be seen that the geographical orientation of Book 5 is somewhat more diverse than that of Books 3 and 4, which were specifically limited to Europe. In part this is due to the anomalous position of Egypt (neither African nor Asian), as well as the complexities of Asia. Its scope is from the Pillars of Hercules to the Thracian Bosporus, which to some extent provides a complementary structure to that of Books 3 and 4 (now the southern coast of the Mediterranean rather than the northern), but there are more digressions into the interior, such as up the Nile (*NH* 5.47–54) or the course of the Euphrates (*NH* 5.83–90), so the underlying linear *periplous* is not as apparent as previously. Pliny's knowledge of Africa penetrated well into its interior, although without much topographical coherence or understanding (Brent D. Shaw, "The Elder Pliny's African Geography," *Historia* 30 [1981] 424–71; Antonio Santana Santana and Trinidad Arcos Pereira, "La descripción de África en Plino (*Nat.* V y VI)," in *Libyae lustrare extrema: Realidad y literatura en la visión grecoromana de África* [Seville 2008] 347–60). Even though Book 5 covers parts

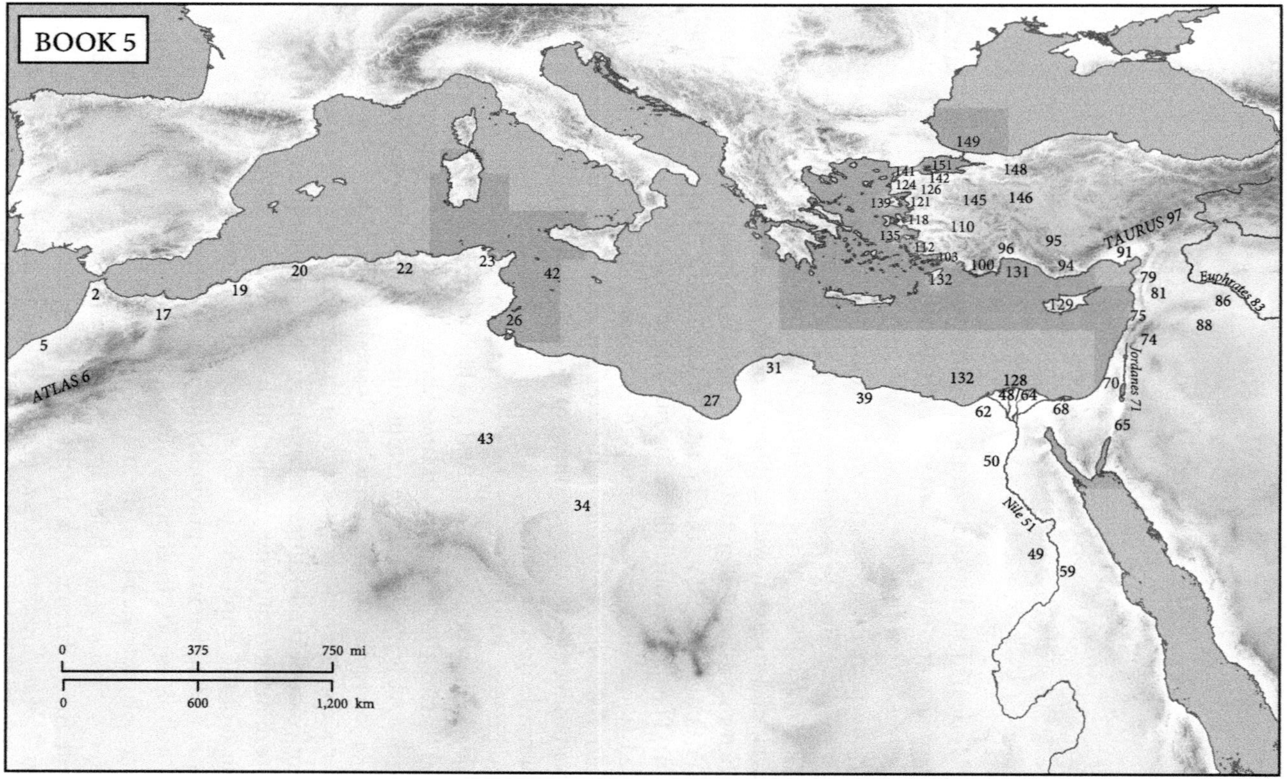

Map 5 Places cited in Book 5 of the *Natural History*, according to chapters (courtesy of Ancient World Mapping Center).

of two continents, interior Arabia and Aethiopia are reserved for Book 6 (*NH* 6.142–97). These regions were seen to be more connected with the world of the Indian Ocean and Mesopotamia than the Mediterranean.

Introduction, Mauretania, and the Atlantic Coast

5.1. The Greek toponym Libya (Libye) was originally an ethnym (Libyes), perhaps first documented in connection with the foundation of Cyrene in the seventh century BC and referring to the inhabitants of Africa west of Egypt and outside the Cyrenaica (*SEG* 9.3, line 33; Hecataeus [*FGrHist* #1] F346, 357). By the fifth century BC the term had begun to evolve into a toponym (Herodotus 4.191–9), eventually meaning all of what is today called Africa outside of Egypt.

It was customary in antiquity for seas to be named after the peoples on their coasts; thus there were many localized terms for the Mediterranean. By the first century BC the Libycan Sea referred to the entire southern portion of the Mediterranean, along the Libyan (African) coast, from the western edge of Egypt to the Pillars of Hercules (Strabo, *Geography* 2.5.19; Pomponius Mela 1.21–3).

As the Roman presence increased, the term Africa came to replace Libya for the continent (again, outside of Egypt). This also was an ethnym that had become a toponym, originally referring to the population west of Carthage, but becoming more generalized by the second century BC (Plautus, *Poenulus* 1304; Sallust, *Jugurtha* 17.3). Yet it remained essentially Roman and Latin terminology; Greek writers continued to use Libya, even when discussing Roman matters.

That Africa (or Libya) did not include Egypt was established by the fifth century BC (Herodotus 4.41), with the boundary – at least by Roman times – at the Catabathmos ("Descent"), located at modern Sollum, which is still the boundary between modern Libya and Egypt (Sallust, *Jugurtha* 17.4; Pomponius Mela 1.40). This was a locality where the main road drops down a sharp defile to the coast (Strabo, *Geography* 17.1.14; Roller, *Historical and Topographical Guide* 953).

The contrast between the relatively smooth north coast of Africa – extending for over 3,000 km. – and the highly indented European coast of the Mediterranean was notable (Sallust, *Jugurtha* 17.5). The peculiarities of foreign names were a common complaint of Greco-Roman geographical authors (Strabo, *Geography* 16.4.18; *NH* 3.139). The use of the word *castella* (fortresses) was a comment on the limited urbanization of Africa: the word suggests a defensive settlement or village, contrasted with a traditional town or city (Pliny, *Histoire naturelle* 5.1 [ed. Jehan Desanges, Paris 2003] 79). Needless to say, Africa had its cities (such as Cyrene, Carthage, and Mauretanian Caesarea), and the *castella* were limited to the indigenous population.

5.2. The first region discussed is that of the two Mauretanias (through *NH* 5.21), the northwest portion of the continent of Africa. Pliny presented the name as a plural because of its historic division into two parts. The districts had been local kingdoms, which became extinct by 33 BC. Eight years later Juba II and Cleopatra Selene were placed on the throne of a united Mauretania by Augustus as allied monarchs (Strabo, *Geography* 6.4.2; Dio 53.26; Roller, *World of Juba* 100–6). Juba was succeeded by his son Ptolemy (the grandson of Cleopatra VII and Marcus Antonius) in late AD 23 or early the following year (Tacitus, *Annals* 4.23), who ruled until he fatally tangled with his cousin, the emperor Gaius Caligula, probably because he was showing too much independence and flaunting his Ptolemaic ancestry. The emperor summoned him to Rome and in time he was executed, late in AD 40 (Suetonius, *Gaius* 35, *Caligula* 35; Roller, *World of Juba* 253–5). Yet Caligula was unable to provincialize the kingdom before his death early the following year (Pliny was in error in suggesting that he had done so, but presented a more accurate version of the events at *NH* 5.11). It took several years of Roman military operations – the Mauretanians were hardly pleased at the removal of their king and revolted – and it was actually Claudius who effected the provincialization (Dio 60.8–9), dividing the former kingdom into two (restoring the situation before the accession of Juba II), with Mauretania Caesarensis to the east (essentially modern Algeria) and Mauretania Tingitana to the west (modern Morocco).

In *periplous* fashion, Pliny provided an itinerary from the region of the Pillars of Hercules into the Atlantic and along the coast (through *NH* 5.10), with his primary sources Hanno of Carthage (for whom, see *NH* 2.169), Polybius (see *NH* 5.9), and the map of Agrippa. The account extends almost to the equator. Distances are inconsistent and contradictory, representing the various sources and the effects of a little-known region, with the maximum reported at 336 miles (497 km.), from around the Pillars to Rutubis Harbor (*NH* 5.9). Yet the itinerary goes far beyond this point with no distances provided. Rutubis may be at modern Moulay Abdallah (or Tît) in Morocco (Pliny, *Histoire naturelle* 5.1, 112), but like most of the toponyms on this coast can only be located speculatively. The lack of distances beyond Rutubis may be because this was the limit of the map of Agrippa.

The first portion of the *periplous* is along the strait between the Mediterranean and the Atlantic, particularly noting Tingi (or Tingis, modern Tangier). It was an ancient city with mythological origins (*NH* 5.3–4), lying due south of Baelo in Baetica (*NH* 3.7). The crossing of 30 (presumably miles, or 44 km.) is long – the actual distance is more like 32 km. – but the route may not have been direct, as is the case with ferries crossing the strait today. Yet this was the primary connection between the Iberian peninsula and western Mauretania. At least part of the African coast in this region was considered a portion of Baetica: Zelil (or Zelis, Strabo, *Geography* 3.1.8, probably at modern Asilah) was under its control. Attaching certain parts of Mauretania to Baetica probably occurred during the interregnum between the death of the last indigenous king, Bocchus II, in 33 BC, and the accession of Juba II eight years later.

The *periplous* continues around the northwestern extremity of Africa (Cape Ampelusia, modern Cape Spartel) and along the Atlantic coast. Lixus is one of the few places that can be identified without question, located at modern Larache in Morocco, an important archaeological site with remains from at least the sixth century BC (M. Euzennat, "Lixus," *PECS* 521). It is at the mouth of the modern Loukkos River, which preserves the ancient name, and has a rich mythological history (*NH* 5.3–4).

5.3. Lixus was the home of the giant Antaeus (Antaios), noted for his wrestling skills. As a son of Poseidon, he drew his strength from the earth, and when thrown became even stronger. He decorated his palace with the skulls of his victims. Yet Hercules (Herakles) lifted him up and defeated him by not allowing him to touch the ground (Pindar, *Isthmean* 4.52–7; Apollodorus, *Bibliotheke* 2.5.11). Whether there was any physical evidence surviving in Roman times that was believed to be his palace is unknown. He allegedly founded Tingi (*NH* 5.2) and was buried there. The Roman adventurer Sertorius, effective ruler of the city in 81–80 BC, reported that he had found the grave and excavated the remains (Plutarch, *Sertorius* 9).

The Hesperides ("the Daughters of the Evening") guarded the trees that produced golden apples in the garden mentioned by Pliny. Their acquisition was the last labor of Hercules, who killed the dragon that protected them (Pherecydes [*FGrHist* #3] F17; Sophocles, *Women of Trachis* 1099–1100; Euripides, *Herakles* 394–402). The serpentine turns of the Lixus River – still visible today – were said to represent the dragon. Places associated with Hercules abound on the Atlantic coast of Mauretania; there was a cave and altar of the hero west of Tingi at a locality still called the Grottoes of Hercules (Strabo, *Geography* 17.3.3; Pomponius Mela 1.26; Michèle Coltelloni-Trannoy, *Le Royaume de Maurétanie sous Juba II et*

Ptolémée [Paris 1997] 174–7). The tidal peculiarities of the region were also known to Strabo, with Juba II the probable source.

5.4. Citing Cornelius Nepos, Pliny was dismissive of the mythological tales of both Greeks and Romans, as well as certain statements by Nepos himself. That Lixus was more powerful than Carthage is an exaggeration, but nevertheless it was the most important city on this coast, and its visible remains testify to this. Pliny also took issue with Nepos' use of "lying opposite" (*ex adverso*), a navigational term generally used to mean the relationship of places on opposing coasts (Herodotus 2.34), but here seemingly implying localities on the same latitude (at *NH* 6.199; again quoting Nepos but derived from Hanno, it is the Carthaginian trading post of Cerne that is "opposite" Carthage). Pliny rejected Nepos' statement of the relationship between Carthage and Lixus (although their latitude is similar) in part because of Nepos' report that the distance between Lixus and Tingi was impossible to measure. Yet Pliny had already shown this to be 57 miles (84 km., *NH* 3.2). To be sure, Nepos was probably writing before the map of Agrippa laid out the distances that Pliny used.

5.5. The *periplous* continues south as far as Sala (at Chellah, near modern Rabat in Morocco; M. Euzennat, "Sala," *PECS* 793–4). There are digressions on a number of inland cities, some of which were given colonial status by Augustus, probably during the interregnum of 33–25 BC. The most significant of these is Volubili (more commonly Volubilis), arguably the most important archaeological site in Morocco. It did not become a Roman municipality until the provincialization of the AD 40s (M. Euzennat, "Volubilis," *PECS* 988–9; Pliny, *Histoire naturelle* 5.1, 94). The city actually lies somewhat closer to the Atlantic than the Mediterranean, at the northern edge of the Atlas uplands.

The distances in this section are probably from the map of Agrippa, and are questionable and appear contradictory. From Lixus to Banasa (at Sidi Alibou Jenoun) is about 115 miles (170 km.), and it is 35 miles (52 km.) farther to Volubili. The coastal distance of 50 miles to the mouth of the Sebubus River (modern Oued Sebou) is short; it is about 75 miles (111 km.) today, but the lower course of the river may have changed since antiquity.

The presence of the elephant in this region has been documented from early times: Hanno encountered the animal north of the Lixus (Hanno 4). The local elephant was one of several species known in Greco-Roman times (*NH* 8.32). The North African type had become extinct by late antiquity (H. H. Scullard, *The Elephant in the Greek and Roman World* [Cambridge 1974] 30–1; André Jodin, "L'éléphant dans le Maroc antique," in *Actes du quatre-vingt-douzième congrès national des sociétés savantes* [Paris

1970] 51–64). Juba discussed the animal thoroughly in his *Libyka* (*FGrHist* #275, F47–54), whose extant fragments are scattered through various sources, especially the *Natural History* and Aelian's *On the Characteristics of Animals*.

The Autoleles (whose name may be a Greek descriptive term: "Self-Sufficient People") lay on the route inland from Sala to Mt. Atlas. The mountain is the subject of a lengthy discussion, continuing to *NH* 5.18.

5.6–7. The Atlas is not exactly a mountain, but a range extending across northwest Africa, from western Tunisia to the Atlantic coast of Morocco, a distance of over 2,000 km. On average it is about 250 km. inland, separating the fertile coastal regions of northwest Africa from the western Sahara. The highest part of the range is modern Toubkal, near the western end, rising to 4,167 m.; it was this summit that was probably the Mt. Atlas of antiquity. On both sides its base is low, giving the mountain a particular prominence, and the range is by far the highest in Africa north of the equatorial regions. Mt. Atlas was an object of curiosity in antiquity: it was said to extend to the heavens, in fact to the orbit of the moon. This is less curious than one might think, since until the Alps and Caucasus were explored, it was the highest known peak in the ancient Mediterranean world. Hanno was aware of the mountain (Hanno 7), as were presumably other travellers in the region. The indigenous name is Dyris or Addiris (Strabo, *Geography* 17.3.2; *NH* 5.13), but by the fifth century BC the range was called after the giant Atlas, whose home was near to the ocean that bears his name (Hesiod, *Theogony* 507–19). Pliny's source for the mountain is probably Juba, supplemented by the report of C. Suetonius Paulinus in the AD 40s (*NH* 5.14)

Because of its remoteness, the Atlas was seen almost as a mythical paradise, especially on the north side, where there were hidden fertile valleys, many fresh springs, and refuges of elephants in their old age (Aelian, *On the Characteristics of Animals* 7.2). The locals were elusive, although their fires were visible and their music could be heard at night. Pliny has mixed in material from Hanno about peoples farther south, near the equator (Hanno 14; Pomponius Mela 3.95), although there is also a certain generic quality about the description of strange sounds at night in remote areas. The Aegipani, or Goat Pans, were first mentioned in extant literature by Pomponius Mela, who placed them farther south, implying that he had learned about them from Hanno, although no such term appears in the extant text of the Carthaginian explorer. They are usually associated with Satyrs (Pomponius Mela 1.48), the wild animalistic people of Greek folk tradition; mention of the two may be more of a descriptive gloss than two specific ethnic groups.

Due to their unusual features and the remoteness of the Atlas mountains, they were remarkably prominent in Greek myth. Tales connected with Hercules abound in northwest Africa: the apples of the Hesperides were said to be on Mt. Atlas (Apollodorus, *Bibliotheke* 2.5.11). Perseus' association with the region is somewhat more obscure, but the Gorgons were believed to live in the vicinity of the Hesperides, and the implication is that this is where the hero killed Medusa (Hesiod, *Theogony* 270–94). Pliny's final sentence – that the distance to Mt. Atlas is "immense and uncertain" (*inmensum incertumque*) – demonstrates that at this point he was considering the mountains in their mythological context, not as the geographical feature cited elsewhere in the description.

5.8. Pliny had introduced Hanno the Carthaginian, who lived around 500 BC, at *NH* 2.169. His presence is constant in the early sections of Book 5. Yet Pliny showed too much incredulity at his account: the extant manuscripts of the explorer's text list only six city foundations (Hanno 1–5). Pliny may be referring to the report of three hundred Carthaginian cities on the Atlantic coast that had been abandoned by the first century BC (Strabo, *Geography* 17.3.3). All of these need not have been founded by Hanno – although the opening lines of his report indicate that his expedition began as a major colonizing effort – and the fact remains that there were certainly many Carthaginian outposts along the Atlantic coast, even if the number of three hundred is not believable.

5.9–10. Polybius, better known today as a historian, was an avid explorer, which is how he wanted to be remembered (Pausanias 8.30.8). He was at the fall of Carthage in 146 BC, and was given a ship by the conqueror, P. Cornelius Scipio Aemilianus, in order to investigate the Carthaginian presence on the Atlantic coast of Africa. His report has not survived, but references to it, in addition to *NH* 5.9–10, are at Polybius 3.59.7, *NH* 6.199, and Geminos 16.32 (see also Marijean H. Eichel and Joan Markley Todd, "A Note on Polybius' Voyage to Africa in 146 B. C.," *CP* 71 [1976] 237–43; Pliny, *Histoire naturelle* 5.1, 106–21; Walbank, *Historical Commentary* 3, 633–8; Roller, *Ancient Geography* 137–9). Pliny's account also mixes in material from Hanno (5–7) and the map of Agrippa (the distances in miles are probably from the latter) as well as Juba (Alfredo Mederos Martín, "Los Periplos Atlánticos Norteafricanos de Polibio y Juba II Trasmitidos por Plinio (*N. H.*, V, 9–10)," *Gerión* 31 [2013] 239–68). After Rutulus Harbor there are no more town names or mileages, but only ethnyms and rivers; this may mark the end of the map.

The Darat River was the northern limit of the range of the crocodile, an animal that had been known to exist on this coast since the time of Hanno and his contemporary Euthymenes of Massilia (Aristides 36.85–95). Hippopotami were also first recorded at the same time. If the Darat is the modern Draa, as seems probable, 616 miles (912 km.) would be the approximate distance from the Pillars of Hercules.

The Aethiopians and Gaetulians were ethnyms that were applied over a wide area of northern Africa. Although the former appears in Hanno's text (7, 11), it is unlikely that he used this hellenized term, which must be due to the translator, unless Hanno had access to Greek literature, which is not impossible. But throughout Greek literature "Aethiopian" was a generalization for the people of Africa beyond the northern coastal regions (Homer, *Iliad* 1.423, 23.205–6), with the name carrying a sense of remoteness. By the fourth century BC they were seen as one of the four ethnic groups defining the extremities of the inhabited world (along with the Indians, Scythians, and Celts; Ephorus F30a [= Strabo, *Geography* 1.2.28]). The Gaetulians were generally the people of interior Mauretania, but both ethnyms were loosely applied to the little-known region beyond the Atlas, providing familiar names to describe the locals.

By the middle of *NH* 5.10 Pliny has changed sources, from Polybius to Hanno (although he may have had access to Hanno through Polybius). Theon Ochema, or Chariot (or Support) of the Gods, and Cape Hesperu (Cape of the West, where Pliny created a latinized Greek genitive) are toponyms from the Greek version of Hanno's text (14, 16), although it actually has the Horn (Ceras) of the West, changed by a later source (perhaps Polybius) to Cape. These toponyms are in equatorial West Africa; there is little doubt that Theon Ochema, described by Hanno as an active volcano, is Mt. Cameroon, the only such feature on this coast (P. Schmitt, "A la recherche du Char des Dieux," in *Litterature gréco-romaine et géographie historique: Mélanges offerts à Roger Dion* [*Caesarodunum* 9bis, Paris 1974] 473–9).

The sailing time between Theon Ochema and Cape Hesperu, ten days, is not the figure in the extant text of Hanno and may be from Polybius; Hanno reported four days (Hanno 16), which was also recorded elsewhere by Pliny (*NH* 6.197). In the final sentence of the section the antecedent of *locavit* ("located") is not perfectly clear. The obvious one is Polybius, but the statement has a visually descriptive quality that suggests the map of Agrippa. Regardless, the positioning of Mt. Atlas is not to be taken as a precise geographical note but a generalization: the mountain is roughly in the center of the circuit from the Pillars of Hercules to Theon Ochema. The two comments – that Atlas is in the center and that it is the farthest part (*in extremis*) of Mauretania (here used in the singular and referring only to Mauretania Tingitana) – are not contradictions but two different perspectives of the mountain.

5.11. When Ptolemy of Mauretania was executed, probably in late AD 40, the Mauretanians revolted, led by a certain Aedemon, presumably a senior member of the royal household. The revolt seems to have been widespread, and was put down by various Roman commanders (Coltelloni-Trannoy, *Le Royaume* 60–5; Duncan Fishwick, "The Annexation of Mauretania," *Historia* 20 [1971] 467–87). With provincialization in place, the first governors were of senatorial rank, such as C. Suetonius Paulinus (*NH* 5.14), but in time an equestrian administration was established.

5.12. Pliny, arguing against unnamed personalities, showed a certain tension between the equestrian and senatorial orders, especially those of the former who became senators. There is also a reference to the predatory and luxurious nature of some equestrians, whose primary interest in Mauretania was its wealth. Three luxury commodities from Mauretania are listed. The first is ivory, which had been a valuable product in the Mediterranean world since the Bronze Age: the Greek word, *elephas* (Homer, *Iliad* 4.141, etc.), demonstrates that early encounters with the elephant were to harvest their tusks. Ivory was the most expensive animal product obtained from the land (*NH* 37.204) and was exported to Rome from the elephants of Mauretania for a variety of artistic and decorative uses.

Citrus wood was another important Mauretanian export, processed into luxury furniture as early as the late Republic: it was said that Cleopatra VII used it in her palace (Lucan 10.144–6). But the demand was so great that by Pliny's day the best wood was no longer available, perhaps due to equestrian extravagance.

Finally, Mauretanian purple dye (called Gaetulian purple) was of the highest quality. The industry may have originated in the Carthaginian period, and Juba established a dye works at islands off the Atlantic coast (at modern Essaouira or Mogador); the islands are now joined to the mainland (Juba F43 = *NH* 6.201). Gaetulian purple was in favor during the Augustan period, almost as much as the famous variety from Tyre (Horace, *Epistles* 2.2.181–2; Tibullus 2.3.58; Ovid, *Fasti* 2.319; Pomponius Mela 3.104; *NH* 9.127).

5.13. At *NH* 5.5 the *periplous* had reached the town of Sala (modern Chellah near Rabat). Here Pliny called it Salat, indicating a different source, probably local information (*indigenae*) obtained by Juba or Suetonius Paulinus rather than the Carthaginian and Greek material previously used. This second *periplous* continues down the coast to the Asana (modern Oum el-Rbia) and Fut (modern Tensift) Rivers, where there was a route inland to the Atlas. The rounded distances indicate locally derived approximations rather than measurements, probably converted from travel days. The use of the indigenous name for Mt. Atlas, Addiris, also demonstrates a local source (Vitruvius 8.2.6; Strabo, *Geography* 17.3.2).

The Ivor River cannot be located, and the form of the name is variable in the manuscripts of the *Natural History*. The implication is that it was an affluent of the Fut, but it may be the modern Ksob, farther down the coast, as suggested in the *Geographical Guide* of Ptolemy (4.1.3, in the form Diour). The location of the oasis, whether on the coast or inland, is equally enigmatic; it may be one of the abandoned Carthaginian settlements that were common along this coast (Strabo, *Geography* 17.3.3). Viniculture is documented in this region from at least the fourth century BC and was probably a Carthaginian import (Pseudo-Skylax 112; Dionysios Skytobrachion [*FGrHist* #32] F8 = Diodorus 3.68.2).

5.14–15. In AD 41, C. Suetonius Paulinus was at the beginning of a lengthy and distinguished career that would reach its peak twenty years later when, as governor of Britannia, he put down the revolt of Boudica (Caitlin C. Gillespie, *Boudica: Warrior Woman of Roman Britain* [Oxford 2018] 28–33). Pliny's personal reference may indicate that the two knew each other, and the unusually lengthy section on Mt. Atlas may be due to this friendship. Suetonius' consulship was probably shortly after his service in Mauretania (Tacitus, *Annals* 14.31–7). Pliny had access to his report on Mauretania, which was probably unpublished (*FRH* #83, F1). Although *NH* 5.14–15 specifically refers to his campaign, other details in Pliny's account of the region may also come from his material.

What was notable is that Suetonius actually crossed the Atlas and went far to their south, well into the Sahara. Many suggestions have been made about his route, but all are speculative, although he was able to see Mt. Atlas in summer snow (Pliny, *Histoire naturelle* 5.1, 135–41). He was unusually interested in the flora and fauna of the region, which would have been strange to anyone from the Mediterranean. The tree in question may be the Mt. Atlas cedar (*Cedrus atlantica*), which can reach a height of 40 m. (Pliny, *Histoire naturelle* 13 [ed. A. Ernout, Paris 1956] p. 99).

The only toponym recorded beyond the Atlas was the Ger River (whose name, as is often the case in sub-Saharan Africa, simply means "river," as in Niger: J. Oliver Thomson, *History of Ancient Geography* [Cambridge 1948] 268). It is plausible that this was the Niger but there is no proof. Reports of a major river in the southern Sahara had existed since the fifth century BC, when the routes across the desert became known to the Mediterranean world (Herodotus 2.32). If Suetonius reached the Niger he went a long way: he had marched 10 days to reach the Atlas (perhaps from a point on the Mediterranean coast), and then continued through the desert to the Ger, astonished at the winter heat (probably the winter of AD 41–2). Today the caravan route from the end of the road in southern Morocco to Timbuctu on the Niger takes 52 days, as shown on the famous sign at Zagora. If Suetonius followed this, or its predecessor, his expedition would have lasted several months and he certainly reached the tropics and saw its unusual wildlife. The exotic flora and fauna of southern Mauretania was a frequently reported feature of the region: Strabo provided a long list (*Geography* 17.3.4).

Pliny suggested that the local Canarii (who probably lived just south of the Atlas) were animalistic in nature, perhaps a conclusion merely from the assumption that their name meant "Dog People." Yet it is probably a local ethnym; whether it has any connection with the toponym Canaria, a name applied by Juba (in the Greek form Kynika) to an island off the Mauretanian coast (modern Gomera, one of the Canary Islands) is difficult to determine (Juba F44 = *NH* 6.205). This name was not indigenous, but was given to the inhabitants because of the dogs (Greek *kynes*) on the island. The relationship between the mainland Canarii and the islanders remains uncertain.

5.16. The Perorsi are the only ethnic group mentioned in addition to the Canarii. They were perhaps the same as the Pharusians of Strabo, located on the coast south of the Atlas (Strabo, *Geography* 17.3.7). They were caravan nomads whose range was across much of northwestern Africa (John Ferguson, "Classical Contacts With West Africa," in *Africa in Classical Antiquity: Nine Studies* [ed. L. A. Thompson and J. Ferguson, Ibadan 1969] 1–25), probably encountered both by Juba and Suetonius Paulinus; thus Pliny's source is uncertain. At this point he used "Aethiopian" in its generic sense of the indigenous people of northern Africa.

In his final comments about Mt. Atlas, Pliny referred to Juba's scholarship, especially his pamphlet *On Euphorbion*, which discussed a plant discovered in the mountains, either by Juba himself or his physician Euphorbos, after whom it was named (Juba F42; *NH* 25.77–9; Dioskourides 3.82; Galen, *On the Combining of Drugs According to Place 1* [13.278 Kühn]; Roller, *World of Juba* 178–9). The plant became widely used as a topical curative. The name of Juba's physician is the most visible surviving remnant of the Mauretanian kingdom, due to the botanical family Euphorbiaceae and the genus *Euphorbia*, or spurge.

5.17. Pliny's description of the two Mauretanias concludes with a topographical and ethnographic description, extending from the Pillars of Hercules east to the borders of Numidia (through *NH* 5.21). There are no terminal points recorded for the distance of 370 miles (548 km.); it is probably from Tingi to the Muluccha River (perhaps the modern Moulouya or Kiss, at the border between Morocco and Algeria; Sallust, *Jugurtha* 110.8).

The catalogue of indigenous ethnic groups is from west to east, and includes both Mauretanias. The Mauri, or Maurusii, were the dominant population of the region; hence the use of their name for the territory as a whole. By the mid-first century BC they were involved in the African phase of the Roman civil war as mercenaries (*de bello africo* 3, etc.). A detailed ethnography, probably from Juba, was provided by Strabo (*Geography* 17.3.7). Farther east were the closely related Masaesyli, whose leader in the late third century BC was Syphax (see further, *NH* 5.19). The suggestion that the Mauri and Masaesyli had been wiped out is certainly an exaggeration; to be sure, the populations had been reduced by the time of provincialization, but this was probably more due to assimilation or the pushing of the indigenous people into the hinterland in the face of increasing Roman settlement.

For the Gaetulians, see *NH* 5.10. Three of their sub-groups are listed. The Baniurae may have been Carthaginian mercenaries as early as the late third century BC (Silius Italicus 3.303); for the Autoteles (or Autoleles, or Autololes) see *NH* 5.5. The Nesimi are otherwise unknown.

5.18–19. For the local elephants, see *NH* 5.5. The format of the *Natural History* reverts to a *periplous* in order to identify coastal locations from Mt. Abila (modern Jebel Musa, at the entrance to the Mediterranean: *NH* 3.4) to the Malvane River near the eastern boundary of Mauretania Tingitana. It then continues to the east into Mauretania Caesariensis, but the itinerary is confused, perhaps representing a problematic blending of two sources. The Mulucha (or Muluccha) river is actually west of Siga, not east (*BA*, map 29).

Syphax (or Sophax, Strabo, *Geography* 17.3.9), king of the Masaesyli in the late third century BC, was adept at playing the Carthaginians and Romans against each other. He ended up on the Carthaginian side, marrying the famous Carthaginian princess Sophonisba (or Sophoniba). Eventually he was captured by the Numidian king Masinissa and handed over to the Romans, dying in captivity at Tibur in Italy. Sophonisba committed suicide to avoid being taken (Appian,

Libyka 28; Livy 30.12.15). Syphax's royal capital of Siga (at modern Takembrit in Algeria) was destroyed. Its location opposite Malaca in Spain (for which, see *NH* 3.8) is not literally accurate (the longitude of Malaca is actually over two degrees to the west), but this indicates that there was a sailing route between the two points, probably from Carthaginian times.

Two cousins, Bogudes (or Bogud) II and Bocchus II, simultaneously ruled portions of Mauretania in the mid-first century BC. Bogudes, who was associated with Julius Caesar and had a Greek wife, Eunoe Maura (Suetonius, *Divine Julius* 52; Dio 43.36–8), became involved in the Roman civil war, joining Caesar in Hispania. Eventually he abandoned his kingdom and went to Marcus Antonius in Greece, where he was killed in 31 BC (*de bello alexandrino* 59, 62; Strabo, *Geography* 8.4.3; Dio 50.11.3). Bocchus, on the throne since the 60s BC, died naturally in 33 BC (Strabo, *Geography* 17.3.5; Dio 41.42.7, 49.43.7). Neither king seems to have had an heir, and this resulted in the interregnum that only came to an end when Juba II and Cleopatra Selene became monarchs of both territories in 25 BC (Roller, *World of Juba* 55–8).

5.20–1. The *periplous* continues east along the coast; references to the Roman presence indicate that Pliny's source was produced after the provincialization of the AD 40s, and in part after Vespasian gave colonial status to Icosium (modern Algiers) sometime between AD 69 and 76 (Pliny, *Histoire naturelle* 5.1, 166–9). The account is almost totally topographic, with only two ethnyms, the Macurebi and the Nababes. Both these peoples lived in the eastern part of Mauretania, west of the Ampsaga River (the modern Oued el-Kebir in Algeria), which itself was the boundary between Mauretania and Numidia (Pomponius Mela 1.30; *NH* 5.22). They were obscure and played no role in international events.

The cited distance of 322 miles (477 km.) from Mauretanian Caesarea to the Ampsaga is quite long (the direct measurement is closer to 200 km.), but as always reflects the peculiarities of a sailing or road route. The length of 1,038 miles (1,536 km.) for the two Mauretanias again seems long but with coastal indentations may be reasonable. It is hardly possible to determine the meaning of the width of 467 miles (691 km.): the most reasonable suggestion is from the Pillars of Hercules along the Atlantic coast to a point beyond the Atlas, corresponding somewhat to the coastal distances in *NH* 5.9. All these figures are almost certainly from the map of Agrippa.

Numidia

5.22. After the detailed discussion of Mauretania, Pliny's account of Numidia – the territory to its east – is remarkably brief. The toponym is not indigenous, but from Greek *nomas*, originally meaning "roaming" or "grazing." It came to be applied to non-sedentary peoples such as the Scythians (Herodotus 1.15). Eventually the term evolved into an ethnym (in the plural, Nomades in Greek), specifically applied to the population south and west of Carthage (Polybius 1.19.2).

The Numidians became a recognizable ethnic group by the late third century BC, and were ruled by a series of prominent kings, who, like many in North Africa, balanced their interests between the Carthaginians and the Romans. The most famous was Masinissa (ruled 205–148 BC), who was involved both in the Second and Third Punic Wars as well with Sophonisba, the wife of Syphax (Polybius 14–15; Livy 25–42; Appian, *Libyka* 10–28). He became an early example of the Roman friendly and allied monarch, a phenomenon that would reach its peak with Juba II, Cleopatra Selene, and others in the Augustan and Julio-Claudian periods. In fact, Masinissa was Juba's great-great-great-grandfather (Roller, *World of Juba* 11–17). The last king of an independent Numidia was Juba's father, Juba I, who was a player in the Roman civil war and committed suicide in 46 BC. Numidia was provincialized as Africa Nova and the infant Juba II was removed to Rome, eventually to be placed on the throne of a united Mauretania (*de bello africo* 91–7; Livy, *Epitome* 114; Plutarch, *Cato the Younger* 65–70; Appian, *Civil War* 2.98–9).

Mapalia (or *magalia*) were the dwellings of the local population. The word is indigenous, probably coming into Latin through Punic. Sallust described them as oblong in shape with curved roofs of reeds or rushes, somewhat like the hulls of ships (Sallust, *Jugurtha* 18.8; Silius Italicus 17.88–9; *NH* 16.178). They may also appear in art. Obviously they were light in weight and portable (G. Marcy, "Remarques sur l'habitation berbère dans l'antiquité: à propos des *mapalia*," *Hespéris* 29 [1942] 23–40).

The *periplous* continues to the east, but is interrupted for the mention of three inland towns (again listed roughly west to east). Cirta Sittianorum (modern Constantine in Algeria) was the most prominent and the primary royal seat of the Numidian kings. The account then returns to the coast and continues to the Tusca River (modern Oued el-Kebir in Tunisia), the historic boundary between Numidia and the province of Africa.

Numidian marble, known today as *giallo antico*, came from near modern Chemtou in Tunisia, where impressive quarries and workshops are still visible. The stone began to be excavated by the early second century BC and was soon sent throughout the Mediterranean. It was the most important export product of Numidia, and can be seen today in the Temple of Concord and the Forum of Augustus in Rome (*NH* 36.49; Heinz Günter Horn, "Die antiken Steinbrüche von Chemtou/Simitthus," in *Die Numider: Reiter und Könige nördlich der Sahara* [ed. Heinz Günter Horn and Christoph B. Rüger, Köln 1979] 173–80). Wild beasts had been considered a formulaic (although valid) feature of Numidia, and North Africa in general, since the fifth century BC (Herodotus 4.191; Strabo, *Geography* 2.5.33).

The Roman Province of Africa

5.23–5. The region of Africa proper, where the toponym originated, lies to the east of Numidia. It is described in *periplous* fashion, from the borders of Numidia at the Tusca River to Sabrata, in western modern Libya. The Roman province was closely contiguous with modern Tunisia. Cape Apollo (Ras Sidi Ali el-Mekki) lies just to the east of the longitude of Sardinia; Cape Mercury (Greek Hermaia, modern Cape Bon) is well southwest of Sicily, but it and Lilybaeum, the southwest corner of Sicily, create the Strait of Sicily and thus the narrowest part of the Mediterranean (about 150 km. across). Hippo Dirutus or Diarrhytos ("Divided by Water") is modern Bizerte; the Hipponiensian Bay is an inland lake whose outlet to the sea is what divided Hippo. The two capes – Apollo and Mercury – are the limit of the great bay, usually called the Gulf of Utica (Uticensis Sinus) or the Carthaginian Gulf (Strabo, *Geography* 17.3.16), on which both Carthage and Utica lie.

Pliny's source for Africa was to some extent remarkably out of date and largely reflected the era of the Roman civil war of the 40s BC rather than the Imperial period. It also included material from the original organization of the province after the end of the Third Punic War in 146 BC. Yet at the same time Pliny was personally familiar with the region, and here and elsewhere noted autoptically the farming methods of the Byzacium region, the coastal district at the eastern edge of the province (*NH* 17.41; Healy, *Pliny* 13–14). M. Porcius Cato the Younger (the great-grandson of the Cato who wrote the *Origines*), carried the cause of the civil war to Africa after the death of Pompey the Great in September 48 BC. But his effort was futile, and April 46 BC he committed suicide at Utica. Yet he was remembered as one of the last proponents of the moribund Republic (*de bello africo* 88–93; Plutarch, *Cato the Younger* 71; Appian, *Civil War* 2.98–9).

"Libyphoenician" was originally a generic term for the ethnicity that resulted from the mixing of the Phoenician, Carthaginian, and indigenous populations of North Africa; it was an ethnym already in use by the fifth century BC (Hanno 1; see also Diodorus 20.55.4). Yet it came to mean a specific ethnic group, such as those who allegedly sent envoys to Alexander the Great at Babylon shortly before his death (Diodorus 17.113). It also had a certain toponymic force, referring to the coastal region south of Carthage and toward the Cyrenaica, in the district also called Byzacium (Strabo, *Geography* 17.3.19). The round figure of 250 miles (370 km.) for its circumference is a conversion of 2,000 stadia, probably from Polybius. This was a region of immense fertility; in fact the eastern portion, toward Libya, was associated with the mythological Lotus Eaters, a connection made as early as the fifth century BC (Herodotus 4.177; Strabo, *Geography* 17.3.17; *NH* 5.28, 5.41; Serena Bianchetti, "I Lotofagi nella Tradizione Antica: Geografia e Simmetria," in *L'Africa Romana: Atti del xiii Convegno di Studio, Djerba 10–13 Dicembre 1998*, vol. 1 [ed. Mustapha Khanoussi *et al.*, Rome 2000] 219–29). A distance of 580 miles (858 km.) from the Ampsaga River to Sabrata is reasonable; the width, 200 miles (296 km., converted from 1,600 stadia), "as far as is known," demonstrates the limit of knowledge of the interior, even after the Roman presence.

The existence of two provinces of Africa was brief. The original one had been established after the defeat of Carthage in 146 BC, with Utica the residence of the first Roman officers (Appian, *Libyka* 75). After the defeat of Juba I and the dissolution of the Numidian kingdom in 46 BC, his territory became the province of Africa Nova; the original African province was called Africa Vetus (Appian, *Civil War* 4.53). But by 40 BC they had a common administration, and shortly thereafter they were joined together simply as Africa (Appian, *Civil War* 4.53; Dio 48.20.4). Thus Pliny's use of the distinctive terms Vetus and Nova refers only to a brief period after 46 BC and certainly before the Augustan reorganization of 27 BC. This seems an odd anachronism, especially because of Pliny's procuratorship in the province under Vespasian (Syme, "Pliny" 214–15). Interestingly Pomponius Mela (1.33–8) did not mention the brief division into two provinces, so Pliny's source probably preceded him and is unknown. A possibility is C. Sallustius Crispus (Sallust), the first governor of Africa Nova, who, after a disastrous administration, avoided prosecution and retired from politics to devote himself to scholarship (*de bello africo* 97; Dio 43.9; *in C. Sallustium Crispum oratio* 19–20).

The trench separating the two provinces was dug by P. Cornelius Scipio Aemilianus and the Numidian kings, perhaps some time before the fall of Carthage in 146 BC. It extended inland to Thenae, whose exact location is not known, and which was still visible in Pliny's day (Pliny, *Histoire naturelle* 5.1, 242).

5.26. The Syrtes lie between the province of Africa and the Cyrenaica. The Lesser Syrtis is to the west (the modern Gulf of Gadès). Farther to the east is the Greater Syrtis (the modern Gulf of Sidra). The shallow bay between the two, with an extent of several hundred kilometers, seems to have had no particular name. The Greater Syrtis is the southernmost part of the Mediterranean. Both Syrtes were notorious for their treacherous shallows and tidal currents (Polybius 1.39.1–6; Strabo, *Geography* 17.3.20). Polybius' 300 miles (444 km., converted by Pliny from 2,400 stadia) from Carthage to the Lesser (Western) Syrtis is far too short (Polybius 34.15.8); Pliny may have incorrectly recorded the data and repeated the figure of 300 miles for the circumference, which is reasonable (Pomponius Mela 1.35). The depth of 100 miles (148 km. from 800 stadia), is approximately correct for the Lesser Syrtis.

Routes into and across the Sahara had been known since early times, such as the crossing to the probable Niger River noted by Herodotus (2.32). The territory immediately to the interior of the Syrtes was notorious for its dangerous ophids (Sallust, *Jugurtha* 89.5). In fact the Psylli (see also *NH* 5.27) were famous for their ability to cure snake bites (Herodotus 4.173; Strabo, *Geography* 17.1.44; Molly Jones-Lewis, "Tribal Identity in the Roman World: The Case of the Psylloi," in *The Routledge Handbook of Identity and the Environment in the Classical and Medieval Worlds* [ed. Rebecca Futo Kennedy and Molly Jones-Lewis, Abingdon 2016] 192–209).

The route cited by Pliny was from the territory of the Augiles to that of the Garamantes. In his description the itinerary is somewhat truncated, because there is no mention of the fact that the Augiles themselves were well into the desert, since the oasis of Augila is 400 km. inland from the Cyrenaica or 4 days from the southernmost point of the Greater Syrtis (Strabo, *Geography* 17.3.23). From there the route was 12 days (or 10, Herodotus 4.183) to the Garamantes, who lived at Gerama (modern Germa in the Fezzan region of Libya), a major oasis with striking archaeological remains. Some of Pliny's information may be as early as Herodotus (Mario Liverani, "The Libyan Caravan Road in Herodotus IV 181–185," *JESHO* 43 [2000] 496–520), with other details derived from the report of the expedition of L. Cornelius Balbus, who reached Garama in 20 BC

(Strabo, *Geography* 17.3.19; *NH* 5.36; R. C. C. Law, "The Garamantes and Trans-Saharan Enterprise in Classical Times," *JAH* 8 [1967] 181–200).

5.27. The Psylli lived around Lake Lycomedes, which has not been located but may have been north of Augila, given Pliny's use of "above" ("super"), suggesting that he was looking at the map of Agrippa (his "in the center," or "medio," for the Augilae demonstrates a similar origin). The account is confused, which may be in part due to inadequate source blending of data from Polybius, Sallust, and the map of Agrippa, and the scant information about this little-known region. The distances have been converted from stadia, indicating Polybius was the original source (which the later ones may also have used), but the topography is somewhat vague. The 250 miles (370 km., from 2,000 stadia) is roughly from Meninx to beyond Lepcis Magna. The amounts for the Greater Syrtis (625 by 312 miles, or 925 by 462 km., from 5,000 by 2,500 stadia) are excessive. The former is more representative of both Syrtes together, and the latter the distance from Cyrene to the southernmost point of the Greater Syrtis.

The Cisippades are not mentioned elsewhere in literature; whether they were the mercenaries fighting in the Roman army along the Danube in the late first century BC and later, and who were mentioned in various inscriptions, remains speculative (Pliny, *Histoire naturelle* 5.1, 266). In Pliny's account they are vaguely located, perhaps southwest of Lepcis Magna.

5.28. The implication is that the coast of the Lotophages ("Lotus Eaters") would be on the south side of the Greater Syrtis, but by all other accounts they were on the western coast of the Little Syrtis, although the ethnym may have spread all along this coast. Pliny's source was Greek, as demonstrated by the Greek accusatives, and was probably Eratosthenes (*Geography*, F105), who was cited at *NH* 5.41, where Pliny noted that their traditional home was the island of Meninx (modern Jerba in Tunisia, still a major resort). The Machroae may be the Machlyes of Herodotus (4.178), who were consumers of the lotus and neighbors of the Lotophages. The lotus is probably a type of water lily, abundant on this coast (*NH* 13.104–6).

The Altars of the Philaeni, at the southernmost point of the Greater Syrtis (modern Ras el-Aali), commemorated two Carthaginian brothers who determined the border between their territory and the Cyrenaica, and died on the spot. The altars existed from at least the fourth century BC, and may have been important to navigators, since they were frequently noted (Pseudo-Skylax 109.3; Sallust, *Jugurtha* 79; Strabo, *Geography* 17.3.20; Pomponius Mela 1.38).

The topography continues to be somewhat confused, in part because after describing the province of Africa Pliny abandoned the *periplous* format and the details become more scattered. The coast and hinterland of the Greater Syrtis were not well known, even in Roman times, and rather than continue with a *periplous*, Pliny provided a variety of toponyms to the south and east. Callimachus, from Cyrene, was well informed about this region (at least its status in the third century BC), and his location of Pallantias is not as anomalous as Pliny thought, because the nearer side of the Little Syrtis (from the point of view of the Cyrenaica) and the region between the two Syrtes are essentially contiguous (Callimachus F548; see also *NH* 5.27).

The toponym Borion is a transliteration of Greek Boreion, or Northern, and refers to the small promontory (modern Ras Tauínes) just southwest of Berenice, the starting point of the Greater Syrtis as one travelled south from the Cyrenaica. This province had been created after the Ptolemaic withdrawal from the region in 74 BC (Sallust, *Histories* 2.39; Appian, *Civil War* 1.111). Historically the Cyrenaica continued well to the southwest, to the Altars of the Philaeni, but beyond Boreion there was scant Roman presence, due to the light population and the hazardous nature of the territory.

5.29–30. As was his custom with the Iberian Peninsula and parts of western Europe, Pliny catalogued the urban settlements of the region just discussed, listed in a broad extent from the Ampsaga River (*NH* 5.20–1) to Cape Borion (*NH* 5.28). Most of the communities named (generally as local ethnyms), 41 in all, are in the western portion. As usual, they are listed alphabetically according to status.

In addition, there are 10 ethnic groups – the indigenous population that was not urbanized – and one toponym, Gaetulia. The order in which these ethnyms are presented is not obvious, but most are in the historic region of Numidia (Pliny, *Histoire naturelle* 5.1, 328–42). The Nattabutes were south of Cirta (Ptolemy, *Geographical Guide* 4.3.24) and the Capsitani were presumably around Capsa (modern Gafsa), farther to the southeast, where Jugurtha had a treasury (Strabo, *Geography* 17.3.12). To the east were the Musulami, an extensive confederation of indigenous people who regularly opposed the Romans (Tacitus, *Annals* 2.52, 4.24). The Sarbabares are not otherwise known, although there were peoples with similar names in eastern Numidia. The Massili (or Maesyli) were one of the formative Numidian groups: Gaia (or Gala), the father of Masinissa, took the title "king of the Massili" and was involved in the Second Punic War (Livy 24.48–9). The Nicives (or Nicibes) were southeast of Cirta. The

Vamacures were around Timgad. The Cinithi were on the coast of the Lesser Syrtis near the island of Meninx, and were part of an anti-Roman coalition led by the indigenous insurgent Tacfarinas in the early years of Tiberius (Tacitus, *Annals* 2.52). The Musuni were south of the Musulami, and the Marchubi between the coast and Cirta.

Gaetulia was often a generic name for the northern part of Africa inland from the coastal settlements. The Niger River (whose name means simply "river") is probably the Oued Djedi, which flows from west to east across the northern Sahara in central Algeria, and at a length of 480 km. is the one of the longest rivers in the region and an important landmark. Awareness of it may have come from the expedition of Balbus in 20 BC (Strabo, *Geography* 17.3.19; *NH* 5.36). "Aethiopian" is here used in its generic sense of all interior peoples of Africa west of Egypt.

Cyrenaica

5.31. The Cyrenaica (through *NH* 5.38) lies between the Greater Syrtis and Egypt. Greeks had arrived in this region in the seventh century BC – Herodotus' detailed description is one of the best extant accounts of their overseas settlement (Herodotus 4.150–8) – and it became the only significant area of Greek population west of Egypt. The Cyrenaica came under Roman control in 74 BC (Sallust, *Histories* 2.39; Appian, *Civil War* 1.111). Pliny is the first extant author to use the toponym Pentapolis for the district, based on its five prominent cities, but the Greek term must have been in use much earlier.

The oracle of Hammon (or Ammon) lies about 400 miles (600 km.) from Cyrene, but well to the southeast, at Siwa in Egypt (Ahmed Fakhry, *Siwa Oasis* [Cairo 1973]). It was an ancient cult center, named after the god Ammon, whom Greeks equated with Zeus. It became famous in the Greco-Roman world because of the visit of Alexander the Great in early 331 BC (Diodorus 17.49–51; Strabo, *Geography* 17.1.43; Arrian, *Anabasis* 3.3–4), but by the first century BC it was in decline. The region was also of geological interest because seashells had been found in this area far from the sea; thus it was important for understanding the formative processes of the earth (Eratosthenes, *Geography* F15 [= Strabo, *Geography* 1.3.4]).

The Spring of the Sun was a feature of the oasis. Its temperature changed according to the time of day, and it was part of the local tourist circuit (Herodotus 4.181; Pomponius Mela 1.39). Springs with properties that seemed anomalous were a matter of curiosity in antiquity: Pliny catalogued several, and Lucretius offered a lengthy scientific explanation for the spring at Ammon (*NH* 2.227–8; Lucretius 6.848–78).

Pliny was correct in noting that the location of Greek mythological tales can move. The toponym Hesperides (or Euhesperides) is documented in the Cyrenaica in early sources (Herodotus 4.171, 204; Pseudo-Skylax 108), presumably from a time when this region was a remote western area of Greek settlement. As such it became associated with the last labor of Hercules, which in time moved to the farthest extremity of North Africa (*NH* 5.3). But Pliny reversed the direction, suggesting that the tale came from west to east. The putative Gardens of the Hesperides may be one of the fertile depressions near modern Benghazi (the site of ancient Berenice). The distance of 375 miles (555 km.) from Lepcis to Berenice is the direct sailing route across the mouth of the Greater Syrtis.

5.32. The distances along the Cyrenaian coast were catalogued in *periplous* fashion, taken from the map of Agrippa (*NH* 5.40). Those from town to town are accurate, as expected in this well-known region. Cape Phycus (modern Ras Aamer) is the northernmost place on this coast, the sailing point for anyone coming from the Aegean or Greece; it is due south of Cape Taenarum in the Peloponnesus (*NH* 4.15–16), somewhat closer than Pliny's 350 miles (518 km.), and southwest of the Criu Metopon (*NH* 4.58), the southwest corner of Crete, which is more than the stated 125 miles (185 km.).

East of Apollonia the population dropped substantially and no more towns were recorded. Cherronesus (or Chersonesus, modern Ras et-Tin) is where the coast turns sharply to the south, 88 miles (130 km.) from Cape Phycus. To the east of Chersonesus there was little settlement, and the 216 miles (320 km.) to Catabathmos (*NH* 5.38–9), the border of Egypt, reflects the greater indentation of the coast.

5.33. Through *NH* 5.38 there is a discussion of the interior of the Cyrenaica, with additional material about more remote points in the continent of Africa. Some of this is repetitive of earlier material. The Marmaridae were the indigenous people of the hinterland, said to be scattered all across the region, from the Greater Syrtis as far east as Paraetonium (modern Marsa Matruh), the westernmost major town in Egypt (see *NH* 5.39). Other sources, however, localize them closer to the Catabathmos (Pseudo-Skylax 108; Strabo, *Geography* 17.1.13).

The Acrauceles are probably the Araraukeles of Ptolemy (*Geographical Guide* 4.4.9), east of the region of Barca. The Nasamonians had been known since the early days of Cyrene, living southeast of the Greater Syrtis; they were famous for perfecting the route across the Sahara, from Ammon to the river beyond the desert, where there was a large city inhabited by black people, speculatively Timbuctu on the Niger. The Nasamonians also knew the route to the oasis of Augila (*NH* 5.26); Herodotus provided a brief ethnology of them (Herodotus 2.32, 4.172). Their alternate name, Mesammones ("Those In the Midst of the Sands") is not documented in extant sources before Pliny and may be nothing more than a Greek rationalization.

The interior of the Cyrenaica, from the seacoast to the desert, was divided into three agrarian zones. Along the coast was a strip of 15 miles (22 km.) that had trees, followed by another strip of the same width that was cultivated. Beyond this was the vast area that produced silphium, 30 by 250 miles (44 by 370 km.) in extent, and then the desert. Silphium, or silphion, was a major export product of the Cyrenaian economy. It was widely used in the Greco-Roman world as a spice and also medicinally, but became extinct in the Imperial period: the last known example of the plant was presented to the emperor Nero. It was described in detail by both Theophrastus and Pliny (Theophrastus, *Research on Plants* 6.3.1–6; *NH* 19.38–45; Dalby, *Food* 303–4).

5.34. The Asbytae were in the uplands east of Berenice and were known for their horsemanship (Herodotus 4.170; Strabo, *Geography* 2.5.33). The Macae lived near the head of the Greater Syrtis and are best remembered for assisting the Carthaginians at the end of the sixth century BC in driving out the Spartan Dorieus who had created a settlement on the Kinys River (the modern Wadi Caan in Libya) that was seen as an illegal intrusion (Herodotus 5.42).

The Garamantes are mentioned frequently in this portion of the *Natural History*, demonstrative of their importance and wide geographical extent. At *NH* 5.26 Pliny located them 12 days from Augila (to Herodotus [4.183] it was 10); here the distance is 11 from the Greater Syrtis. Although the differences are slight, they are not truly compatible, and must represent different localities associated with the ethnic group. For a map of the route, see Liverani, "The Libyan Caravan Road" 519).

Water is abundant in the Sahara, if one knows how to find it. But Pliny's account is also based on an assumption common in Roman times (and to some extent surviving into the nineteenth century) that the source of the Nile was in Mauretania (at a certain Lake Nilides) and that the river passed through various lakes or flowed underground until reaching its familiar route south of Egypt. This was a theory promulgated by Juba, in part for political reasons, but also based on his own research and certain preconceptions about the ability of rivers to flow underground (Juba F38a; see further, *NH* 5.51–4).

Salt was an important commodity in the Sahara and along the caravan routes. Herodotus (4.181–5) reported that there was a deposit of salt every ten days; the regular intervals suggest that these were man-made depots. There were also dwellings built of salt, which, as Herodotus noted, could only survive in a region without rain.

According to the text of the *Natural History*, the Trogodytae lived southwest (toward the winter sunset) of the Garamantes. This passage is problematic because by all other accounts they lived on the coast of the Red Sea, which would be to the southeast (Agatharchides 5.62; Strabo, *Geography* 16.4.17). This may be simply an error on the part of Pliny or the later manuscript copyists, since no Trogodytae are known west of the Garamantes, or the common confusion of "Trogodytae" and "Troglodytae," the latter meaning "Cave Dwellers" (see also *NH* 5.45). The reference to Aethiopia is too vague to assist in locating them. Carbuncles are garnets, and it may be that the Garamantes kept the places of origin of their trade items secret, indulging in misdirection so that others could not find the sources, a common technique in such circumstances. In fact, the Garamantes themselves may have been misled by those bringing the items to them. Moreover, it is odd that carbuncles are the only gems mentioned: Balbus' expedition seems a lot of effort to obtain something that is not particularly interesting as a precious stone. It is even possible that Balbus himself withheld information about what he actually found (Lisbet Thoresen, "Archeogemmology and Ancient Literary Sources on Gems and Their Origins," in *Gemstones in the First Millennium* AD: *Mines, Trade, Workshops and Symbolism* [ed. Alexandra Hilgner *et al.*, Mainz 2017] 195–8; see also *NH* 5.37). The entire passage, with toponyms ranging across North Africa, is too uncertain to be analyzed in detail.

5.35. Phazania was a region in the uplands south of the Lesser Syrtis. Its subjugation was probably the work of L. Cornelius Balbus in 20 BC. Pliny may have used two different sources for the expedition, one in Greek that provided the accusative of Aele (Aelen), which was perhaps rather condensed and did not name all the localities on his route, and the official account preserved at *NH* 5.36–7. Mt. Ater, the Black Mountain, is probably Hamada al-Hamra, the uplands east of the Phazanian territory.

5.36–7. Using the *Acta Triumphorum (NH* 1.5), Pliny gave an account of the expedition of Balbus to Garama (see also *NH* 5.26). He also used a visual report of Balbus' triumph. Balbus (*PIR* C1073), from Gades in Iberia, was a member of one of the distinguished families of the first century BC: his uncle of the same name had been consul in 40 BC (Broughton 2, 378). The younger Balbus was an associate of Julius

Caesar and eventually proconsul in Africa, triumphing in 19 BC after his expedition (Pietro Romanelli, "La campagna di Cornelio Balbo nel Sud Africano," in *Mélanges offerts à Léopold Sédar Senghor* [Dakar 1977] 429–38). A few years later he commissioned a theater in Rome, some of whose remains are still visible (Richardson 381–2). For a suggested outline of his route to the Garamantes, see Jean-Philippe Lefranc, "La Géologie, Pline l'Ancien et l'Itinéraire de Cornelius Balbus (20 avant Jésus-Christ): Nouvelles Identifications," in *Histoire et archéologie de l'Afrique du Nord: actes du IIIe colloque international réuni dans le cadre du 110e congrès national des sociétés savantes, Montpellier, 1–15 avril 1985* [Paris 1986] 308).

For another spring with peculiar properties, see *NH* 5.31. Having mentioned Cidamus and Garama previously (*NH* 5.35), Pliny excluded them from the list taken from the *Acta Triumphorum*, although they surely must have appeared. The list is a mixture of town names and ethnyms, as well as geographical features, provided without comment except for Mt. Giri, which was the source of gems. The place has not been identified despite intense speculation (Pliny, *Histoire naturelle* 5.1, 409–10): if the toponyms are in geographical order it should be close to Garama, but it may be farther west. The fact that it was singled out may hint at the purpose of Balbus' expedition, to determine the source of the precious stones whose trade the Garamantes controlled (*NH* 5.34). Moreover, there was a placard (*titulus*) carried in the triumph recording that Mt. Giri was the source of the gems. Yet it is more probable that the mountain was a rendezvous point where gems from a wide area, even beyond Africa, were made available to the Romans (Thoresen, "Archaeogemmology" 197–8).

5.38. The Garamantes did not react positively to the arrival of the Romans, and made the route that Balbus used impassable by filling in the wells, a stratagem practiced by Saharan peoples into modern colonial times (Pliny, *Histoire naturelle* 5.1, 412). It was not until AD 70 that another route was discovered, perhaps due to the activities of C. Calpetanus Rantius Quirinalis Valerius Festus (*PIR* C184), who commanded a legion in North Africa and tangled with the Garamantes during the period of instability growing out of the Roman civil war of AD 68–9.

The Oeensi were the people of Oea (at modern Tripoli in Libya). They were involved in a dispute with their neighbors at Lepcis Magna, and called upon the Garamantes for allies, who were eventually routed. A by-product of this incident was the discovery of the new route to Garama. Its descriptive name, By the Head of the Rock, suggests that it may have stayed more in the uplands than the previous one, and perhaps went directly from Oea (or Lepcis Magna) to Garama, thereby reducing the overland portion by four days (*BA*, Map 36).

For the Catabathmos, see *NH* 5.32. This was the boundary between the Cyrenaica and Egypt, 1,060 miles (1,569 km.) from the Lesser Syrtis, probably reflecting the coastal sailing route but one that cut across the mouth of the Great Syrtis. A distance of 910 miles (1,347 km.) for the inland extent would be approximately that from Carthage or Utica to Garama by a direct land route.

5.39. Although discussion of Egypt was reserved for *NH* 5.47–64, Pliny provided a *periplous* along the coast from the Catabathmos to Paraetonium (the distance between the two is somewhat farther than the 86 miles [127 km.] reported). To Pliny, Paraetonium was suspended between Libya and Egypt, although the Catabathmos was regularly considered the boundary (Strabo, *Geography* 17.3.22; Pomponius Mela 1.40). Apis (modern Zawyet Umm el-Rakham) is misplaced, since it lies on the coast just west of Paraetonium and is much less than 62.5 miles (93 km.) away. Pliny, who was using Eratosthenes at this point, may have failed to convert stadia to miles. But the 200 miles (296 km., probably from 1,600 stadia) to Alexandria and the 525 miles (777 km.) from Cyrene to Alexandria are accurate. The distances were probably converted from Eratosthenes' 1,600 and 4,200 stadia.

5.40. In concluding the material on Africa, Pliny made it clear that his major sources were Eratosthenes, Polybius, and the map of Agrippa (F58). He also cited two additional ones, probably used only for distances: Artemidorus of Ephesus (F86), from around 100 BC, and Isidorus of Charax (F9), from the Augustan period. It was common for Pliny to end a broad geographical section with comparative measurements.

The two distances for the width of North Africa vary greatly, as one might expect, because of their length. Isidorus' measurement, 3,697 miles (5,472 km., perhaps converted from 30,000 stadia), covers a shorter extent that that of Agrippa (3,080 miles or 4,558 km.). Although the west end of both is probably the same, Isidorus' terminus at the Canopic (western) mouth of the Nile contrasts with Agrippa's inclusion of all of Lower Egypt, thus to a point well east of the Canopic mouth. The distances attributed to Polybius (34.15.6) are derived from Eratosthenes (*Geography* F103), and were converted by Pliny from 8,800 and 13,500 stadia. Mention of Polybius before Eratosthenes (in reverse chronological order) suggests that Pliny accessed the latter through the former. Isidorus and Artemidorus were frequently mentioned in tandem, raising the possibility that Isidorus summarized the work of the earlier scholar. Such large distances are impossible to critique, but the use of the Canopic Mouth of the Nile, just east of Alexandria, as well as Tingi and Carthage, demonstrates that these are coastal sailing routes.

Islands off North Africa

5.41. There is a brief discussion of the islands off the coast of North Africa, probably almost totally from Eratosthenes (*Geography*, F103, 105). Particular attention is paid to Meninx (modern Jerba in Tunisia), the alleged home of the Lotus Eaters (*NH* 5.28). Its length of 25 miles (37 km., converted from 200 stadia) is actually the circumference. The island lies just off the African coast (1.5 miles or 2.2 km., converted from 12 stadia). To the north are the two adjacent islands of Cercina and Cercinitis (modern Grand Kerkennah and Gharbi); again the "length" of the former (25 miles) is actually a circumference. The suggestion that Cercinitis faces Carthage is, at the very least, peculiar, since any relationship of the islands to Carthage is odd topographically, because both are nearly 300 km. away with the bulge of the African coast intervening. The only explanation is the vagaries of sailing routes, but this is speculative, and the more likely possibility is that there has been an error, either by Eratosthenes or someone later. The two islands are about 500 m. apart today, and there are visible remains of the ancient bridge, which existed from at least the third century BC (Pliny, *Histoire naturelle* 5.1, 439).

5.42. This section is probably also derived from Eratosthenes, with the exception noted below. Lepadusa, Gaulos, Galata, and Cossora (as Cossyra) have been mentioned previously (*NH* 3.92), but the variant spelling indicates two different sources, with the former from a Roman catalogue of toponyms. Most of the islands are between Africa and Sicily, but the list is highly selective: Gaulos (modern Gozo) is cited but not the adjacent island of Melita (modern Malta). The distance, 50 miles (74 km. or 400 stadia), is that from Cercina to Lepadusa, and the other islands are scattered in an arc around to Carthage. The list probably reflects a shipping route between Cercina and Carthage.

The Altars (Arae) are anomalous on this list. The source is probably still Eratosthenes, but they are hardly known, and their location between Sicily and Sardinia does not fix the context and puts them far away from the other islands. The only extant source earlier than Pliny to mention them is Vergil (*Aeneid* 1.108–12), but he seems to have them more between Sicily and Carthage, which would fit Pliny's location of the other islands. Vergil knew Eratosthenes (Roller, *Eratosthenes* 14), and may have used the Arae for his own needs. But Pliny's location remains peculiar, and any further explanation – not assisted by the fact that the islands had disappeared by his time – is difficult to make. Eratosthenes' knowledge of the western Mediterranean was limited (Eratosthenes, *Geography* F131 = Strabo, *Geography* 2.1.41), and this may have contributed to the confusion and perhaps caused a misplacement of the Arae.

The matter of scorpions is probably not from Eratosthenes, but a Roman point of view. Scorpions were considered endemic to North Africa and the adjacent islands; Pliny discussed them in detail at *NH* 11.86–91. They were said to die on Clupea and Galata because of the unique quality of the soil, and thus it was exported to other localities in order to kill the animal (*NH* 35.202; Kitchell, *Animals* 165–6).

Interior Africa

5.43. The final section on Africa (through *NH* 5.46) is about its interior. For the most part it is a catalogue of ethnic groups, some with unusual characteristics. Such anatomically strange people could either be a misunderstanding of local customs, dress, or appearance, or even the pervasive idea that remote people are different.

The account, with some exceptions and confusion, is roughly east to west. The first groups are those who live beyond the Gaetulians and the desert, starting with the Libyan Egyptians, a generic term which seems to mean little more than those in the interior west of Egypt. By late antiquity they had localized around a large lake that was said to flow into the Nile, perhaps Lake Chad (Orosius 1.2.32).

The Leucoe ("White") Aethiopians were said to adjoin the Libyan Egyptians (Pomponius Mela 1.23). The Nigritae and Pharousians were often mentioned together and lived south of the Atlas; they moved into the coastal regions as Carthaginian power declined (Strabo, *Geography* 17.3.3, 7; Ferguson, *Classical Contacts* 21–2). The Pharusians were caravan nomads who wandered across much of northwest Africa, as far east as Cirta. Some were called Gymnetes ("Naked"), perhaps because they wore few clothes (*NH* 6.190). For the Perorsi, see *NH* 5.10, 16.

Having crossed Africa ethnically from Egypt to the Atlantic coast, Pliny reversed direction and mentioned the peoples more to the east, all of whom have been cited previously (*NH* 5.26, 34). Yet the prevalence of the terms "Aethiopia" and "Aethiopians," found all across Africa, reflects the tendency to use the words generically for the populations in and south of the Sahara. This appeared as early as the Greek translation of Hanno, perhaps in the fifth century BC (Blomqvist, *Date*). Hanno, if he was familiar with Homer – not impossible – may have actually used the ethnym. Homer had divided the Aethiopians into two parts (*Odyssey* 1.23–4), and this provided authority for a belief that there were "Western" Aethiopians far to the west of the traditional land of Aethiopia on the upper Nile, even though Homer may have meant nothing more than that the river divided the peoples. There was a further idea that all tropical peoples – even as far east as India – were Aethiopians (Herodotus 3.94, 7.70; Pomponius Mela 3.67).

5.44–6. The Nigris River is not certainly identified, since its name is generic and means "river." It could be the Niger, as implied by Herodotus and Suetonius Paulinus (Herodotus 2.32, *NH* 5.14), or one of the West African rivers flowing into the Atlantic, such as the Draa or Senegal. Large rivers that were reported in the tropics tended to be equated with the Nile, in character if not in actuality. This was in part because the Nile was the only large tropical river known to the early Greek world; even Alexander the Great thought that the Indus was in some way connected to it (Strabo, *Geography* 15.1.25). Around 500 BC Euthymenes of Massilia went south along the African coast and came to a large river – not certainly identified but most probably the Senegal – and thought that it was the Nile (Roller, *Through the Pillars* 15–19).

West Africa was noted for its abundance of reeds, some of which had a capacity of about eight liters, perhaps actually sugarcane (Strabo, *Geography* 17.3.5). Papyrus was said to grow both in West Africa and at the source of the Nile, strengthening the idea that the river began in the west (Pomponius Mela 3.96). One of the reasons that Euthymenes thought his river was the Nile was the presence of crocodiles. Although the various species of flora and fauna varied from place to place, these similarities gave credence to a connection between the Nile and the West African rivers, a point of view promoted by Juba (F38a = *NH* 5.51), who may be Pliny's source for some of this material, since there are some Greek accusatives and Juba wrote in Greek. The Aethiopian Tarraelii are otherwise unknown, but presumably were near the Oechalichae, who lived south of the Greater Syrtis. Their town of Magium has not been located.

The rest of *NH* 5.44, and through 5.46, enters the world of fantastic peoples and their anatomically unlikely characteristics. Pliny's phrase, "if we are to believe it" (*si credimus*), which follows similar diction by Pomponius Mela (1.23), indicates that both authors looked upon this material as questionable, although much of it was derived from Herodotus (4.180–8). The name of the Atlantes was from Mt. Atlas; they were the Atarantes to Herodotus, who located them at one of the way stations on the route taken by the Garamantes across the desert. Solar cults were documented in this region (Herodotus 4.188).

For the Aegipani, see *NH* 5.7. The Blemmyae are probably the unnamed headless people of Herodotus 4.191. They were first cited by name in Ptolemaic times and were believed to live at the source of the Nile (Theokritos 7.114; Strabo, *Geography* 17.1.2). The Gamphasantes were said to be particularly elusive because of their lack of weapons; the name may be a variant of Garamantes (Pomponius Mela 1.23, 47; Herodotus 4.174).

The Satyrs were the partially human wild men who frequented remote mountainous regions. The Himantopodes ("Strapped-Foot People") may have practiced binding of the feet (Pomponius Mela 3.103). The Trogodytae are probably not those who lived on the Red Sea coast, who were far more civilized than implied (Roller, *World of Juba* 200–1). Those cited here were presumably primitive peoples west of the Nile who lived in caves, and Pliny may have confused *troglodyte* ("cave dweller") with the ethnic group on the Red Sea, perhaps misreading Herodotus 4.183; see also *NH* 5.34.

Marriage customs that were seen as promiscuous were also believed to be a feature of primitive peoples: Herodotus cited several examples from North Africa (Herodotus 4.172, 176, 180). The religious practices of the Augilae were described in greater detail by Pomponius Mela (1.46): the source is not Herodotus but probably the expedition of Cornelius Balbus. Visual representations of the Aegipani may have derived from Suetonius Paulinus, but would have been of dubious accuracy since they were virtually never seen.

Confusion of the Pharusii and Persians was in part based on onomastic similarity, but also the mythology of Hercules, who was believed to have had Persians in his entourage who eventually settled on the Atlantic (Sallust, *Jugurtha* 18.4; Erich Gruen, *Rethinking the Other in Antiquity* [Princeton 2011] 275–6). The Pharusians were said to have used scythed chariots, something implausible that was perhaps a conflation of their supposed Persian ancestry with their reputation as warriors (Strabo, *Geography* 17.3.7). Moreover, Persians were not unknown on the Atlantic coast of Africa in the early fifth century BC: for the voyage of Sataspes, see *NH* 3.8 and Herodotus 4.43.

Egypt and the Nile

5.47. Having completed Libya, Pliny next provided an extended section on Egypt and the Nile (through 5.64). It begins with the distances along the west side of the continent of Asia, from the Canopic Mouth of the Nile (near Alexandria) to the Tanais River (modern Don), the boundary of Europe (*NH* 4.78). As usual, comparisons were made from several sources, but interestingly the map of Agrippa is not mentioned, so these may be distances established in Ptolemaic times from Alexandria to points on the Black Sea. The sources are presented chronologically, beginning with Timosthenes of Rhodes, the senior naval officer for Ptolemy II in the first half of the third century BC. He recorded 2,638 miles (3,904 km., from 21,000 stadia) from the Canopic Mouth to the mouth of the Pontus, or Black Sea, at the northeast end of the Thracian Bosporus. This is excessive (it is hardly more than 1,500 km.), and represents the twists and turns of a sailing route (Timosthenes F15 Roller).

Next is Eratosthenes, half a century later, who gave the distance across the Black Sea to the mouth of Lake Maeotis, at the end of the Cimmerian Bosporus (Eratosthenes, *Geography* F115 = *NH* 6.3). His distance was 12,360 stadia, which Pliny recorded as 1,545 miles (2,287 km.). Again this is not a direct route (which would be about 725 km.) but reflects a coastal sailing. Finally there are Artemidorus (F81) and Isidorus (F10), mentioned in tandem as usual, from the first century BC, who recorded the full distance as far as the Tanais River as 40,000 stadia, converted by Pliny to 5,013¾ miles (7,420 km.). This contrasts with 6,375 miles (from 51,000 stadia, equal to 9,435 km.), Pliny's measurement from unspecified sources for the entire length of Asia (*NH* 6.209).

It is interesting that Pliny noted there are many seas named after the peoples living on them – something generally true – but the Pontus, or Black Sea, is the one major exception to this rule, taking its name from the Greek word for sea, *pontos*, rather than any of its coastal inhabitants.

5.48. It is possible that some of the account of Egypt is autoptic, and a heavily restored inscription from Arados in Phoenicia may mention Pliny and implies an Egyptian posting (*OGIS* 586; Michel Malaise, "Pline l'Ancien a-t-il séjourné en Égypte?" *Latomus* 27 [1968] 852–63). The discussion of Egypt begins with an examination of the Delta, from where the Nile divides into its various channels down to the mouths. Pliny only named the left (Canopic) and right (Pelusiac) mouths, which enter the Mediterranean 170 miles (252 km.) apart, an accurate distance following the coastline. There were actually seven traditional mouths (Strabo, *Geography* 17.1.18). This was the place where the word Delta was first used as a toponym (Herodotus 2.13), but by Hellenistic times it had become a generic term for the mouths of other rivers. The division of the river, where the Delta begins, is just below the site of Memphis. The

Canopic channel, which flows just east of Alexandria, is slightly shorter than the Pelusiac.

5.49. The territory was traditionally divided into Lower and Upper Egypt, with the former the Delta, and then, after a gap of about 400 km., the latter around the ancient city of Thebes in the district called the Thebais. Pliny called the administrative units "town prefectures" (*praefecturae oppidorum*), his translation of the Greek word *nomos*, which originally meant "pasture" (Homer, *Iliad* 2.475 etc.). By at least the fifth century BC the term had been applied to Egyptian districts (Herodotus 2.166). The number of them varied, and over half were in the Delta. Pliny listed a total of 45; there may be some redundancy because the names changed.

There is no rigid geographical order to Pliny's list, with the only region specifically noted that around Pelusium at the eastern mouth of the Nile. The numerous Greek names demonstrate that the nomenclature continued to evolve in the Ptolemaic period. The list is presented without comment except to note that the Hammoniac nome extended to the oracle of Hammon (or Ammon, *NH* 5.31). Alexandria was outside the nome system (and technically not even in Egypt because it was west of the Delta). Lake Mareotis (modern Mariout) was to the south of the city, connected to the Canopic Mouth by canals (Strabo, *Geography* 17.1.7) No source is provided for the list of nomes, but the most likely is a Roman administrative document.

5.50. Lake Moeris (modern Birket Qarun) lies about 50 km. west of the Nile in the modern Fayum region. It was a natural lake (despite Pliny's assertion) but had been modified extensively since earliest times, and was the most fertile region of Egypt. Its circumference varied substantially with its rise and fall. It was large when Strabo visited it in the 20s BC (Strabo, *Geography* 17.1.35), but according to Pomponius Mela (1.55) in the following century it was merely 20 miles (30 km.) around. A few years later, perhaps in the AD 60s, C. Licinius Mucianus recorded that it was 450 miles (666 km.) in circumference (*HRR* F6). Pliny, if he is to be taken literally, reported that it was dry. The name Moeris was believed to come from the throne name of the Twelfth Dynasty king Amenemes III (ruled 1853–1806/5 BC; Tusculum 5, 164).

Memphis, 62 miles (92 km.) to the northeast of Moeris (by a route along the canal to the Nile and then downriver), was the royal capital of Egypt, even after the establishment of the Ptolemaic dynasty, lasting as such until Alexandria became the administrative capital of the Roman province (Günther Hölbl, *A History of the Ptolemaic Empire* [tr. Tina Saavedra,

London 2001] 77–8). From Memphis one could go west 12 days to the oracle of Hammon, about 500 km. away. The numeral 15 at the end of the section is without units and cannot easily be understood: from Memphis to the beginning of the Delta is about 50 km., but 15 miles would be 22 km. The text implies days, which makes no sense.

5.51. There follows a lengthy discussion of the Nile (through *NH* 5.59), arguably the most famous river of classical antiquity. The major topics discussed are its source, course, and flooding. Pliny's account of the river is complex and often contradictory, given the many previous reports that he used (Andy Merrills, *Roman Geographies of the Nile From the Late Republic to the Early Empire* [Cambridge 2017] 279–97). For the upper river he relied on Juba's *Libyka* (F38a), providing the fullest extant discussion of his theory that the river originated in Mauretania, far to the west of its lower course, a point of view that was still promulgated to some extent until 1862, when John Hanning Speke (1827–64) discovered the actual source in the lake that he named Victoria Nyanza (John Hanning Speke, *Journal of the Discovery of the Source of the Nile* [New York 1868] 416–29).

The source of the Nile had long been one of the most persistent questions of Greek geographical theory. Pliny may have reflected a certain military bias in his suggestion that the failure to discover its source was due to the lack of military expeditions up the river (Beagon, *Roman Nature* 188). Ptolemaic explorers had gone far upriver (*NH* 6.183) but had not reached the source. The Carthaginians seem to have believed that its origins were in West Africa, and Juba, researching the Carthaginian and Ptolemaic data and adding his own preconceptions, equated the Nile with the great rivers of West Africa (Juba F38); Euthymenes of Massilia had already advanced such a theory (Roller, *Through the Pillars* 15–19). Connecting the Nile to Mauretania allowed Juba to create a geographical allegory of the connection between Egypt, the homeland of his wife Cleopatra Selene and an implementer of his research, and their new territories.

Lake Nilides – whose location is unknown – probably has no onomastic connection with the Nile unless it was a name given to a lake after it was thought to be the source of the river. Juba's theory included the idea that similar fauna meant the same river, and cited three types of fish and the crocodile as proof. The *alabeta* (or *alabes*) is the *Labeo niloticus*, and the *coracinus* (or *korakinos*) a tilapia. The *siluris* is a catfish (Strabo, *Geography* 17.2.4; Roller, *Historical and Topographical Guide* 977). All three are found in the Nile.

To provide physical documentation of his theory, Juba placed a crocodile from Lake Nilides in the Temple of Isis at his capital, Mauretanian Caesarea, which was still visible in Pliny's day. He may have been influenced by Sobek, the sacred crocodile of Egypt, and thus made another linkage between Mauretania and his wife's home (Jane Draycott, "The Sacred Crocodile of Juba II of Mauretania," *AClass* 53 [2010] 211–17). This was a direct imitation of Hanno, who exhibited tokens of his expedition at Carthage. Juba also wrote a treatise on Hanno's cruise (*NH* 6.200; Athenaios 3.83 c). Another factor influencing him was a seeming correlation between snow in Mauretania and the Nile flood, but this was of little significance since snows in central Africa were in fact the cause of the flooding.

5.52. The river was said to be alternatively on the surface and underground, at one point for as much as twenty days; the use of travel time indicates that Juba's explorers traced its assumed course. The Nigris Spring is another example of the generic hydrological name found in Africa. Underground rivers were considered a normal phenomenon in Greco-Roman theories of what lay below the surface of the earth: Eratosthenes and Strabo knew of several examples (Eratosthenes, *Geography* F140; Strabo, *Geography* 8.8.4). Most of these are far more reasonable geologically than the suggested course of the Nile, extending over hundreds of kilometers. Yet the Sahara has many oases, watercourses, and lakes, and with proper preconceptions it would be possible to create a route for the Nile from Mauretania to its known portion in Aethiopia (Thomson, *History* 268).

5.53. Eventually, according to Juba, the river reached familiar territory in Aethiopia above Meroë. He was a linguist, among his other scholarly achievements, writing a treatise titled *Resemblances* (*Omoitetes* in Greek) on the subject of comparative linguistics (Juba F14 = Athenaios 4.170), and thus he provided the translations for the various indigenous river names, although he seems to have missed the fact that "Gir," like "Niger," meant "river." The Astapus is probably the Blue Nile and the Astaboras the modern Atbana. The Astasapes (Astasobes) is perhaps the White Nile, with "asta" the local word for "river" (Strabo, *Geography* 17.2.2). The Blue and White Nile join at modern Khartoum, and the Atbana comes in farther downstream below Meroë, at which point it was said that all the parts of the river were together and (in antiquity) the Nile actually began, although this was merely a matter of nomenclature. "Nile" as a toponym is not documented until Hesiod (*Theogony* 338); "Triton" (*NH* 5.54) was perhaps an attempt to connect a certain Lake Tritonis and its associated river, located in eastern Libya, to the Nile system (Herodotus 4.179, 191; Apollonios 4.269).

5.54. The animistic account of the river's flow may be a feature of Juba's style. Since Pliny's report is oriented downstream, which is unusual but a necessary element of the structure of Juba's *Libyka*, the "last cataract" is actually that conventionally known as the first, or northernmost, at modern Aswan. After this point, the river downstream is more gentle. Pliny transliterated Juba's descriptive term for a cataract, *katadoupa*, and turned it into a Latin toponym, but also provided the normal Latin word for the feature, *catarracta*. He later returned to a discussion of the river, still based on Juba, at *NH* 6.175–80.

5.55–6. The Nile flooding (discussed through *NH* 5.58) was also a perennial intellectual problem for Greeks and Romans. It was considered by Thales of Miletus (Seneca, *Natural Questions* 4a.2.20–2) and discussed in detail by Herodotus (2.20–2), who summarized various theories about the phenomenon (see also Diodorus 1.38–41). In Pliny's day there were as many as eight possibilities under consideration (Tusculum 5, 168–70), but by late Hellenistic times it had already come to be realized that there were mountains in central Africa – Polybius' exploration in the 140s BC had shown this (*NH* 5.9–10) – and thought began to coalesce toward the correct explanation, which was precipitation far to the south. The other persistent theory, originally advanced by Thales, was that the Etesian winds, blowing from the northwest in summer, caused the river to back up and be unable to flow into the Mediterranean. This point of view seems to have been favored by Roman authors (Lucretius 6.712–18; Pomponius Mela 1.53; Lucan 10.239–47). It may also be connected with the idea of Euthymenes of Massilia, who believed that an oceanic (perhaps tidal) flow into the rivers of West Africa eventually reached the Nile (Seneca, *Natural Questions* 4a.2.22; Aristides 36.85–95).

Yet Pliny described the theory of the mathematical and astrological author Timaeus, little known today, of uncertain late Hellenistic date. Pliny mentioned him a few times (see also *NH* 2.38; Jones, "Timaios (Astrol.)," *EANS* 810). It is not clear why he reported Timaeus' theories in detail, which he considered arcane. The toponym Phiala is otherwise unknown but may be derived from *phiale*, the Greek word for "bowl," and, as Pliny hinted, is probably a geological feature.

Timaeus believed that the flooding was due to the heat of the sun at the time of the rising of the Dog Star (Sirius) and when the sun enters Leo, events of summer: the heat would draw up the river and cause the floods. Pliny, however, rejected this theory, since the sun was actually farther north in summer, and moreover the absence of shadows in the tropics could not affect the river since this happened all year. A point of interest is that these theories about the flooding do not presume a Mauretanian source of the river, as proposed at *NH* 5.51–4.

5.57. The flooding of the Nile began after the summer solstice (thus late June or into July) and then, after reaching a peak, diminished through late summer and into autumn. The length of the flood varied according to the sources, from the hundred days of Herodotus (2.19) to as few as forty days (Strabo, *Geography* 17.1.4); this may be due to different ways of reckoning the flooding period.

Nilometers, or Niloscopes, had long existed on the river. Diodorus (1.36.11–12) saw one at Memphis and reported that the observers sent readings throughout Egypt. Strabo (*Geography* 17.1.48) saw an example at Elephantine, which was particularly important since this was where the river entered Egypt. They were simple devices, yet because records had been kept for a lengthy time, it was possible to predict the size of the next harvest (John Peter Oleson, "Irrigation," in *Handbook of Ancient Water Technology* [ed. Örjan Wikander, Leiden 2000] 198–200).

5.58. This section is an important testimonium on the varying rises of the river, and the effects of too much or too little flooding. The recorded amounts of the rise are from five to 18 cubits (roughly 2.2 to 8 m.), with 16 (7.2 m.) the best. But Pliny's account covers only little more than the century before his time and thus is likely to be limited to Roman reports. It is hardly unexpected that the amount of the rising could have psychological effects on the people of Egypt, given that the flooding was essential to their agricultural survival. Moreover the floods could be interpreted as portents: the poor flood of 48 BC was related to the events of that summer and autumn, especially to supporters of Pompey the Great in the Roman civil war, who was defeated by Caesar at Pharsalus in the summer and then killed in Egypt a few weeks later.

5.59. The river entered Egypt from Aethiopia at Syene, just below the First Cataract (Juba's last cataract: see *NH* 5.54). Syene was on an island in the Nile, Elephantine, with a commercial area on the east bank (the Arabian side, at modern Aswan). This was an international market where goods from Aethiopia and interior Africa came to Egypt (D. J. Blackman, "Syene," *PECS* 871). There were several islands in the area, slightly upriver, including Philae (Strabo, *Geography* 17.1.50). The region was famous to scholars as the place where Eratosthenes determined the circumference of the earth (*NH* 6.171). Because of the cataract, Elephantine was the head of Egyptian navigation on the river. Aethiopian vessels would come north to just above the cataract and their goods would be carried around it over a portage, which Strabo traversed in the 20s BC (Strabo, *Geography* 17.1.50).

The distances down the Nile varied in the reports, from 400 miles (592 km., from 3,200 stadia) to 600 miles (888 km., from 4,800 stadia). Since the river had been carefully measured by Ptolemaic surveyors in the third century BC (Martianus Capella 6.596–8), these differences are surprising, and Pliny rejected the three earlier authors and provided his own distance (from an unspecified source) of 585 miles (866 km.).

5.60. Amasis (reigned 570–526 BC) was the Twentieth Dynasty king whose hellenophilia made him an important figure in Greek history, as described in detail by Herodotus (2.172–82), who was the source for the report that Egypt had 20,000 cities. Those listed by Pliny (through *NH* 5.61) seem to be from a Greek tourist guidebook extolling the local places to visit: the single distance provided, 7.5 miles (11 km.), is converted from 60 stadia. They are described downstream from the region of Thebes.

The hundred gates of Thebes (at modern Luxor and Karnak) were first reported by Homer (*Iliad* 9.383), perhaps a reference to the elaborate and extensive architecture still visible at the site. Coptos (at modern Qift) was where the Nile comes closest to the Red Sea, which is 150 km. away along a well-built road. It was where goods from Arabia and India, which had come up the Red Sea and then across on the road, began their transport down river (Steven E. Sidebotham, *Berenike and the Ancient Maritime Spice Route* [Berkeley 2011] 125–74). Abydos (at modern el-Araba el-Madfuna) is west of the river and was a major tourist site: Strabo visited it in the 20s BC (Strabo, *Geography* 17.1.42).

5.61. The touristic account of Egypt continues downriver, with a passing reference to the pyramids. The labyrinth (near modern Hawwara, between the Nile and Lake Moeris), had been built by Amenemes III (1853–1806/5 BC) as an administrative and mortuary center. It had been visited by Herodotus (2.148), Diodorus (1.66.3–6), and Strabo (*Geography* 17.1.37). Its complexity – 3,000 rooms – was what gave it its popular name.

5.62. Alexandria was founded by Alexander the Great in 331 BC west of the Canopic Mouth at a site known as Rhocotes (Rhakotis), an early anchorage (Diodorus 17.52; Strabo, *Geography* 17.1.6–10; Plutarch, *Alexander* 26; Arrian, *Anabasis* 3.1.5). It was located on a narrow spit of land between the sea and Lake Mareotis. Pliny attributed the construction to an architect named Dinochares (see also *NH* 7.125). This is probably an error for Dinocrates (Deinokrates), a Macedonian, whose ambitious projects included a cremation platform in Babylon for Alexander's companion Hephaisteion and a colossal statue of the king on Mt. Athos, neither of which was implemented (Vitruvius 2.pref.1–4). Yet his name, Master of Marvels, sounds suspicious, and the real personality behind these projects is difficult to identify (Heckel, *Who's Who* 106).

The figure of 15 miles (22 km., from 120 stadia) is essentially the circumference of the urban area, much of which was royal property. The comparison with a Macedonian cloak demonstrates that the city was essentially a rectangle with rounded corners, a shape usually described as a *chlamys*, an outer garment worn by horsemen, such as is visible on the Parthenon frieze. This analogy was well known (Strabo, *Geography* 17.1.8). The royal palace – which in Pliny's day had become, in part, the Roman administrative center – occupied a vast area adjoining the harbor in the northeast part of the city. Its precinct also included the Museum and Library, and the royal tombs, including that of Alexander and the Ptolemaic rulers from Ptolemy I to Cleopatra VII (Judith McKenzie, *The Architecture of Alexandria and Egypt c. 300 BC to AD 700* [New Haven 2007] 64–71).

5.63. Lake Mareotis (or Mareia, modern Mariout) lies to the south of the city and was connected to the Canopic Mouth by canals. It served as a holding basin for ships, especially those that had come down the Nile. Its remnants today are much smaller than its size in antiquity. The dimensions of the lake are from a rare citation of the *Histories* of the emperor Claudius, of which only six fragments exist, all but one preserved by Pliny (*FRH* #75). He did not provide Claudius' units of measurement, which cannot be miles, unless the figure of 250 is an error, since at its largest the lake was about 30 by 65 km. In fact the figures in the section cannot be understood, and the unattributed length of 40 *schoeni* is equally impossible. The *schoenus* ("rope") was common in Egypt, but was variable in length, ranging from 30 to 120 stadia (6 to 24 km.), causing much difficulty to those calculating overland distances, especially in points east of the Mediterranean, where it remained in use into the Roman period (Herodotus 2.6; Isidorus, *Parthian Stations* 1, etc.; Strabo, *Geography* 17.1.24).

5.64. Pliny's final section on Egypt concerns the mouths of the Nile. Previously he had mentioned only the easternmost and westernmost (*NH* 5.48), but now he added the traditional five others, as well as naming several towns in the Delta (Tusculum 5, 175–80), and one additional mouth, the Naucratitic or Heracleotic.

Arabia

5.65. The next region examined is Arabia. The account begins at the Pelusiac (eastern) mouth of the Nile, and continues along the coast to

Gaza, the border of Palestine (*NH* 5.66), as well as including the territory to the southeast along the Red Sea and its hinterland. The term Beata (Fortunate) is a translation of the Greek Eudaimon, in use as early as the fifth century BC (Euripides, *Bacchants* 16), and referring to the wealth and prosperity of the region, especially in regard to the frankincense-producing territory of the south. The word Beata came into vogue in the Augustan period; Pliny also used an alternative, Felix (Horace, *Odes* 1.29; *NH* 12.84). From the earliest Greek knowledge of Arabia, there was a distinction between the desert in the north, populated by nomads, and the fertile agricultural region of the south (Eratosthenes, *Geography* F95; Strabo, *Geography* 16.4.2).

Frankincense (Pliny's "fragrance"), *Boswellia sacra*, and to a lesser extent myrrh (*Commiphora myrrha*), had been luxury products in the Levant and the Greco-Roman world for some time (Genesis 37:25; Herodotus 1.183). Frankincense, at least, reached the Macedonian court in the fourth century BC (Plutarch, *Alexander* 25.4–5; Plutarch, *Sayings of Kings and Commanders* 179 f). The first account of the Arabian peninsula was by Eratosthenes, although frankincense itself had been described in detail by Theophrastus (*Research on Plants* 9.4; see also *NH* 12.54–65).

The three ethnic groups mentioned are a limited selection of those known in Arabia. The Catabanes (Kattabanes) were in the south, with their capital at Tamna (modern Hajar Kuhlan in Yemen), and were one of the frankincense-harvesting groups (Jean-François Breton, *Arabia Felix From the Time of the Queen of Sheba* [tr. Albert LaFarge, Notre Dame 1999] 43–5). The Esbonitae may have been around Esbous (biblical Heshbon, Numbers 21:25–30), at the other end of Arabia northeast of the Dead Sea. Scenitae ("Tent Dwellers") is a Greek generic ethnym for many nomadic populations, applied to peoples in a broad swath from Arabia to the Euphrates and upper Mesopotamia, this last perhaps reflected in Pliny's comment about the Syrian borderlands (Juba, F66 = *NH* 13.34; Strabo, *Geography* 16.1.26–7).

Mt. Casius (Kasion, modern Ras Qarun on the Egyptian coast east of the Delta, perhaps biblical Baal-Zephon [Exodus 14:2, 9]) is an isolated hill (90 m. in height) and the only notable feature on the route between Egypt and the Levant (Herodotus 2.6; Strabo, *Geography* 16.2.33). Pliny's suggestion that it was the only prominent point in Arabia is hardly true, but indicates that his thoughts were limited to the coastal route through its extreme northwest, perhaps viewing the region from the point of view of a *periplous*. The Canchlei and Cedrei presumably lived inland from this coast; the latter may be the biblical Kedar (Psalm 120:5).

The Nabataeans were the famous traders of interior northern Arabia, who had been known to the Greek world since the fourth century BC and were noted for their spectacular city of Petra (Diodorus 19.95–8; Strabo, *Geography* 16.4.21). By the first century BC their territory extended from Damascus south to the Red Sea, and west to the coast around Gaza.

The two bays of the Red Sea are conspicuous today, the modern Gulf of Suez and the Gulf of Aqaba to its east. The route from Aelana (modern Aqaba) to Gaza was well travelled, and used to transport Arabian aromatics to the Mediterranean. The accurate distance of 150 miles (222 km., from 1,200 stadia) is probably from Eratosthenes: the route went north from Aelana to Petra, and then across to Gaza. The second distance, Pelusium to Arsinoë, although from the map of Agrippa, is also Greek in origin (125 miles, or 185 km. from 1,000 stadia). Arsinoë is near modern Suez, and the route followed the ancient canal from the Mediterranean to the Red Sea. An interesting point is that in Agrippa's day Cleopatra VII was renamed Arsinoë Kleopatris (a designation still in use in the fifth century AD), but this was hardly politically correct for the Augustan period and Agrippa put the old early Ptolemaic name on his map (Cohen, *Syria* 308–9).

Pliny's comments on Arabia conclude with a statement about climate and life zones, noting the changes from the Mediterranean to the Red Sea. Today the annual rainfall at Gaza is 148 mm. while that at Suez is only 20 mm. No source was recorded; the most likely are Eratosthenes or Juba, whose work pervades this portion of the *Natural History*.

Phoenicia and the Southern Levant

5.66. The narrative moves north along the eastern coast of the Mediterranean toward Asia Minor. Pliny laid out the extent of the region to be discussed (through *NH* 5.90): north along the coast to Cilicia, and east to Babylonia and Mesopotamia, including areas north of Mesopotamia into the southern Taurus. His model may have been Eratosthenes, who divided the inhabited world into units that he called sealstones; Pliny's geographical extent is similar to Eratosthenes' third sealstone (Eratosthenes, *Geography* F66, 84).

5.67. There are two distances defining the extent of the region laid out at *NH* 5.66. Sources were not provided, but the most probable is Eratosthenes, with 470 miles (696 km., from 3,760 stadia) an accurate figure from essentially the northeastern corner of the Mediterranean to the

Phoenician–Arabian border beyond Gaza. The second measurement, 175 miles (259 km., from 1,400 stadia), was more specifically defined: from Seleucia Pieria (at modern Kapısuyu in Turkey) on the coast near the mouth of the Orontes River to Zeugma (at Kavunlu in Turkey) on the Euphrates, also an accurate measurement, as would be expected on this well-travelled route. Another source, perhaps the map of Agrippa, lays out the various territorial districts in this region, including the normal pattern of naming a sea after the people on its coast.

As early as the fifth century BC, if not before, it was believed that the Phoenicians had invented the alphabet and that it was brought from them to the Greeks (Herodotus 5.58). Although the tradition improbably put this event in the Bronze Age, there is little doubt that early Greek connections with the Phoenicians facilitated this exchange: the Phoenicians were in the Aegean as early as the end of the Bronze Age (Duane W. Roller, "Phoenician Exploration," in *The Oxford Handbook of the Phoenician and Punic Mediterranean* [ed. Carolina López-Ruiz and Brian Doak, Oxford 2019]) 646–8). Yet the actual way in which Greek alphabetic writing developed remains obscure (Rudolf Wachter, "Alphabet: The Greek Alphabet," *BNP* 1 [2002] 530–41). The Phoenicians were also credited with the invention of astronomy and navigation (related disciplines), which were attributed specifically to the people of Sidon. They were said to be the first to observe the stars for navigational purposes and to build the first merchant ships, allegedly an invention of Hippus of Tyre (Strabo, *Geography* 16.2.24; *NH* 7.208–9). All this is simplified and in many ways problematic, but nevertheless the Phoenicians contributed significantly to emergent Greek culture in the early first millennium BC. Moreover, even though Greeks and Romans could be diffident about cultures not their own, respect for Phoenician accomplishments was a notable exception (Gruen, *Rethinking* 138–9).

5.68. A *periplous* extends from Pelusium as far as Phoenicia, along the Mediterranean coast. Chabrias was an Athenian adventurer who was in the service of the Egyptian king in the early fourth century BC (Diodorus 15.29). According to Strabo his artifact was a defensive wall, perhaps marking the boundary of Egypt, which was still visible in the first century BC (Strabo, *Geography* 16.2.33).

Pompey the Great arrived in Egypt shortly after his defeat at Pharsalus in August 48 BC, but misjudged his expected welcome and was killed upon landing on the orders of Potheinos, the minister of Ptolemy XIII. There are varying reports as to the disposition of his body: Strabo reported twenty years after the event that it was still at Mt. Casius, a short distance west of

Ostracine. According to Plutarch and Appian, his head was sent to Julius Caesar, who arrived in Egypt a few days later (Strabo, *Geography* 16.2.33 Plutarch, *Pompeius* 80; Appian, *Civil War* 2.86). The body was either cremated or buried, and a monument was erected.

Pliny's account of the coastline east of Pelusium is somewhat confused. The 65 miles (96 km.) from Pelusium would be the distance to Rhinocolura, effectively the boundary of coastal Arabia. But Idumaea (biblical Edom) is an inland territory, in the uplands west of the southern Dead Sea, with little direct access to the coast. Palestina should be beyond (north of) the short coastal stretch of Arabia, yet Pliny's account reverts west to Lake Sirbonis (modern Sabkhet el-Bardawil), which extended from Pelusium to Ostracine, by all accounts part of Egypt. This is probably a failure to blend sources: the 150-mile (222 km., from 1,200 stadia) circumference of the lake is probably from Eratosthenes (see his *Geography* F15). It was already marshy by the fifth century BC (Herodotus 2.6).

Anthedon is essentially on the coast, and Iamnea (modern Yibna) is a few kilometers inland. It and its seaport, Iamneia Paralios (modern Yavnah Yam) are the two Iamneae to which Pliny referred. Neither of these "inland" towns, both so close to the sea, negates the concept of a *periplous*.

5.69. Jope (modern Jaffa, within the urban region of Tel Aviv) was a rare example of a city on the southern Levantine coast that was connected to Greek mythology. It was an ancient town, famous for the story of Andromeda, tied to a local rock to be devoured by a sea monster but rescued by Perseus, as told in detail by Ovid (*Metamorphoses* 4.604–803). Ceto (Greek *ketos*) is merely a common Greek word for a large fish (Homer, *Odyssey* 12.97) but eventually it became a proper noun applied specifically to the creature that plagued Andromeda (Manilius, *Astronomica* 5.600). Its skeleton was brought to Rome by M. Aemilius Scaurus and displayed during his aedileship in 58 BC, when he gave games that were noted for their splendor (*NH* 9.11; Broughton 2, 195). It was 40 feet long (about 12 m.); such enormous skeletons are usually explained as those of prehistoric animals or whales, although it is difficult to imagine a whale coming so far into the Mediterranean. Yet an exceedingly large skeleton (30 m. long and perhaps three m. high) was reported to Posidonius as having been found in the Makros Plain southeast of Arados in Phoenicia (Posidonius F244). Pliny's comment that Ceto was worshipped suggests a primitive seafaring cult that may have assisted in giving rise to the story of Andromeda. The marks of her chains, probably rock cuttings, whether natural or human-made, were still visible in the late first century AD (Josephus, *Jewish War* 3.420).

Strato's Tower (Stratonopyrgos), a Phoenician outpost founded by Straton I of Sidon in the early fourth century BC, was absorbed into Caesarea (Kaisareia), the great city built by Herod the Great in the late first century BC (Duane W. Roller, "The Problem of the Location of Straton's Tower," *BASOR* 252 [1983] 61–6). Vespasian renamed Caesarea Colonia Prima Flavia Augusta. Just north of Caesarea, where the ridge of Mt. Carmel reaches the sea and creates a major topographical barrier, was the boundary between Palestine and Phoenicia.

At this point Pliny diverged from his *periplous* (not continued until *NH* 5.75) and named a few towns in the interior, specifically in Samaria, the inland region to the east between Jope and Caesarea. This was the seat of the ancient Samaritan kingdom, with its major city at Samaria, which had been named Sebaste by Herod. In Pliny's day it was part of the Roman province of Judaea, with Neapolis (modern Nablus), just to the southeast, the administrative center.

5.70. Pliny defined Judaea in a broad sense (through *NH* 5.73), from the Syrian border north of the Sea of Galilee to the Dead Sea and Idumaea. The location of Peraea is confused: it is actually the territory east of the lower Jordan, but not in the south as Pliny implied, although it would join Arabia (in its most expansive sense) on its east. Few distances were provided, but the presence of the map of Agrippa is apparent in the use of the word "below" (*NH* 5.73), implying the orientation of a map on a wall, and the material may ultimately have come from Herodian sources (Nikos Kokkinos, *The Herodian Dynasty: Origins, Role in Society and Eclipse* [Sheffield 1998] 61).

A toparchy was a local district, a term that may have originated in Hellenistic Egypt but which was especially common in the Levant (1 Maccabees 11:28; Josephus, *Jewish War* 2.98). Josephus, writing at about the same time as Pliny, recorded a similar list of toparchies, but with a total of eleven, corresponding to Pliny in eight of them. Such districts were subject to change topographically, especially in the uncertain conditions after the political convulsions of the AD 60s (Anthony Keddie, *Class and Power in Roman Palestine* [Cambridge 2019] 29–30).

Hiericous, or Jericho, was long famous for its rich agricultural environment, still visible today. The region was a metaphor for luxury (Horace, *Epistles* 2.2.184; Strabo, *Geography* 16.2.41; Josephus, *Jewish War* 4.459–75). There was no actual town at Herodion, which was the site of the palace and tomb of Herod the Great, located in an isolated region 12 km. south of

Jerusalem, although it is probable that a village grew up in the area (Josephus, *Jewish War* 1.419–21; *Jewish Antiquities* 15.323–5). Yet Pliny's explicit mention of a town (*oppidum*) may refer to the elaborate Herodian complex. In fact, the eulogistic tone of this passage – describing Jerusalem as by far the most famous city (*longe clarissima urbium*) of the East, and Herodion as an illustrious town (*oppido inlustri*) – may be further evidence of Herodian sources and the effect of his magnificent building program (Daniel R. Schwartz, "One Temple and Many Synagogues: On Religion and State in Herodian Judaea and Augustan Rome," in *Herod and Augustus* [ed. David M. Jacobson and Nikos Kokkinos, Leiden 2009] 391).

5.71. The Jordan River flows from its source at the spring of Paneas (modern Baniyas), which emerges from a cave. This was the site of a Hellenistic cult of Pan, perhaps initiated by the Ptolemies in order to superimpose a Greek shrine over a local one (Zvi Uri Ma'oz, "Banias," *NEAEHL* 136–7). The river drops sharply to the Genesara (the Sea of Galilee) and then is more languid in its course to Lake Asphaltitis (the Dead Sea; see *NH* 5.72). Few rivers that are so short (251 km.) have had such a rich cultural history. The environment of the Genesara is particularly pleasant, especially on its west side. Tarichaea was famous for its fish-salting industry (hence its name) and was the center of the Roman presence in the region at the time Pliny was writing the *Natural History* (Nikos Kokkinos, "The Location of Tarichaea: North or South of Tiberias?" *PEQ* 142 [2010] 7–23). The length of 16 miles (24 km.) is accurately measured and probably came from the map of Agrippa.

5.72. Lake Asphaltitis (the Dead Sea) was a peculiar phenomenon to the Greco-Roman world. It first came to Greek notice at the time of Alexander the Great, and was called Asphaltitis because it was a source of bitumen (Aristotle, *Meteorologika* 2.3.359a; Diodorus 19.98–9; Strabo, *Geography* 16.2.42). The dimensions (probably from the map of Agrippa) are problematic (at least as the sea is today): rather than over 100 miles (148 km.) long, it is only half that, and thus, as is often the case, the figure may be the circumference. The width (75 miles or 111 km.) is again far too great; it is never more than 25 km. across and, given its topographical situation, could hardly have been wider. At the peninsula of Lisan, which protrudes from the east near its south end, it narrows to only a few kilometers across. The south end was an area of regular volcanic activity, reflected in the biblical tale of Sodom and Gomorrah (Genesis 19). Greek geographers had difficulty with this unusual region: the accounts of both Eratosthenes (*Geography* F17–18), and Strabo (*Geography* 16.2.41–4) are confused. Yet its high degree of salinity remains a curiosity today.

5.73. The Essenes were one of the most famous ascetic groups of the southern Levant. The major sources for them are Philon, *Every Good Man is Free* 75–91, and Josephus, *Jewish War* 2.119–61. They were not totally withdrawn from the world, since one of their number, a certain John, was involved in the uprising of the AD 60s (Josephus, *Jewish War* 2.567), and some of them, at least, settled in urban areas (Philon, *Hypothetika* 11.1; Josephus, *Jewish War* 2.124). Pliny's notice is important because it provides a place for at least one of their major groups, probably north of the town of Engada on the west shore of the Dead Sea, although the exact location of their settlement has long been disputed (Jean-Paul Audet, "Qumrân et la notice de Pline sur les Esséniens," *RBi* 68 [1961] 346–87; Christoph Burchard, "Pline et les Esséniens," *RBi* 69 [1962] 533–69). He also implied that they were regularly receiving converts. For many years these Essenes have been associated with the residential structure at nearby Khirbet Qumran and the caves above it where the Dead Sea Scrolls were found, but this has also been questioned (Duane W. Roller, Review of Hirschfeld, *Qumran in Context*, BASOR 340 [2005] 94–5). Both Engada (modern Ein Gedi) and Jerusalem were said to be "a heap of ashes," the effects of the uprising of AD 66–70. Pliny's account, having followed the Jordan from its source, then continues south to Masada and the southern boundary of Judaea.

5.74. The Decapolis was a loose confederation of ten cities, located between the Sea of Galilee and the Dead Sea, on both sides of the Jordan. It was not topographically cohesive, since Damascus, far to the north, was a member, and Pliny referred to the tetrarchies scattered among the cities. Despite its name the number of members varied. There is no evidence of any common administration, and it probably was merely a descriptive unit, coming into existence at the time of the Roman organization of Syria in 64–63 BC. Pliny is the first extant source to use the term and to list its members (S. Thomas Parker, "The Decapolis Reviewed," *JBL* 94 [1975] 437–41).

The Scythian connection with Judaea is problematic, seemingly only based on the name of one of the members of the Decapolis, Scythopolis (at modern Beth Shean). The site has been occupied since early times (Joshua 17:11; Judges 1:27; 1 Kings 4:12). When it became a Ptolemaic possession it was renamed Scythopolis, for reasons that remain obscure; the secondary name Nysa also appeared at this time (Cohen, *Syria* 290–9). But there is no evidence for Scythians (the nomadic peoples north of the Black Sea) in this region, and the name Nysa, that of the nurse of Dionysos (Father Liber), seems to have generated local myths about the two and the placement of them on their coinage (Head, *Historia Numorum* 803).

A tetrarchy was originally an administrative unit that was part of a group of four, perhaps originating in Thessaly during the fourth century BC (Theopompus F208; Demosthenes, *Third Philippic* 26; see also Hellanikos [*FGrHist* #4] F52). It eventually came to mean a petty state not worthy of kingship, with the sense of "four" no longer meaningful. The concept was common in the Levant, most notably when Augustus awarded tetrarchies to two of the sons of Herod the Great in 4 BC, thereby denying them kingship (Josephus, *Jewish War* 2.94). Pliny's tetrarchies were local states scattered east of the Sea of Galilee.

5.75. Pliny returned to the *periplous* that he had left at *NH* 5.70, starting from Crocodilon, north of Caesarea, and continuing as far as Cape Album (modern Ras el-Abyat), just north of the current Israeli–Lebanese border. There was a major glass industry along the coast north of Ptolemais (modern Akko), located on a beach that only extended for a few hundred meters. The locals believed that the production of glass was invented here, but this is unlikely, yet this region saw innovations in the industry in the first century BC, perhaps at Sidon (Strabo, *Geography* 16.2.25; *NH* 36.190–3; E. Marianne Stern, "Glass Production," in *The Oxford Handbook of Engineering and Technology in the Classical World* [ed. John Peter Oleson, Oxford 2008] 520–47).

5.76. Tyros (Tylos), or Tyre (modern es-Sur), was of particular interest. It was the southernmost and most famous of the Phoenician cities, sending forth settlers into the west, and most notably founding Carthage and Gades (Roller, "Phoenician Exploration" 645–53). It was originally an island, but Alexander the Great built a causeway to it in 332 BC (Arrian, *Anabasis* 2.16–24), and ever since it has been attached to the mainland. Pliny's 700 paces (1 km.) corresponds to the four stadia reported by Diodorus (17.40.4). It was famous for its purple dye industry, due to an indigenous snail, probably *Murex purpurea*, described in detail elsewhere by Pliny (*NH* 9.125–38). The dye and the textiles colored by it were highly prized, especially in Rome. Palaetyros (Old Tyre) was actually several kilometers south of Tyre on the mainland. The first figure, 19, must be miles (28 km.) if it is to include Palaetyros, but the next is in stadia (probably taken from Eratosthenes or Posidonius): 22 stadia (4.4 km.) for the circumference of the town is reasonable.

Sidon (modern Saida) was the next major Phoenician city north of Tyre. Its assumed relationship to Boeotian Thebes is through a variant of the usual tale that Cadmus actually came from Tyre to Central Greece

(Herodotus 2.49), but the Sidonian version was known by the fifth century BC (Euripides, *Phrixos B* F819).

5.77. Returning to the interior, Pliny described the two mountain ranges of the region, Libanus and Antilibanus, with the former closer to the coast. They lie parallel to one another with a deep valley in between, known as Coele ("Hollow") Syria. Zimyra, or Simyra (Pomponius Mela 1.67), is not in Coele Syria but a townsite near the coast, perhaps at modern Tell Kazel (*NH* 5.78; H. Sader, "Tell Kazel and Ancient Simyra," *Berytus* 38 [1990] 16–22). The 1,500 stadia (300 km.) is probably the length of the Libanus range, although far too long (it is more like 150 km.). Placing the Decapolis, tetrarchies, and Palestine beyond Antilibanus is loose terminology, as they are well to the south, with the exception of the Decapolis city of Damascus.

5.78. Returning to the *periplous*, the account continues to the north (here Zimyra seems to be near the coast, beyond the north end of Mt. Libanus). Arados (modern Arwad or Rouad in Syria) is more like three km. offshore, rather than merely 200 paces (about 300 m.) Again there is a mixture of units of measurement, with stadia for the circumference of the city (7 stadia or 1.4 km.; see also Strabo, *Geography* 16.2.13), probably from Posidonius. At this point the *periplous* is beyond (north) of the Libanus and Antilibanus, but there are still rugged (but lower) mountains just interior of the coast. Nevertheless the coastal plain widens to a maximum of about 20 km.

Syria

5.79. The boundary between Phoenicia and Syria was just north of Arados in the modern Syrian province of Latakia. The *periplous* continues along the Syrian coast as far as the mouth of the Orontes River (modern Nahr el-Asi) and Seleucia Pieria (at modern Kapısuyu), just to its north (José M. Blázquez Martínez, "Siria en la *Historia Natural* de Plinio," in *"Debita verba": Estudios en Homenaje al Profesor Julio Mangas Manjarrés* [ed. Rosa María Cid López *et al.*, Oviedo 2013] 1, 413–28). There is a diversion to Antiochia (Antioch, modern Antakya), about 25 km. inland. This was the most famous of many cities with that name, often called Antioch-on-the-Orontes or Antioch Near Daphne for clarity, and was the administrative center of both the Seleucid empire and the Roman province of Syria. Daphne (modern Harbiye) was a resort in the mountains just south of Antioch (Strabo, *Geography* 16.2.6).

5.80. The local Mt. Casius (Kasion, modern Jebel el-Akra) is an isolated summit (1,759 m. high) just to the southwest of Antioch. The celestial phenomena noted are an example of the Greco-Roman fascination with what was perceived as unusual solar activity, especially from mountain summits. In this case it would be dark in the west while the sun rose in the east. The best example of such a curiosity was on Mt. Athos in Macedonia (Strabo, *Geography* 7.F15). The 19-mile (28 km.) road to the summit was probably the distance from Seleucia or Antioch. The mountain is hardly four miles high (5,920 m.), but mountain heights were poorly calculated in antiquity and generally exaggerated (Keyser, "Geographical Work" 353–72).

The narrative returns to the mouth of the Orontes River and continues north along the coast to the border between Syria and Cilicia at Mt. Amanus (modern Nur Dağları). This is the high ridge that is at the northeastern corner of the Mediterranean and which forms a natural boundary, the beginning of Asia Minor. The Syrian Gates are a pass through the Amanus north of Antioch (the modern Belen Pass at 750 m. elevation), and the Rhosian Mountains (modern Kara Dağ) are the southwestern end of the Amanus on the coast. Yet the Taurus are far to the north (see *NH* 5.97) and are not truly part of this system.

5.81. The vast interior of Syria, as far as the Euphrates, is the next topic. The account shows vestiges of three alphabetical lists of toponyms, ethnyms, and tetrarchies, mingled together. The source is probably official Roman reports of the population of Syria.

The lists have no details beyond the names except for a brief statement on the religious history of Bambyce (modern Membid), where there was an ancient cult of Atargatis that had been revived by Stratonike, the wife of the Seleucid king Antiochus I, in the early third century BC (Lucian, *On the Syrian Goddess* 21–4; Cohen, *Syria* 172–8). The Azetae, Gindareni, and Gabuleni were all in the vicinity of modern Aleppo, where there are still remnants of the toponyms. The Hemeseni were the inhabitants of Hemesa, or Emesa (modern Homs). The Hylatae are otherwise unknown, but their name suggests a Greek descriptive term, "Woodland Dwellers." The Ituraeans lived in the southern part of Hollow Syria, and were ruled by a prominent local dynast in the first half of the first century BC, whose name, Ptolemy, demonstrates the historic connections between this region and Egypt (Strabo, *Geography* 16.2.10; Richard D. Sullivan, *Near Eastern Royalty and Rome, 100–30* BC [Toronto 1990] 70–2). The Baaethaemi are otherwise unknown, but may be a Semitic name for the Hylatae (Tusculum 5, 195–7).

5.82. The Mariamnitani were probably the inhabitants of Mariamme, which the dynast Straton of Arados gave to Alexander the Great in early

332 BC (Arrian, *Anabasis* 2.13.8). Presumably it lay inland from Arados at modern Mariamin. The Cardytenses may be the inhabitants of the Syrian town of Kardytos mentioned by Hecataeus (*FGrHist* #1, F279).

The ethnyms in the final list (all inhabitants of towns) have Greek names, reflecting Seleucid foundations. With the exception of the Beroeenses, from Beroia (modern Aleppo), all are on or near the upper Orontes (Cohen, *Syria* 106–8, 116–17, 119, 263). As a final comment, Pliny once again mentioned the Greco-Roman prejudice against strange-sounding foreign toponyms (*NH* 5.1).

The Euphrates

5.83. A feature of the Syrian hinterland is the Euphrates River, and Pliny provided a lengthy account of its course (and the adjoining districts), from its source into Mesopotamia (through 5.90). At its closest the Euphrates is less than 200 km. from Antioch, so this was an appropriate place to discuss the river. It is about 2,800 km. long today but was shorter in antiquity since there had not been as much siltation at the head of the Persian Gulf, yet it was one of the longest rivers known. Pliny's report is from a mixture of sources, and also includes divergences to localities not actually on it but within its immediate region. Parts of the river were known better than others, due to the topography, and distances were provided where the river was navigable. In its upper portions it had long served as a major boundary and was seen as the eastern limit of Asia Minor.

Its source, like that of the Nile (*NH* 5.51), was a matter of disagreement, although since early times it was known to be somewhere in Greater Armenia (Herodotus 1.180). Pliny's recent eyewitnesses were not specified, but were probably Domitius Corbulo and Licinius Mucianus, mentioned immediately thereafter. Caranitis (Karanitis) is the district around modern Erzurum in Turkey. There are several streams in this and the surrounding regions (including the modern Fırat, Kara, and Murat Su), and it was not certain which of them should be considered the actual Euphrates.

Cn. Domitius Corbulo (*FRH* #82, F3; see also *NH* 2.180, 6.23), who had been suffect consul in AD 39, led a number of campaigns against the Armenians for several years beginning in 55 BC, a period of local instability. Pliny used him as a source three times. Mt. Aga is probably the Mt. Abos of Strabo (*Geography* 11.14.2), which he identified as the source of the river. C. Licinius Mucianus was in the region after his suffect consulship of AD 64. He believed the source to be at Mt. Capotes (probably modern Kurtlu Tepe), a high prominent summit near one of the affluents of the river. It cannot be determined if his distance of 12 miles (18 km.) to Zimara is correct.

5.84. There seems to have been an accurate survey of the river from Zimara to Elegea (near modern Kömürhan), perhaps due to the Roman activities in the region. It was navigable through this section, but below Elegea it enters a deep canyon as it passes through the Taurus Mountains; this portion is largely inaccessible even today (Timothy Bruce Mitford, *East of Asia Minor: Rome's Hidden Frontier* [Oxford 2018), vol. 1, 123–48).

5.85. Downstream from the gorge the river became better known, although there was one more canyon to traverse. Pliny changed sources, with measurements now in *schoeni*, a unit of variable length. Yet this does not necessarily mean an indigenous report, since Eratosthenes (*Geography* F27 = *NH* 12.53), Strabo (*Geography* 11.11.5), and Isidorus of Charax all used the unit. The latter's *Parthian Stations* included material about Syria and upper Mesopotamia. Writing in the Augustan period, he was perhaps more relevant to Pliny's needs.

Given the uncertain length of the *schoenus*, it is impossible to determine the reported width of the river below the gorge, but it can be assumed that it spread across a wide area. The bridge mentioned has not been located but may have been around Barsalium (modern Kıllık) or even above the gorge, where the road from upper Mesopotamia to Melitene crossed the river. After passing through a second gorge it reached the important town of Samosata (modern Samsat), 40 miles (59 km.) below the head of navigation into Mesopotamia. The account has reverted to miles and thus is probably from a Roman source.

5.86. Although the course of the Euphrates continues to be the focus of the narrative, places far from the river on both sides are included. Arabia may seem like an unusual term for the region below the second Taurus gorge (the district is normally called Osrhoene), but there was a tradition that the locals were Arabians, which presumably meant Nabataeans (*NH* 6.125), and King Abgar II (Greek Ariamnes) of Osrhoene, who opposed M. Licinius Crassus, was said to be Arabian (Plutarch, *Crassus* 21). The region was known as Arabia as early as the late fifth century BC (Xenophon, *Anabasis* 1.5.1), and it is possible that a group of Nabataeans had moved into the area and had retained some of their original identity into the first century BC or even later.

Crassus, the consul of 70 BC and a member of the unofficial triumvirate of 11 years later, made an ill-advised expedition against the Parthians in 55 BC, going well beyond the Euphrates and meeting his death at Carrhae in

53 BC (Plutarch, *Crassus* 19–33; Dio 40.25). Carrhae, biblical Harran (Genesis 11:31), at modern Altınbaşak in Turkey), is about 80 km. east of the river.

Another group of Arabians, the Praetavi, were around Singara (probably modern Balad Sinjar in Iraq), east of Carrhae. The Imeni are otherwise unknown. Below Samosata the river entered the upper part of Mesopotamia. There was a major crossing at Zeugma (modern Belkis in Turkey). A bridge (Greek *zeugma*) was built here by the Seleucids, probably at the time of Seleucus I, and this became the familiar name of the town, officially Seleucia-on-the-Euphrates (*NH* 5.82). Remains of the bridge may still be visible. The distance of 72 miles (107 km.) from Samosata is accurate.

5.87. The account continues down the Euphrates as far as Sura (modern Suriyya in Iraq), where it makes its sharp bend to the east. There are textual difficulties in this section, since there is a lack of grammatical coherence (Tusculum 5, 202). The name of the Rhoali survives in modern Ruala. The Arabian Scenitae ("Tent Dwellers") were the local nomads, extending to the south, some of whom were known to the Greek and Roman world as plunderers and brigands (Strabo, *Geography* 16.1.26–7). They are a fine example of a descriptive adjective becoming an ethnym.

5.88. Pliny's narrative left the river to describe the desert south to Arabia, with a brief encomium on Palmyra. He triangulated the location of the oasis and its city, with distances along routes converging there, perhaps from the map of Agrippa. As reported, Seleucia-on-the-Tigris was 337 miles (499 km.) to the southeast, the nearest point on the Syrian coast 203 (300 km.) to the west, and Damascus 176 (260 km.) to the west. The only figure that is problematic is the one to the Syrian coast; the nearest point is Arados, not an effective destination from Palmyra and far closer than the reported distance. It is more likely that it is to Antioch or Seleucia Pieria, following a main route across to the Orontes and then downstream.

Pliny emphasized the significance of Palmyra due to its location between the two great empires (*imperia summa*) of Rome and Parthia; thus it had a "private fate" (*privata sors).* But its location was exceedingly remote, and the locals took every possible means to create routes to the city and to keep them safe (Andrew M. Smith II, *Roman Palmyra: Identity, Community, and State Formation* [Oxford 2013) 1–2; Gary E. Young, *Rome's Eastern Trade* [London 2001] 136–8).

5.89. Pliny's overview of this lightly populated region (*solitudo*) continues south from Palmyra as far as Elatium (presumably the Aelana of *NH* 5.65), at the head of the eastern gulf of the Red Sea. This followed an ancient trade route south from Damascus. No specific distances were provided – perhaps none appeared on the map of Agrippa – and Petra is far more than merely two-thirds of the distance from Damascus to the Red Sea.

The account then returned to Sura and the Euphrates, noting the neighboring town of Philiscum, presumably on the opposite (left, or north) bank, since it was in Parthian territory. From here one could sail to Seleucia-on-the-Tigris and Babylon, which are about 60 km. apart, in 10 days, covering a distance of 400–500 km.

5.90. The lower Euphrates is a tangle of channels and canals, intermingling with the Tigris (which in antiquity came to within 25 km. of the Euphrates, near the future site of Baghdad). Today this region is significantly different hydraulically. From Zeugma it was 594 miles (879 km.) to Masica, an unidentified location which is probably where the various channels began.

In order to complete his material on the Euphrates, Pliny briefly considered the matter of its flooding. This was similar to that of the Nile, but did not excite as much interest. It flooded in early summer, began to fall in late August, and returned to normal in late September. The flooding of the three great rivers of antiquity – the Nile, Euphrates, and Indus – was seen as the benevolence of the natural world for the sake of human endeavor (Cicero, *de natura deorum* 2.130).

Southeastern Asia Minor

5.91. The account returns to where it had left the coast after *NH* 5.80, and goes around the northeast corner of the Mediterranean, continuing as a *periplous* along the south coast of Asia Minor, with some mention of inland localities. The summaries as far as Lycia (*NH* 5.97) are brief. This begins a long narrative along the southern and western coast of Asia Minor as far as the Thracian Bosporus (*NH* 5.151), which occupies most of the rest of Book 5.

After noting some final locations in coastal Syria beyond the Amanus Mountains, the account enters Cilicia through the pass known as the Cilician Gates, an ancient route between Syria and Asia Minor (Strabo, *Geography* 12.2.7). Cilicia, or Kilikia, was the southeastern part of Asia Minor, and had been an object of Roman interest since the early first century BC. In Pliny's day it was a Roman province.

The *periplous* moves along the Cilician coast, occasionally venturing slightly inland to record the towns in the region known (in its eastern part) as the Aleian Plains, and, more generally as Cilicia Pedias, or Level Cilicia. At least two sources were used – Tarsos was mentioned twice (see also *NH* 5.92) and described in slightly different terms – but neither is identified. The most likely possibilities are the map of Agrippa and a Roman demographic catalogue.

5.92. Beginning again at Tarsos, the account names additional towns in Cilicia Pedias, and continues farther west into the other main district of Cilicia, Tracheia ("Rough"), an extremely rugged mountain region that drops down steeply to the coast. The narrative goes as far as Cape Venus, or Aphrodisias, about 100 km. from Cyprus; the island is described in greater detail at *NH* 5.129–30.

5.93. After noting the westernmost localities in Cilicia, Pliny cited five inland towns scattered over a wide area, from Castabala (probably the site at modern Bodrum Kale), the easternmost town in Cilicia, to Iconium (modern Konya), far inland and the major city in Lycaonia. The Anazaberi are probably the citizens of Anabarzos (modern Anavarza) in the midst of Cilicia Pedias. The toponyms seem to be listed roughly east to west, and may come from a third source that examined localities in south central Asia Minor.

5.94. Isauria was the mountainous region between Cilicia and Pamphylia. It was not documented before the Hellenistic period (Diodorus 18.22), and had little independent history, generally being at the mercy of its neighbors. In Pliny's time it was part of the Roman province of Galatia, to its north (David Magie, *Roman Rule in Asia Minor to the End of the Third Century After Christ* [Princeton 1950] 288–90).

The Omonandes (or Oroandes) cannot be specifically located, but are probably the small wealthy population mentioned by Polybius in the context of events of the 180s BC (Polybius 21.41.7; Walbank, *Historical Commentary* 3, 155–6). This region of western Cilicia, and beyond toward Pamphylia, was wild and remote, and one might expect a large number of isolated fortresses. Pliny made only passing mention of Pisidia, the next region to the west. It too had little independent history, and became part of the province of Galatia.

5.95. Lycaonia was a small region in the southern interior plains of Asia Minor, with Iconium its major city (Sviatoslav Dmitriev, "Observations on the Historical Geography of Roman Lycaonia," *GRBS* 41 [2000] 349–75). It had tended to be unorganized and subject to outside rulers since Hittite times. Lycaonia was officially attached to the province of Galatia, but its people may have looked more to the west and the cities of Roman Asia for trade and commerce. The five ethymns all refer to a series of towns located on a relatively fertile strip on the borders of Phrygia at the west end of Lycaonia and along the main road from the Aegean through the Cilician Gates to Syria (Magie, *Roman Rule* 455–6).

The tetrarchy mentioned may refer to the rule of part of Lycaonia by Amyntas of Galatia, who had been placed on the throne of that territory in 36 BC by Marcus Antonius, surviving for the next 11 years (Sullivan, *Near Eastern Royalty* 171–4). Pliny's terminology is somewhat anachronistic, since Galatia, even though it had a long history of tetrarchs, had been unified and raised to a kingdom by Deiotaros in 42 BC.

5.96. Pamphylia was limited to a small district on the coast south of Pisidia, essentially the alluvial plain of the Kestros River (modern Aksu). It was no more than 60 km. across and 30 deep. There was little in the way of any local identity, but its coastal orientation made it of early Greek interest. The alleged ancient name, Mopsopia, comes from the early wandering prophet, Mopsos, and it was believed that the region had been settled by Trojan refugees (Herodotus 7.91; Strabo, *Geography* 14.1.27, 14.4.3). Roman interest began in the early first century BC, and eventually it too was attached to the province of Galatia.

The Taurus

5.97. West of Pamphylia was Lycia, an important region of southwest Asia Minor (see *NH* 5.100–2). But Pliny digressed to discuss the Taurus Mountains, the main range of southern Asia Minor (through *NH* 5.99). It had long been believed that the Taurus separated the Asian part of the inhabited world into a northern and southern portion. To be sure, the name Tauros strictly applies to the mountains in Asia Minor, which reach up to 3,000 m. high in Lycia, but it was thought that the range continued eastward across southern Asia and all the way to India. In India, the local Hemodi Mountains (*NH* 6.56; the modern Himalayas, which preserve the name) were said to be its eastern end. Eratosthenes (*Geography* F47) was the first to develop this concept, seeing the mountains as a line that divided the inhabited world, a theory that was to some extent geographically sound. There were many ways of defining the Taurus in Asia Minor, but to Pliny their western end was at Cape Chelidonium (modern Gelidonya Burnu), one of its most southern points.

The western part of the greater Taurus range, as far east as the modern Hindu Kush, runs roughly west to east, but beyond it the Himalayas curve toward the south, passing through as much as 10° of latitude. Pliny's language is rather obscure and animistic, yet the implication is that if the eastern end of the range continued west on the same latitude it would enter the ocean and thus not continue toward Asia Minor. This issue was solved by the northerly trend of the range, which eventually placed it in a direct line to cross southern Asia and Asia Minor, past the various seas mentioned. But the catalogue of seas is problematic, because the Caspian and Hyrcanian are the same (as Pliny knew at *NH* 6.15), and Lake Maeotis (the modern Sea of Azov) is far to the north and has little relationship to the Taurus. The almost poetic rendering of this section suggests that the source may be a philosophical account of the nature of the inhabited world rather than a geographical treatise.

5.98. With additional animistic terminology, Pliny catalogued the mountain ranges of southern Asia, linking them with the Taurus. The Ripaean ("Tossing") Mountains were a mythical concept in the far north, perhaps vaguely connected with early knowledge of the Alps, an idea developed no later than the sixth century BC (Aristotle, *Meteorologika* 1.13.350b; Eratosthenes, *Geography* F8; Strabo, *Geography* 7.3.1).

There are three lists. The first cites 11 ranges from west to east. The Imaus and Emodus are local variants of the same name, preserved in the modern Himalayas, and the Taurus in this list is the west end of the range. Then there is the Caucasus, far to the north, the highest mountains known in classical antiquity (reaching 5,642 m.), although it is not certain that this was actually realized, and moreover there was no perception of the true height of the Himalayas. The third list cites three assumed branches of the Taurus, all of which are promontories or peninsulas that protrude into the Mediterranean from southern Asia Minor. All put together, there are 15 mountains mentioned, creating essentially a full catalogue of the various ranges of southern Asia in a broad swath from India to the Aegean.

5.99. Further features of the greater Taurus include the fact that despite being a major barrier there were several passes through it. Pliny used the Latin Portae, a translation of Greek Pylai, a term that since at least the fifth century BC had described a low route through otherwise impassable territory, most famously at Thermopylai in Central Greece. Implicit in the animistic account (see also *NH* 5.97) is a certain tension between the mountains and the seas.

The Parihedrians were around Mt. Pariades, the Lesser Caucasus of Armenia (*NH* 6.25). The Moschians were to their northwest in interior Colchis (*NH* 6.29). The Amazons were the semi-mythical warrior women of northern Asia Minor; Mt. Amazonius may be one of the ridges above the Themiscyra River in northern Pontus. The Coraxians were at the northeast corner of the Black Sea, and Mt. Corax may be part of the western Caucasus. The Scythians, in the most general sense, were the nomadic people north of the Black Sea (Ephorus F30a), but there is no Mt. Scythicus otherwise known. Pliny's basic argument is valid, that the influence of the greater Taurus was so vast that many populations were named after various portions of the range. This marks an important understanding of the relationship between toponyms and ethnyms, but the argument is circular, and has led modern topographers to seek the toponyms that are allegedly based on the ethnyms where there may be none.

Ceranius (Greek Keraunia, "Thunderer") is a common toponym for mountain peaks, usually conspicuous ones. Here Pliny may be referring to the modern Dagestan Mountains of the eastern Caucasus (Pomponius Mela 3.39), but it is hard to understand Ceranius as a synonym for the entire Taurus or even the Caucasus, even despite such an implication by Pomponius Mela (1.109).

Lycia

5.100. The account returns to the *periplous* of the coast of Asia Minor and where it left off at *NH* 5.96. Lycia (through *NH* 5.102) is west of Pamphylia, and was a federation of cities that was generally independent of the major powers. The region was the last area of Asia Minor to come under Roman control, when local instability led Claudius to provincialize the territory in AD 43 (Suetonius, *Claudius* 25).

The narrative remains close to the coast, passing through most of Lycia. Its eastern part was active volcanically and gave rise to the story of the Chimaera, the monster that was part lion, goat, and snake, and which breathed fire. She was known to Homer (*Iliad* 6.178–83; Strabo, *Geography* 14.3.5), and her activities are still visible today in the Yanar Taş region.

5.101. The reduction in the number of Lycian towns is probably as much due to Roman reorganization as depopulation, and may reflect a different Roman view of the demography of the region after provincialization. The list of towns shows vestiges of alphabetization and may be from Roman documents. Lisa, Melanoscopium, and Telandrus are west of Telmesus and thus beyond Pliny's stated boundary of Lycia, lying across the Bay of Telmesus, but probably culturally if not politically Lycian. Only three of the nearly thirty Lycian cities named were inland, all close together in the district known as Cabalia.

5.102. The southern protrusion that is Lycia marks a natural division in the topography of Asia Minor as well as in the nature of the adjoining seas. Rhodes, the first of the many islands that are a feature of the Aegean, is directly southwest of Lycia; from here along the coast one is never out of sight of islands until the end of the Aegean at the Thracian Bosporus. Moreover, on the land, the Gulf of Telmesus (the homonymous city is on its east side) was the historic boundary between Lycia and Caria (*NH* 5.103). When the Roman province of Asia was created, Caria, at least in part, seems to have been included (Magie, *Roman Rule* 154–5), which is presumably what led Agrippa (F25–6) to see the Lycian–Carian boundary as a significant dividing point, separating Asia (i. e. Asia Minor as a whole) into two parts, a point of view perhaps reinforced by the continuing independence (in Agrippa's day) of Lycia. Agrippa's line between the two Asias ran roughly to the northeast, between Phrygia and Armenia, and presumably west of the province of Pontus, although Pliny's description is not clear on this point. This is the first specific mention of Agrippa since Arabia (*NH* 5.65), but one suspects that his map was a constant source of information for Pliny.

The measurements provided by Agrippa lack any specific points, and thus are difficult to understand. The length (east–west) of the western part of Asia, 475 miles (703 km.), would be approximately that from the Aegean to the center of Asia Minor, east of Phrygia; 575 miles (851 km.) for the eastern part would perhaps be to the Euphrates. The two widths (north–south) are the same (320 or 325 miles, 474 or 481 km.), and are essentially the shortest distance across Asia Minor, from Pamphylia to the Black Sea coast east of the Thracian Bosporus. There are different figures at *NH* 6.24, based on a Greek source.

From *NH* 5.103 through 5.140, Pliny's *periplous* moves along the Aegean coast of Asia, with some inland incursions. The coastal islands are also listed. There are numerous topographical and orthographical problems; many have been discussed in J. M. Cook, "Pliny On Icarian Shores," *CQ* 9 (1959) 116–25.

Caria

5.103–4. Caria (though *NH* 5.109) is the southwestern portion of Asia Minor. It was an ancient region that had its own cultural prominence, especially during the fourth century BC under the Hekatomnid dynasty of Mausolos and his siblings. In time it became more involved in the politics of the Hellenistic world and was absorbed by Rome in the second half of the second century BC.

The first part of the account is a standard *periplous* from the Lycian border following the complex indentations of the coast, and passing the closest mainland point to the island of Rhodes – said to be 20 miles (30 km.) away but actually closer – as well as moving along the prominent peninsula of Cnidus. Doris is not a town or region, despite the implication, but the large and deep bay (modern Gökova Körfezi) that is surrounded by Caria, with Cnidus at its southern entrance and Halicarnassus at its northern (*NH* 5.107). It was believed to be an area of ancient Dorian settlement (Herodotus 1.144).

5.105–6. Pliny catalogued the people of inland Caria according to Roman jurisdictions – the same pattern he had used in Hispania – using the Latin term *conventus*, referring to a group of communities who came to a central point for legal proceedings. The three mentioned – Cibyratic, Synnas, and that at Apamea – are all in what Pliny called the *rursus* ("back") part of Caria, in other words, its far interior, toward Phrygia (in fact Cibyra, at modern Horzum, was actually in Phrygia). A prominent geomorphic element of interior Caria was its rivers; unlike many other parts of the eastern Mediterranean, western Asia Minor was exceptionally well-watered, and a total of eight rivers are mentioned.

The story of Marsyas and Apollo was localized around Apamea (at modern Dinar). The city was a Seleucid foundation at the ancient site of Kelaina on the Marsyas River (modern Dinar Su). Marsyas challenged the god Apollo to a music contest and paid a horrific penalty for losing: the tale is a typical example of conflict between religions, and is best recorded by Ovid (*Metamorphoses* 6.382–400). Musical innovation is an element of the tale, and it seems probable that this part of the Phrygian–Carian borderlands was where such developments took place (Strabo, *Geography* 10.3.15; Athenaios 14.624b; Landels, *Music* 153–9).

5.107. The narrative returns to the Carian coast (connecting with the end of *NH* 5.104) and continues around the Bay of Doris to the end of the peninsula on which Halicarnassus sits, and into the Iasian Bay (modern Asın Körfesi), as far as Bargylia (modern Varvıl). In early times there was a local Dorian federation of six cities, whose name is not known (Herodotus 1.144). The six were Cnidus and Halicarnassus in Caria, Lindos, Ialysus and Camirus on Rhodes (*NH* 5.132), and the island of Cous or Cos (also Ceos; *NH* 5.134). The context is the sixth century BC, and its existence may be what led Pliny to suggest, in a vague way, that Doris was a separate political entity (see also *NH* 5.103).

Alexander the Great conquered Halicarnassus in the autumn of 334 BC and appointed Ada, sister of the famous Mausolos, as satrap. He consolidated the territory under her rule (Arrian, *Anabasis* 1.23.6–7). The six towns mentioned form a circuit within 20 km. of Halicarnassus.

5.108–9. There are several catalogues of interior Carian towns, with nearly thirty listed. There is little reflection of the Roman organization, although it was noted in passing that Alabanda (modern Araphisar) was the center of a *conventus*, without reference to what towns were included. There seems to be hardly any geographical coherence (Alinda and Alabanda, only a few kilometers apart, are in separate lists), and eight towns are described by local ethnyms rather than toponyms. The account may be Seleucid in origin, given the citation of several toponyms associated with that dynasty.

Pliny noted the distinction of many of these towns, but without providing details. Mylasa (modern Milas) was an ancient cultic locality that was the home of the famous Hekatomnid family, who came to rule all Caria (Strabo, *Geography* 14.2.23–4). Nysa (modern Sultanhisar) was an educational center that numbered among its students the geographer Strabo (*Geography* 14.1.43–4). Trallis (modern Aydin) was the home of Pythodoris, who became queen of Pontus and an important personality of the early first century AD (Duane W. Roller, *Cleopatra's Daughter and Other Royal Women of the Augustan Era* [Oxford 2018] 99–120). Aphrodisias was a major artistic and cultural city, with an ancient oracle of the goddess and her indigenous predecessor. Other Carian towns had their own prominence.

Pliny often positioned towns in terms of rivers. The names are provided without any other comment except for the odd statement that Pygmies lived around Thebais, which sounds more like a district than a town. It is not certain where it was located. There was a Thebai on the south side of Mt. Mykale in Ionia, near Priene, the closest known use of the toponym to Caria (Theodor Wiegand and Hans Schrader, *Priene* [Berlin 1904] 469–74). A founder of Priene was a certain Philotas from Boeotian Thebes (Strabo, *Geography* 14.1.3), yet this becomes a convoluted argument to address the question of Pygmies in Caria, although they were said to be in many places beyond their traditional central African home, including as far north as Thrace (*NH* 4.44).

Lydia

5.110. Lydia was north of Caria, across the Maeander River. It lay inland from Ionia and was notable as the land of Croesus, described in great detail by Herodotus (1.6–94). Its central city was the famous Sardis (at modern Sart), on the Pactolus River (modern Sart Çay).

The earlier name for the region, Maeonia, was used by Homer (*Iliad* 2.864–6). Lydia as a toponym first appears in the account of Herodotus. "Hyde" is dubious and is found only at *Iliad* 20.385 as a locality near Mt. Tmolus, but by Hellenistic times it was believed to be an ancient term for the entire district (Strabo, *Geography* 13.4.6; Roller, *Historical and Topographical Guide* 781).

5.111. Having covered the early toponymic history of Lydia, Pliny as a contrast turned to its current status under the Roman administration, an account centered on Sardis (Francesca Gazzano, "La Lidia di Plinio il Vecchio," *Maia* 70 [2018] 260–79). The ethnyms are the inhabitants of various towns in the region, some of which show evidence of Roman reorganization.

Cadi, or Kadoi (modern Eski Gediz), in the borderlands between Lydia and Phrygia, probably had a settlement of Macedonian veterans. The Maeonii may merely be an attempt to recreate the ancient ethnym, and Antoniopolis was a renaming of ancient Apollonia (near modern Yenice), presumably reflecting the local presence of the triumvir Marcus Antonius in the 40s BC (Christian Habicht, "New Evidence on the Province of Asia," *JRS* 65 [1975] 83–5). The Apollonihieritae lived around a sanctuary of Apollo, and Mysotimolitae sounds like another created name for Roman purposes, referring to a group near Mt. Tmolus.

Ionia

5.112. Ionia is north of Caria and west of Lydia. Its distinguished cultural and intellectual history may be what inspired Pliny to discuss it at length (through *NH* 5.120), but there is little coloration to the narrative, which is almost exclusively toponymic. No sources are cited, but the use of both stadia and miles suggests at least two, one Greek, perhaps from Miletus, and the other probably the map of Agrippa.

The account begins at the Iasius Bay (*NH* 5.107) and continues north along the coast to Miletus, in antiquity on the south side of the Maeander estuary. The Branchidae were the family in charge of the oracle of Apollo, but in time their name became toponymic and virtually interchangeable with Didyma, the actual location of the shrine (Herodotus 1.92; Strabo, *Geography* 14.1.5). The site is slightly inland (20 stadia or 4 km.), and from there to Miletus it is about 15 km., at variance with Pliny's 180 stadia

(36 km.), which may be a textual error. The route went along the well-known Sacred Way, whose remains are still visible.

One of Miletus' ancient names, Pityusa ("Piney"), may reflect the extensive local pine forests, which, among other uses, would have provided timber for shipbuilding, thus aiding in the extensive overseas settlement for which the city became famous (Alan M. Greaves, *Miletos: A History* [London 2002] 13–14). Many of the numerous Milesian outposts encircled the Black Sea and its approach from the Aegean. Pliny's 90 settlements is an excessive number, as far as is known, and may represent an effusive Milesian source, or include secondary foundations. Seneca had 75 (*de consolatione ad Helviam matrem* 7.2), and perhaps 50 can be identified today. Nevertheless there is little doubt that the number far exceeded that of any other Greek city, and from the eighth or seventh century BC until its decline late in the following century due to Persian activities, Miletus was the most important state in the Greek world (A. J. Graham, *Colony and Mother City in Ancient Greece* [Manchester 1964] 98). Its citizen Cadmus (*FGrHist* #489), oddly the only person mentioned in Pliny's account of Ionia other than Alexander the Great, was generally accepted as one of the inventors of prose writing. There is no extant mention of him before the first century BC, a curious point in itself, and his career remains vague and essentially undated beyond an assumption that he was before the fifth century BC (Diodorus 1.37.3; Strabo, *Geography* 1.2.6; *NH* 7.205).

5.113. The Maeander River (modern Menderes) is the longest in western Asia Minor (548 km.), and as such had its own personality and mythology. Its twists and turns produced the word "meander." Pliny outlined its entire length, from its source in southern Phrygia at Lake Aulocrene (or Aulotrene, modern Pınarbaşı) to the Aegean. Today the coast is much farther to the west than it was in antiquity, due to the immense siltation

The *periplous* continues from Miletus around the head of the Maeander estuary: the towns mentioned are all inland today. It then proceeds from Mt. Mykale (not mentioned) to the central gathering place of the Ionians. Pliny called this Panionia, but strictly speaking this is the term for the festival that was held there; the site itself was the Panionion (Herodotus 1.148).

5.114. Phygela (or Pygela: Strabo, *Geography* 14.1.20; Pomponius Mela 1.88) and Marethesium were farther along the coast from Panionia. Pomponius Mela equated the name with Greek *phyge*, "banishment" or "exile" (Pomponius Mela 1.88), but Strabo derived it from *pygalias*, or "buttock pain," an affliction suffered by the rowers of Agamemnon when they were in the region. Magnesia (on the Maeander) lies about 20 miles

(30 km.) inland from Phygela and 15 miles (22 km.) southeast of Ephesus. Today the site is on the Maeander but in early times it was on or near the sea. The Derasidae Islands (*NH* 2.204) have vanished with the siltation of the Maeander estuary.

5.115. Thyatira (at modern Akhisar) seems peculiarly out of place, since it is about 30 km. north-northwest of Sardis in Lydia (rather than Ionia). Pliny may have left it out of his description of Lydia. Ephesus (at modern Selçuk) is the next town on the coast after Phygela, and the most famous in Ionia after Miletus. Its alleged Amazonian foundation is part of a general belief that Amazons were involved in the early history of the region, perhaps a point of view put forth by Ephorus of Cyme (in nearby Aeolis: *NH* 5.121) in the fourth century BC, whose city was said to be one of their foundations (Strabo, *Geography* 12.3.21, 12.8.6). The region around Ephesus was also subject to extensive siltation, although not as much as on the Maeander, yet the island of Syrie (modern Kurutepe) has been a knoll on the north edge of the plain since Hellenistic times. The Temple of Diana is the famous shrine to Artemis.

5.116–17. The account continues along the lengthy peninsula that extends west between Ephesus and Zmyrna (Smyrna), ending at Mt. Mimas (Boz Dağ), the westernmost part of Asia south of the Troad. The distance of 150 miles (222 km.) to Mt. Mimas is probably the sailing route around the peninsula from Ephesus to Zmyrna.

This long peninsula was a hindrance to navigation: the lengthy route around it was contrasted with two isthmuses, the shorter of which (the more western) was only 7.5 miles across (11 km., from 60 stadia). Alexander the Great planned to construct a canal, but the project is not well documented and may have been little more than a proposal (Pausanias 2.1.5). More successful was the reattaching of the Chytrophoria islands to the mainland, at Clazomenae (modern Klazumen). There had been a causeway connecting the city to the mainland, two stadia (400 m.) long, but the locals removed it when the Persians were threatening, probably during the Ionian revolt at the beginning of the fifth century BC. Alexander eventually rebuilt it (Pausanias 7.3.9).

5.118. Zmyrna (more commonly Smyrna, modern İzmir) was another of the cities allegedly founded by an Amazon. The nature of Alexander's restoration is not clear, and it may not even have happened. The city had been of little importance in the fifth and fourth centuries BC, and any rejuvenation may not have begun until somewhat later (Strabo, *Geography* 14.1.37; Pausanias 7.5.1–3; Cohen, *Europe* 180–3). It was 12 miles (18 km.) from some unspecified place on the coast, but there is no obvious candidate (the previous coastal town was Clazomenae, about 35 km. away)

Starting with Mt. Mastusia (modern Dikmen Dağ), just west of Zmyrna, Pliny described a chain of mountains that runs to the east as far as the perceived beginning of the Taurus (see *NH* 5.97). This is to some extent an artificial construct. One can easily link Mastusia, Termetis, Olympus, Draco, and Tmolus into a joined sequence for about 100 km., essentially from just west of Zmyrna along a line south of the ancient and modern road from Zmyrna to Sardis, and somewhat beyond. But east of Tmolus there are no contiguous mountains over a distance of about 50 km. until Cadmus (modern Babadağ), which is an isolated summit south of the upper Maeander. Then it is another 200 km. to what was considered the beginning of the Taurus, across the uplands of southern Phrygia and Pisidia. Yet the chain that Pliny presented, even if questionable, is a fine example of structuring topography to fit geographical theory, supporting the idea of an east–west range of mountains going all the way across Asia.

5.119. As with the Maeander (*NH* 5.113), Pliny outlined the course of the Hermus River (modern Gediz Çay), the second longest in western Asia Minor (401 km.), from its source near Dorylaeum in Phrygia (modern Şarhöyük) south and then west to the Aegean. Although siltation was not as pronounced at its outlet than farther south, the lower course has changed since antiquity. Phocaea (modern Foça), at the end of the peninsula near the estuary, was the northernmost town in Ionia.

5.120. As a final note on Ionia, Pliny described the two Roman *conventus* that made up the territory, one at Ephesus and the other (which extended north of Ionia into Aeolia, or Aeolis) at Zmyrna. The Ionian members of this *conventus* were along the Hermus River, northeast of the city, where there were Macedonian settlers. Those in the Ephesian *conventus*, insofar as they can be identified, were between Mt. Tmolus and the Maeander.

Aeolia

5.121–3. Aeolia, or Aiolis, was north of Ionia, extending to Troas, or the Troad. Despite the implication, Aeolis and Mysia were not the same geographical region. The former, named after one of the major ethnic and linguistic groups of Greece, was essentially the coastal strip between Ionia and Troas, including some of the adjacent islands. The Aeolians were first mentioned by Hesiod, suggesting that they had arrived in the region during the population movements after the Trojan War (Hesiod, *Works and Days* 636; Herodotus 1.6 etc.; Strabo, *Geography* 13.1.8). Mysia was more inland, and the Mysians were Trojan allies, although their exact location is difficult to determine (R. Hope Simpson and J. F. Lazenby, *The*

Catalogue of the Ships in Homer's Iliad [Oxford 1970] 176–7). Pliny's assumption that the Mysians were in the region earlier than the Aeolians is thus correct, but the two populations were not in the same location.

The account is a *periplous*, extending from Phocaea (*NH* 5.119), the northernmost Ionian town, to the north and around the Gulf of Adramyttium (the large bay on the south side of Troas, the modern Gulf of Edremit), and as far as Lectum at the southwest point of Troas (modern Balaburnu, the westernmost point of the continent of Asia). There are no distances and the account rarely ventures far from the coast. Yet the tone of the narrative is one of devastation everywhere: 15 towns were either abandoned or destroyed. This probably reflects the effects of the Mithridatic Wars, which lasted for over 25 years in the early first century BC and which had been especially intense locally. Although the exact date of Pliny's lists is not certain, the Mithridatic destruction was still noticeable decades after the death of Mithridates VI in 63 BC, since Herod the Great visited Chios in 14 BC and helped rebuild a stoa that had been destroyed in the wars and was still in ruins (Josephus, *Jewish Antiquities* 16.17–19). The impact of the wars continued into the Julio-Claudian period (Charles Brian Rose, *The Archaeology of Greek and Roman Troy* [Cambridge 2014] 236–7). Such disruption would have caused movement of populations and the abandonment of villages.

Adramytteos (at modern Karataş), at the head of its homonymous gulf, was the commercial and administrative center of the region. Pliny's list of urban ethymns covers a wide region, including localities on the Rhyndacus River (modern Koca Çay), whose source is over 100 km. to the northeast and which flows away from Adramytteos to an outlet on the Propontis. Assuming that this is the river he meant, it is difficult to imagine those living along it going a vast distance across mountainous terrain to Adramytteos, but there is no further explanation. It is significant that although the city was the seat of a *conventus*, the ethnyms are not listed as belonging to it, but rather people would come for *negotium* (business matters).

Troas and Mysia

5.124. Troas, or the Troad (through 5.127), is the northwestern corner of Asia Minor and the westernmost part of the continent. The account continues along the west coast of Troas to the entrance of the Hellespont at Sigeum (modern Yenişehir). There is only the slightest allusion to Homeric issues, noting that four rivers mentioned in the *Iliad* can no longer be found, and a passing reference to Ilium (Troy) itself, 2.5 miles

(3.7 km, from 200 stadia) inland. Pliny chose not to enter into the complex arguments about the actual location of the town that was the site of the Trojan War (Strabo, *Geography* 13.1.26–7), and his account of the Trojan Plain remains unclear (J. M. Cook, *The Troad: An Archaeological and Topographical Study* [Oxford 1973] 181–2).

5.125. The unnamed bay is the broad entrance to the Hellespont, and most of the sites mentioned are along its southern coast. Alleged tombs of those who fought at Troy are a common feature of the Trojan Plain, most of which are from an earlier era than the war itself. The Tomb of Achilles is a Late Neolithic settlement four km. southwest of Troy, which was monumentalized in the third century BC. The Tomb of Ajax was destroyed through sea action by the early second century AD and rebuilt by Hadrian. Its construction is still visible, locally called İn Tepe; what Pliny saw or read about is not known (Strabo, *Geography* 13.1.30; Philostratos, *On Heroes* 8.1; Rose, *Archaeology* 190–2, 247–8).

Teuthrania is misplaced and oddly described, assuming that it is the well-known town on the lower Caicus River at modern Kalerga, close to the coast and nowhere near the source of the river. It is probably the same as the Teuthranie of *NH* 5.126, and may have been left out of the description of Aeolis.

5.126. The towns listed are all in Mysia, around Pergamum or in the uplands to its east, as far as they can be identified. Pergamum was certainly the most important city in northwestern Asia Minor in Hellenistic times, due to the culturally prominent dynasty of the Attalids. Even after it had become Roman territory it continued to be a major player in the Mithridatic Wars and a Roman administrative center. The 12 peoples (all urban ethnyms) that came to the city for legal and other purposes were mostly on the upper Caicus (modern Bakır Çay), and farther to the southeast toward Lydia, in the watershed of the Lycus River (modern Gördük Çay).

5.127. To complete the topographical account of Troas, Pliny returned to Rhoeteum (*NH* 5.125) on the Hellespont and continued as far as Cape Trapeza, or the Dardanian Cape (modern Kepez Burnu), where the Hellespont narrows. The current is from the Black Sea to the Aegean.

Continuing the theme of depopulation, he cited five ethnic groups mentioned by Eratosthenes (*Geography* F126). These were not limited to northwestern Asia Minor: only the Lelegians seem to have been local (Homer, *Iliad* 10.429). The Solymi were thought to be from Lycia or Phrygia (Herodotus 1.173; Strabo, *Geography* 14.3.10). The Bebrykes were Mysian (Strabo, *Geography* 12.3.3). The Colycantii and Tripsedi are not otherwise known. Isidorus of Charax (F19) named two more ethnic groups, but they cannot be located. The town of Apamea, formerly called Damea, is mentioned nowhere else, and Isidorus' positioning of it covers most of eastern Asia Minor. Apamea (Apameia) was a common toponym in the Seleucid world, since King Seleucus I, the companion of Alexander the Great, named cities after his wife, the Bactrian princess Apame (Cohen, *Europe* 375).

Islands off the coast of Asia Minor

5.128. Pliny's account leaves the mainland to examine various islands off the coast of Asia Minor (through *NH* 5.140). The narrative begins at the Canopic mouth of the Nile (considered the beginning of Asia) and continues east and north into Phoenicia. This coast has few islands, and only four are mentioned over several hundred kilometers.

Pharos, the site of the famous lighthouse of Alexandria, is now joined to the mainland. Its position was a major topographical problem in antiquity, since Homer made it clear that when Menelaus was stranded there for twenty days it was a day's sail from the coast (*Odyssey* 4.354–62). This is difficult to reconcile with the fact that in historic times the island was immediately offshore. Eratosthenes (*Geography* F10) was concerned about the discrepancy but offered no solution. Strabo (*Geography* 1.2.23) reasonably suggested that Menelaus was exaggerating his perilous situation.

The lighthouse was built in the early third century BC and survived until after AD 1300. It was one of the Seven Wonders of the World, a catalogue of technological marvels from earliest times to the Hellenistic period (Beresford, *Ancient Sailing Season* 201–2). The channels mentioned are the routes of access into the harbor, where there were numerous rocks and reefs that required careful navigation between Pharos (to the west) and Cape Lochias (to the east). The complexity of these hazards appears on early modern maps of the city, especially that by Gaston Jondet from 1911–1915 (McKenzie, *Architecture* 19–30).

Paria is no longer visible as an island, but presumably is one of the features, either on the land or under the sea, at modern Jaffa in Israel. Pliny had already discussed the story of Andromeda (*NH* 5.69). Arados (see *NH* 5.78) lies about three km. offshore; the channel between it and the mainland was noted for a freshwater spring at a depth of about 25 m. (Lucinius Mucianus F8). This mingling of fresh and salt water was of particular interest (Lucretius 6.890–9; Strabo, *Geography* 16.2.13).

5.129. From Arados to Cyprus there were no significant islands. In discussing Cyprus, Pliny cited an unusually detailed list of sources (six in all), from Philonides (*FGrHist* #121, F1), dating to the era of Alexander the Great, to Isidorus of Charax (F13) in the Augustan period. This large number suggests that Pliny may have been using a recent handbook on the geography of the island.

Cyprus is the third largest island in the Mediterranean (9,251 sq. km.; only Sicily and Sardinia are larger), and had been an independent entity under a number of local kingdoms – nine is the canonical number (Diodorus 16.42.4; Pomponius Mela 2.102) – until succumbing to Ptolemaic and then Roman control. The varying distances around the island depend on the sailing routes. Timosthenes (F16 Roller) reported 427.5 miles (633 km., from 3,400 stadia), which is the best (the official figure is 780 km.). That of Isidorus (375 miles or 555 km., from 3,000 stadia), calculated 250 years later, may have used more direct routes. The 162.5 miles (240 km, from 1,300 stadia) of Artemidorus (F117), from Cape Clidae (Greek Kleides, modern Cape Apostolos Andreas), at the northeast corner, to Acamas (modern Cape Arnaoutis) at the northwest, along the north shore, is close to the official distance of 227 km; that of Timosthenes reflects a different route (200 miles or 396 km., from 1,600 stadia).

5.130. There is no indication of the Roman organization of the island (after 30 BC it was its own province), and the list of fifteen towns and the three no longer surviving may be from one of the sources at *NH* 5.129. Anemurium is the closest point on the Asian mainland, 50 miles (74 km., from 400 stadia) away. The Aulon Cilicius ("Cilician Trench") refers not to the depth of the ocean but the trench-like nature of the channel between Cyprus and the mainland, a narrow passage with mountains on either side. All the islands listed off Cyprus are small and close to the coast.

5.131. Although the Cilician coast has practically no islands, to the west, from Lycia to the Aegean, they are numerous, and Pliny recorded nearly thirty along this stretch, commenting that there were others he did not name. No source is cited, but it may be Timosthenes, whose navigational manual would have had such detail, including items such as noting the dangers of the Chelidoniae (perhaps "Swallow Islands," the modern Beşadalar). These are a line of reefs and islets extending directly south from the mainland and across the coastal shipping route. Ships rounding the mainland promontory of Cape Gelidonya would be subject to disaster, as demonstrated by the many shipwrecks in this area (Lionel Casson, *The Ancient Mariners* [second edition, Princeton 1991] 26–7).

5.132. Rhodes, lying just off the Carian coast, had an unusual political and cultural history. One of the leading maritime powers of the Hellenistic world, it managed to maintain a measure of independence into the Roman period. It was a major intellectual center: Strabo listed 16 Rhodians of note, more than from any other city in the eastern Greek world (Strabo, *Geography* 14.2.13). Pliny's eulogistic opening statement reflects its fame as the outstanding example of Greek culture in the early Imperial period, and many famous Romans, including Cicero and Julius Caesar, studied there (Plutarch, *Cicero* 4.4; *Caesar* 3.1).

As always, the varying numbers for the circumference of the island may reflect different sailing routes. The first figure (125 miles or 185 km., from 1,000 stadia), unattributed, may be from Timosthenes, although it is surprisingly small (the actual amount is 270 km.) and thus may only be the distance between two Rhodian ports.

Rhodes was well positioned in the Mediterranean, which enhanced its role as an important maritime center. The sailing route to Alexandria (a direct distance of 325 nautical miles or 600 km.) could be covered in 3 days (Appian, *Civil War* 2.89; Casson, *Ships* 293). The distances cited by Pliny (equal to a span of 740 to 863 km.) reflect several possible routes between the two points: one could go straight south from Rhodes to the African coast and then east to Alexandria, or take various coastal routes east past Cyprus and the Levant. The most problematic was to go directly southeast from Rhodes and hope to gain the Egyptian shore near Alexandria. The distance to Cyprus (176 miles or 260 km., perhaps 1,400 stadia) is unattributed and in fact is little more than half the actual amount. It is improbable there would be such an error on this well-travelled route, and it is likely a conversion or manuscript problem.

5.133. Except for Carpathos and Casos, which are southwest of Rhodes toward Crete, the over twenty islands listed are northwest of Rhodes, toward Cnidus and Cous. Some cannot be identified, and may only be rocks or tiny islets. This, and the notice of welcoming harbors on the minor island of Symi (and also the record of its circumference), suggests that Pliny was continuing to follow, at least in part, a navigational guide such as that of Timosthenes.

5.134. Cous is the normal Latin form for the island known in Greek as Cos (Cicero, *Letters to Atticus* #176, from 49 BC, etc.). It lies 15 miles (22 km., from 120 stadia) southwest of Halicarnassus, and often had a political connection with the Carian city. Like Rhodes, it continued to have a certain degree of independence into the Roman period, until it was incorporated into the province of Asia at the time of Augustus. It was famous as the home of the physician Hippokrates and for the local Asklepieion.

Most of the other toponyms mentioned, except those associated with Cous, are islands in the vicinity, and between it and Halicarnassus. Some of these cannot be identified, and, as before, were probably islets or rocks, again suggesting that Pliny was continuing to use a navigational guide. Oddly Caryanda and Passala are mainland locations near Halicarnassus, with the former already mentioned (*NH* 5.107); to be sure, both are seaports and would feature in a mariners' handbook. Pliny may have consolidated a list of islands and seaports into a single list presumed to be of islands.

Cinaedopolis (Greek Kinaidopolis) is otherwise unknown, and the name is that of a city, not an island. It would mean something like Catamite City, which may be a slander. It is by no means certain that Alexander the Great is meant: the form "King Alexander" is unique in Book 5 of the *Natural History* and Pliny generally used Magnus Alexander (e. g. *NH* 5.62, 107, 116). Nevertheless, it probably does refer to Alexander the Great (although there were other kings with that name), and the anomalous usage suggests a local negative view outside the historical mainstream.

5.135. Pliny's itinerary north from Samos seems to reflect two routes to Chios. One of them keeps away from the coast and goes to the Trageae, Corseae, and Icaros before returning west to Chios. The other is the direct route north along the Asian coast, past Miletus. Samos is the third largest island in the Aegean and thus has numerous topographical features. This is the only citation of Aristotle (F570) in Book 5 of the *Natural History*, and it is not known what the context is.

5.136. Chios was said to be 94 miles (139 km., probably from 750 stadia) north of Samos, but the islands are much closer than that, and the distance is probably the sailing route from Samos city to Chios city. Chione was the daughter of Boreas, the north wind, and as such her name, Snow (from *chion*), was perfectly appropriate (Hecataeus of Abdera [*FGrHist* #264] F12; Apollodorus, *Bibliotheke* 3.15.2–4). The marble from Chios (known today as *portasanta* because it was used in the doors of St. Peter's in the Vatican) is veined and colored, and became common in Rome in the first century BC due to the efforts of L. Licinius Lucullus (consul in 74 BC), who spent many years in the region (*NH* 36.49–50). The unattributed perimeter of Chios (125 miles, or 185 km, from 1,000 stadia) is perhaps from Timosthenes, and is reasonable, given the highly indented coast.

5.137–8. Pliny listed over fifty islands in the northeastern Aegean, essentially from Ephesus north to Troas. Fewer than twenty can be identified today, which, in some respects, shows the detail of the navigational guide that was his source (H. T. Wade-Gery, "The Islands of Peisistratos," *AJPh* 59 [1938] 470–5). The extensive siltation on this coast also means that many former islands are today merely knolls or hillocks in the coastal plain. There is evidence for four lists: one of islands around Chios, a second around Ephesus, a third around Zmyrna, and the last north toward Troas.

The oddity in the list is Teos, not an island and thus hardly in the open sea (*in alto*). It was one of the 12 Ionian cities, but not particularly famous (*clara*). Although it was involved in the military activities of the region, it was only notable intellectually as the home of the poet Anakreon (Strabo, *Geography* 14.1.30). Moreover, Teos is not 72.5 miles (107 km. or 580 stadia) from Chios by any route. The emendation by Wade-Gery (above) to the island of Tenedos has a great deal of merit, since the distance of 72.5 miles would thus be reasonable, and although close to the shore, Tenedos is relatively isolated. Moreover, it was known to Homer as an island sacred to Apollo (*Iliad* 1.452, etc.), giving it a certain amount of fame. In Pliny's day it might have been even better known because Vergil mentioned it several times in the *Aeneid* (2.21, etc.), calling it *notissima fama.* To be sure, Tenedos was mentioned by Pliny at *NH* 5.140, but this is from another source with its own problems.

5.139. Lesbos, the second largest island in the Aegean, lies about 60 km. due north of Chios; the 65 miles (96 km., from 520 stadia) reported by Pliny is the sailing distance from Chios city to Mytilene, the major city on the island. Methymna is on the north side of the island and is closer to, and more connected to, the cities on the mainland, which lies across a channel only 10 km. wide. The 37 miles (55 km., probably from 300 stadia) is the distance to Adramytteos (*NH* 5.122), the most important city on this part of the mainland.

It is not certain what the source is for the figure of 1500 years for the history of Mytilene. The town was not mentioned by Homer (although the island of Lesbos, which had been captured by Achilles, was frequently cited: *Iliad* 9.129, etc.). There are prehistoric remains near Mytilene, especially at the site of Thermi, about 12 km. north of the city (Dorothy Leekley and Robert Noyes, *Archaeological Excavations in the Greek Islands* [Park Ridge, N. J. 1975] 20–1), but Mytilene only seems to have become important in Archaic times (Herodotus 1.160, 2.135). Pliny's 1500 years would assume prominence several generations before the Trojan War, conventionally dated in antiquity to around 1200 BC (Herodotus 2.145; Eratosthenes [*FGrHist* #241] F1a), and his statement remains enigmatic.

5.140. As usual, the circumferences provided (168–195 miles, or 249–289 km.) are larger than expected and reflect sailing routes. The distance to the mainland (7.5 miles, or 11 km., from 60 stadia) is accurate. Continuing north, Pliny reached Tenedos (modern Bozca Ada, perhaps the Teos of *NH* 5.138). Its distance of 56 miles (83 km.) is not that from the closest point on Lesbos, but from Mytilene. The 12.5 miles (18 km., from 100 stadia) to Sigeum, to its northeast, is accurate. Tenedos is the closest significant island to the Aegean entrance to the Hellespont.

Northwestern Asia Minor

5.141. With a typically animistic description of the Hellespont violently forcing its way along and creating the separation between Asia and Europe, Pliny returned to the region north of the Trojan Plain and Cape Trapeza (*NH* 5.127). His account continues through the Hellespont and into the Propontis, the sea between it and the Thracian Bosporus. The places named are all on the south (Asian) side of the Hellespont and Propontis, and there is some confusion between the Granicus (modern Biga Çay) and Aesepus (modern Gönen Çay) Rivers, since the former is the more western.

5.142. Cyzicus (Greek Kyzikos, modern Belkis) was a Milesian settlement of at least the early seventh century BC, and a major player in the political and military events of the Classical and Hellenistic periods. It remained a prosperous commercial center in the Roman period (Strabo, *Geography* 12.8.11; Boardman, *Greeks Overseas* 245–6). The city was on an island just off the Asian coast of the Propontis. Alexander the Great would have been here in the spring of 334 BC, but nothing more is known about his engineering efforts, and it is possible that the story has been confused with his activities at Tyre.

The account continues along the coast of the Propontis east of Cyzicus, with references to mountains visible in the region. Artace is repeated from *NH* 5.141 and is now misplaced between Placia and Scylace. The Rhyndacus River (essentially the modern Koca Çay) was the dividing point between the provinces of Asia and Bithynia.

5.143. The narrative moves east along the Asian coast of the Propontis. The derivation of the Homeric Halizonians (probably more correctly Alizonians), or Salt People, may be correct (Homer, *Iliad* 2.856, 5.39), but they are generally placed much farther east, in Paphlagonia (Strabo, *Geography* 12.3.20–1; Hope Simpson and Lazenby, *Catalogue* 176). The use of the toponyms Iuliopolis and

Germanicopolis (at modern Tahtalı) is indicative of a Roman source; in the latter case, the ancient town of Helgas was renamed, probably when Germanicus, the nephew of Tiberius, was in the region in AD 18 (Tacitus, *Annals* 2.54). Although there is still a certain coastal orientation to the account, it is no longer primarily based on the Hellenistic *periplous* that has been Pliny's primary source for the circumnavigation of Asia Minor.

5.144. The account continues to Cios (at modern Gemlik), at the head of the bay of the same name, the more southern of the two deep indentations of the Propontis. It was an important trade outlet for the peoples of the interior and the regions to the east.

5.145. Pliny broke off his coastal description (returning at *NH* 5.148) to provide a brief discussion of two of the regions of interior northwestern Asia Minor. Phrygia was an ancient district: the Phrygians, who were believed to have come from Europe, were said to be Trojan allies (Homer, *Iliad* 2.862–3). It was later famous as the home of King Midas of Gordium (Herodotus 1.14). After the collapse of his dynasty, it had little central organization and tended to be appropriated by its neighbors, eventually being incorporated into the province of Asia. Pliny's list of towns is alphabetical, but essentially does not go beyond the letter C, indicating its probable partial nature. It suggests a Roman administrative document, but no details were provided.

5.146–7. Galatia, to the northeast of Phrygia and created out of part of its territory, was an anomalous region. Its population was not one of the ancient peoples of Asia Minor, but Celts from Europe who migrated into the area in the early third century BC, giving the Greek form of their name to the district (Pausanias 10.19.5–12). There were originally three ethnic groups (the Tolostobogii, Trocmi, and Teutosages), who did not differ in any cultural way but were probably distinguished by family relationships. The Voturi and Ambitouti were little-known sub-groups of the Tolostobogii. Pliny's account is from an early source, since by the first century BC the Galatians had consolidated under a single king. The kingdom survived until provincialized by Augustus in 25 BC (Stephen Mitchell, "The Galatians: Representation and Reality," in *A Companion to the Hellenistic World* [ed. Andrew Erskine, Oxford 2005] 280–93). The 12 ethnyms are all local peoples in towns in central Asia Minor (not limited to Galatia, but borders were fluid). The priests of the Mother of the Gods were called Galloi, which implied a Galatian origin (Strabo, *Geography* 13.4.14).

5.148. Returning to the coast at Cios (*NH* 5.144), Pliny concentrated on Bithynia, the extreme northwesternmost region of Asia Minor. It emerged as an independent kingdom in the fifth century BC and remained so until King Nikomedes IV willed it to Rome in 74 BC and it became a Roman province (Sullivan, *Near Eastern Royalty* 33–5). Hannibal, the famous Carthaginian leader, spent his last years in the region, but whether he was involved in the building of Prusa (at modern Bursa) is uncertain (Cohen, *Europe* 403–4). Yet there is no doubt that he lived on an estate at Libyssa, on the coast west of Nicomedia, where he committed suicide in 183 BC (Livy 39.51; Strabo, *Geography* 12.4.3; Plutarch, *Flamininus* 20; Appian, *Syriaka* 11).

From Cios to Nicaea (modern İznik) is 25 miles (37 km., from 200 stadia), as reported. The emphasis continued to be on towns that were no longer in existence, probably another example of the ravages of the Mithridatic Wars, as elsewhere in northwestern Asia Minor. In fact, the provincialization of Bithynia in 74 BC was one of the major causes of the Third Mithridatic War, and there was much fighting in the region. No source is provided for Pliny's account, but the autoptic reference to Hannibal's tomb may suggest that he saw the effects of the destruction at first hand.

5.149. Nicomedia (modern İzmit), the capital of Bithynia, lies at the head of the deep Astacenus Bay, the more northern of the two eastern bays of the Propontis. The narrative heads west along the north shore of the bay to its western end at Cape Leucatas (modern Yelken Kaya Bursa), 37.5 miles (56 km., from 300 stadia), and then along the Propontis itself to Calchadon (normally Chalcedon, modern Kadıkoÿ), 62.5 miles (93 km., from 500 stadia) from Nicomedia. The locals at Calchadon had long been ridiculed because they chose an unfavorable site only seven stadia across the Bosporus from where Byzantium would be founded 17 years later in a more propitious location (Herodotus 4.144; Strabo, *Geography* 7.6.2). Nevertheless Calchadon retained periods of independence until it was incorporated into the province of Bithynia in 74 BC. Pliny's sources are mixed: the comment about its freedom must come before that date, but the town of Juliopolis suggests an Augustan or later foundation.

5.150. The final section discussing the west coast of Asia Minor goes as far as the entrance to the Thracian Bosporus, joining with *NH* 4.74–5. The Bosporus begins 12.5 miles (18 km., from 100 stadia) from Calchadon, and its actual narrows are 8.75 miles (13 km., from 70 stadia) farther. The Thyni and the Bithynians are probably variants of the same ethynm.

From Calchadon to Sigeum, at the Aegean end of the Hellespont, was reported at 322.5 miles (477 km.), probably the coastal route along the Asian side. With this Pliny concluded his account of western Asia, which had begun at *NH* 5.100 when he began his discussion of Lycia. The narrative continues at *NH* 6.4 with the south side of the Pontus Euxinus, or the Black Sea.

5.151. As a final note to Book 5, Pliny cited over twenty islands in the Propontis. These were listed without comment except that Elaphonnesus, or Proconnesus, was the source of Cyzicene marble. Although he implied otherwise, the two islands are probably not the same: the former is most likely modern Ekinlik Adası, northwest of Cyzicus, and the latter Marmara Adası to its northeast, whose modern name indicates its historic role. The local stone was white with blue veins, and was used in the Hekatomnid palace at Halicarnassus. It appeared in Rome in the second century AD and became popular in late antiquity, most notably visible at Ravenna and in Hagia Sophia (Vitruvius 2.8.10; *NH* 36.47).

Demonnesus (also known as Chalkitis, modern Heybeli Ada) was a major mining center, producing *kyanos* and *chrysokolles*. These were found in association with copper, which was used for statues and medicinal reasons (Theophrastus, *Lithika* 25; *On Marvellous Things Heard* 58). *Chrysokolles* is probably malachite (David A. Scott, *Copper and Bronze in Art* [Los Angeles 2002] 102–7).

BOOK 6

The Remainder of Asia

Book 6 discusses the portion of the continent of Asia that had not been examined in Book 5, beginning at the Thracian Bosporus, where Book 5 ended (5.151). Now Pliny continued from the Bosporus around the Asian shore of the Pontus Euxinus, or Black Sea (1–24), with some diversion into the interior of Asia Minor. He then moved into Armenia and the Caucasus (25–35) and the Caspian region (36–49). After a consideration of the extreme north and east of the known world (50–5), the *Natural History* focused on India, Taprobane, and Ariana (56–106). Returning west, the account moves into the Persian plateau and Mesopotamia (107–141), and then the part of Arabia not previously discussed (142–77), followed by Aethiopia (178–97), and concluding with various islands (198–205). The final portion of the geographical books of the *Natural History* is about terrestrial parallels (206–20).

Continuing with the ideological perspective apparent in all the geographical books, Book 6, the last of them, mostly examines regions that were not Roman territory at the time Pliny wrote, the antithesis of the highly romanized world of Book 3. Only the first few chapters (1–24) are about Roman territory (perhaps reflecting a perceived need to finish the coasts of the Black Sea before moving to other regions). But beginning with 6.25, he left Roman territory for the remainder of the geography, circling around the fringes of the known world in a clockwise fashion from north to south (most of the western edge of the inhabited world was Roman territory and had already been discussed).

Thus the geographical books of the *Natural History* demonstrate an ideological progression, describing the world of Vespasian from those areas that are most Roman (Iberia) through Italy to those least Roman, regions that few Romans had seen, such as Taprobane and interior Africa. Significantly the last place discussed is the Fortunate Islands (6.202–5), the modern Canaries, associated with the Blessed Islands (Makarioi Nesoi) of myth, known since earliest times as the fertile paradise that was the home

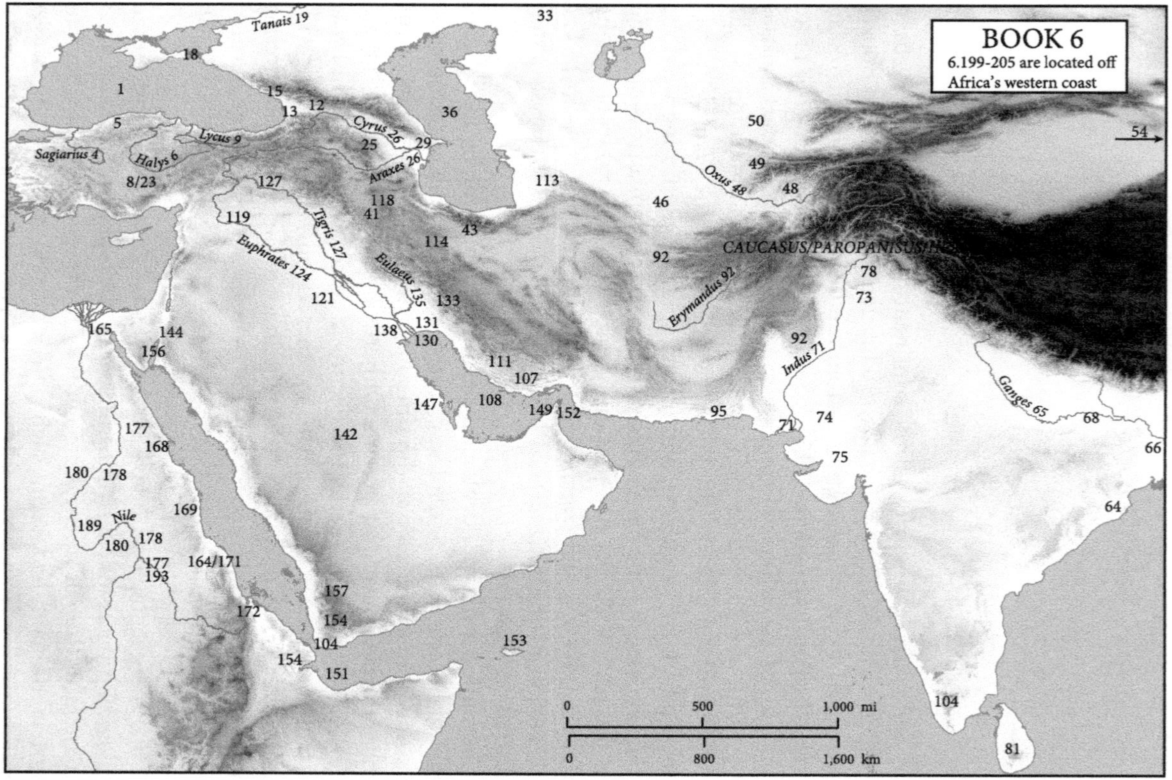

Map 6 Places cited in Book 6 of the *Natural History*, according to chapters (courtesy of Ancient World Mapping Center).

of the heroes (Hesiod, *Works and Days* 166–73; Pindar, *Olympian* 2.68–72). By ending the geography with these islands – and other remote ones – Pliny's text takes on a mythic tone, a fitting end to his account of the Roman world.

The Southern and Eastern Coasts of the Black Sea

6.1. What is today called the Black Sea was the Euxinus Pontus in antiquity: *euxeinos* seems to be a Greek translation of Persian *aesana*, meaning "dark" or "sombre," since it was inhospitable in nature (West, "Most Marvellous" 151–67). As was his custom, Pliny described it in animistic terms, telling how the ocean, as if a living being, needed to force its way into various portions of the inhabited world, breaking apart Europe and Asia at the Pillars of Hercules. Thus it created the Black Sea and Maeotis (modern Sea of Azov) as it thrust out of the Aegean. Such geographical features – where land and sea seemed to be in conflict – were considered natural wonders (Cicero, *Tusculan Disputations* 1.45; Murphy, *Pliny* 46–7; Evans, "The Cruel Sea?" 105–19).

6.2. Pliny drew attention to the two straits called the Bosporus, at opposite ends of the Black Sea. The Thracian Bosporus, which connected the Black Sea to the Aegean, was only 875 paces in width (1.3 km., from seven stadia) at its narrowest, just south of its midpoint. The standard explanation of the name Bosporus, "Cow Ford," was applied to the Cimmerian Bosporus and was mythologically said to be because of the passage of Io in bovine form (Aeschylus, *Prometheus Bound* 729–35), although this does not take into account that there were two widely separated places with the same name. Pliny's onomastic explanation, that both were so narrow they could be forded by cattle, is quite possible, even if not literally true. But they were the narrowest ocean passages known, and thus were objects of curiosity. Since they both divided Europe and Asia, it was considered peculiar that sounds – human or otherwise – could be heard from one continent to the other (Pliny's "two worlds" or *duos orbes*).

6.3. Pliny provided a rare instance of his criteria for providing distances, measured "in our own time" (*in aevo nostro*), although there were neither any further details nor any exact definition of the phrase. But he may have wished to emphasize reports from recent Roman expeditions, such as that of Cn. Domitius Corbulo to Armenia in the late AD 50s (*NH* 6.40). The map of Agrippa, although nearly a century old, was also a constant source. Pliny's phrase may have been intended to distinguish Roman from Greek sources, but nevertheless many of his distances were converted from stadia into miles, and thus were based on Greek information.

Despite his stated reliance on recent sources, the first cited is Eratosthenes (*Geography* F115), nearly 300 years old in Pliny's day. His distance (1,438½ miles or 2,129 km., from 11,500 stadia) conflicts with a previous citation of the same source (Eratosthenes, *Geography* F116 = *NH* 5.47), and may demonstrate that Pliny did not consult Eratosthenes directly, or used different methods of conversion. Even the numbers provided by Agrippa (F51) – 1,000 and 360 miles, or 1,480 and 533 km. – were converted from stadia (8,000 and 2,880). Yet the direct distance across the Black Sea is less than 300 km., so the figures were from coastal sailing routes, and the disagreement Pliny reported may be from different itineraries.

6.4. The account moves east along the north coast of Asia Minor from the northern mouth of the Bosporus, in the form of a *periplous*. The Sangaris (Sagarius or Sangarios, modern Sakarya) is the most important river of northwestern Asia Minor (824 km. long). Pliny called it "famous" (*inclutis*) presumably because it was known to Homer and Hesiod (Homer, *Iliad* 3.187, 16.719; Hesiod, *Theogony* 344).

The Mariandyni were a small ethnic group living in coastal northwest Asia Minor, and were immigrants who had come from Thrace but who were eventually absorbed by the Bithynians (Herodotus 1.28; Strabo, *Geography* 12.3.4). There are several species of aconite, a poisonous plant growing in this region; the eponymous village, Acone, was presumably on the coast east of Heraclea but has not been located. The plant and its poison were said to have been created from the drooling of Cerberus, who was believed to have come up from the underworld here, probably at the nearby Acherusia cavern (Theopompus F181b; Dioskourides 4.76–7; Strabo, *Geography* 12.3.7; *NH* 27.4–5). The cavern itself, two stadia (400 m.) deep, had been visited by Xenophon on his return from Mesopotamia (Xenophon, *Anabasis* 6.2.2). At the Billis River (or Billaios, modern Filyos Çay) the account enters Paphlagonia.

6.5. Paphlagonia was a minor district of northern Asia Minor, squeezed between Bithynia and Pontus, and often coveted by both. The Eneti were another Homeric topographical problem. Homer recorded that Paphlagonia was "the land of the Eneti" (*Iliad* 2.851–5), but there was no remnant of them except the obscure village of Eneti, a name that seemed spurious. This led the Homeric scholars of the Hellenistic period to emend Homer's text to fit the known toponyms of the region (Strabo, *Geography* 12.3.8). Moreover, it was thought that the Eneti of Homer became the Veneti at the head of the Adriatic, which conveniently removed them from having any historic presence in Paphlagonia (Cornelius Nepos F7; Strabo, *Geography* 5.1.4).

6.6–7. Cape Carambis (modern Kerembe Burnu) was the closest point to the north shore of the Black Sea. Pliny's stated distance across the sea to Criometopon in the modern Crimea (*NH* 4.86) is not actually to this point but to the mouth of the Cimmerian Bosporus, an obvious shipping route to access the towns along the strait. The distances (325/350 miles, or 481/518 km., to the Thracian Bosporus, and 312.5 miles, or 462 km., to the Cimmerian) are approximately accurate.

The narrative then reaches Sinope (modern Sinop), the most important city on this coast, near the narrowest part of Asia Minor. The distance of 200 miles (396 km., from 1,600 stadia) is too short; it is about 500 km. from the Cilician coast to the Black Sea. Nevertheless there was an early conception that this was a narrows that made Asia Minor into a peninsula, documented from the fifth century BC. It was an ancient trade route that allowed the people of Syria and Mesopotamia to access the sea, and it was said that one could make the crossing of the peninsula in five days (Herodotus 1.72, 2.34). The account continues east along the coast to the mouth of the Halys River (modern Kızılırmak), the longest river completely in Asia Minor (1,355 km.). It was the traditional cultural boundary between the Aegean and Mesopotamian worlds (Aeschylus, *Persians* 865; Herodotus 1.6, 75; Strabo, *Geography* 12.3.12).

The Dorians, Ionians, and Aeolians were the three primeval ethnic and linguistic stems of the Greek peoples. Pliny's statement contrasting these with an otherwise barbarian population had its origins in the ethnology of Ephorus of Cyme (F162), who enumerated 13 barbarian groups in addition to the Greeks. The unusual ethnic diversity of Asia Minor was a matter of interest to the Greco-Roman world (Strabo, *Geography* 14.5.23).

6.8. Cappadocia was the inland portion of central Asia Minor, a territory with a long and rich history, having Hittite and Assyrian influences. In early Hellenistic times it came under the control of the major powers, but eventually was ruled by local dynasties that lasted until provincialization in AD 17. Pliny's source for Cappadocia was recent because of the references to the emperor Claudius and the town of Neocaesarea, which was named by Vespasian (Tusculum 6, 161). Samiramis, or Semiramis, was the legendary Assyrian queen (Herodotus 2.184; Diodorus 2.4–20), who was mythically associated with many city foundations in central Asia Minor; such sites are often prehistoric mounds.

6.9. The districts mentioned are all minor regions of eastern Asia Minor, generally north and east of Cappadocia. The Leucosyri ("White Syrians") lived in northern Cappadocia and perhaps as far as the Black Sea, and were probably remnants of an Assyrian trading presence. The adjective "white" cannot be explained. Syrians are documented in this region as early as the fifth century BC (Pindar F173; Strabo, *Geography* 12.3.9).

6.10. There is a list of both coastal and inland towns, and the major rivers that empty into the Black Sea. Ziela (more commonly Zela, modern Zile) was an ancient city sacred to the goddess Anahita that was little known outside its region until C. Valerius Triarius was defeated there by Mithridates VI of Pontus, allegedly losing 7,000 men (Appian, *Mithridateios* 89; Plutarch, *Lucullus* 35; Dio 36.12–13). Twenty years later Julius Caesar was victorious over Mithridates' son Pharnakes II, describing it through his famous phrase "veni, vidi, vici" (*de bello alexandrino* 72–7; Suetonius, *Divine Julius* 37).

6.11. The Genetae and Chalybes lived farther east along the coast. The former are little known (although in the first century BC there was a toponym Genetes in this region: Strabo, *Geography* 12.3.17). The Chalybes (seemingly "Iron People") were known from early times in this iron-producing region, where vestiges of early mining are still visible (Aeschylus, *Prometheus Bound* 714–16; Herodotus 1.28; Strabo, *Geography* 12.3.19; Boardman, *Greeks Overseas* 245).

Several other ethnic groups are listed along this stretch of coast, essentially from Amisum (modern Samsun) to Trapezus (modern Trabzon). In Pliny's day this region had only recently become Roman territory and there is no evidence in his account of Roman organization. The list, structured geographically, is probably from the period of the reconstituted Pontic monarchy of Queen Pythodoris, who ruled for half a century in the era of Augustus and Tiberius, and who claimed the territory east of Pontus as well as Colchis. She had a close relationship with Livia Drusilla, the wife of Augustus, which may explain the single Roman-inspired toponym of Liviopolis, which has not been located (Roller, *Cleopatra's Daughter* 108).

6.12. The account continues as a *periplous* from Trapezus around the southeast corner of the Black Sea to the famous land of Colchis, the home of the Golden Fleece and the goal of Jason and the Argonauts. The ethnym Heniochi ("Chariot Dwellers") refers to several peoples in the eastern and northern Black Sea regions (see also *NH* 6.14), describing the common wagon nomads of this region.

Hiberia (or Iberia) seems misplaced, since it is usually a district on the Caspian side of the Caucasus (*NH* 6.29), but in these little-known regions, the location of ethnic groups may reflect trade networks rather than actual homelands. The narrative keeps strictly to the coast, and continues to be derived from a source connected to the Pontic world of the early Imperial period, since Pythodoris was also queen of the Colchians: hence the statement that the Phasis River (modern Rioni in Georgia), the primary stream of Colchis, was actually in Pontus (see also Strabo, *Geography* 11.2.18).

6.13. The Phasis, the largest river entering the Black Sea on its eastern side (327 km. long), had been known to Greeks since earliest times, probably in the context of the story of Jason and the Argonauts (Hesiod, *Theogony* 340). It was once believed to be the division between Europe and Asia, and was also thought to be connected to the External Ocean (Herodotus 4.45), but this view was essentially obsolete by the Hellenistic period (David Braund, *Georgia in Antiquity* [Oxford 1994] 16–18). Aea (Aia) was the home of Medea's father Aietes (Strabo, *Geography* 1.2.39). Pliny's description of this region is particularly focused on its rivers, with emphasis on their navigability. The Phasis was a route for bringing Colchian products to the Black Sea for export: the tale of the Golden Fleece shows the local importance of precious metals.

6.14. It is common for ethnyms in this region, and north to the Cimmerian Bosporus, to have Greek descriptive names, indicating that the populations were little known (perhaps only when their representatives appeared at the trading centers). The name of the Phthirophagi ("Lice Eaters") may be a local descriptive term translated into Greek. There were also more of the ubiquitous Heniochi; this particular group was thought to be descended from the charioteers of the Dioscuri, who travelled with the Argonauts. Their memory was strong in this region (Pomponius Mela 1.111). For Sebastopolis, see *NH* 6.16.

6.15. Pliny digressed from the coast to mention briefly the Caucasus region east of the Black Sea, still using the term Pontus in its expansive sense that included Colchis. The Caucasus Mountains extended from the Black Sea southeast to the Caspian or Hyrcanian Sea, with the west end believed to be pointing toward the mythical Ripaean Mountains of the far north (*NH* 4.78). The locals in this region were known to the Greco-Roman world for their violence and fierceness, as both Mithridates VI and Pompey the Great learned, confronting them with difficulty in the 60s BC (Appian, *Mithridateios* 101; Plutarch, *Pompeius* 34–6). Nevertheless Coraxian wool (or garments made from it) had come into Ionia by the sixth century BC (Hipponax F2).

Dioscurias (modern Sukhumi in Georgia) – another toponym associated with the Dioscuri – was the major mercantile center of this region, having been founded in the fifth century BC. The 300 peoples with different languages may be an exaggeration (Timosthenes F27 Roller), but there is no doubt that the city received traders from a vast region of the Caucasus and beyond. It was also considered (somewhat inaccurately) the farthest point one could reach in the Mediterranean system (Strabo, *Geography* 11.2.16; David Braund, "Greeks and Barbarians: The Black Sea Region and Hellenism Under the Early Empire," in *The Early Roman Empire in the East* [ed. Susan E. Alcock, Oxford 1997] 121–36). Pliny did not specify exactly in what period 130 interpreters were necessary, but it was probably before the founding of Sebastopolis in the early Imperial period.

6.16. The assumed presence of Ampitus and Telchius, the charioteers of the Dioscuri, helped to explain the prevalence of chariot nomads in the region. Pliny implied that Dioscurias and Sebastopolis were two separate places, a status confirmed by Arrian, sailing along this coast half a century later (Arrian, *Periplous of the Euxine Sea* 11). Sebastopolis was identified as a fortress, and it probably was a construction of the Pontic queen Pythodoris in the Augustan period, one of many examples of toponymic honoring of the emperor, here using the Greek form of his name, Sebastos (Suetonius, *Augustus* 60). The suggestion that the two localities were separate (the distance of 70 has no units attached, so it cannot be determined whether it is miles or stadia) implies that there was a realignment of the urban zone, perhaps reflected in the statement that Dioscurias was now deserted (*NH* 6.15), contradicting what Arrian saw not much later. Archaeology is of little help because of a high water table (J. G. F. Hind, "Greek and Barbarian Peoples on the Shores of the Black Sea," *AR* 30 [1983–4] 92).

The various ethnic groups lie on the coast beyond Dioscurias and Sebastopolis. The indigenous name of the Achaeans sounded like that of the inhabitants of the northern Peloponnesus, and there were a number of appropriate myths to explain their presence along this coast (Strabo, *Geography* 11.2.12, 14). The Mardi cannot be located (they may have been peoples migrating from the Mardi in Media: Strabo, *Geography* 11.13.3), and the Cercetae were still farther along the coast. They were believed to be Indian in origin, perhaps a vestige of trade connections with that region (Strabo, *Geography* 11.2.14; Hesychios, "Kerketai") The Seraci and Cephalotomi are otherwise unknown; there is no obvious explanation for the peculiar name of the latter, "People With Their Heads Cut Off."

Pityus (at modern Bichvinta or Pitsunda, about 50 km. from Dioscurias) was another wealthy trading center that was subject to hostile incursions from the hinterland (Strabo, *Geography* 11.2.14). The Epagerritae were part of the large ethnic group called the Sarmatians or Sauromatae (the distinction that Pliny made is more linguistic than ethnic). They had been known to the Greek world since the fifth century BC and controlled trade routes from the cities on the northern coast of the Black Sea into the little-known interior (Herodotus 4.21).

6.17. Mithridates was made king of Bosporus by Claudius in AD 42, during a period of local instability, but was soon overthrown by his brother Cotys I. Mithridates fled into the interior but was eventually turned over to the Romans and in time was executed by the emperor Galba (Tacitus, *Annals* 12.15; Plutarch, *Galba* 15; Dio 60.8.2). Pliny implied that he wrote some sort of memoir, but nothing more is known.

Since the time of Alexander the Great it was believed that the Caspian Sea was connected to the External Ocean. The supposed outlet was discovered by the Seleucid explorer Patrocles in the early third century BC, probably a misinterpretation of the massive delta of the Volga River, at the north end of the sea, as well as certain preconceptions (Patrocles F4a; Strabo, *Geography* 2.1.17; *NH* 6.36).

The *periplous* continues as far as the Cimmerian Bosporus (modern Kerch Strait). The relationship of the distances to the topography can only be speculative since the analysis becomes circular. The Cimmerian Bosporus was the extremity of Asia in this direction; Pliny had approached it from the European side at *NH* 4.87.

6.18. The final portion of Asia in this region is the peninsula situated between the Black Sea and Lake Maeotis (modern Sea of Azov). The 67.5 miles (100 km., from 540 stadia) seems long, but the initial point cannot be determined. The stated width of no less than two *iugera* is perhaps literally true, but makes little sense, because using *iugera* as a linear measure (it is normally an area) would produce a distance of only about 150 meters. Pliny may have misunderstood the nature of the measurement, since the peninsula is penetrated by numerous inlets and waterways, or the number may be at fault in the manuscripts. Lake Maeotis would technically be split between Europe and Asia, but Pliny placed it in the former.

6.19. The Maeotis region is the subject of an astonishingly lengthy discussion (through *NH* 6.22), mostly catalogues of ethnic groups, with some topographical features (Viktor Burr, *Nostrum Mare* [Stuttgart 1932] 37–40). No source is provided, but the most probable is a report about the various peoples who came to the coastal trading centers, such as Dioscurias or Panticapaeum. These populations are generally unidentifiable.

The Tanais River (modern Don) flowed into the farthest extremity of the Maeotis. It is 1,870 km. long and provided access to the remote regions north of the Black Sea (Pliny's "rough wilderness with forested valleys"). It was also the dividing line between Europe and Asia and had been known to the Greek world since the fifth century BC (Herodotus 4.20–1). Yet even in Roman Imperial times the upper river was hardly understood, although Archaic Greek pottery has been found 400 km. upstream (Strabo, *Geography* 11.2.2; Boardman, *Greeks Overseas* 244).

The Sarmatae, perhaps originally from the Persian plateau, were noted for the prominence of their women (Herodotus 4.110–17); the name Gynaecocratumenians is based on Greek *gynaikokratia*, a society ruled by women. This term was developed by Greek political theorists in the fourth century BC to describe a situation in which women held power, something believed to have pernicious effects on the local economy (Aristotle, *Politics* 2.6.6–7, 5.9.6; Diodorus 2.45). In time it was applied to non-Greek cultures such as the Amazons. Yet the term was certainly appropriate for the region north of the Black Sea, where there are numerous tombs of wealthy royal women (Anna A. Trofimova, ed., *Greeks on the Black Sea: Ancient Art From the Hermitage* [Los Angeles 2007]). For the Ripaean Mountains, see *NH* 4.78.

6.20–2. There are further lengthy lists of ethymns, with over fifty named. These are probably not limited to the Tanais and Maeotis region, but from a broad extent of the hinterland of the eastern and northern Black Sea: mention of the Caucasus and the Ceraunian Mountains (the east end of the Caucasus) demonstrates this. The emphasis on river travel is apparent, as these were the peoples who came down the rivers to the sea to trade their wares. The Panticapaeans have been mixed into the Tanais ethnyms, but Panticapaeum, on the European side of the Cimmerian Bosporus, was the capital of the Bosporanian kingdom and would have been one of the trade centers that these peoples reached. The many ethnic groups give some credence to the vast number of indigenous peoples said to come merely to Dioscurias (*NH* 6.15).

Interior Asia Minor

6.23. At this point, having covered the entire coast of the Mediterranean system, Pliny ended his *periplous*, although there would be further coastal itineraries, especially regarding the routes between India and the Persian Gulf and Arabia (*NH* 6.92–111). Yet most of the remainder of Book 6 is about the interior of Asia.

Pliny's next topic is an examination of some portions of interior Asia Minor that were not previously discussed (through *NH* 6.25). His major source was the expedition of Cn. Domitius Corbulo, starting in 58 BC (see also *NH* 5.83), and there were also reports from various allied kings and their children who came to Rome. There was a long tradition of foreign royalty coming to the city, either for education, as hostages, or to escape instability at home. The first documented is Ariarathes V of Cappadocia, who visited before taking the throne in 163 BC (Livy 42.19.3–6). Given the citation of Domitius Corbulo, Pliny was probably referring to kings from the region and period of his command, such as Tigranes VI of Armenia, who lived in Rome for many years before being placed on the throne (Tacitus, *Annals* 14.26).

6.24. Continuing to take a broad view of the kingdom of Pontus, Pliny now included Cappadocia, to its south, as part of it, a point of view that was part of the expansionist policies of Mithridates VI in the early first century BC but no longer valid in Pliny's day. He had already discussed the status of Cappadocia in his own era at *NH* 6.8. Here the description is oriented on someone travelling from Pontus and the Black Sea south, with Greater and Lesser Armenia, as well as Commagene, on the left (east) side. The itinerary continues as far south as Cilicia and Antioch (presumably the town in Syria on the Orontes). This overland route was said to be 640 miles (947 km.) in length, a much more realistic figure than the 200 miles of *NH* 6.7, but still rather long. The stated length (east–west) of Asia Minor, 1250 miles (1,850 km., from 10,000 stadia) is probably a circumference, or the route from Pontus to Antioch by sea.

Armenia and East of the Black Sea

6.25–6. Armenia is the rugged upland country of the interior of northeastern Asia Minor. Pliny had already discussed the Euphrates, whose source was in the territory, and also the Orroean Arabians (*NH* 5.83–5). Armenia was an ancient region: both its highest peak, Mt. Ararat, and the source of the Euphrates are mentioned in Genesis (2:14, 8:4). In Hellenistic times a local kingdom developed; eventually there were two separate states, Greater Armenia, the larger and stronger, to the east, and Lesser Armenia, to the west, which tended to be appropriated by its neighbors. Yet Greater Armenia retained its independence and became a major power under Tigranes II (95–55 BC), the son-in-law of Mithridates VI of Pontus. The two states came under increasing Roman control, especially with the activities of Domitius Corbulo during the reign of Nero.

Pliny's description is mixed chronologically. Caesarea (it is not known which of the many cities with this name is meant) can only have been founded after 27 BC, but Tigranocerta, the capital of Tigranes II, was destroyed in the 60s BC. There are also several references to the Cyrus River (modern Mikvari or Kura, in Azerbaijan and Georgia), which flows from northwest to southeast between Armenia and the Caucasus (David Braund, "The River Cyrus and Its Tributaries: Routes Through and Beyond the Caucasus," in *Topoi* [forthcoming]). This provides Pliny a transition to the region of the Caspian Sea.

6.27. To conclude his discussion of Armenia, Pliny cited two recent sources. Aufidius Bassus (*FRH* #78, F3) was an older contemporary who wrote a history of his own era, although the source of his estimate of the circumference of Armenia is uncertain. The figure, 5,000 miles (7,400 km.), is far too great; Pliny or Aufidius may have failed to convert from stadia.

The emperor Claudius (*FRH* #75, F4) wrote a history, which Pliny used regularly (*NH* 6.31, 128; 12.78). As with Aufidius, the distance (1,300 miles or 1,924 km.) is excessive: from Dascusa (modern Pağnik) on the upper Euphrates to the Caspian is perhaps 800 km., but through rugged country. From Tigranocerta (abandoned but still used as a reference point) to Hiberia (Iberia), on the upper Cyrus, is perhaps 500 km. Again the context is not certain, but given the extensive Roman involvement in Armenia during the first century AD, there would have been several sources. One possibility for both Aufidius and Claudius is the Armenian expedition of C. Caesar, the grandson of Augustus, which began in late 2 or early 1 BC.

6.28. The Cepheni may be the people mentioned by Herodotus (7.61) as a Persian group; no population with that name is known in the region under discussion. The Adiabeni are the inhabitants of Adiabene (*NH* 6.27). The Menobardi are otherwise unknown, and the Moscheni are probably the Moschi of *NH* 6.29. For the Medes and Media, see *NH* 6.43.

6.29. This section is topographically loose. The focus is on Albania and Hiberia (Iberia), but some of the ethnyms seem to belong in northern Armenia. Albania was a district on the lower Cyrus River, extending to the Caspian Sea and located in the northern part of modern Azerbaijan. It was outside the Persian and Seleucid worlds, and was only vaguely known (Eratosthenes, *Geography* F110–11) until the expedition of Pompey the Great in the 60s BC (Strabo, *Geography* 11.4.1–8). Hiberia was to its west, on the upper Cyrus. It had been under Seleucid control, but like Albania was little known until the time of Pompey (Strabo, *Geography* 11.3.1–6; Appian, *Mithridateios* 101–3; Braund, *Georgia* 161–7).

The Armenochalybes (perhaps "Armenian Iron People") seem connected with the Chalybes (*NH* 6.11), but if so are misplaced. The Moschi are not certainly located, although they were under Persian control in the sixth and fifth centuries BC and seem to have been somewhere north of Colchis (Herodotus 3.94, 7.78; Strabo, *Geography* 11.2.14). The Sacasani, or Saka, were a Scythian group that had penetrated south into Armenia and beyond (Strabo, *Geography* 11.8.4). The Macerones, Silvi ("Woodland People"), and Sodi are otherwise unknown, and the Lupeni lived at the mouth of the Cyrus, with the Didani or Didouri inhabiting the Caucasus (Ptolemy, *Geographical Guide* 5.9.22).

6.30. The Caucasian Gates were one of the major passes through the Caucasus. Identification is difficult, in part because of the tendency – still common today – to call them the Caspian Gates, a feature actually south of the Caspian Sea that had been used by Alexander the Great (Eratosthenes, *Geography* F48; Strabo, *Geography* 11.12.4–5; P. M. Fraser, *Cities of Alexander the Great* [Oxford 1996] 108). Some of the passes across the Caucasus had been known to the Greek world since the sixth century BC (Herodotus 1.104), but connecting them to specific events is difficult. The most likely attribution for the Caucasian Gates mentioned by Pliny is the modern Dariel or Dariali Pass, one of the more eastern ones, which took a week to cross (Braund, *Georgia* 225–6; Eberhard Sauer, *Dariali: The 'Caspian Gates' in the Caucasus From Antiquity to the Age of the Huns and the Middle Ages* [Oxford 2016]).

6.31. The area between the Black Sea and the Caspian was considered an isthmus, something that could be crossed by a canal. But the reported distances are greatly in variance, demonstrating how poorly known this region was. The unattributed 375 miles (555 km., probably from 3,000 stadia) may be from the map of Agrippa, and is probably the most accurate. Cornelius Nepos had 250 miles (370 km., from 2,000 km.), and Claudius (F5) 150 miles (222 km., from 1,200 stadia) from the Cimmerian Bosporus to the Caspian Sea, all far below the approximate distance of 790 km. The figure of 200 miles (296 km., from 1,600 stadia) from the Caucasian Gates to the Black Sea is perhaps the most accurate, yet even this is far too short. But there were even smaller estimates for the isthmus; Posidonius (F206) reported a figure of 1,500 stadia (300 km.) and believed that it was little different than the distance from the Mediterranean to the Red Sea (Strabo, *Geography* 11.1.5).

Claudius, whose interest in hydraulics was deep and learned, is the only source for the canal project of Seleucus I, the founder of the Seleucid dynasty, who was killed by Ptolemy Ceraunus, a son of Ptolemy I of Egypt, in 281 BC. The gratuitous notice of the death of Seleucus provides a moral context: this was the fate of those who attempted to make grandiose changes to the natural world (Daniel Ogden, *The Legend of Seleucus* [Cambridge 2017] 266). There is no reason to believe that construction of the canal moved beyond the planning stage.

6.32. Before continuing with the Asian interior, Pliny listed several islands in the Black Sea (a region with few of them), including some near the outlet of the Thracian Bosporus, near the mouth of the Danube, and off the coast of the Pontic kingdom. The story of the aggressive birds of Chalceritis is one of several such tales in Greek culture: other examples include the birds of Diomedes off the Adriatic coast of Italy (*NH* 10.126–7), and most famously, those of Lake Stymphalus in Arcadia, the object of the fifth Labor of Hercules.

6.33. In Pliny's time it was thought that most of the coast of the External Ocean had been explored. The only major exception was from Scandinavia to India, a stretch where the Ocean had only been touched (it was believed) at the mouth of the Caspian Sea (*NH* 6.36). It was felt that one could sail from this point to India, but from the Caspian to Scandinavia the Ocean was unknown (Patrocles F4b; Strabo, *Geography* 11.11.6). Pliny, always interested in coastal navigation, and using an unspecified Greek source, outlined a potential itinerary that crossed the mythical Ripaean Mountains of Europe (*NH* 6.15) and accessed the Ocean somewhere in the far north, proceeding along it toward India. He divided this route into three parts, giving names to the oceanic segments: first the Scythian, named after the peoples of the northern part of the inhabited world (Ephorus F30a), the Eoan (simply Greek for "Eastern"), and the Indian (not the modern ocean with that name, but the Bay of Bengal). The Eoan Ocean is clearly a construct for the unknown central part of this coast. Pliny also provided some generic material about the nature of this unknown coast, noting that there would be bays and local names, something not based on any empirical evidence. The cold temperatures of the north were always a matter of curiosity to Greeks and Romans: it was baffling that people could live where vines could not grow and where the sea froze, where the population was limited and the countryside endlessly forested and wild (Strabo, *Geography* 2.1.16).

6.34–5. Aquilo was the north-northeast wind, and thus the Scythians (a term used generically for the people of the north) are here localized at that extremity of the inhabited world. Beyond them were the Hyperboreans ("Those Beyond the North Wind;" see *NH* 4.89). Cape Lytharmis and the Carambucis River are positioned in Celtica (and west and northwest), which suggests somewhere in Scandinavia (Stephanos of Byzantium, "Elixoia"; Bridgman, *Hyperboreans* 129–31).

The Arimphaei may be the same as the Arimaspaeans, the strange one-eyed people of the far north (Herodotus 3.116; Strabo, *Geography* 1.2.10). They were described with the conventional terminology of remote northern populations: relatively passive, living in the woods and eating berries, and having a subsistence economy of hardy simplicity. This was similar to the people that Pytheas of Massilia encountered in the British Isles in the late fourth century BC (Diodorus 5.21); in fact, Pytheas, well known to Pliny, may have been one of his sources (Pytheas T19–25).

Other peoples associated with the far north are listed without comment. The Cimmerians were known to Homer (*Odyssey* 11.14) and gave their name to the Cimmerian Bosporus. They had penetrated into Asia Minor, and beyond, in the eighth and seventh centuries BC and were the biblical people of Gomer (Genesis 10:2; Ezekiel 38:6; Herodotus 1.15). Except for the Amazons, the other ethnic groups are hardly known, although some of them seem to have Greek or Latin descriptive names (the Anthi are probably "Flower People" and the Georgi "Farmers").

6.36. Caspian and Hyrcanian were interchangeable names in antiquity, reflecting two different ethnic groups on the coast of what is known today as the Caspian Sea. It was known to the Greek world from the sixth century BC (Hecataeus of Miletus F290; Herodotus 1.202–4). Pliny cited several sources for his data, beginning with Clitarchus (*FGrHist* #137, F12), the popularizing historian of Alexander the Great. He was generally considered unreliable (Strabo, *Geography* 11.1.5), yet provided an astute statement about the relative size of the Caspian and the Black Seas. Today the latter is somewhat larger but the Caspian has shrunk since antiquity.

Eratosthenes (*Geography* F111) recorded a distance for the extent of the sea (certainly obtained from Patrocles), moving clockwise. The total is 12,600 stadia, which Pliny (or his Roman source) converted to 1,575 miles (2,331 km.), or exactly eight stadia to the mile, proof of the method of conversion Pliny normally used. The final source is Artemidorus of Ephesus (F110), from the end of the second century BC, who made a slight adjustment to Eratosthenes' figures.

6.37. In addition to the names for the Eastern Ocean listed at *NH* 6.33, Pliny added another toponym, the Sericus Ocean, somewhere east of the Caspian, as shown on the map of Agrippa (F23). This ocean would be connected with the Seres people, for which see *NH* 6.54. Agrippa's measurements for the Caspian essentially conform to those of Eratosthenes, as expected, since no one had sailed its full extent since Patrocles, but the excessive figure of 2,500 miles (3,700 km., or 20,000 stadia), from an unnamed source, cannot be explained.

6.38. Varro provided a description of the supposed connection between the Caspian and External Ocean as well as naming those living along the coasts of the sea, certainly based on Patrocles. The most likely explanation for the connection to the Ocean is that Patrocles saw the broad delta of the Volga River (ancient Rha, but a name not documented until the second century AD: Ptolemy, *Geographical Guide* 5.9.12–21, 6.14.1–9), or perhaps that of the Ural River. Moreover, he was driven by his preconceptions – or perhaps misled by local reports – and assumed that this was the connection to the External Ocean he had been seeking, which those with Alexander the Great said existed (Patrocles F4a; Strabo, *Geography* 2.1.17; Arrian, *Anabasis* 7.16.3; Roller, *Ancient Geography* 115–17). The various ethnic groups, some of which have been mentioned previously, again come from Patrocles' report, and are otherwise little known, except for the Scythians, Sauromatae, and Albani. The Nomads ("Pastoralists") have a generic Greek name.

The memory of Jason and the Argonauts was strong in these regions, especially between the Black Sea and the Caspian, particularly in Albania and to some extent in Colchis, where the anchor of the *Argo* was still visible in the second century AD (Justin, *Epitome* 42.3.5; Arrian, *Periplous of the Euxine Sea* 9).

6.39. Pliny had already mentioned the Albanians at *NH* 6.29, but here provided a more thorough account of them, based on the map of Agrippa (F53). For the Sarmatians see *NH* 6.16. The Aorsians were east of the Maeotis (Strabo, *Geography* 11.5.8; Marek J. Olbrycht, "Die Aorser, die Oberen Aorser und die Siraker bei Strabon. Zur Geschichte und Eigenart der Völker im nordostpontischen und nordkaukasischen Raum im 2. – 1. Jh. v. Chr.," *Klio* 83 [2001] 425–50). The Aroteres ("Plow People") were agriculturalists. For the Coraxici see *NH* 6.26. The southwest coast of the Caspian – where the eponymous Caspi were – is precipitous, interrupted only by the river deltas, although Agrippa's 425 miles (629 km.) may be excessive.

6.40. Pliny provided a more complete discussion of the confusion between the Caspian and Caucasian Gates than had appeared at *NH* 6.30. The issue originated at the time of Alexander the Great, who manipulated the topography in order to enhance his reputation (Braund, *Georgia* 216, 226). Aeschylus in the *Prometheus Bound* (422, 717–21) had made the Caucasus known to the Greek world as the highest of mountains, which reached nearly to the stars. They were an obvious goal for Alexander, but he went nowhere near them, passing south of the Caspian Sea through the Caspian Gates (modern Sar Darreh, east of Tehran; see J. F. Standish,

"The Caspian Gates," *G&R* second series 17 [1970] 17–24; Roller, *Ancient Geography* 102–4). But if these gates were actually in the Caucasus, Alexander would have crossed those mountains, a manipulation that was easy in these little-known regions. Even those with Domitius Corbulo (see *NH* 6.23) hundreds of years later remained confused. Nevertheless Pliny's notice of maps (*depicti*) sent back to Rome by the expedition is a rare reference to such documents in a military context (Klaus Geus, "Hellenistic Maps and Lists of Places," in *Hellenistic Astronomy: the Science in Its Contexts* [ed. Alan C. Bowen and Francesca Rochberg, Leiden 2020] 234).

South and East of the Caspian Sea

6.41. The account moves on to the Persian plateau, south of the Caspian and north of the Persian Sea (Gulf). The Parthian state came into being in the mid-third century BC, created by King Arsaces I in the district called Parthyaea east of the southern Caspian Sea. The last Persian king, Darius III, had died in 330 BC after being defeated the previous autumn by Alexander at Gaugamela (modern Tell Gomeh) in the Arbilitis district of northern Mesopotamia. Although the Parthians and Persians were related both ethnically and onomastically, the two kingdoms were quite separate from one another.

6.42. Pliny followed the ancient route from Artaxata, the capital of Greater Armenia, to Ecbatana, the Median capital, heading south near Ninos (ancient Nineveh, whose location was probably lost) in upper Mesopotamia and then east into the Persian plateau. Mygdonia was a toponym that moved extensively: it was originally a region of Thrace, whose people migrated across the Hellespont to Bithynia (Herodotus 7.123–4; Strabo, *Geography* 12.8.10). There was also a Mygdonia in Phrygia (*NH* 5.145), and the one in northern Mesopotamia, so named because its landscape was similar to that of Thracian Mygdonia (Strabo, *Geography* 16.1.23). No source is provided for the road distances; the one from Artaxata to Nisibis (750 miles or 1,110 km., from 6,000 stadia) is far too long, but that to Ecbatana (450 miles or 666 km., from 3,600 stadia) is reasonable. For the founding of Ecbatana, see *NH* 6.43.

6.43. Ectabana (modern Hamadan in Iran) existed long before Hellenistic times, and was known to the Greeks through Herodotus' detailed description of the Median empire (1.95–122). It was evidently refounded by Seleucus I Nicator; Pliny's statement is supported by the existence of a royal mint from his era (Cohen, *East* 206–8). Seleucia

Magna is probably Seleucia-on-the-Tigris (*NH* 6.122), but 750 miles (or 1,110 km., from 6,000 stadia) is a major error (the actual distance is about 300 km.), probably due to a conversion or manuscript problem, and perhaps a duplication of the same figure in *NH* 6.42.

This is the most complete extant description of the Caspian Gates, which were believed to pass through the Taurus Mountains, as broadly defined (Strabo, *Geography* 11.12.5). Alexander went through them in the summer of 330 BC (Arrian, *Anabasis* 3.19–20). Since they were seen as the limit of the civilized world, they were used frequently as a topographical locus by Eratosthenes (*Geography* F37, etc.; Standish, "Caspian Gates" 17–24). The narrows, eight miles (12 km., from 64 stadia) long, are a passageway from the highlands of the Iranian plateau to the coastal regions of the Caspian Sea. Pliny's description is certainly an eyewitness report, perhaps from someone with Alexander, and implies that the pass had been enhanced through human effort.

6.44. The reference to the Tigris indicates that the citation of the Carduchi or Cordueni is out of place and probably belongs at *NH* 6.42. They lived on the south slopes of the Taurus and had caused much difficulty for Xenophon (Xenophon, *Anabasis* 4.1.8–11).

The Roadside (Pliny used the Greek, *par'hodon*) Pratitae were at the Caspian Gates – nowhere near the Carduchi – and presumably guarded them, probably exacting tolls from those passing through. After the Gates, a traveller would reach the Caspian littoral and the Parthian heartland. Their capital was said to be Hecatompylos ("Hundred Gates," a name not explained), a city of Greek origin. The founder is uncertain, perhaps Alexander or Seleucus I. Its location is perhaps at Shahr-i Qumis in Iran (Cohen, *East* 210–15).

6.45. The importance of the Caspian Gates as a topographical datum is emphasized by the number of distances calculated from it, all of which translate to even numbers of stadia (one is actually preserved in stadia). The source is derived from those with Alexander who recorded the distances along his path: Pliny made use of at least two of them, Baeton and Diognetus (*FGrHist* #119, 120; see *NH* 6.61, 69). There was also a work titled *Asiatic Stopping Points*, not mentioned by Pliny but used by Eratosthenes and Strabo (Eratosthenes, *Geography* F78; Strabo, *Geography* 15.2.8), which may have been the published record of the distances Alexander travelled. Yet not everything in this passage is from sources connected to Alexander, since he never reached the Cyrus River; the additional material may be from Eratosthenes.

6.46. The ethnic groups listed are all around the southern edge of the Caspian Sea, and the source is probably Patrocles through Eratosthenes (*Geography*, F108), although the Staures are not cited elsewhere. The Hyrcanians gave their name to the portion of the sea that they lived on, which Pliny implied was only the part beyond the Sideris River (not certainly located but probably one of the streams flowing into the sea from the east). Nevertheless the two names Hyrcanian and Caspian remained totally interchangeable throughout antiquity.

Margiane is the region in modern Turkmenistan around the ancient oasis of Merv. The 1,500 stadia (300 km.) refers to the extensive circuit wall built by the Seleucid king Antiochus I (reigned 281–261 BC), although such a length is questionable.

6.47–8. The Bactrians were the people of Bactria, which is largely in modern Afghanistan. They had been known to the Greek world since Persian power had extended this far, by the early fifth century BC (Aeschylus, *Persians* 306, 318, 732; Herodotus 1.153), and Alexander was in the region in the winter of 329–8 BC and again in 327 BC. In Hellenistic times it was ruled by a local hellenized dynasty (Strabo, *Geography* 11.11.1–2). The Mardians are normally placed much farther west, in Persis (Herodotus 1.125; Strabo, *Geography* 11.6.1, 11.13.3), but given the citation of the Margus River in Bactria, which flows through the oasis of Merv, there may be another group with a similar name.

The foundations of Alexander and Antiochus I at Merv remain elusive (Strabo, *Geography* 11.10.2; Fraser, *Cities* 116–18). M. Licinius Crassus was defeated and killed by the Parthians in 53 BC at Carrhae in upper Mesopotamia; Orodes II was the contemporary Parthian king but was not present at the battle (Dio 40.27).

The ethnic groups in the first list are hardly known, although some were recorded by Pomponius Mela (1.13) and Ptolemy (*Geographical Guide* 6.11.6). There is a second list of 17 ethnyms, presumably of peoples farther to the east. Mention of the Gandari, who are normally associated with the upper Indus (Herodotus 3.91; Strabo, *Geography* 15.1.26), suggests that the list covers a broad region from Merv to India. The Pariani, Arasmi, Marotiani, Syrmatae, Oxyttagae, Moci, and Bateni are otherwise unknown. The Zarangae were among the peoples at the eastern extremity of the Persian empire (Herodotus 3.93), and the Gaeli, Matiani, and Dribices seem misplaced, as they were said to be around the Caspian Sea (Pomponius Mela 1.13, 3.39; Strabo, *Geography* 11.5.1, 11.13.2). The Saraparae were in Media and were particularly violent and aggressive; their name was said to mean "Beheaders" (Strabo, *Geography* 11.14.14).

6.49. Sogdiana was north of Bactria and was the farthest limit of Persian control (Herodotus 3.93, 7.66; Strabo, *Geography* 11.11.2). Panda is otherwise unknown, but it is a widespread name (see *NH* 6.76). The Alexandria in question is Alexandria Eschate ("Farthest Alexandria"), which Alexander developed in 329 BC. It did not last long because of local incursions and has become the most obscure of his foundations. It may have been at modern Khogend in Tajikistan (Arrian, *Anabasis* 4.1.3; Cohen, *East* 252–5). The site was on or near the Iaxartes River (modern Syr Darya), and was considered the end of the world. Part of Alexander's self-image was that he had gone farther than anyone else, including the primeval divine travellers, Hercules (Herakles) and Father Liber (Bacchus or Dionysos). There were said to be pillars at some point (generally located in India) marking the limit of their exploits, corresponding to the Pillars of Hercules at the other end of the inhabited world (Diodorus 17.95; Arrian, *Anabasis* 5.29.1; Strabo, *Geography* 3.5.5). Cyrus the Great of Persia and Semiramis of Babylon were also said to have come this far; Cyrus certainly had been east of the Caspian, where he was killed by the Massagetae in 529 BC (Herodotus 1.214), but it is unlikely that either monarch made it to India, although this was a persistent tradition (Ctesias F1b; Xenophon, *Kyropaideia* 1.1.4; Strabo, *Geography* 15.1.5).

Another example of Alexander's manipulation of topography was calling the Iaxartes River the Tanais. The latter, which flowed into the upper end of Lake Maeotis, was considered the boundary between Asia and Europe (Herodotus 4.45; *Airs, Waters, and Places* 13). Alexander never came anywhere close to the river, and thus in order to assert that he had gone beyond the limits of Europe, his people said that the Iaxartes was actually the Tanais, even using botanical data to prove their point (Polykleitos of Larisa [*FGrHist* #128] F7; Strabo, *Geography* 11.7.4).

Demodamas (*FGrHist* #428, F2), probably from Miletus, was sent by the Seleucids in the early third century BC to investigate this region. This account by Pliny is the only extant record of his exploration of the far northeast of the inhabited world. He published his report, and it was probably the source for the long list of ethnyms in *NH* 6.50. He crossed the Iaxartes and probably went farther in this direction than anyone in Greek antiquity, erecting altars to the patron divinity of his home town, Apollo of Didyma. There were undoubtedly commercial aspects to the expedition, as well as an attempt to define the potential limits of Seleucid territory (M. Rostovtzeff, *Social and Economic History of the Hellenistic World* [Oxford 1941] 174).

6.50. Reverting to the common use of "Scythian" to mean all peoples of the north, Pliny listed a large number of ethnic groups. Many of the names probably came from the report of Demodamas, but quite a few are peoples west of the Caspian Sea, and some are connected with the expedition of Pompey the Great to the region in the 60s BC. These were transmitted to Pliny through Varro.

Aramii as an alternate name for the Scythians is not documented elsewhere, nor are the two alleged Scythian names, the Chorasari and Croucasis: the latter may be a local toponym recorded by Pompey. The Sacae – clearly a variant of Scythian – were Persian allies in the fifth century BC (Herodotus 7.64), and the Massagetae were well known in the Greek world as the object of the last and fatal campaign of Cyrus the Great of Persia (Herodotus 1.201–6). Many of the other ethnic groups are otherwise unknown.

Among those that can be identified, the Dahae are probably the Daans who lived east of the Caspian (Strabo, *Geography* 11.7.1). The Essedones were near Colchis (*NH* 6.21). The Rumnici were probably the Rhymmoi of Ptolemy (*Geographical Guide* 6.14.10), who lived between the Caspian and the External Ocean. The Pestici were on the east coast of the Caspian (Pomponius Mela 3.39), and the Camacae were near the Essedones. The Euchatae may be the same as the Auchatai, a Scythian group, and the Cotieri were nearby (Herodotus 4.6). The Arismaspi were the semi-mythical one-eyed people of the far northeast (Herodotus 3.116; *NH* 4.94).

6.51. Pliny may, in part, have been alluding to the manipulation of topography by those associated with Alexander, but there is little doubt that the world beyond the Caspian was hardly known, a vast area populated by many different peoples. In fact, Pliny reported that it was the most poorly understood region in the world, a reasonable supposition given the topographical issues that developed at the time of Alexander as well as its remoteness (Bolton, *Aristeas* 106). The salinity of the Caspian is about one-third that of the ocean, and Pliny provided what is still the generally accepted reason, that there are several large rivers, notably the Ural, Volga (ancient Rha), and Kura (ancient Cyrus), whose substantial inflow reduces the salinity. Since the Caspian was considered part of the External Ocean, this was a matter of bafflement. Pliny has changed from Hellenistic sources such as Patrocles and Demodamas to Varro, who supplied information about Pompey's campaign to the Caucasus, reported by his chronicler, Theophanes of Mytilene (*FGrHist* #188), whom Pliny did not cite but who may have been used through Varro.

6.52. It is highly unlikely that Pompey sent people to Bactria: his Caucasus campaign was already a deviation from his commission, to track down Mithridates VI, who had retreated from Pontus to the Bosporanian region. Moreover, Pompey ran into great difficulty in the Caucasus, due to hostile locals and the weather (Plutarch, *Pompeius* 35; Appian, *Mithridateios* 103). What Varro described is a lengthy trade route beginning in India and ending at Phasis on the Black Sea, in Colchis. It had probably long existed, but certain details remain vague or even erroneous, most notably the assumption that the Oxus River (modern Amu Darya) flowed into the Caspian. Today it actually empties into the Aral Sea. It is certainly possible that its course was different in antiquity, but given the general confusion about topography in this region, it is difficult to determine the exact path of the route (J. R. Hamilton, "Alexander and the Aral," *CQ* 21 [1971] 106–11). It is significant that while travel times are listed for the western and eastern ends, there are none for the middle, the part in question. It was known that there were islands in the Caspian Sea, but they seem to have been largely uninhabited and of little interest. Zazata is otherwise unknown unless it is the Talge of Pomponius Mela (3.58) or the Talka of Ptolemy (*Geographical Guide* 6.9.8); Talge was a sacred island known for its fertility.

6.53. Another trade route is described, in this case from west to east. It diverged from the Indian route and headed more to the north and east, an early example of what came to be the Silk Road to China. The Scythian Cape, not identifiable, seems to have been the diverging point between the two routes (Pomponius Mela 3.59).

It is obvious that there was little known about this route until it reached its eastern terminus, the Seres (*NH* 6.54), although it was believed never to have been very far from the External Ocean. The reference to snow suggests that it passed through the high country north of the Himalayas. The only toponym is Mt. Tabis, said to be on the coast (Pomponius Mela 3.60), which makes locating it impossible. Much of the description is a generic account of the terrors of unknown country, with its lack of cultivation, threatening beasts, and, worst of all, cannibals (Anthropophagi), who were assumed to be common among remote peoples (Strabo, *Geography* 7.3.9; *NH* 4.88, 7.11–12).

6.54. Eventually the land of the Seres ("Silk People") was reached. Although the conventional wisdom considers these the Chinese, their identity is far more nuanced. It is more probable that at the time Pliny wrote they were merely traders in silk located somewhere between Bactria,

India, and China (John Ferguson, "China and Rome," *ANRW* 2.9.2 [1976] 582–5; Roller, *Ancient Geography* 119).

The Mediterranean world became aware of eastern silk in the third century BC, when Greeks saw it in India (Strabo, *Geography* 15.1.20). The toponym China (as Thina) was known in Pliny's day as a great inland city that was somewhere beyond India (*Periplous of the Erythraian Sea* 64), but as yet there seems to have been no connection between Thina and the Seres, other than that both were in the remote and barely comprehended northeastern corner of the world.

Pliny's description of the processing of silk (given in greater detail at *NH* 11.76–8) has nothing to do with the Chinese product. It is an account of the indigenous industry on the island of Cous (Cos) in the Aegean, taken from Aristotle (*Research on Animals* 5.19.551b), and thus in existence from at least the fourth century BC and using a different species of animal (*Saturnia pyri* and *Pachypasa otus*) from that known in China (*Bombyx mori*; Forbes, "Silkworm" 22–6). The comment on transparent clothing is a moralism of the early Imperial period, which also referred to Coan silk (Horace, *Satires* 1.2.101–3). The final comments on the Seres are an interesting statement about their nature and trade techniques: Pomponius Mela (3.60) recounted how they merely left their goods in a remote place and were not personally involved.

6.55. Most of the toponyms and ethnyms cannot be identified, but presumably were located in the far northeast, beyond Bactria and India. There may be little topographical coherence to the list and it perhaps is simply a repertory of toponyms learned at some eastern trading center, perhaps in India. Cape Chryse is usually located beyond the mouth of the Ganges and associated with southeast Asia (Roller, *Ancient Geography* 175), and the description of the local climate is more applicable to the tropics than to central Asia. For the Hyperboreans, see *NH* 4.89. Amometus (*FGrHist* #645, F2) is hardly known, and his topic is not clear in Pliny's text, but he and Hecataeus of Abdera (*FGrHist* #264, T6b) both wrote on geographical paradoxes and fantasies, a subtle indication by Pliny that the places he has been describing do not belong to the real world of terrestrial geography. Yet this is less true than he believed, in part because his account is lacking in the telltale indicator of anatomically impossible people.

The Attacori are otherwise unknown. It is remotely possible that the Thuni are the people of Thina, or China (*Periplous of the Erythraian Sea* 64). The Toarci, or Tochari, lived beyond Bactria, and were part of a Chinese-led coalition that brought the Hellenistic kingdom of Bactria to an end in the second century BC (Strabo, *Geography* 11.8.2; W. W. Tarn,

The Greeks in Bactria and India [revised third edition, Chicago 1997] 270–311). The Casiri, Ciconae, and Brisari are otherwise unknown, but the reference to "here" (also in *NH* 6.56) in regard to the Indian nomads shows that this is a report from some unspecified location, probably a trading emporium visited by these peoples somewhere in Central Asia on what came to be called the Silk Road.

India

6.56. Pliny's next topic is India, one of the most detailed sections of his geography (through *NH* 6.80). His initial direction of approach was from Bactria, following the trade route southeast past the Hemodi Mountains (the Himalayas). Here the Eoan (Eastern) Sea means the Bay of Bengal, but this does not contradict the previous account (*NH* 6.33) that it continued well north of India toward the Caspian Sea. The other ocean mentioned, the Indian, conforms to the usage of today, but was reported from the perspective of the mouth of the Indus and thus seen to be in the south.

Pliny relied on Eratosthenes (*Geography* F70) for the distances along the Indian coast, which are presented from east to west (Eratosthenes' own perspective for his account of Asia). Eratosthenes himself used Megasthenes (*FGrHist* #715), who was at the court of the Mauryan king Chandragupta at the end of the fourth century BC (see further, Eratosthenes, *Geography* F69, for a more detailed account of his own measurements). The east coast of India was said to be 1,875 miles (2,775 km.) as far as the "bend," its southernmost point (modern Cape Comorin). From there it was 2,475 miles (3,663 km.) to the mouth of the Indus, for a total of 4,350 miles (6,438 km. or 35,000 stadia), close to the actual distance from the mouth of the Ganges around to that of the Indus. Pliny's comment that the Indus was at the western border of India indicates that the context of this passage is before the arrangement made between Chandragupta and Seleucus I around 305 BC that allowed the former to extend his possessions west of the river (Karttunen, *India and the Hellenistic World* 260–4).

6.57. Again reverting to commercial information, Pliny noted that it took 40 days to sail the entire length of India, presumably from the Ganges to the Indus. But this figure was variable, and contrasts with Pomponius Mela's 60 days (3.61). The north–south distance, 2,850 miles (4,218 km.), has no source, and does not conform to the other figures. Those of Agrippa (F32) are 3,300 miles (4,884 km., from 26,400 stadia) east–west and 1,300 miles north–south (1,924 km., from 10,400 stadia); these differences show how little was known about the size and shape of the subcontinent (see further, *NH* 6.59). The lack of defined endpoints makes it difficult to correlate the numbers to actual distances.

Posidonius (F212) had India extend from the summer to the winter rising of the sun, or roughly northeast to southeast, which contrasts with the earlier view that the orientation was east–west, as noted in *NH* 6.56. Posidonius was correct but tended to be ignored, and the east–west view persisted into the second century AD (Kidd, *Posidonius* 2, 756–7). He was thus aware that India faced the west wind (Favonius), which was a reason for its healthy climate. The wind also helped sailing from the Red Sea to India (*NH* 6.101). Placing India "opposite" (at the same latitude as) Gallia seems an odd statement, since the two localities differ by about 30° of latitude, but his point was that both faced the same way.

6.58. Pliny's source was probably still Posidonius: astronomy and climate were areas of his interest. After the time of Alexander, the unusual qualities of India were first recorded in detail by Megasthenes (F8) at the end of the fourth century BC, and repeated by Eratosthenes (*Geography* F75), both of whom were particularly interested in its strange fauna and flora, especially the ability to have two harvests, something generally alien to the Mediterranean world (Strabo, *Geography* 15.1.20–2). Because India was farther south than most Greeks had been, the celestial phenomena were different. The south end of India is at 8° north latitude, and the Great Bear – the mainstay of astronomical observation – begins to disappear at the Tropic of Cancer (Dicks, *Geographical Fragments* 126–7, 172–4).

The vast number of ethnic groups in India had long provoked comment: Herodotus (5.3) had noted that it was the most populous region on earth. This is Pliny's first mention of the sources for India. Patrocles (*FGrHist* #712), the Seleucid explorer of the Caspian, has been apparent, who provided data on the dimensions of India, which were extensively used by Eratosthenes (*Geography* F22, 47, 50, 69, 73). Megasthenes was perhaps the most important writer on India, especially the regions beyond the Indus that Alexander never visited. His *Indika* in (probably) four books was the most thorough analysis of the region before the Roman period, and was extensively used by Eratosthenes, Strabo, and Pliny. Dionysius (*FGrHist* #717) is hardly known except for Pliny's comment in this passage: that he was sent to India (probably to the Mauryan court) by Ptolemy II of Egypt (reigned 281–246 BC) to report on matters of interest, particularly manpower.

6.59. Pliny was well aware of the variable and inconsistent information about India. The comparison with the Aegean island of Cous (Cos or

Ceos) seems peculiar, but was part of the tradition (Strabo, *Geography* 15.1.3), and probably relates to the period of 309–308 BC when Ptolemy I of Egypt was in residence on the island and his son and successor Ptolemy II was born there. The account of Ptolemy I (*FGrHist* #138), who had been a companion of Alexander the Great, was a major source for events of the latter's career, and he may have inserted the item about Cous into his narrative (Diodorus 20.27.3; Theokritos 17.58–76; A. B. Bosworth, *A Historical Commentary on Arrian's History of Alexander* [Oxford 1980–95], vol. 2, 318–19). Those with Alexander tended to exaggerate details about India, something well known by late Hellenistic times (Strabo, *Geography* 15.1.33). Onesicratus (*FGrHist* #134, F6) is the source that India was one-third of the world.

The idea that the Indians did not leave their own territory is a slight misrepresentation of a statement by Megasthenes (F14), who reported that they did not go from their homeland to make war, perhaps a slander at Alexander. Megasthenes was also the source for the Indian king list – although his total was 6,042 years – which ran from Dionysos to Chandragupta (ruled 318–294 BC), a somewhat hellenized version of the official account at the Mauryan court.

6.60. The Indian rivers were astounding to the Greeks, not so much for their length but their volume, which was said to be greater than that of either the Istros (Danube) or the Nile (Arrian, *Indika* 4.13). Alexander descended the Indus from February to June of 325 BC, but his journey was not continuous, so the comment about the necessary time for it is somewhat misleading. The Ganges (2,525 km.) is actually shorter than the Indus (3,180 km.), but the upper course of neither was comprehended.

The *Indika* of Seneca is hardly known, and may never have been completed; Pliny, as a contemporary, may have had access to notes. The figures provided probably originated with Megasthenes (F9a), who had a detailed discussion of the rivers of India. Mention of the Caucasus is another reflection of the manipulation of topography by those with Alexander, who called the mountains that he crossed in his approach to India (the modern Hindu Kush) by that name, so that he could say he had crossed the Caucasus (Eratosthenes, *Geography* F23; Arrian, *Anabasis* 5.3).

6.61–2. Pliny's comment about following in the footsteps (*vestigii*) of Alexander is both literal and generalized: the course of his journey was recorded in detail by those noted below, and his memory influenced all later thought about the route to India (Grant Parker, *The Making of Roman India* [Cambridge 2008] 97–8). Diognetus (*FGrHist* #120, F1) and Baeton (*FGrHist* #119, F2a) kept the record of the distances he travelled. The former is only known through a few references by Pliny, of which this is the most complete. Baeton is somewhat better documented, and wrote a work called *The Stages of Alexander's March* (Athenaios 10.442b). The technical term for both was *bematistes*, "one who measures in paces," a word that may have first been used by them. The process by which they worked, or whether they operated separately or together, is not known. Even though specific citations of them are rare, their data probably pervade the material on Alexander and India, and they also provided important information about Alexander's actual route. The itinerary that they created runs from the Caspian Gates to the Hypasis River, the least-known region that Alexander passed through (since the Caspian Gates were thought to be the end of the civilized world). The Hypasis (modern Beas) was where a revolt in June of 326 BC forced Alexander to turn back, and where he erected an altar to the Twelve Gods (Diodorus 17.95.1; Arrian, *Anabasis* 5.25–9).

Alexander's papers are not extant, but were among his personal effects discovered at Babylon after his death (Badian, "A King's Notebooks," 183–204). Whatever they were, they indicate that the king kept his own record of the journey, and that this material was available to historians. Pliny noted that he himself consulted multiple copies of a given text and that they did not always agree, especially where numbers were concerned, a perennial problem in a world where all texts were transmitted by hand.

6.63. Alexander went no farther than the Hypasis, but the route across to the Ganges and as far as its mouth had been recorded by the early Seleucid period. The Mauryan kings built a measured road as far as Pataliputra (Palibothra, their capital), and Baeton, in a later stage of his career, may have recorded some of these distances (*NH* 6.69). From there to the mouth of the Ganges the route was not well known (Strabo, *Geography* 15.1.11). There is no reason to believe that Seleucus I went any farther than he had gone with Alexander, but Pliny's meaning must be that his agents at Pataliputra reported the data back to him, either using existing Mauryan information or their own bematists.

6.64. Pliny's phrase, "after the Hemodi," indicates that his report was still oriented around the trade route from Bactria to India. The peoples mentioned seem to be located south of the mountains and north of the Ganges. Hemodi and Imaus are variants of the same toponym; Pliny's derivation from "snowy" is correct (Karttunen, *India and the Hellenistic World* 106–8).

The list of various ethnic groups – from west to east – is probably from Megasthenes, reflecting information that he learned at Pataliputra, although some of the names may be from sources connected to Alexander (Richard Stoneman, *The Greek Experience of India: From Alexander to the Indo-Greeks* [Princeton 2019] 23). Most of them are not otherwise known. The Bragmanae may be the social class of the Brahmans, but if so are out of place in this list. For the Calingae see *NH* 6.65–6. The Malli may be the ones who battled with Alexander (Diodorus 17.98; Arrian, *Anabasis* 6.5–6), but would have gone far afield to do so.

6.65. The source of the Ganges was not precisely known to the Greco-Roman world (Strabo, *Geography* 15.1.72); it rises in the Himalayas in the Indian state of Uttarakhand. It had a large number of tributaries (Megasthenes F9a; Arrian, *Indika* 4.2–7). After the river leaves the mountains, the flow is gentle for the rest of its route, dropping only about 100 m. over a thousand km. The width was worthy of note, recorded by Megasthenes (F9b) at 100 stadia (20 km.), which seems excessive but may reflect a flood stage. The Gangaridae Calingae lived on its lowest part: how these were related, if at all, to the Calingae mentioned at *NH* 7.30, a population probably in the south of India with strange habits, cannot be determined.

6.66. This is one of several reports of the manpower of various Indian kingdoms (see also *NH* 6.67–8). They go back to the time of Alexander, and, as always, are probably exaggerated, with the numbers varying in the sources. Moreover, the information obtained by Alexander has been mixed into later data, such as the accounts from the envoy of Seleucus I, Dionysius (*NH* 6.58). It is unlikely that the Gangaridae Calingae mobilized and went nearly 2,000 km. to oppose Alexander, but this was consistently reported (Diodorus 17.93.2; Plutarch, *Alexander* 62.2). Nevertheless the material, even if its exact context is questionable, gives an interesting insight into what was available to the Indian kings, especially their extensive elephant forces.

The classes of Indian society were another matter of interest to the Greek world. Pliny listed farmers, military, administrators, philosophers, and hunters (primarily elephant handlers). This is either a condensed summary of Megasthenes' seven classes (F4, 19) or from a different report, and it remains difficult to believe that the remote Gangaridae Calingae were the ultimate source. There is also a passing mention of the practice of ritual suicide, but again lacking any details, although similar to the account of Megasthenes (F34). This was best documented in the instance of Kalanos, from Taxila, who returned with Alexander to Pasargadai in Persis, where he committed suicide because of illness, a story that became one of the most famous episodes of Alexander's expedition (Diodorus 17.107; Strabo,

Geography 15.1.68; Plutarch, *Alexander* 69.3; Arrian, *Anabasis* 7.3). Another example was Zarmanochegas, who was part of an Indian embassy that met Augustus at Daphne in Syria in 20 BC. He travelled as far as Athens, where he immolated himself: Strabo saw his tomb there with its commemorative inscription (Strabo, *Geography* 15.1.73).

Skilled elephant handlers were an essential part of Indian warfare at this time (Scullard, *Elephant* 55–9). Megasthenes (F20) described in detail the military use of the animals, as well as their care and training; Pliny provided a highly condensed description. Elephants were little known in the Greek world until the time of Alexander (Pausanias 1.12.3–4).

6.67. The Modogalinga may have lived on one of the many islands in the extensive Ganges estuary. The ethnic groups are otherwise unknown, although some of them may appear in the Indian text known as the *Aitareyabrahmana* (7.18; Pliny, *Histoire naturelle* 6.2 [ed. Jacques André and Jean Filliozat, Paris 2003] 96–7). The additional manpower figures may come from Dionysius' report (*NH* 6.58).

The Dardae, or Derdai, lived in the northwest of India, and were familiar because of their association with the famous gold-digging ants, which were probably small mammals but were nevertheless considered one of the marvels of the region. They were known to the Greek world as early as the fifth century BC (Herodotus 3.102–5; see also *Mahabharata* 2.48.4). The Dardae themselves were first mentioned by Megasthenes (F23b; Strabo, *Geography* 15.1.44). The Setae were probably in the same region: silver was said to be common in India (Onesicritus F21; *NH* 6.74; Karttunen, *India and the Hellenistic World* 248).

6.68. The best-known kingdom of Early Hellenistic India was that of the Mauryans, centered at Pataliputra (Greek and Latin Palibothra or Palimbothra, modern Patna) on the Ganges. This was the territory of the Prasii, whom Alexander had heard about but had had no personal contact with (Quintus Curtius 9.2.3). In the late fourth century BC Chandragupta Maurya established a powerful dynasty that continued into the reign of his son Bindusara (or Amitrochates), and his son, the famous Asoka, and which came to control much of northern India as far as the Indian Ocean. They also had close relations with the Seleucids and perhaps other Greek kingdoms, and published their royal decrees in Greek and other languages. The empire lasted until its collapse after the death of Asoka in the late 230s BC (Karttunen, *India and the Hellenistic World* 253–71). Greek envoys were regularly at the Mauryan court, most notably Megasthenes (in the time of Chandragupta) and Daimachos (*FGrHist* #716) in the time of Bindusara, as well as Dionysius. It is through their writings, largely preserved by Strabo and Pliny, that the Greek impression of the Mauryans survives.

6.69. The Monaedes and Suari are otherwise unknown, but were near the Prasian territory. The astronomical information that follows may be from Megasthenes (F7b), but mention of Baeton, the bematist of Alexander (F4), suggests that he may have had a further career in India after his time with the expedition.

The data about Mt. Maleus are repeated from *NH* 2.184. The mountain would be farther south than one would expect any of the Greeks with Alexander or with the Mauryans to have gone, well into the southern part of India and certainly south of the Tropic of Cancer, where the Septentriones (Great Bear) begin to disappear. Details were presumably reported to those at Pataliputra (Megasthenes F7; Daimachos F3; Karttunen, *India and the Hellenistic World* 127–8; Pliny, *Histoire naturelle* 6.2, 146–8). Dramasa is perhaps from *dhruvamsa*, "a fixed pole" (Pliny, *Histoire naturelle* 6.2, 100), but it may not have meant the same as the concept of "pole" to Greeks. The south pole was completely theoretical in Greek astronomical thought, but the fact that India was so far south may have led to some consideration of it.

6.70. The idea that the dark complexion of people in the south was due to their location had, to some extent, long been a part of Greek anthropological theory, although it was difficult to explain the differences between Indians and Aethiopians, the two most tropical people known (Arrian, *Indika* 6.9). The reference to the Indus River is probably an error for the Ganges or another river; the Indus is nowhere near the Prasians. The Pygmies were one of many anatomically unusual people believed to live in India: Megasthenes (F27) and Daimachos (F5) provided lists. Any connection with the Homeric pygmies (*Iliad* 3.6) is improbable.

The distance reported by Artemidorus (F108) between the two rivers is probably an error, although it is difficult to tell which affluents of either were used for the measurements. But the 2,100 miles (3,108 km.) may have been 210 miles (311 km.), or the figure may be stadia (about 420 km.), which would be close to the actual distance between them (about 500 km.).

6.71. The Indus, the longest river in India, was better known to the Greek world than the Ganges, because of Alexander's travels on it. Pliny described it and the peoples living around it at length (through *NH* 6.80). No sources were cited, but Pliny had a variety of reports from the era of Alexander, which is presumably the context of much of the account; the many ethnic groups mentioned that were not directly on the rivers were names learned during the journey. The environment of Alexander is demonstrated by the misplacement of the Caucasus, reflecting the manipulation of topography of that period (Roller, *Ancient Geography* 102–4), although there are also some later details. Alexander's expedition was on the river for several months in 325 BC, and thus had every opportunity to collect information. The most thorough extant account of the cruise is that of Arrian (*Anabasis* 6.1–18). The Indus was noted for its large number of affluents, many of which come together in what is now central Pakistan. It was thought to be narrower and shallower than the Ganges, but could be as wide as 100 stadia (20 km.) when in flood (Strabo, *Geography* 15.1.32).

6.72. The official length of the Indus today is 3,180 km., of which 1,240 miles (1,835 km.) was considered navigable in antiquity. This is substantially more than the length of Alexander's journey on it, so presumably there was received information about the upper river. Its primary direction is to the southwest.

Before considering the Indus region proper, Pliny recorded a summary *periplous* around India from the mouth of the Ganges. This cannot be from the era of Alexander, or from the same source as *NH* 6.56, since the distances are quite different. The toponyms are generally unidentifiable, although Cape Perimula was known for its pearls (*NH* 9.106), and, as the location of the best-known market in India, was probably at one of the major trading centers on the west coast, where Mediterranean and Indian goods were exchanged. Patala, near modern Hyderabad, was the last city on the Indus, located on an island (*NH* 6.80).

6.73. The first region discussed lies between the upper Indus and the Jomanes (modern Jumna), which is the main tributary of the Ganges. This is roughly from eastern Pakistan as far as Delhi, with the Great Indian Desert to its south. The various ethnic groups are otherwise unknown. Tigers (in this case the Bengal tiger, *Panthera tigris tigris*) became known to the Greeks at the time of Alexander but remained an exotic and unusual animal into the Roman period (Strabo, *Geography* 15.1.37; Kitchell, *Animals* 183–5).

6.74. The area examined may be the Western Ghats, where mountains parallel the west coast of India for several hundred kilometers. None of the ethnyms can be identified. Mt. Capitalia is probably an indigenous toponym, and may refer to Mt. Anamudi in the Western Ghats, at 2,695 m. the highest point in India outside the Himalayas. Alternatively, it may be the isolated peak of Mt. Abu (1,220 m. elevation) in Rajasthan. Gold had been associated with India since well before the time of Alexander (Herodotus 3.94, 106), and alluvial gold was common (Megasthenes F27b; Karttunen, *India and the Hellenistic World* 249–50). For silver, see *NH* 6.67.

6.75. There are more ethnic groups not known from other sources, but presumably located on or near the coast and east and south of the mouth of the Indus. The Oratae may be the Oreitai of Arrian (*Indika* 23.5), but this group is west of the Indus and thus seemingly in a different region. If they are the same, this was where Alexander placed a Macedonian, Apollophanes, as satrap, but he soon succumbed to local hostility (Heckel, *Who's Who* 41–2).

Crocodiles (the Indian variety, *Crocodylus palustris*) were common in the Indus system; Alexander saw them in the Hydaspes, and believed that this was proof of a connection with the Nile, although that crocodile is a different species, *Crocodylus niloticus* (Strabo, *Geography* 15.1.25; Kitchell, *Animals* 41). Automula, otherwise unknown but perhaps near modern Broach, was another of the market centers of the western Indian coast that traded with the Mediterranean world.

6.76. The Pandae are more in the world of mythology than ethnology. They were descended from a daughter of Herakles/Hercules named Pandaia, who became ruler of her territory, with the forces listed. In part, the story is in the category of women-dominated societies at the edge of the known world, but Panda was a widespread ethnym, representing a group that had passed through India in the ninth or eighth century BC (Klaus Karttunen, *India In Early Greek Literature* [Helsinki 1989] 202–7). The tale is probably an attempt to reconcile the presumed role of Herakles with a tradition of local women rulers.

The toponyms are not otherwise known, and are probably in the south of India. The distance, 1,925 miles (2,849 km.), from the Caspian Gates is probably the addition of the bematists' measurements to the length of the Indus cruise.

6.77. Having completed a catalogue of the ethnic groups east and south of the mouth of the Indus, Pliny now proceeded upriver, listing the ethnyms. The first nine – otherwise unknown – are on the lower river, probably below where it splits into various channels (a point about 250 km. upstream), and where there are a number of islands. Then there is a stretch of 250 miles (370 km., from 2,000 stadia) across the western edge of the Great Indian Desert, interrupted by a small inhabited region of three otherwise-unknown ethnic groups. Leaving the desert, the route continues upstream to the junction of the Indus and the Hydaspes, and then up the latter, following Alexander's route in reverse, but perhaps not totally dependent on data from his era. It passes through six ethnic groups and seventeen cities (all unidentified) to the town of Bucephala, perhaps near modern Jalilpur in Pakistan, named after Boukephalos, Alexander's horse, who had been with him since childhood and died at this point, and who is most famously represented on the Alexander mosaic in Naples (Cohen, *East* 308–12; Andrew Runni Anderson, "Bucephalas and His Legend," *AJPh* 51 [1930] 1–21).

6.78. Continuing to show his debt to sources connected to Alexander, Pliny called the mountains north of Bucephala the Caucasus. The ethnic groups mentioned cannot otherwise be identified. Still following Alexander's route in reverse, the narrative crosses over to the Indus, which lies about 200 km. from Bucephala. Despite Pliny's assertion that his account was going down this river, the five unidentifiable ethnic groups are presumably in the foothills northwest of Bucephala, where the ancient city of Taxila (Taxilla) was located (near modern Rawalpindi). There are four other ethnic groups somewhere in this general region. For the four satrapies, which are west of the Indus, see *NH* 6.92–5; the inclusion of them in India reflects the agreement at the end of the fourth century BC between Seleucus I and Chandragupta, where the former yielded much of his far eastern territory.

6.79. To conclude his discussion of India, Pliny provided some mythical information associated with the region. There were a number of localities that seemed to give credence to a local presence of Dionysos (Father Liber or Bacchus). A town named Nysa was believed to replicate the one in Asia Minor where the god had been born (Homer, *Iliad* 6.130–5; Arrian, *Anabasis* 5.1.1–6). Mt. Meros seemed to be the Greek world for "thigh," reminiscent of the circumstances of the god's birth. Moreover, the flora and fauna among the Aspagani (perhaps the Astakenoi of Strabo, *Geography* 15.1.27), especially the production of grapevines, seemed similar to those of the god's homeland. As a descendant of Dionysos, Alexander promoted such comparisons, and the presence of some early Greek settlements in Bactria – refugees from Ionia transplanted by the Persian king Darius I (Herodotus 6.9) – may have meant that elements of Greek cult were found closer to India than one might expect. For Taprobane, see *NH* 6.81–91.

6.80. The final comments on India are about various islands off its coast. As far as can be determined, most of these are in the Indus delta, near the mouth of the river, or just to its west. Chryse and Argyra have Greek toponyms descriptive of their primary products (gold and silver). Pliny's information on Chryse (generally thought to be somewhere on the Malay peninsula, perhaps misplaced in Pliny's list) is expectedly vague, although in his day it was already a major trading center and was considered the farthest east point known to the ancient world (*Periplous of the Erythraian Sea* 63). Crocala and Bibaga (Bibakta) were encountered shortly after the fleet under Nearchus and Onesicritus headed west from the Indus (Arrian, *Indika* 21.6, 11). The distances (20 miles or 30 km., and 12 miles, or 18 km.) are from the sailing record.

Taprobane

6.81. Taprobane (modern Sri Lanka) has an unusually long description in the *Natural History* (through 6.91), probably because an extensive report on the island came to Rome in Pliny's lifetime, through the agency of the freedman of Annius Plocamus and the embassy of Rachias to Rome (Murphy, *Pliny* 105–8; D. P. M. Weerakkody, *Taprobanê: Ancient Sri Lanka as Known to Greeks and Romans* [Turnhout 1997]). The island was not known to the Greek world before the time of Alexander. The two sources mentioned by Pliny, Onesicritus and Megasthenes, provided the earliest information. Onesicritus (*FGrHist* #134) was the pilot of the fleet Alexander sent from the mouth of the Indus to the Persian Gulf, and seems to have been the first to hear about the island (F13). What he learned is replete with superlatives and has a quality of local promotion, but also a certain amount of discreet vagueness, such as implying that there was only one river in Taprobane, a place with over a dozen rivers longer than 100 km. Nevertheless the Taprobanian elephant was a distinct subspecies, *Elephas maximus maximus*, larger and more powerful than those on the mainland (Aelian, *On the Characteristics of Animals* 16.18; Scullard, *Elephant* 72). *Margaritae* (pearls) and gold were also common throughout the island and were local export products (see *NH* 6.89; Ptolemy, *Geographical Guide* 7.4.1), although in modern times only small amounts of alluvial gold, not economically viable, have come from the island (C. B. Dissanayake and S. W. Nawaratne, "The Occurrence of Gold in the Gem Beds of Sri Lanka," *Economic Geology* 76 [1981] 733–8). Further information, more topographically and ethnographically oriented, was provided by Megasthenes (F26). The ethnym Palaeogoni may sound like a Greek descriptive term ("Those Born Long Ago") but is actually a hellenization of an indigenous name for the island or one of its towns, Palaesimundum (*Periplous of the Erythraian Sea* 61; *NH* 6.85–6).

Because of its location as essentially the farthest south region known in antiquity, below 10° north latitude, Taprobane was long thought to be part of another continent (Hipparchus, *Against the Geography of Eratosthenes* F5; Dicks, *Geographical Fragments* 115–16). It was considered to be the theoretical Antichthones ("Opposite Land"), where the Antipodes ("Those With

Their Feet Opposite") lived; these hypothetical terms compensated for the fact that all of the known world was north of the equator, and it was expected that there should also be lands to its south (Anna-Dorothee von den Brincken, "Antipodes," in *Trade, Travel, and Exploration in the Middle Ages* [ed. John Block Friedman *et al.*, New York 2000] 27–9). Eratosthenes' measurements (*Geography* F76) for the island (7,000 by 5,000 stadia, or 1,400 by 1,000 km.) are far too large and are generalized (the maximum distance across is about 350 km.), and may reflect the confusion about its situation.

6.82. Pliny's account moved from the era of Eratosthenes to an unspecified later period when there had been improvements in seamanship. The reported orientation of Taprobane – east to west – shows how little was known about the island (as well as southern India) in Hellenistic times. A sail of 20 days from the Prasians (at Pataliputra) would include the several hundred kilometers down the Ganges to its mouth, and then along the east coast (that of the Eoan Sea) to Taprobane; this information must be from Megasthenes. Onesicritus (F12) also gave a figure of 20 days' sail from the mainland (presumably the mouth of the Indus), but this need not be any confusion of data, since the length of the two routes is similar. He too would have learned this by hearsay. The voyage of seven days is from Eratosthenes (*Geography* F74), but the information is still confused, and it is not clear what "our ships" refers to, although it should be Pliny's own era. Nevertheless the strait between Taprobane and the mainland was notoriously shallow and full of shoals – most notably the feature known today as Adam's Bridge – and dangerous to navigate.

The difference between the boats used around India and those of the Mediterranean world was of interest to the Greeks: it was said that the local ones were poorly rigged and had other problems, something noticed by Onesicritus, an experienced seaman (F12; Strabo, *Geography* 15.1.15; Pliny, *Histoire naturelle* 6.2, 158–9). The number of amphoras a ship could carry was the standard measure of its capacity when laden with liquid cargo (generally wine or olive oil), but the variable size of the amphora meant this did not allow a precise measurement (Casson, *Ships* 170–3). Yet in 43 BC a ship with a capacity of over 2,000 amphoras was considered a large one (Cicero, *Letters to His Friends* 12.15.2).

6.83. It was hardly true that the stars were not used locally for navigation, but in these low latitudes, where they were so different, understanding them was problematic or even impossible for Greek seamen. It is an exaggeration that Septentrio (or the Septentriones, the Great Bear) is not visible in Taprobane, but it did set, and that also was a strange phenomenon to the Greeks: Homer (*Odyssey* 5.273–5) had noted it was the only constellation that was always visible. But south of the Tropic of Cancer it would begin to disappear, and would set entirely at times at the latitude of Taprobane (Eratosthenes, *Geography* F67; Dicks, *Geographical Fragments* 127).

The use of birds to locate land was a famous technique, familiar from the story of Noah (Genesis 8:6–12), yet would be rarely needed in coastal sailing. Nevertheless they could be useful during fog or other situations where sight of the shore was lost. Avoiding sailing for a hundred days after the summer solstice was because of the dangers of the monsoon season.

6.84. The account of the unnamed freedman of Annius Plocamus is one of the most interesting passages in the geographical section of the *Natural History*. It is a rare example of direct Roman contact with a culture at the ends of the earth, rather than through the highly derivative medium of a trade route. Annius Plocamus is almost certainly the Popilius Annius Plocamus whose name appears on graffiti from AD 6 along the road from Coptus (Koptos) in Egypt to Berenice on the Red Sea: Pliny reported that he had mercantile dealings on the sea. His family was prominent in the business community of Puteoli on the Bay of Naples (Sidebotham, *Berenike* 72). Sometime shortly after AD 41, he sent his freedman – who may be the Lysas named on the graffiti – to investigate new markets, since the route to northern India was already well used, as the extant and contemporary *Periplous of the Erythraian Sea* demonstrates. The freedman used the familiar excuse of being driven off course to explain why he ended up at Hippuros, an unidentified location probably in northwest Taprobane. Even though the toponym sounds Greek, it is probably local (Pliny, *Histoire naturelle* 6.2, 113–15). He was hospitably welcomed and stayed for at least six months.

6.85. After a somewhat exaggerated and idealized account of how the unnamed local king was impressed by Roman ways, especially the integrity of their coinage, nothing more is heard about Annius Plocamus or his freedman. The narrative shifts to the embassy sent to Rome by the king, led by a certain Rachias (or Rachia), who provided his own exaggerated and idealized account of his homeland, but with additional valuable information about the land of the Seres, who had been mentioned in a different context at *NH* 6.54. Rachias may be the Raki named in Sri Lankan inscriptions, who was the son-in-law of the ruler and thus would have been well positioned to lead a lengthy foreign expedition (Ananda Abeydeera, "Raki's Mission to Romanukharattha. New Evidence in

Favor of Pliny's Account of Taprobanê, *N. H.* 6.84–91," *Habis* 40 [2009] 145–65).

Envoys from the far east to Rome were not a new phenomenon when Rachias and his entourage arrived in the city, sometime between AD 41 and 54. Augustus reported that Indian embassies came regularly to him (Augustus, *Res gestae* 31; Suetonius, *Augustus* 21.3); these included the well-equipped one that came in 20 BC to Daphne in Syria with a vast number of unusual gifts (Strabo, *Geography* 15.1.73; Dio 54.9.8–9). An embassy was even said to have come from the Seres, but this is so vaguely documented, and there are so many issues concerning the Seres, that it is problematic (Florus 2.34.62).

6.86. Despite Rachias' expansive and detailed account, too little is known about contemporary Taprobanian topography to relate his description to the landscape. The king's territory was centered at the city of Palaesimundum, which, since it seems to be a name for the entire island, may suggest that he ruled it in totality. The city may be in the northwest at modern Anuradhapura (Ptolemy, *Geographical Guide* 7.4.10).

6.87. The use of the first person (see also *NH* 6.84) raises the possibility that Pliny himself had personal contact with the envoys. The trees in the ocean around Taprobane were probably mangroves, which Greeks had first encountered when Androsthenes of Thasos was sent by Alexander the Great down the Arabian coast of the Persian Gulf (*FGrHist* #711, F5; Theophrastus, *Research on Plants* 5.4.7–8). It would seem to the Greco-Roman world that the tropical oceans were filled with forests.

The suggestion that the moon was only visible for about a week is obviously a misunderstanding, perhaps of its phases. Canopus (α Carinae; see *NH* 2.178) would have been high in the sky in Taprobane. The reversal of shadows was a curiosity encountered by Greeks and Romans as they ventured farther south, first noticed on the upper Nile (Eratosthenes, *Geography* F41). But the idea that the sun's motion was reversed is another peculiar misunderstanding, made more inscrutable by depending on the viewer's perspective, and perhaps based on the idea that everything in the far south was reversed.

6.88. As at *NH* 6.81, the distance and position of Taprobane is exaggerated and erroneous. A length of 10,000 stadia (2,000 km.) may actually be the perceived distances along the east coast of India, since the figure is immediately followed by a reference to the Hemodi Mountains (Himalayas) and the Seres. The report on them is part of the information obtained in Rome from the visit of Rachias, whose father travelled to a major trading emporium on a river at some unknown point, presumably north of the Himalayas, where he encountered certain Seres. Some Roman merchants had also been to this market, and this is probably the same trading post described in the roughly contemporary *Periplous of the Erythraian Sea* (65), located on the borders of Thina (China), where there was an annual rendezvous.

What is of particular interest is that Rachias' father described the physical characteristics of the Seres. That their language was otherwise unknown and said to be harsh is an expected characteristic of remote peoples, but that they were large and had red hair and blue eyes means that these people were not the Chinese. Presumably they were Indo-Europeans of central Asia who were middlemen somewhere between India and China and who traded in Chinese silk (Samuel Lieberman, "Who Were Pliny's Blue-Eyed Chinese?" *CP* 52 [1957] 174–7). Given the role of silk as a luxury product in Rome, Pliny could not resist another derogatory comment about excess, although there is no evidence that Rachias' father felt the same way.

6.89–90. The complaints about luxury carry over into a discussion about the products of Taprobane. Gold and silver are to be expected in this region, and the tortoise-shell marble may be a type of alabaster, if not the shell itself (Lionel Casson, *The Periplus Maris Erythraei* [Princeton 1989] 156; Karttunen, *India and the Hellenistic World* 213–14). Pearls (*margaritae*) were a major luxury item in Rome from the late second century BC (Pierre Schneider, "Erythraean Pearls in the Roman World," in *The Indian Ocean Trade in Antiquity* [ed. Matthew Adam Cobb, Leiden 2018] 135–56) and were the most famous product of the region, valued more than any precious stone other than diamonds; the pearl industry was first described by Megasthenes (F13a; *Periplous of the Erythraian Sea* 59; *NH* 9.106–23). Coinage of Cleopatra VII shows her wearing what appear to be pearl earrings and perhaps a pearl necklace, and there is the famous story, perhaps apocryphal, of her dissolving a pearl in vinegar (*NH* 9.119–21; Macrobius, *Saturnalia* 3.17.14; Duane W. Roller, *Cleopatra: A Biography* [Oxford 2010] 132, figs. 11d, e). Gems remain a major product of Sri Lanka (C. B. Dissanayake *et al.*, "The Geology, Mineralogy and Rare Element Geochemistry of the Gem Deposits of Sri Lanka," *Bulletin of the Geological Society of Finland* 72 [2000] 5–20). There are also a few general comments about Taprobanian society, presumably still from Rachias' report, but synthesized in such a way as to reflect contemporary Roman attitudes. Their alleged cult of Hercules was connected with the pearl industry, as he was said to have discovered the product (Megasthenes F13a; Arrian, *Indika* 8.10).

It is not possible to determine the accuracy of all the material attributed to Rachias about the nature of the Taprobanian government. In fact, some of the comments seem suspiciously directed to a Roman audience, and are a critique of the state of affairs in Rome during the Julio-Claudian period. In Taprobane, supposedly, the price of grain was never inflated, there was no litigation, and the kingship was elective as a defense against hereditary monarchy (Chester G. Starr, "The Roman Emperor and the King of Ceylon," *CP* 51 [1956] 27–30). These comments also sound like a report from a world idealized because of the controls on its rulers. Rachias came to Rome in the time of Claudius, when over three years of the erratic Gaius Caligula had severely tested the Imperial system, and upon whose death (by assassination) the Senate made its last attempt to reassert power, to no effect. The royal dress – otherwise unexplained – is another attempt to link Greek culture to the far east, especially the world of Dionysos (Father Liber).

6.91. The account of the Taprobanian lifestyle, also somewhat idealized, probably contains some accurate details. The grapevine was not introduced to Taprobane during antiquity, and was essentially limited to northern India (Karttunen, *India in Early Greek Literature* 207–10). Tortoise shell was being exported from the island at the time of Rachias' visit, and the hawksbill turtle (*Eretmochelys imbricata*) was a common local feature (*Periplous of the Erythraian Sea* 61; Casson, *Periplus* 231). The use of large tortoise shells as homes also existed in Carmania (*NH* 6.109). The long lives of people at the fringes of the known world was a typical idealization: elsewhere Pliny credited this report insofar as Taprobane was concerned to the geographical author of the late second century BC, Artemidorus of Ephesus (*NH* 7.30).

From India to the Persian Gulf and Egypt

6.92. The account moves west from India toward the Iranian plateau, with a discussion of the four satrapies west of the Indus that had been mentioned at *NH* 6.78. Capisene and its city Capisa were on the upper Kophen River (modern Kabul), but there is no credible evidence that either Cyrus or Samaramis (Semiramis) got this far, although it was persistently reported that they had reached India (Ctesias [*FGrHist* #688] F1b.16–19; Strabo, *Geography* 15.1.5). Arachosia was to the south in the vicinity of modern Kandahar. The various ethnic groups named are generally unidentifiable. An exception is the Arii, who lived near Alexandria-of-the-Areians, an obscure city that is usually suggested to be around modern Herat in

Afghanistan (Cohen, *East* 260–1). The Paropanisidae (Paropamisidai) lived in the mountains of the same name (the modern Hindu Kush), and were one of the four satrapies (Tarn, *Greeks* 95–102). The fourth satrapy, Gedrosia, is seemingly reserved for *NH* 6.94–5. The variants in the spelling of the toponyms from *NH* 6.78 indicate that this account is from another source.

6.93. The narrative moves farther to the south into Ariane and the desert regions closer to the coast. The Alexandria in question is Alexandria-in-Aria, probably the same one mentioned at *NH* 6.92, although the location and history of towns named Alexandria in this region can be inscrutable (Cohen, *East* 260–1). Artacabene may be at the same place, or nearby. Antiochus is probably the third of that name (ruled 222–187 BC), who was active in the eastern Seleucid empire.

6.94. The account is a mixture of rivers and ethnyms scattered over a wide area between India and Persia. The Euergetae ("Benefactors") were so designated because of services rendered to Cyrus the Great of Persia (Diodorus 17.81.1; Strabo, *Geography* 15.2.10; Arrian, *Anabasis* 3.27.4–5). Except for them, and perhaps the Gedrusi (see *NH* 6.95), the ethnic groups cannot be identified. The names may reflect the route of Alexander from India to Persia (although the lack of known ethnyms makes this improbable), or perhaps trade routes across the region. Some of the names, such as the Suares, refer to locations close to the sea, and others were much farther inland, such as the Euergetae and the district of Daritis, in Media, so there is probably a conflation of routes.

6.95. The itinerary continues to the west, although if the Gedrusi are actually the Gedrosians, they are the inhabitants of the fourth satrapy of *NH* 6.78 and were to the east, close to India. Ichthyophagi ("Fish Eaters") is a generic name for primitive coastal populations: the members of Alexander's fleet gave it to the people west of India, and there was another unrelated group on the coast of the Red Sea (Strabo, *Geography* 15.2.2, 16.4.4) They were seen as the epitome of a society functioning at a subsistence level in a hot environment (Diodorus 3.15–17; *NH* 7.30). Alexander's prohibition of their eating fish may simply reflect Greek incredulity at people with such habits (Truesdell S. Brown, *Onesicritus: A Study in Hellenistic Historiography* [Berkeley 1949] 51–2). But it also may have been an attempt to force agriculturalism on a population that did not practice it, and moreover who seemed to have an anomalous diet. This was not the only time Alexander indulged in social engineering (Plutarch, *On the Virtue of Alexander* 5.328 c; Tarn, *Alexander* 254).

The Arbii lived along the Arbis River (modern Hab in Pakistan, the first major stream west of the Indus (Strabo, *Geography* 15.2.1). Carmania and Persis are territories along the east side of the Persian Gulf; thus the itinerary skips over nearly 1,000 km. of intervening territory, reserving that for the following account of the sea voyage from the Indus to the Persian Gulf.

6.96. There is a lengthy discussion of the various sea routes between India, the Persian Gulf, and the Red Sea (through *NH* 6.106). There are several layers of reports. The earliest are those of Nearchus (*FGrHist* #133, F13) and Onesicritus (*FGrHist* #134, F28), who commanded the fleet that Alexander sent from the mouth of the Indus to the Persian Gulf, and who wrote separate accounts of the cruise. Then there is material from the *On Arabia* of Juba II of Mauretania, from the early first century AD, which updated the earlier commentary and was Pliny's primary source (F28; Roller, *Scholarly Kings* 130–40). There is also a source between the time of Alexander and Juba, which may be Posidonius, and finally, there is material from Pliny's own era, which parallels the information in the *Periplous of the Erythraian Sea*, but is probably not derived from it.

Nearchus and Onesicritus made their cruise in the last months of 325 BC and into the following year. Pliny's recension emphasizes ethnyms and toponyms; the version by Arrian (*Indika* 18–42), from half a century later, is more thorough and has much more ethnography, indicating this was the focus of the original report. It is interesting, however, that even by Pliny's day the location of a number of places identified by Nearchus and Onesicritus had been forgotten.

6.97. Alexandria is probably the town founded at Rhambakia (Diodorus 17.104.7; Arrian, *Anabasis* 6.21.5; Cohen, *East* 297–301). Leonnatus, a Macedonian, was the bodyguard of Alexander and thus a senior member of his staff, and was active in the manoeuvering after the king's death (Heckel, *Who's Who* 147–53). For the Ichthyophagi, see *NH* 6.95; the fact that it took 30 days to sail past their territory indicates the ethnym is a generic term for the people along this coast.

Solis Island (Island of the Sun) was a locality where sailors either disappeared or were turned into fish by the nymphs who lived nearby at the Nymphorum Cubile (the Bed of the Nymphs). Eventually the sailors were returned to human form by the Sun and became the ancestors of the Ichthyophagi (Pomponius Mela 3.71; Arrian, *Indika* 31). This was a typical sailors' tale in the manner of Circe and Kalypso. The island may be modern Astaku, the only important one on this coast and thus a hazard to shipping.

6.98. The Hyctanis, with its gold sands, has not been certainly located (Pliny, *Histoire naturelle* 6.2, 130). Since leaving the mouth of the Indus, Nearchus and Onesicritus had been sailing due west just north of 25° north latitude, but as they reached Carmania the coast began to turn north and familiar stars began to be seen. It is not quite accurate to say that the Septentriones (the Great Bear) first appeared at this point – they had been theoretically visible at least part of the time at the mouth of the Indus (*NH* 2.185) – but around the turn of the year, the period of this portion of the cruise, the constellation rises in the northern sky and would have become more obvious. Arcturus (α Boötis, the Guardian of the Bear), was in a similar situation.

Carmania (essentially the modern Kerman region in Iran) was the effective eastern end of the Achaemenid Persian empire (Strabo, *Geography* 15.2.1), and for Nearchus and Onesicritus there was probably a sense of returning to more familiar territory when they reached this point. This was a region of rich mineral resources: arsenic, in its form as a sulphide, and cinnabar, or mercury sulphide, were used as pigments and medicinally (Theophrastus, *Lithika* 58; Dioskourides 5.94, 105).

Cape Carmania is probably modern Ras al-Kuh, where the coast turns sharply north at the entrance to the Persian Gulf. The Macae lived primarily on the Arabian side of the entrance (the modern Strait of Hormuz). Fifty miles (74 km., from 400 stadia and repeated at *NH* 6.152) is the distance across the narrowest part of the strait. Oracta, or Dorakta, is probably modern Qeshm, off the east coast and the largest of several islands in the gulf (Strabo, *Geography* 16.3.7; Arrian, *Indika* 37.1–2).

Large sea creatures, presumably whales, were frequently reported as common in these waters. Greeks had little experience in sailing the External Ocean, and thus would find them frightening (Diodorus 17.106.6–7; Juba F3; Strabo, *Geography* 15.2.12; Arrian, *Indika* 30, 39). Twenty cubits is about nine meters.

6.99. The narrative moves up the Persian side of the gulf into more familiar waters, and with an emphasis more on navigation and commerce than exploration, presumably due to Juba's updating. The island Gyani are mentioned only here. It was not possible to go upstream from the Persian Gulf all the way to Pasargadai (which lies in an inland basin), but the mouth of the Sitioganus (or Silakos) was presumably the point of embarkation for the journey. Bitumen, or asphalt, had a variety of uses, largely as a sealant and medicinally, and was common in this region: Alexander the Great notoriously almost burned down his headquarters at Babylon experimenting with it (Strabo, *Geography* 7.5.8; Dioskourides 1.73; Plutarch,

Alexander 35). Eventually the marshy region at the head of the gulf was reached.

6.100–1. The following sections (through *NH* 6.106) are one of the best extant accounts of Roman contacts with India; only the *Periplous of the Erythraian Sea*, from a generation earlier than Pliny, has similar detail, but it is focused more on commodities. Pliny heavily condensed the report of Nearchus (preserved more fully by Arrian throughout his *Indika*), and has confused the Tigris with the Pasitigris, which is actually the route to Susa. But the head of the Persian Gulf has changed greatly since antiquity, and even in Hellenistic and Roman times extensive siltation was apparent. Charax, a trading center originally founded by Alexander at the head of the gulf, was 75 km. inland by the Augustan period (*NH* 6.138–40)

Alexander, going overland from the mouth of the Indus, arrived at Susa before Nearchus and Onesicritus. Since he took seven months, the three months of the cruise must only be the time at sea, and thus they would have been often on land. The arrival of the ships at Susa was in early 324 BC, when Alexander was holding a festival to celebrate the marriages between the Macedonians and Persians, including his own to Stateira, the oldest daughter of Darius III, the last king of Persia (Diodorus 17.107.6; Plutarch, *Alexander* 70).

At this point in the narrative, with the cruise of Nearchus and Onesicritus having come to an end, Pliny turned to later modifications of the route to India. The first is no earlier than the late second century BC and is probably from Posidonius, although not specifically attributed to him. This was when the discovery of the Hippalus wind (or perhaps the activities of a seaman named Hippalos) made direct travel to India possible (Strabo, *Geography* 2.3.4; *Periplous of the Erythraian Sea* 57). Cape Syagrum is one of the promontories on the south side of the Arabian peninsula (perhaps modern Ras Fartak), where ships coming out of the Red Sea would leave the coast and head across the Indian Ocean to India, a distance calculated at 1,332 miles (1,971 km.). This was a rare cross-ocean sailing route in Greek antiquity (as opposed to ones along coasts), and would have been difficult, but it was nonetheless traversed regularly (Nigel Groom, "The *Periplus*, Pliny and Arabia," *AAE* 6 [1995] 180–95).

The second modification, probably reported by Juba, changed the port of landfall in India from an unspecified point to Cape Zigerus, perhaps the Melizeigara of the *Periplous of the Erythraian Sea* (53), at modern Jaygarh south of Bombay. The final route, still active in Pliny's day and originating sometime in the first century AD, was shorter than the previous one, but no details were provided beyond the fact that it allowed the voyage to be made annually. Yet the ships had to be armed because of persistent piracy, especially in the Red Sea (Diodorus 3.43.5; *Periplous of the Erythraian Sea* 20).

The adverse balance of payments between Rome and east Asia was a regular problem: commodities to the value of one hundred million sesterces went to India, Arabia, and the Seres, in return for luxury goods. Half of this amount was to India, and it was believed that Indian products were sold to the Romans at a immense profit (*NH* 12.84; J. Innes Miller, *The Spice Trade of the Roman Empire* [Oxford 1969] 23). Roman coin hoards have frequently been found in southern India (Mortimer Wheeler, *Rome Beyond the Imperial Frontiers* [Harmondsworth 1955] 165).

6.102–3. The account now covers the trade route from Alexandria Next to Egypt to India, going up the Nile and across to the Red Sea. This itinerary is from the Roman period, as the name Juliopolis indicates. Juliopolis is a toponym otherwise undocumented, but its location – two miles or three km. from Alexandria – makes it probable that it was Nikopolis, the victory city that Octavian built after he defeated Marcus Antonius and the Ptolemaic army at the site on 1 August 30 BC, resulting in the effective end of the Ptolemaic empire (Strabo, *Geography* 17.1.10). Boats set sail here and headed up the Nile to Coptus, 309 miles (457 km.) upstream. Coptus (Koptos, modern Qift) was where the Nile comes closest to the Red Sea, and the overland journey to the seaport of Berenice began. Pliny's account is the most detailed extant. There was was a difficult trek for 257 miles (380 km.) across a hot and rugged country; it was necessary to make the journey in summer – when it would be particularly arduous – to coincide with the Etesian winds that would allow the sail to India and to make the round trip within a year. The road and many of the way stations are still visible, as well as several parallel routes (Sidebotham, *Berenike* 125–74; Melinda Székely, "Tra Coptos e Berenice (Plin. *Nat. Hist.* 6.26.100–106)," *AAntHung* 50 [2010] 63–9). The number of locations named Hydreuma (Watering Place) is significant.

6.104. The departure from Berenice would be in early July with the rising of the Dog Star (*Periplous of the Erythraian Sea* 39, 49, 56; Casson, *Periplus* 289–91). There might be a stopover at Ocelis (or Acila, *NH* 6.151), at the southwest point of the Arabian peninsula in the frankincense-producing region: Kane and Muza were other ports in the area. These locations (Ocelis in particular) would have been the last landfall and watering stations before heading across the open sea (*Periplous of the Erythraian Sea* 25). The brief digression on the frankincense region is not really relevant to the Indian route, and is to some extent a summary of the more detailed discussion at *NH* 12.51–84, also from Juba's *On Arabia* (F30–3). For Hippalus, see *NH* 6.100.

The various Indian toponyms are all on the southwestern coast, in the modern state of Kerala. Muziris (near modern Cranganore) was a major trading port of the region (reported in detail at *Periplous of the Erythraian Sea* 53–4). It was also described in a Tamil poem of the second or third century AD as a place where Westerners arrived with gold and left with pepper (Casson, *Periplus* 296). Muziris had a Roman mercantile colony with its own Temple of Augustus, as shown on the late antique Peutinger Map. Offloading cargo into boats would have made it particularly susceptible to pirates as well as the winds and seas of the monsoon, since the ships were not sheltered. Caelobothras, probably the same as Keprobotos, whose indigenous name was something like Cheraputra, had probably been on the throne for thirty years or more when Pliny was writing (*Periplous of the Erythraian Sea* 54; Casson, *Periplus* 217).

6.105. The Neacyndes lived at Nelkynda (*Periplous of the Erythraian Sea* 53–5), 500 stadia (100 km.) from Muziris. Pandion was the ruler of the Pandya kingdom (in Greco-Roman accounts of India the names of rulers and their states are often the same). The kingdom had been powerful since the time of Alexander (*NH* 6.76; Arrian, *Anabasis* 8; *Periplous of the Erythraian Sea* 59) and had sent an ambassador to Augustus (Strabo, *Geography* 15.1.4). Cottonara is still a major pepper-producing region (Casson, *Periplus* 221). The local ship was a *monoxylis* ("single-wooder"), a generic term for a dugout; the Greek word had been in use since at least the fifth century BC (Xenophon, *Anabasis* 5.4.11; Casson, *Ships* 8). At some time in this narrative Pliny ceased to use Juba's *On Arabia* and relied on mercantile sources of his own era, as demonstrated by the citation of the Egyptian calendar at *NH* 6.106.

6.106. Pliny is the only source for the westbound journey from India. The northeast wind on this cruise was less threatening than the southwest one of the outbound journey, so the departure date for the return was flexible within a month or even more, yet it was still possible to make the entire round trip within a year.

6.107–8. There is an exceedingly brief discussion of Carmania (see also *NH* 6.98), the region near the mouth of the Persian Gulf, a summary taken from Nearchus (F14). To those coming from India, this cultivated region would have appeared to be the first civilized area that they encountered. Its fertility was notable, although there were no olives (Strabo, *Geography* 15.2.14).

A discussion of the Persian Gulf region follows. Here Pliny called it the "Red Sea," in contrast to *NH* 6.103, where Red Sea means the same as the modern term. This confusion is because the toponym (Erythra Thalatta in Greek) was at first a general word for the northwest part of the Indian Ocean, and today's Red Sea was originally the Arabian Gulf (Herodotus 2.11, etc.). It and the Persian Gulf were seen merely as inlets of the greater Red Sea (hence Pliny's statement that the sea twice bursts into the land). As more came to be known about the perimeter of Arabia the various names became localized to what they are today (as Pliny explained at *NH* 6.108), but a certain amount of confusion remained.

King Erythras was a Persian or Egyptian who had settled on the island of Tyros (modern Bahrain) in the Persian Gulf (Strabo, *Geography* 16.3.5; Pomponius Mela 3.79; *NH* 7.206). But there was general disagreement as to whether the Red Sea (in its broadest sense) was named after him, or from the color of its waters (Agatharchides of Cnidus, *On the Erythraean Sea* [ed. Stanley M. Burstein, London 1989] 42–5).

Eratosthenes (*Geography* F93) recorded a circumference of 31,250 stadia for the Persian Gulf, which Pliny converted to 2,500 miles (3,700 km.), information obtained from the cruise of Androsthenes of Thasos (*FGrHist* #711, F2), who was sent by Alexander some time between 325 and 323 BC to explore the coast of the gulf. The figure for the coast of Arabia, 1,500 miles (2,220 km., converted from 12,000 stadia), is probably from Juba, since the explorers of Alexander failed to make its circuit (Arrian, *Indika* 7.20.7). The distances are in round numbers of stadia and presumably are estimates. The source of the comparison of the gulf to a human head is not known, but was also cited by Pomponius Mela (3.73).

6.109. Pliny completed his discussion of the regions between India and the Persian Gulf with a number of scattered toponymic details (through *NH* 6.111). After providing the total distance from the Indus to Babylon – as reported by Nearchus (F15) and Onesicritus (F29) – at 1,700 miles (2,516 km., from 13,600 stadia), there is a comment on the Chelonophagi, or Turtle Eaters, who lived on the Carmanian coast; Pliny discussed their habits in greater detail at *NH* 9.35–6.

6.110. The various toponyms in this section are presented in a confusing manner, and may simply be material that Pliny left out of his previous accounts. They extend from the outer Persian Gulf east to India, seemingly with an eastbound orientation, which suggests a source other than Nearchus and Onesicritus. The Harmozae (whose name survives in the Strait of Hormuz) are at the entrance to the Persian Gulf, but the other toponyms, insofar as can be determined, are near the mouth of the Indus (Tusculum 6, 214).

6.111. The toponyms are from the region of the Persian Gulf, providing a transition to the following account (*NH* 6.112–16) of the Persian or Parthian empire, which Pliny saw as merely a sequential name change (see also *NH* 6.41). This is inaccurate, but a point of view perhaps generally accepted in the Rome of his era, since one could believe that the Achaemenid Persian empire, which Alexander had brought to an end, had seamlessly evolved into the Parthian empire that emerged several generations later, although there was actually no real continuity.

Parthia

6.112. Pliny's next area of interest is the populations of the Iranian plateau. The Parthian kingdom came into existence in the mid-third century BC, originating in their homeland southeast of the Caspian Sea. By the Augustan period they had spread west to the Euphrates and east toward India, and a wary but peaceful coexistence with Rome was established in 20 BC (Augustus, *Res gestae* 29, 32–3).

The division of Parthian petty kingdoms is information found only in the *Natural History*, as well as the separation into upper (northern) and lower (southern) kingdoms; this may be visual evidence from the map of Agrippa. Parthyaea is the latinized Greek term for their homeland, although it is hardly possible that it could border all the remaining kingdoms.

6.113. Pliny made a circuit around the core Parthian territory, beginning in the east. For the Arii, see *NH* 6.78. The Median Pratitae, who must be in the northwest, are mentioned only here, unless they are the Pratitae of *NH* 6.44, who seem farther west. For the Hyrcanians, see *NH* 6.36. The suggestion that Parthian territory was surrounded by deserts is somewhat of an exaggeration: it was true in regard to the north and east, but not so much the south (toward the ocean) or the west (toward Mesopotamia). Nomad is a Greek descriptive term ("Pastoralist"), often used for remote agricultural or transhumant populations. Alexandropolis in Nisiaea is only documented here. The name of the city is suspicious (very few of the alleged foundations of Alexander have this form), and there is no other record of the king creating a city in this region; if it existed at all, it was probably established later (Cohen, *East* 203).

6.114. Media was southwest of the Caspian Sea in the highlands of northwestern Iran. It was a powerful state that originated in the ninth century BC, described in detail by Herodotus (1.95–122). The Medes were absorbed by the emergent Persian empire in the mid-sixth century BC, and after the time of Alexander were subject to the major states, including the Parthians. Yet memory of their importance and power was such that "Medes" and "Persians" virtually became synonyms in both Greek and biblical sources. There is a circuit of their borders, starting from the east. For the Caspi, see *NH* 6.39.

6.115. The homeland of the Persians was Persis (the region of Fars in modern Iran). It was here that a local dynast, Cyrus, revolted against the Medes around 560 BC and established the Persian empire, events that are the main topic of Book 1 of Herodotus' *Histories*. Pliny's derivation of the name Persian Gulf is accurate, following the habit of naming seas from the inhabitants on their coasts.

Climax Megale (Greek for "Great Staircase") is presumably part of the route from the coast at Ceribobus (which is unidentified) to the upland valley in which Persepolis, the Persian capital, is located (at 1,634 m. elevation). Alexander was there in the spring of 330 BC, and although the literary tradition is emphatic about its destruction, the extent of the fire is still debated, since the site continued to be occupied (Diodorus 17.72; Strabo, *Geography* 15.3.6; Plutarch, *Alexander* 38; Arrian, *Anabasis* 3.18.11–12; Bosworth, *Historical Commentary* 1, pp. 330–3). Pliny is the only source for a Laodicea in Persis; neither its location nor its founder (perhaps Antiochus I) can be identified (Cohen, *East* 189).

6.116. Phasargis (more commonly Pasargadai) lies about 40 km. northeast of Persepolis. The tomb of Cyrus is probably one of the several visible royal tombs around the city. The Magi are familiar in both Greek and biblical literature as a religious group with political overtones, but were also one of the original Median ethnic groups (Herodotus 1.101).

The reference to Ecbatana (at modern Hamadan), the ancient Median capital about 750 km. northwest of Pasargadai, is confused. The city was far older than the Persian kingdom. It may be that something has dropped out of the text and that one of the Persian kings named Darius transferred something from Ecbatana to Persepolis. The Paraetaceni were a Median ethnic group (Herodotus 1.101).

Mesopotamia

6.117. A lengthy discussion of Mesopotamia follows (through *NH* 6.141). The Greek toponym was originally applied to the northern part of the plain between the Euphrates and Tigris, probably when Alexander the Great passed through the region in the summer and autumn of 331 BC. The earliest known citation of the term is by Eratosthenes (*Geography* F83).

Eventually it became more generalized, and spread south to the head of the Persian Gulf.

The Macedonian redistribution of the population was probably after the time of Alexander (Tusculum 6, 216). For the Orroei, see *NH* 6.129; the Arabian Mardani are otherwise unidentified. Arabian Antioch and its founder remain elusive: the name Nicanor is common in the early Hellenistic period, and, moreover, it may be an error for Nicator, in other words, Seleucus I, the companion of Alexander and founder of the Seleucid empire. There are too many little-known cities named Antioch to determine which was meant (Cohen, *East* 61–7).

6.118. The general context is upper (northern) Mesopotamia and the regions to its north. The Eldamari, Salmani, and Masei are not otherwise mentioned, although any group believed to be Arabian would be south of the right bank of the Euphrates. Gordyaea was on the upper Tigris, south of Armenia (Strabo, *Geography* 16.1.8), and was generally part of the Armenian empire. The final four ethnic groups are otherwise unidentified but are probably in the region south of Gordyaea and into the northern Mesopotamian plain.

6.119. The narrative has moved into the plains and Mesopotamia proper. Nicephorion (probably at modern Raqqah in Syria) was said by Pliny to be a foundation of Alexander, but there is also a tradition attributing it to Seleucus I (Appian, *Syriaka* 57), which is more sensible, since when Alexander passed through the region in the summer of 331 BC he had no recent victory to celebrate (Cohen, *East* 85–8). Nevertheless Pliny's account essentially follows the route of Alexander across northern Mesopotamia that summer. Since Caphrena is not mentioned elsewhere, its role as a tax collection center cannot be explained. It may have been at modern Ank Köy, about 15 km. northeast of Zeugma. If Pliny's figure is correct the town would have been over a kilometer across.

6.120. This section covers a broad area, almost the entire extent of Mesopotamia, from the north (where Thebata and Oruros were presumably located) to Babylon. Pompey the Great was involved in diplomatic activities with the Parthians in the 60s BC, during his extraordinary command against Mithridates VI of Pontus. Ever since L. Cornelius Sulla had been the first Roman known to have met with the Parthians, in the late 90s BC (Plutarch, *Sulla* 5.4–6), the boundary between Rome and Parthia was effectively at the Euphrates, until the balance of power was destroyed by the invasion of M. Licinius Crassus in the 50s BC.

Gobares is otherwise unknown but it is probably a version of the common Persian aristocratic name Gobryas. Some holders of that name held positions in Babylonia (Pierre Briant, *From Cyrus to Alexander: A History of the Persian Empire* [tr. Peter T. Daniels, Winona Lake, Ind., 2002] 988).

6.121. Babylon (near modern Hillah in Iraq), the most famous city in Mesopotamia, had existed since the second millenium BC, and became the headquarters of Alexander the Great, where he died in June 323 BC. The Chaldaic (Chaldaean) people were a local group in southern Babylonia, but eventually the ethnym became synonymous with the Babylonians. To the Greeks, however, as early as the fifth century BC they were the scholars of Babylon, who lived in a separate quarter of the city (Herodotus 1.181–3; Diodorus 2.29–31; Strabo, *Geography* 16.1.6).

The walls of Babylon were one of the Seven Wonders of the World. The list was probably collated in the second century BC or slightly earlier (*Greek Anthology* 8.177, 9.58). The dimensions of the walls, probably exaggerated, were taken by Pliny from the calculations of Herodotus, as well as the notice about the difference in the length of the foot. One of the unusual features of the city was that the Euphrates flowed through it. The most visible remnant of the walls is the Ishtar Gate, now in Berlin. The Temple of Jupiter Belus (Zeus Baal) occupied the center of one of the halves of the city. The Babylonian role in astronomy was well known to the Greco-Roman world, and had affected Greek thought since the beginnings of Greek intellectualism.

6.122. Seleucia-on-the-Tigris (at modern Tell Umar) was founded by Seleucus I about 60 km. north of Babylon, probably around 300 BC (Cohen, *East* 157–73). There were three major ethnic groups living in the city: Macedonians, Greeks, and Syrians (Josephus, *Jewish Antiquities* 18.372). It was intended to replace Babylon as the major urban center of the region, and to create a Seleucid presence in Mesopotamia; as such, it was considered one of the great cities of the Hellenistic world (Sallust, *Histories* 4.60.19). Pliny's population figure supports this (although perhaps not including everyone who lived in the city; Rostovtzeff, *Social and Economic History* 498). Its prosperity was in part due to astute placement on several trade routes and in an especially fertile location.

Ctesiphon, across the Tigris from Seleucia, was not a Parthian foundation – it had existed from the late third century BC (Polybius 5.45.4) – but in later years, as Parthian control spread over southern Mesopotamia, it became a royal residence, although not necessarily their capital (Strabo, *Geography* 16.1.16; Cohen, *East* 100–1). Vologesus I of Parthia (reigned AD 51–76/80) founded Vologescerta, probably a few miles downriver.

6.123. Five Babylonian towns are listed (the ethnyms are all urban ones). The only one that can confidently be identified is the town of the Orcheni, who were the inhabitants of Uruk (at modern Warka), the famous Sumerian city, located about 175 km. downriver from Babylon. The town underwent a renaissance in Hellenistic times (as Orchoi), and became a scholarly center (Cohen, *East* 174–7). Hipparenum was perhaps ancient Nippur (modern Nuffar), located between Babylon and Uruk. Babylonian documents support Pliny's statement that these cities were centers of learning, especially astronomy (Mathieu Ossendrijver, "Hellenistic Astronomy and the Babylonian Scribal Families," in *Hellenistic Astronomy: the Science in Its Contexts* [ed. Alan C. Bowen and Francesca Rochberg, Leiden 2020] 427–9).

6.124. Pliny returned to Nearchus (F1 h) and Onesicritus (F30), as well as Juba (F29), for a series of distances on the lower Euphrates. Because of the extensive siltation in the region, it is not possible to equate these with the modern landscape. Such changes even in antiquity are reflected in the differences in the distance from Babylon to the coast at the time of Alexander (412 miles or 610 km.) and that from Juba, 300 years later (adding 50 miles to his figure since that was the distance Charax was from the coast at that time: see *NH* 6.139), 225.5 miles or 334 km. Juba spent time at Charax (see further, *NH* 6.139), which explains the frequent mention of this trading center in these chapters (Roller, *World of Juba* 233–4).

Overland measurements were surprisingly inconsistent in antiquity before the Roman mile established a universal constant. Even then it is not always possible to tell whether distances expressed in miles were converted from other units, and what the perceived length of those units was. The *schoenus*, parasang, and stadion were all used in the Greek world, and in Greek literature, and remained in use in Roman times, but the relationship between them varied. The *schoenus* ("rope"), Egyptian in origin, could be anywhere from 28 to 60 stadia (perhaps 6 to 24 km.: Strabo noted how its length changed when one entered the Thebais in Egypt [*Geography* 11.11.5, 17.1.24]). The parasang (*parasangos*) was a Persian unit, and became known to Greeks as early as the fifth century BC (Herodotus 2.6), most famously used by Xenophon in the *Anabasis*. It was normally 30 stadia (6 km.). But the length of both the *schoenus* and the parasang could not easily be converted into stadia, because the stadion itself varied. This is one of the rare citations of parasang in Latin, although Pliny may have used it more than is apparent, unrecoverable because of the layered conversions to Roman miles.

6.125. The context is the lower Euphrates. A lack of chronological grounding means that the relationship of Charax to the head of the Persian Gulf is uncertain (see further, *NH* 6.138). The lower river has always been a region infested by brigands and robbers, but the Attali are mentioned only here. The Scenitae ("Tent Dwellers") and Nomads are peoples of the desert given generic Greek names (*NH* 5.87); the account has returned to the east–west course of the Euphrates in northern Mesopotamia.

6.126. Various measurements are provided, without attribution. The figures are complex: 1,125 miles (1,665 km., from 9,000 stadia) for the length of the Euphrates is clear enough, and 320 miles (474 km.) on the Tigris from the Red Sea (here meaning the Persian Gulf) is probably to Seleucia on that river (Pliny ignored the small distance of the crossover). Presumably the 724 miles (1,072 km.) to Zeugma is from the unspecified starting point on the Euphrates. The 175 miles (259 km., from 1,400 stadia) from Zeugma to Syrian Seleucia is to the town at the mouth of the Orontes, on the Mediterranean, following a well-established route. The 944 miles (1,397 km.) for the width of the Parthian empire is presumably from Zeugma to some unspecified point.

At the end of the section Pliny began a discussion of the Tigris, through *NH* 6.141.

6.127. The source of the Tigris was said to be in Lesser Armenia. Its official modern length is 1,850 km., flowing roughly southeast, coming within 25 km. of the Euphrates around Baghdad. On its final stretch before the Persian Gulf there is a complex network of channels and canals, some of which go back to earliest times. Today the Tigris and Euphrates join around modern Basra.

The sources that Pliny used for the Tigris included the map of Agrippa (*NH* 6.136) and the emperor Claudius (*NH* 6.128). For the lower river there were also Isidorus of Charax and Juba (*NH* 6.141). The meaning of the word "tigris," said to be Median for "arrow," was generally accepted (Eratosthenes, *Geography* F89; Strabo, *Géographie* 8 [ed. François Lasserre, Paris 2003] 125). Lake Aretissa, or Aresene (modern Lake Erçek in Turkey), is not part of the Tigris system, but is close to the upper river and was thought to be connected to it, especially since it was believed that parts of the river ran underground (*NH* 6.128).

6.128. As with the Nile and other rivers, parts of the Tigris were said to be underground, a belief due to the difficulty of identifying its many tributaries in this remote mountainous region. The emperor Claudius had a deep interest in hydraulics (such as his draining of the Fucinus Lake in Italy and the construction of the harbor at Ostia), but the context of this passage cannot be specified beyond one of the Roman expeditions in the region (Claudius F6).

Pliny reverted to the Euphrates, with an item that may have been left out of his discussion of the upper parts of that river (*NH* 5.83–5). The Arsanias River (the modern Murat) is one of its major eastern tributaries. Rivers that maintained their individual characteristics after merging with others were another interest of Greek and Roman hydrological studies.

6.129. The route of the Tigris is followed from Armenia to Mesene (the southernmost part of the plain of the Tigris and Euphrates), but in a greatly condensed fashion, giving the impression that the Gordyaean Mountains (part of the Taurus system) and Mesene were adjacent, when in fact they are hundreds of kilometers apart. The Cauchian Plains are otherwise unknown, but it is unlikely that any channel of the river would head in a northerly direction. The Pasitigris ("Universal Tigris") was the name applied to the lower river, where many of its channels come together.

6.130. The description is centered on the lower Tigris. The Choaspes River is one of its main eastern tributaries. A width of 62 miles (92 km.) may represent the river at flood stage. Siltation and hydraulic works over many centuries make it impossible to reconcile Pliny's description with the current state of the region, but at some time the Orcheni (see *NH* 6.123) diverted the Euphrates for irrigation, one of many reports regarding the relationship between it and the Tigris over their lowest course (Strabo, *Geography* 15.3.4).

6.131. The region under discussion is north of the left bank of the Tigris. Parapotamia is a common Greek toponym for territory along a river (Strabo, *Geography* 16.2.11). Mt. Zagrus, which retains its ancient name, is the mountainous region northeast of Mesopotamia. Chalonitis is nearly equidistant between the Caspian Sea and Syria, but the 380 miles (562 km.) reported to Persis is only slightly more than half the actual distance.

6.132. Sittacene is the region north of Seleucia-on-the-Tigris. Pliny's implication is that it was an area of Greek habitation, and in fact Sittace, the major town, although possibly existing in early times, was used by Xerxes to settle people transported from Central Greece (Diodorus 17.110.4; Cohen, *East* 104–5). They retained their cultural identity into the first century BC. The Greek local presence was supplemented by the founding of an Apamea by Antiochus I (ruled 281–261 BC), which became topographically notable because of its relationship to the Tigris and Archous Rivers (Cohen, *East* 125–8).

6.133. Susiane lies east of the Tigris and north of Mesene, at the edge of the plains. The city of Susa (near modern Shustar) was actually an ancient

Elamite center, but had been destroyed by the Assyrians in the seventh century BC. Darius I of Persia reestablished it because it created a major Persian city between Persis and Babylonia, and also because this was where the conspiracy occurred that put him on the throne in 522 BC (Herodotus 3.70).

Since Barbitace – which seems to have a Greek name – is not otherwise mentioned, the attitude of the locals about gold cannot be confirmed, but it is one of Pliny's several statements about the negative effects of luxury (see, for example, *NH* 6.88–9). The Oxii and the Mizaei are not mentioned elsewhere but presumably were in the mountains northeast of Susa.

6.134. The toponyms and ethnyms in this section are in the vicinity of Susa. For the Mardi, see *NH* 6.47; the Saitae, presumably in the same region, are otherwise unknown. Elymais is the Greek name for the ancient territory of Elam, extending from Susa to the east into the mountains and toward Persis (Strabo, *Geography* 16.1.17). The 250 miles (370 km., from 2,000 stadia) from Susa to the Persian Gulf might have been reasonable in Roman times by an indirect river route, but in Alexander's day the city was much closer to the gulf. Pliny's sources for much of this material are Nearchus and Onesicritus, who sailed up the Pasitigris in early 324 BC and eventually reached Susa (*NH* 6.100).

The Cossiaei were in the mountains north of Susa and may have been descendants of the Bronze Age Kassites (Strabo, *Geography* 16.1.18). Pliny's source extended the toponym Caucasus into this region, and noted a trade route to Bactria, which could hardly have been an easy journey.

6.135. The Eulaeus River (see *NH* 6.100) was yet another example of a river that was thought to run underground, like the Nile and the Tigris. It is probably the system of either the modern Karkeh or Karun, mostly in Iran, and one of the longer river systems in the Median territory. Susa was refounded as Seleucia-on-the-Eulaeus, probably a military settlement, sometime in the third century BC; its Temple of Diana (Artemis) shows the persistence of Greek culture, which is documented as late as the first century AD (Cohen, *East* 194–9). The rivers named are otherwise unknown, but presumably are northern tributaries of the Tigris.

6.136. Elymais extends east of Susa, generally in the mountains, toward Persis, although the territory included the coastal marshlands southeast of Susa and to the coast. For the rate of siltation in this region, see *NH* 6.139–40. The narrative moves southwest of Susa to the important center of Charax, the southernmost major town in the Tigris and Euphrates valley. Before considering it, however, Pliny provided data from the map of

Agrippa (F30–1), which he used more widely than implied, although it has not been mentioned since India (*NH* 6.57).

6.137. Using the map, Pliny gave an overview of the territory covered since the Indus River, defining its perimeter and extent. The distance of 1,320 miles (1,954 km.) from the Indus to the Tigris is astonishingly accurate, and may have come from the logs of Nearchus and Onesicritus. The north–south figure, 840 miles (1,243 km.), is much less so, and is a route over rugged country that would be difficult to measure.

Pliny, and thus the map of Agrippa, defined Mesopotamia in its broader sense, rather than limiting it to merely the stretch between the two rivers in the north along the route followed by Alexander (*NH* 6.117). The context uses essentially the modern definition, extending from the Taurus to the Persian Gulf (a distance of 800 miles, or 1,184 km., probably 64,000 stadia, a reasonable figure). The east–west calculation (360 miles, or 533 km.) conforms to where the rivers are their greatest distance apart, in the north where the toponym was originally applied.

6.138. Pliny provided a lengthy discussion of the town of Charax (through *NH* 6.141), because of one his major sources, Isidorus, was from there, and another, Juba, was a colleague of Isidorus' and spent time in the city (John Hansman, "Charax and the Karkheh," *IrAnt* 7 [1967] 21–58). It was a trading emporium on the lower Tigris. Pliny reported most of what is known about the city, which was perhaps at modern Jebal Khayabir in Iraq. It may have been founded as an Alexandria by Alexander the Great, but the evidence is dubious despite Pliny's explicit statement (Cohen, *East* 123–4). Charax was located on a man-made mound near the confluence of the Eulaeus and the Tigris, and settled by disabled veterans and refugees, with a district named after Macedonian Pella.

6.139. By slightly over a century later, Alexander's city had been destroyed by floods, and a King Antiochus founded an Antiochia on the site (Cohen, *East* 109–17). Despite Pliny, this is unlikely to have been Antiochus V (reigned 164–162 BC), who ruled for less than two years and was a child; it is more probably the vigorous Antiochus IV (ruled 175–164 BC), who was active in the region.

This town was also eliminated by flooding, and rebuilt by a certain Spaosines, or Hyspaosines, who was a local ruler in nearby Arabia and a Seleucid satrap (Juba [F30–3] need not have been in error on this). It is not certain when Spaosines' construction took place, or which Antiochus he served, but by the 120s BC coins were minted at the new town, now called Spaosinou Charax, or "The Fortress of Spaosines" (Head, *Historia Numorum* 823). The name Charax indicates that the new city was fortified in a way to protect it from the floods, through the building of levees, perhaps as much as six miles (9 km.) in extent. At some time during its early history, Charax was either on the coast or only 10 stadia (2 km.) from it, but when Juba was there siltation had made it 50 miles (74 km, from 400 stadia) away. It too became a major trading center; by the first century AD Roman merchants were visiting it regularly (*nostri negotiatores*, *NH* 6.140).

The Porticus Vipsania in Rome was where the map of Agrippa was displayed; its location is not known but was probably near the Quirinal (Richardson 319–20).

6.140. The name Spaosinou was essentially forgotten, and Charax flourished as a trading post, but became increasingly distant from the coast. In the 75 years between Juba and Pliny, siltation had moved the shore an additional 70 miles (104 km.).

Since the Persian Gulf was part of the External Ocean, it was tidal; in fact, Seleucos of Seleucia, of the second century BC, who wrote the first treatise on the tides, was from this region (Hipparchus, *Against the Geography of Eratosthenes* F4; Duane W. Roller, "Seleukos of Seleukeia," *AntCl* 74 [2005] 111–18). Tidal activity was so limited in the Mediterranean that when Greeks and Romans encountered the External Ocean, they found the tides enigmatic.

6.141. The manuscripts of Pliny read "Dionysium" for the geographical scholar from Charax, who is otherwise unknown. Given Pliny's effusive statement, an emendation to "Isidorum" has been almost universally accepted, since Isidorus of Charax was a well-known geographer (*FGrHist* #781, T1; Fraser, *Cities* 88–93).

Gaius Caesar, Augustus' grandson and adopted son, was sent on a broad commission to the East, which set forth in late 2 BC or early 1 BC, and lasted until his death at Limyra in Lycia in early AD 4 (Roller, *World of Juba* 212–26). The exact motives for the expedition remain unclear; it may ultimately have been designed to assert a Roman presence near Parthian territory (Pierre Schneider, "*Fauces Rubri Maris*: the Greco-Roman Bab al-Mandab (5th Century B.C. – 2nd Century A.D.)," *OT* 12 [2014] 202–3). Among his advisors were Isidorus (who may not have accompanied the expedition but merely made a preliminary reconnaissance for Gaius) and Juba, who did take part in at least the Arabian portion. Portions of his *On Arabia* were a report on his travels. It survives in 19 fragments, 12 of which are from the *Natural History* (Roller, *Scholarly Kings* 107–66), including the account that follows (F1, 30–3), through *NH* 6.159.

6.142. Embarking on his summary of the topographical material from Juba's *On Arabia*, Pliny discussed the extent of Arabia, defining it as beginning at the southern edge of the Taurus mountains and positioned between Syria and the Euphrates, noting that Tigranes II of Armenia (ruled 95–55 BC) settled people in this region, and as far as Egypt. The four ethnic groups are not mentioned elsewhere, but may have been obtained from the map of Agrippa.

6.143. The comparison of Arabia to Italy may also be from the map, which Juba probably saw when he visited Rome after becoming king. Arabia is not at the same latitude as Italy (all of the former is farther south), which may demonstrate something about the layout of the map. Pliny's mention of the fortunate (*felix*) siting of Arabia is a reference to the common toponym, Arabia Felix, which actually refers not to its location but the wealth of its products (see also *NH* 5.65).

The rest of the section is not from Juba. The comments are Pliny's own editorial remarks, situating Arabia within the context of the geography of the *Natural History*, and referring back to his previous citations of the region (*NH* 5.65, 86; 6.125). This related Arabia to Syria and the Euphrates, which had already been discussed. For Palmyra, see *NH* 5.88. Tents of goat hair have been used since earliest antiquity and remain common in this region today.

6.144. This is one of the earliest extant descriptions of the famous Nabataean city of Petra (see also Strabo, *Geography* 16.4.21, which may be slightly earlier than Juba's account). Greeks had known about the city since the late fourth century BC (Diodorus 19.95–8), but this was before it became the trading center for which it was to be famous. Juba's description is clearly autoptic.

The section is oriented on Gaza, the Mediterranean port that was the outlet for Arabian trade, and also the starting point of the land portion of Gaius' expedition. The distances seem to have been reversed: 600 miles (888 km.) would be appropriate from Petra to the Persian Gulf, and 135 miles (200 km.) that from Gaza to Petra. Petra was at a major crossroads, on the route from Gaza that continued to the east (the route which Gaius and Juba took), and that from Palmyra south to the Red Sea.

6.145. The itinerary heads east from Petra to Charax, reflecting Juba's route. The Omani seem misplaced, since they should be in southeastern Arabia, where the ethnym survives (*NH* 6.149). Although constructions attributed to Semiramis, the Assyrian queen, were noted regularly in western Asia (*NH* 6.8), mention of her at this point may be Juba's reference to his little-known treatise *On Assyria* (F4, 46, 59; Tatian, *Oration to the Greeks* 4; *NH* 8.115; Roller, *Scholarly Kings* 46–7).

There were two main routes to Charax. One came from the west to Forat (otherwise unknown) and then went 12 miles (18 km.) on the Tigris, and the other came from Parthia (probably Persis) to a seaport on the Persian Gulf named Teredon (Strabo, *Geography* 16.3.2). Evidently in Juba's day Charax had a certain autonomy, with a local dynast styling himself "king" who ruled territory beyond the city (see also *NH* 6.146). For the Chaldaei, see *NH* 6.121.

6.146. The toponyms cannot be identified. To speak of anywhere as a 10-day sail from landlocked Petra is incomprehensible, and there may have been too much condensation of Juba's material. Barbatia and Dumatha are otherwise unknown, but the implication is that they were points on the Tigris where traders who had come from Petra could sail to Charax; the use of days rather than distances indicates a source from the camel caravans.

The Apamea in question is probably the one cited at *NH* 6.129 and 132, located where the Tigris split and flowed around the city, which happens today at modern Kut al-Imara (Cohen, *East* 125–8). For the "king" of the Characenes see *NH* 6.145. The region was subject to invasion from downstream (see also *NH* 6.152), perhaps from pirates as much as Parthians, but the flood-control weirs inhibited these activities.

6.147. Juba's *periplous* of the Arabian coast begins at Charax and moves down and around it to the mouth of the Red Sea. Not all the information is from Juba; Pliny may have collated toponyms from various sources – including oral testimony by seamen – and the names may not always appear in geographical order. Attempts to reconcile Pliny's account with the actual topography have been intense and contradictory since the nineteenth century (S. B. Miles, "Pliny's Geography of the East Coast of Arabia," *JAS* 10 [1878] 157–72; Daniel T. Potts, *The Arabian Gulf in Antiquity* 2 [Oxford 1990] 302–13). The initial parts of the coast had already been explored by Androsthenes of Thasos (*FGrHist* #124, F2), who was commissioned by Alexander the Great in the winter of 325/4 or 324/3 BC; some of his report was preserved by Eratosthenes (*Geography* F94). Antiochus IV (Epiphanes) sent out explorers in the second century BC, although Pliny was incorrect in considering this the first such Seleucid expedition: Antiochus III had also been in the region (Polybius 13.9), It is unlikely that Juba travelled these coasts himself, but made use of the earlier accounts as well as trade and mercantile information obtained at Charax. The first segment is from Charax to the trading center of Gerrha, located at least 150 miles (222 km., from 1,200 stadia) down the coast, which had an interesting salt-based architecture (Strabo, *Geography* 16.3.3).

6.148. The *periplous* continues past the island of Tylos or Tyros (modern Bahrain) and its companion Arados (unnamed by Pliny, at modern Muharraq; Strabo, *Geography* 16.3.4). Tylos was described in greater detail at *NH* 12.39. It is only a short distance from the coast, and the quoted figure of 50 miles (74 km.) is probably a sailing route from an unidentified point. The large islands beyond include Tabora (modern Qatar), which is actually a peninsula, but confusion between peninsulas and islands was a common feature of coastal sailing into modern times.

6.149–50. The *periplous* continues down the Arabian coast of the Persian Gulf, with a vast number of toponyms and a few ethnyms, almost none of which can be identified.

Part of the text is missing near the end of *NH* 6.149. The Omani are mentioned again (see *NH* 6.145), but this is essentially their home region. Some of Juba's information has been supplemented by mercantile details from a period close to Pliny's era, although the two emporia that are said to be the busiest in the Persian Gulf, Homna and Attana, are not mentioned elsewhere. As is common along the coast of the Indian Ocean and its bays, many of the peoples are characterized as Ichthyophagi ("Fish Eaters"; see also *NH* 6.95).

As a linguist, whose *Resemblances* discussed the relationship between Latin and Greek (Roller, *Scholarly Kings* 19–31), Juba would have been intrigued in hearing about the strange writing that was reported on the unnamed island (see also *NH* 6.174). It presumably was a local script. Hieroglyphics have been suggested (Tusculum 6, 238), but Juba would not have found them unfamiliar since his wife Cleopatra Selene had imported examples of them to their capital of Mauretanian Caesarea, such as on the statue of Petubastes IV, now in the Cherchel Museum (S75).

6.151. By this point Juba and Pliny have reached the southern coast of Arabia, although the sequence of toponyms in this and the following section is somewhat confused. The Chelonitis ("Tortoise") Islands perhaps reflected the common tortoise harvesting near the mouth of the Red Sea. Eventually the account reaches Acila, or Ocelis, at the southwest corner of the Arabian peninsula, one of the points where those sailing from the Red Sea to India left the coast and headed east (*NH* 6.104).

6.152. After the entrance to the Red Sea, the *periplous* returns to south-eastern Arabia, the home of the Macai (*NH* 6.98), who lived on the Arabian side of the Strait of Hormuz, the entrance to the Persian Gulf, 50 miles (74 km., from 400 stadia) across from Carmania.

The story of Numenius is otherwise unknown; the Antiochus in question is probably Antiochus III (ruled 222–187 BC) or Antiochus IV (ruled (175–164 BC), although this is far from certain. Presumably Numenius, as satrap of Mesene, put down either a local revolt or a Parthian invasion, but the story is of interest because it offers proof of a Seleucid naval force in the Persian Gulf, as well as the role of the tides (see also *NH* 6.140, 146; Potts, *Arabian Gulf* 2, 12–14). The Seleucid expedition established a presence in Arabia, and the city of Gerrha honored Antiochus in a decree and gave him lavish presents (Polybius 13.9; Rostovtzeff, *Social and Economic History* 458).

6.153. The island of Ogyris was beyond the mouth of the Persian Gulf and had been visited by Nearchus (F27; Eratosthenes, *Geography* F94; Strabo, *Geography* 16.3.5). The location reported in the text, 125 miles (185 km.) or 1,000 stadia from the mainland, is a suspiciously round number and does not conform to any known island; its large size (112.5 miles, or 166 km.), another round number (900 stadia), is also dubious. King Erythras, the eponym of the Erythraian Sea, was a Persian or Egyptian who had settled on the island and was said to have been one of the inventors of seamanship (Strabo, *Geography* 16.4.20; Pomponius Mela 3.79; *NH* 7.206). The distance to Dioscuridu (modern Socotra), 280 miles (414 km.), is measured from Cape Syagrum (*NH* 6.100), on the south coast of Arabia; this may be the first stage of the voyage to India.

Returning to the Arabian mainland, the account has reached its southwestern portion, at the edge of the frankincense-producing district. The Gebbanitae controlled the export of the aromatics (*NH* 12.63–68). Thomna (the Thomala of *NH* 6.154) was a collection point for the incense, which was then sent to Gaza, a route of 1,487½ miles (2,201 km.), in caravans that took 65 days.

6.154. Although most of the account is oriented on southwest Arabia, the Trogodytae were on the east coast of the Red Sea, between it and the Nile (Strabo, *Geography* 16.4.17). The other nine ethnic groups were all in the Arabian peninsula: the Chatramotitae were producers of myrrh (Strabo, *Geography* 16.4.4). The Sabaei were the best-known population in the region, as Pliny noted. They lived in Saba, the Sheba of the Bible (1 Kings 10, etc.), and were one of the major producers of frankincense and myrrh. The aromatic industry was described in detail by Juba (*NH* 12.51–72). The other ethnic groups and the eight towns mentioned would be in the same region. Variants of the toponyms indicate multiple sources: Thomala is the Thomna of *NH* 6.153, and Sabbatha is repeated as Sabota at *NH* 6.155. Juba was certainly a major one, as well as, perhaps,

Agatharchides of Cnidus, whose *On the Erythraian* [*Red*] *Sea* was written in the second century BC (Agatharchides, ed. Burstein, 12–18). And, as always, there was an extensive repertory of oral information from merchants, traders, camel drivers, and seamen, available at centers such as Gaza, Petra, and Charax.

6.155. Other localities in the aromatic-producing region were noted. The Atramitai were part of the Sabaeans, living around Sabota, or Sabbatha, modern Shabwa in Yemen. Sources are consistent in reporting its political and cultural importance (Strabo, *Geography* 16.4.2–3; *Periplous of the Erythraian Sea* 27; *NH* 12.52). Pliny's 60 temples may not have been an exaggeration, and the site has notable remains (Casson, *Periplus* 161–2). This district was repeatedly mentioned as an important center of the industry, where the aromatics were collected for transport. Marrebbata, or Mariaba (modern Marib), was another major center (Strabo, *Geography* 16.4.19). Pliny provided no units for his figure of 94; it is most likely stadia (about 19 km.), but even so the inland location of Marrebbata and the lack of bays on the Arabian coast create a problem. Another possibility is that the figure is miles (139 km., from 750 stadia), and it is a misplaced but accurate reference to the depth of the Gulf of Aqaba, discussed in *NH* 6.156.

The narrative turns to localities on the east coast of the Red Sea. The Aelamitae lived at Aela, or Aelana (modern Aqaba at the head of the gulf of the same name). This may have been as far as Gaius Caesar went, since he only glimpsed (*prospexit*) Arabia (*NH* 6.160). The toponyms and ethnyms are probably near the Aelamitae. There were a number of places that did not allow dogs; the most famous was the island of Delos in the Aegean (Strabo, *Geography* 10.5.5).

6.156. All the toponyms except Agra are variants of the place known most commonly as Aelana. The large number of different forms – which seems to have intrigued Pliny – may be due to the antiquity and cultural diversity of the site. There are four sources cited: in addition to Pliny himself and Juba, there are Artemidorus of Ephesus, from the late second century BC, and a fourth unnamed.

At this point the circumference of Arabia, from Charax to Aelana, has been completed, and the material derived from Juba comes temporarily to an end. Pliny concluded the *periplous* with the usual measurements, contrasting Juba's total of 4,000 miles (5,920 km., from 32,000 stadia) with another figure (4,765 miles, or 7,052 km.) from an unnamed source, perhaps the map of Agrippa. Heroum is probably Heroonpolis, east of the Egyptian delta, or a point at the head of the Heroonpolitic Gulf (modern

Gulf of Suez). Arabia is not its widest at the north (although it depends on the definition of the territory), but nevertheless the southern width of the peninsula is somewhat more than the northern.

6.157. Through *NH* 6.159 there is a discussion of the interior of Arabia. No sources are named beyond *veteres* ("old") and *nunc* ("now"), probably the Hellenistic geographers and then mercantile and trading sources from close to Pliny's day.

The toponyms and ethnyms cover a wide swath of western Arabia, although most are otherwise unknown. The Nabataeans (*NH* 6.144) were centered at Petra, in the north, and the Minaei and the town of Maribba (*NH* 6.155) were in the southwest, in the aromatic territory. But the Avalitae were probably around Avalitu Bay, on the west coast of the Red Sea (*NH* 6.174). There is no other evidence for any connection between King Minos of Crete and the Minaei beyond a perceived relationship between the names.

6.158. There are additional toponyms and ethnyms from unidentified parts of interior Arabia. Like Minos and the Minaei (*NH* 6.157), the only connection between the Rhadamaei and Minos' brother Rhadamanthus is the similarity of names. The Homeritae (Himyarites) lived nine days inland from the southwest corner of Arabia and were led by King Charibael, who also ruled the Sabaei (*NH* 6.154). Thus they were in the heart of the aromatic region, and probably the most powerful regional state, as well as the most populous. Charibael was a local person of importance in the mid-first century AD, and sent embassies and gifts to at least two Roman emperors (*Periplous of the Erythraian Sea* 23; Casson, *Periplus* 149–51).

The inhabitants of Mesala (or Mesalum) controlled the export of a unique variety of myrrh, which was white in color (*NH* 12.69). Riphearina seems to be a combination of two local words, "refa" and "arama," or "granary" (Tusculum 6, 232–3). For the Autaei, see *NH* 6.167. Aenuscabales may be related to the local word for "camel"; the location of springs would be essential knowledge for the camel caravans.

6.159. For the most part, the toponyms and ethnyms cannot be identified. Ampelome (or Ampelone), "Vine City," was said to have been a Milesian settlement; it was more likely primarily a Ptolemaic endeavor, from the early third century BC, and was probably one of the Red Sea seaports (Cohen, *Syria* 307). But Miletus was under Ptolemaic control during much of the century and thus could have been involved in the foundation (P. M. Fraser, *Ptolemaic Alexandria* [Oxford 1972], vol. 2, 301–2). Mariba may be the same as Mariaba (*NH* 6.157).

There is another example of an underground river (see *NH* 6.128, 135), although any connection from the Euphrates to western Arabia is more problematic than most of the assumed examples of the phenomenon. There are no significant rivers on the west coast of Arabia.

Ampelome is one of several towns listed with Greek names situated on the Arabian side of the Red Sea. Unless these are local toponyms that sounded Greek, they were probably all Ptolemaic foundations. Except for Ampelome, which is an unusual toponym, they have names of actual Greek towns, and probably included settlers from those localities. None can be located today.

6.160–1. Noting that Gaius Caesar had only a glimpse of Arabia, from Ailana (Roller, *World of Juba* 224–5), Pliny reported on the expedition of L. Aelius Gallus, prefect of Egypt, who made an Arabian incursion in 26–25 BC. He was sent by Augustus, probably to learn about more direct access to the aromatic regions and to insure the cooperation of the locals. Strabo was on Gallus' staff (although he did not join the expedition), and provided a detailed report (Strabo, *Geography* 16.4.22–4). His account was more topographically and militarily oriented than Pliny's. The expedition did not do well, due to Gallus' inexperience and a failure to understand the rigors of the Arabian landscape, and thus there were a large number of Roman casualties (Roller, *Historical and Topographical Guide* 935–9).

Many of the places and peoples mentioned by Pliny have been cited previously, such as Mariaba (*NH* 6.155), the Homeritae (*NH* 6.158), the Minaei (*NH* 6.157), Agraei (*NH* 6.154), Chatramotitae (*NH* 6.154), and Sabaei (*NH* 6.154). These show that Gallus did reach the aromatic districts, and he was able to make ethnographic notes about the regions he passed through. The local gold resources were also investigated.

Although Gallus obtained a vast amount of information, Augustus was not pleased with the difficulties that the expedition encountered, and in the following year (24 BC) Gallus was recalled from Egypt and retired from politics and the military. He devoted himself to scholarship, writing on pharmacology and aromatics, using information that he had learned while on the expedition (John Scarborough, "L. Aelius Gallus," *EANS* 34–5).

6.162. Some of this material may have come from the expeditions of either Gallus or Gaius Caesar. Trading and raiding have always been closely intertwined, and perceptions of which is which depend on the perspective of those affected. The wealth of the Arabians had long been proverbial in the Greek world, from at least the fifth century BC (Euripides, *Bacchants* 16).

The Red Sea

6.163. The narrative moves to the west coast of the Red Sea, using Timosthenes of Rhodes (F13 Roller) and Artemidorus of Ephesus (F103) as primary sources, and perhaps, to some extent, Juba (see *NH* 6.170). Understanding this coast was a Ptolemaic priority, since it was both part of the route to India and to the elephant-hunting grounds of East Africa; elephants were an essential part of the Ptolemaic war machine. Timosthenes synthesized the reports of Ptolemaic explorers into the region, and his material was handed down to Artemidorus and Juba.

Yet Timosthenes' figures, as transmitted, are confused: the length of 1,750 miles (2,590 km, from 14,000 stadia), reported by Artemidorus, is reasonable, but this could hardly have been covered in 4 days (the original number may have been 14). The width of 2 days is accurate. The straits are the Bab el-Mandeb at the south end of the sea; a crossing of 7.5 miles (11 km., from 60 stadia) depends on the actual place of measurement, given the many islands in the region. Eratosthenes' figure (*Geography* F97) of 1,200 miles (1,776 km., from 9,600 stadia) may represent a perceived direct length rather than one derived from coastal sailing.

6.164. Ptolomais (Ptolemais) is probably Ptolemais Epi Theras ("Ptolemais For the Hunt"), the embarkation point for the elephant hunts (Strabo, *Geography* 16.4.7; Cohen, *Syria* 341–3). Since it has not been located, the accuracy of Artemidorus' figure (1,187.5 miles, or 1,758 km., from 9,500 stadia) cannot be determined. The 1,732 miles (2,563 km.) of Agrippa (F55) for the length of the sea was probably taken from Artemidorus (who had 1,750 miles: see *NH* 6.163), with a slight difference in rounding. Various unnamed sources (presumably those previously cited) have a width of 475 miles (703 km., from 3,800 stadia), which is far too wide (it is actually about 300 km.), but may be a sailing distance between two ports on opposite sides. As at *NH* 6.163, the width of the Bab el-Mandeb at the outlet of the sea depends on consideration of the many islands in the straits.

6.165–7. The narrative returns to the Laeanitic Bay (modern Gulf of Aqaba) at Aelana, connecting with the account of coastal Arabia that ended at *NH* 6.156. Although Juba was not mentioned until *NH* 6.170, he probably continued to be Pliny's primary source (Juba F34).

The places and peoples are around the Aea or Heroonopolis Bay, the modern Gulf of Suez. The various populations are otherwise unknown. The initial attempts to build a canal from the Mediterranean to the Red Sea may have begun as early as the Bronze Age, although attribution to the mythical King Sesostris makes this difficult to evaluate (Herodotus 2.137; Diodorus 1.57). The first certain attempt was by King Necho II (ruled 610–595 BC), which was allegedly completed by Darius I of Persia (Herodotus 2.158; Strabo, *Geography* 17.1.25). In fact, construction continued intermittently into the Roman era, and there is physical evidence of several canals over various periods (Carol A. Redmount, "The Wadi Tumilat and the 'Canal of the Pharoahs'," *JNES* 54 [1995] 127–35). The Ptolemy in question is Ptolemy II, whose construction was in 270/69 BC (Diodorus 1.33.9–12; Pliny, *Histoire naturelle* 6.4 [ed. Jehan Desanges, Paris 2008] 46). Sources disagree as to whether he actually finished the project, and the fear of joining seas not naturally connected was persistent: similar reservations existed about cutting a canal through the Isthmos of Corinth (Eratosthenes, *Geography* F16; Strabo, *Geography* 1.3.11). The worry of polluting the fresh water of the Nile was perhaps a more valid concern, since it was an important Egyptian commodity, exported as far as Syria (Herodotus 3.6).

There were three land routes from the Mediterranean to the Red Sea. The first and most western was from Pelusium (*NH* 5.68), at the eastern edge of the Delta, more or less straight south to the Red Sea. It was difficult with minimal road construction, and regularly covered by blowing sand. The second went from Mt. Casius, about 50 km. east of Pelusium, and ran roughly southwest through the territory of the Autaei for 60 miles (89 km., from 480 stadia) until it joined the Pelusium route. There was no anchorage at Mt. Casius and ships had to be offloaded two stadia (400 m.) from the land, making this the least favorable route (Diodorus 20.74.3). The third went from Gerrhum (or Gerrha), just east of Pelusium. This was the shortest, but it went through the waterless uplands of the western Sinai.

All three ended at Arsinoë at the head of the western inlet of the upper Red Sea, the modern Gulf of Suez. Arsinoë was a foundation of Ptolemy II and named after his sister-wife; it was renamed Kleopatris by Cleopatra VII some time before 30 BC, which means that Pliny's itineraries represent the period between the mid-third century BC and 30 BC (Cohen, *Syria* 308–9). Ptolemy II sent out a number of expeditions to the upper Nile and western Red Sea coast; the results of some of these are discussed at *NH* 6.169–77.

6.168. The account continues south from Arsinoë along the west side of the Red Sea, essentially as far as the seaport of Berenice, the sailing point for India. Many of the toponyms and ethnyms cannot be identified (although they would be along the narrow coastal strip): the major exception is Myos Hormos (Mussel Anchorage, at modern Quseir al-Qadim), which served as the primary Indian port before being replaced by Berenice (Berenike), probably in the Augustan period. It was 370 km. farther south, and thus ships returning from India avoided adverse winds for that distance. Moreover, it was closer to the Nile. Both ports were foundations of Ptolemy II (Strabo, *Geography* 16.4.5; *NH* 6.103; Casson, *Periplus* 96–7). Berenice has been the site of major excavations for many years, which have greatly illuminated the only example of a trading center on the Red Sea that is archaeologically documented (Marek A. Woźniak *et al.*, "Ptolemaic Berenike: Resources, Logistics, and Daily Life in a Hellenistic Fortress on the Red Sea Coast of Egypt," *AJA* 125 [2021] 247–81). This is one of several cities called Berenice, named either after the first Berenice, the mother of Ptolemy II, or the second, the wife of Ptolemy III (Dee L. Clayman, *Berenice II and the Golden Age of Ptolemaic Egypt* [Oxford 2014] 133).

6.169. Trogodytice was between the Nile and the Red Sea (Agatharchides 5.62; Strabo, *Geography* 16.4.17). Most of this region was unknown to Greeks until the explorations commissioned by Ptolemy II in the third century BC. The prevalence of Greek toponyms demonstrates that the Ptolemaic explorers essentially wrote the map of the region.

Topazos Island was the source of *topazos*, or peridot, a green gem that the Ptolemaic explorers discovered and presented to Queen Berenice I, the wife of Ptolemy I and mother of Ptolemy II, probably in the early 270s BC (Juba F74–5; *NH* 35.39, 37.107–8). The island in question is modern Zabargad or St. John's, 50 km. southeast of Berenice, where ancient and modern mining workings are visible (James A. Harrell, "Discovery of the Red Sea Source of *Topazos* (Ancient Gem Peridot) on Zabargad Island, Egypt," in *Twelfth Annual Sinkankas Symposium: Peridot and Uncommon Green Gem Minerals* [ed. Lisbet Thoresen, Fallbrook, Calif. 2014] 16–30). The significance of this region to the Ptolemaic economy is demonstrated by the assignment of a royal prefect to it.

The bay full of islands is probably the Gulf of Malsain, or Foul Bay, near the south end of the sea (Pliny, *Histoire naturelle* 6.4, 64). There follows a brief mention of the land of the Candaei, the local ethnym for the people called in Greek the Ophiophagi ("Snake Eaters"). This may be a generic term for several local populations, because the *topazos* island was also called Ophiodes, or "Snaky" (Strabo, *Geography* 16.4.6).

6.170. Pliny made a rare statement about the quality of manuscript copying, indirectly indicating that he probably had copies of Juba's *On Arabia* and *Libyka* available to him. He respected Juba's scholarly diligence but did not hesitate to criticize him when necessary (*NH* 6.139). The problem of homonymic toponyms, especially in the Hellenistic period, was persistent, and there are at least six known places with the name Berenice, all from the early Ptolemaic period, including the three mentioned that are on the west coast of the Red Sea.

The island of Citis, or Cytis, was another locality that produced *topazos*, or peridot, presumably at or in the straits at the outlet of the Red Sea (*NH* 37.107); it may be a doublet for the *topazos* island mentioned previously, or perhaps modern Kod Ali island (Lisbet Thoresen and James A. Harrell, "Archaeogemmology of Peridot," in *Twelfth Annual Sinkankas Symposium: Peridot and Uncommon Green Gem Minerals* [ed. Lisbet Thoresen, Fallbrook, Calif. 2014] 39–40).

6.171. Digressing from Juba, Pliny discussed the astronomical realities of being at a low latitude, to some extent repeating and summarizing material from *NH* 2.183–5. The lower Red Sea lies between 10 and 15 degrees north latitude, and thus the lack of shadows around the time of the summer solstice was noticeable. The explorer Philon (*FGrHist* #670) was the first to record local data about the height of the sun and its effects, on an expedition up the Nile during the time of Ptolemy II (Eratosthenes, *Geography* F40; Strabo, *Geography* 2.1.20). As the Red Sea became better known, it was obvious that the same phenomena applied. Half a century after Philon, Eratosthenes, using his material (and that of others), and collating solar angles at different latitudes, was able to determine the circumference of the earth, publishing his conclusions in his *Measurement of the Earth* (translation and commentary in Roller, *Eratosthenes* 263–7). Pliny's eulogistic statement about Eratosthenes' achievement – "here the world was grasped" (*mundo ibi deprehenso*) – actually applied to Syene on the Nile, not the lower Red Sea, but does not diminish the incredible nature of Eratosthenes' feat.

6.172. The account continues beyond the confines of the Red Sea along the modern Somali coast and into the Azanian Sea (*NH* 6.153), the modern Gulf of Aden, which widens toward the Indian Ocean. Cape Hippalum, not located, is a reoccurrence of the name associated with the route to India. It may originally have been that of the seaman who discovered the Indian route (*Periplous of the Erythraian Sea* 57), but it was also applied to the wind that allowed such travel (*NH* 6.100, 104), and, in this case, a topographical feature. These need not be mutually exclusive: Hippalus the seaman may have had the wind and the place named after him. Sea tortoises were common in this region (*NH* 6.151). The town of the Adulitae is Adulis, discussed in greater detail at *NH* 6.173; the account has moved back into the Red Sea, perhaps due to Pliny's failure to blend sources.

6.173. Adulis was at the head of a deep bay, the modern Gulf of Zula in Eritrea (described in detail at *Periplous of the Erythraian Sea* 4; Casson, *Periplus* 102–6). In the middle of the first century AD it was a major trading post. There is a list of products that arrived at Adulis from the hinterland. Ivory had been known to the Greeks since the Bronze Age, even before they learned about the elephant itself (Homer, *Iliad* 4.141, etc., as well as countless material cultural examples). By Hellenistic times elephants were used both for military purposes and as a source of ivory (Casson, *Periplus* 108).

Although the rhinoceros had been vaguely known since the fourth century BC, perhaps due to reports from Alexander's expedition (Aristotle, *Research on Animals* 2.1.499b; *Parts of Animals* 3.2.663a), it was probably not seen in the Mediterranean world until Ptolemy II collected a number of exotic animals for display in Alexandria (Athenaios 5.201 c; Kitchell, *Animals* 161–3). Its horns came to be prized for folk medicine remedies and luxury oil flasks (Juvenal 7.130). The hippopotamus was noted for its thick hide, which was so tough that spear shafts could be made from it (Herodotus 2.71; Kitchell, *Animals* 87–8). Tortoise shells were a luxury product, used for furniture inlay (*NH* 9.35–9). *Sphingia* were a type of ape, labelled on the Nile mosaic from Praeneste. They seem to have been limited to the region inland from Adulis; their name is a diminutive of "sphinx," which they were said to resemble (Agatharchides F73; Diodorus 3.35.4; Pomponius Mela 3.88; *NH* 8.74; Kitchell, *Animals* 176). The final item on the list is slaves; those from this region were said to be of especially high quality (*Periplous of the Erythraian Sea* 13, 36).

The Aethiopian Aroteres ("Plowmen") would have been agriculturalists, presumably inland from Adulis. The unexplored bay is probably one of the inlets to the south of Adulis, perhaps the "deep bay" of the *Periplous of the Erythraian Sea* (5), modern Hauachil Bay (Casson, *Periplus* 109). As Pliny noted, it is difficult to believe that it was unexplored, unless shoals and other issues meant that penetration was not worth the effort. The Cucios Spring is probably on one of the promontories to the south, which would have been notable as a fresh water supply close to the shore. The *cuci* (Greek *kouki*) is the doum palm (*Hyphaene thebaica*), presumably prevalent around the spring (Theophrastus, *Research on Plants* 4.2.7; *Periplous of the Erythraian Sea* 33; *NH* 13.62).

6.174. The *periplous* continues to the end of the Red Sea and perhaps beyond, with references to a regional coastal trade network (Schneider, "*Fauces*" 229–30). Trogodytic myrrh was considered superior to the

Arabian variety (*NH* 12.69–70). The stele with unknown lettering may be the Pillars of Sesostris reported by Strabo, one of several monuments attributed to this mythical Egyptian king (Herodotus 2.102–4; Diodorus 1.55.7–9; Strabo, *Geography* 16.4.4), whose conquests were said to have ranged from Aethiopia to Thrace and India. In fact, since Baragaza is normally the name of a town in India, one of the main goals of the Indian trade (*Periplous of the Erythraian Sea* 14) – although there may have been a locality of the same name on the Aethiopian coast – it is probable that the comments about Sesotris and Baragaza refer to India, not Aethiopia.

Cinnamon does not come from northeast Africa, but from Taprobane, or even beyond, but it was handled by middlemen on the Somali coast, which came to be known as the Cinnamon-Bearer Territory (Eratosthenes, *Geography* F57; Strabo, *Geography* 2.5.35; Dalby, *Food* 87–8).

East Africa, the Upper Nile, and Aethiopia

6.175–6. Pliny returned to Juba for material about coastal East Africa and the upper Nile, this time taken from his *Libyka*, completed by 2 BC (F35–7). The statement that it was impossible to sail in this region because of the heat is presumably a sailors' tale. Pliny's summary begins at the Mossylitic Cape, on the modern Horn of Africa, perhaps Cape Guardafui, the easternmost point of the continent. Juba's belief that this was the beginning of the Atlantic Ocean represented his ideology that it was not far from Mauretania, a view supported by the alleged finding of Iberian shipwrecks on this coast, including some at the time of the expedition of Gaius Caesar (Strabo, *Geography* 2.3.4–5; *NH* 2.168–9).

Juba's distance, 1,875 miles (2,775 km., from 15,000 stadia), is from Lepta Acra ("Small Cape") or Drepanum ("Sickle"), the cape at the end of the bay where the most important Berenice was located (*NH* 6.103, modern Ras Banas), as far as the open sea, presumably at the Mossylitic Cape. It is about twice the actual distance, and may include a sailing route down the East African coast. The Ascitae Arabians were so named because of their use of hides (*askoi*). Some of the local ethnyms are Greek generic terms, such as the Therothoae ("Beast Hunters"), and more Ichthyophagi. There is a list of seven ethnyms, none of which can be identified, although the Zanganae suggest the district of Zanzibar, in a region known to the author of the *Periplous of the Erythraian Sea* (15–16), where a vigorous trading industry had been established.

6.177. The next topic is the upper Nile, essentially from Syene (at the First Cataract) to Meroë. The latter is at modern Bagrawiya in Sudan, and was the ancient center of the Meroitic kingdom, known to the Greeks since the fifth century BC (Herodotus 2.29). Juba's account is contrasted with that of Bion (see *NH* 6.178). Juba's emphasis is on an Arabian presence in this region – perhaps a point of view obtained when he was on the Arabian expedition of Gaius Caesar – and is based on the issue of the boundary between the continents of Africa and Asia, something debated since at least the fifth century BC (Herodotus 4.45). The Nile seemed the obvious line, since this avoided the matter of how to classify the region between the Egyptian Delta and Phoenicia, but the same people lived on both sides of the river, and thus it was difficult to see it as an actual division (Strabo, *Geography* 1.2.25). Yet the meaning of an Arabian presence – and thus an Asian one – around the Nile, and even west of it, continued to be debated in Pliny's day, although he avoided taking sides. The sentence about people living on the banks because of access to water – obvious and banal – is probably misplaced or an intrusion (Pliny, *Histoire naturelle* 6.4, 104–5).

6.178. Bion of Soloi wrote an *Aithiopika*, probably in the early third century BC (*FGrHist* #668, F4), based in part on autopsy (apparent with the cultic comment about Radata). Six fragments of the work survive, four of which are from these sections of the *Natural History* (Thibodeau, "Bion of Soloi," *EANS* 193). He provided a list of 26 towns and 2 ethnic groups on the east side of the Nile, between Syene and Meroë. Some of the toponyms can be equated with modern ones, but none with certainty.

6.179. Juba's list (F37) is more thorough than Bion's (39 toponyms and one ethnym), perhaps indicating greater knowledge of the region in his day and after nearly 300 years of Ptolemaic rule. A number of the toponyms are Greek, which Pliny transliterated directly, creating latinized Greek case endings. There are few correspondences between the two lists, and again identification of any of the names is difficult. There is only one editorial comment, that Corambis was a source of bitumen; petroleum seeps and lakes were a feature of the region (Pomponius Mela 3.88; *NH* 31.17).

6.180. The next list is from the left ("African") bank of the Nile, still between Syene and Meroë. It is not clear whether Juba or Bion was the source, and it may be Pliny's collation of both. There are 25 town names, of which few can be equated with any modern location. The reference to elephant country beginning at Zamnes suggests a toponym obtained from Juba, since elephant hunting was probably not well established in Bion's day. Elephants were encountered around the fifth cataract of the Nile, but could not be transported down the river because of the cataracts, and thus the Red Sea coast became the primary hunting ground, from which they could be sent by sea (Stanley M. Burstein, "The War Elephants East and West," *World History Connected* 17 [2020] 1–14).

6.181–2. Pliny then turned to information supplied by Roman expeditions above Syene. For the one commissioned by Nero, see *NH* 6.184–8. Publius Petronius was the successor to Aelius Gallus as prefect of Egypt, and was in office 24–21 BC, having previously served as Gallus' second-in-command. During Gallus' Arabian expedition, the Aethiopians attacked the Roman garrison at Syene. Petronius retaliated, and went 870 miles (1,288 km.) up the river beyond Syene, as far as Napata, a major Aethiopian center. This was farther south than any Roman army had previously gone (Augustus, *Res gestae* 26; Strabo, *Geography* 17.1.54).

The comment that the Roman army did not make a desert (*solitudinem*) in Aethiopia brings to mind Tacitus' famous words attributed to the Caledonian chieftain Calgacus (Tacitus, *Agricola* 30–5), written perhaps twenty years after the *Natural History*. Tacitus would have been familiar with Pliny's work, but the parallel diction may suggest that it was a formulaic statement about the predatory nature of the Roman army on campaign, which Pliny, the career military officer, was quick to deny.

The early fame of Aethiopia (here used in its more limited sense of the lands on the upper Nile beyond the First Cataract) was an essential part of Greek mythology. Mentioned frequently by Homer, Aethiopia was originally seen as a powerful kingdom at the edges of the earth, a role established in the opening lines of the *Odyssey* (1.22–7). Under their King Memnon, the Aethiopians came to Troy as allies of King Priam; their activities were the subject of the lost epic *Aithiopis*. Another Aethiopian king, Cepheus, was the father of Andromeda, as outlined in Euripides' *Andromeda*, and since her fate was associated with the region of Jope on the Levantine coast (*NH* 5.69), it was assumed that Aethiopian power once extended that far.

6.183. In an unusually detailed list of sources, Pliny cited five Ptolemaic explorers who went up the Nile. He may have had available a work about the river, such as the ones by Eudoros of Alexandria and by a certain Ariston (*FGrHist* #649, 650; Strabo, *Geography* 17.1.5).

The five explorers are hardly known beyond what Pliny reported. They are probably listed in chronological order, and all most likely fall in the century after the establishment of the Ptolemaic kingdom at the end of the fourth century BC. Dalion, presumably the earliest, went well south of Meroë, probably the first Greek to do so, and wrote an *Aithiopika* (*FGrHist* #666). Aristocreon (*FGrHist* #667) and Bion (*FGrHist* #668) also went beyond Meroë, probably somewhat later; Bion also wrote an *Aithiopika*

(F1). Basilis (*FGrHist* #718, F2) seems to have been primarily interested in India but also wrote on the upper Nile.

But the Ptolemaic envoy who learned the most about the upper river was Simonides (*FGrHist* #669, T1), although ironically he is only known through this entry by Pliny. Yet he lived in Meroë for five years, probably as the Ptolemaic representative. Timosthenes of Rhodes may not have explored the upper river, but as the naval chief of staff of Ptolemy II, he would have collated all the available information (F12 Roller). He noted that it was a 60-day journey from Syene to Meroë, a fact far more important to the Ptolemaic military command than the actual distance. Half a century after Timosthenes, presumably with more information and less of an interest in military movements, Eratosthenes recorded the distance as 5,000 stadia (625 miles or 925 km.). Artemidorus' slight difference (F94), 4,800 stadia (600 miles or 888 km.), may be a matter of conversion. Both Eratosthenes and Artemidorus gave the full distance from Alexandria to Meroë as 10,000 stadia (1,250 miles or 2,000 km.). Sebosus Statius was a little-known geographical writer of probably the Julio-Claudian period (Keyser, "Sebosus Statius," *EANS* 729), and refined the earlier totals to 1,675 miles (2,479 km.).

6.184. Not long before Pliny wrote the *Natural History*, the emperor Nero commissioned an exploration of the upper Nile, probably the first Roman expedition to travel above Meroë. The journey was widely reported, with the fullest account by Pliny's contemporary Seneca (*Natural Questions* 6.8.3–4; Merrills, *Roman Geographies* 206–15). It was led by two centurions, who went beyond Meroë into the tropics, where they reached a marshy region so entangled with vegetation that it was almost impossible to penetrate, probably the Sudd of modern Sudan. Eventually they came to a rocky place where water poured forth, perhaps the Veins of the Nile mentioned elsewhere by Seneca (*Natural Questions* 4a.2.7). This may be Murchison or Kabalega Falls in Uganda. Regardless of how far they went, it was a greater distance up the Nile than anyone in antiquity, and into areas not seen again by Europeans until the nineteenth century. They may have created a map (*Aethiopiae forma*) that they presented to Nero, although the terminology is ambiguous (*NH* 12.19).

The first item reported from Nero's explorers is the itinerary from Syene to Meroë, 976 miles (1,444 km.) to the south. As is often the case in ancient texts, the sum does not equal the total of the individual parts. The expedition recorded the flora and fauna they encountered: the *psittacus* is a parakeet, probably the ring-necked variety (*Psittacula krameri*; Arnott,

Birds 201). For the *sphingion*, see *NH* 6.173. The *cynocephalos* ("Dog Head") is the hamadryas baboon (*Papio hamadryas*; Kitchell, *Animals* 42–3).

6.185. The expedition observed the lusher environment around Meroë, where palm trees became more common. The city was allegedly on an island, although this is not apparent today, despite a consistent tradition (Strabo, *Geography* 17.2.2; Stefan Radt, *Strabons Geographika* [Göttingen 2002–11], vol. 8, 502), and the "island" may actually be the promontory created by the junction of the White and Blue Nile 190 km. to the south (at modern Khartoum). Meroë had long been prosperous, since perhaps the eighth century BC, largely due to the mineral resources of the kingdom. It remains a major archaeological site (P. L. Shinnie, *Meroe: A Civilization of the Sudan* [London 1967].

6.186. Candace is a dynastic name: Queen Amanirenas was on the throne in the Augustan period and was probably the one who opposed the prefect Petronius (Dio 54.5.4–5; Strabo, *Geography* 17.1.54; *NH* 6.181–2). There were several queens from her time to that of Nero's expedition (*BNP Chronologies* 54). Evidently the kingdom was in decline when Nero's people visited, hinted at by the deserted villages downstream, perhaps due in part to the Roman presence (*NH* 6.181). Yet even though the demographic figures are probably exaggerated, there is no doubt that the Meroitic kingdom was powerful and extensive through Hellenistic times (Eratosthenes, *Geography* F98; Strabo, *Geography* 17.1.2, 17.2.2). The large number of artisans (3,000, 4,000 in some manuscripts) were probably mostly involved in metalworking, presumably the basis of the royal economy. The 45 kings that still existed "today" (*hodie*) were presumably petty dynasts and local rulers scattered across the region (Jehan Desanges, "Des éléphants, ou artisans, et des quarante-cinq rois des Ethiopiens (*Pline l'Ancien, vi, 186*)," in *Orbis Aethiopicus: Studia in honorem Stanislaus Chojnacki* [Albstadt 1992] 83–7).

6.187. Having concluded his account of Nero's expedition, Pliny discussed the mythological aspects of the history of the region. The various historical ethynms for the Aethiopians, cited only here, are associated with the distant regions of the inhabited world. Aetheria (the daughter of Night: Hesiod, *Theogony* 124) has a sense of remoteness, but the name can also mean fire (Euripides, *Trojan Women* 1079). Atlantia connects the Aethiopians with the Atlas Mountains and the Western (Atlantic) Ocean, and it was consistently believed that there were Aethiopians in the far west (Hanno 7, 11; see also *NH* 6.197). Aethiops, the son of Hephaistos (Vulcan), a little-known eponym, provides an understanding of the role of fire in the hot regions of low latitudes. There is also the concept that anatomically unusual people live at the fringes of the world. Moreover, the sun was said to be more powerful in these southern places – certainly true – which created the strange fauna of the region, from crocodiles to hippopotami and elephants (Diodorus 2.51.3–4). It was also generally believed that people differed according to the climate of their location (*Airs, Waters, and Places* 5, 13).

Pliny's account of people with unusual physical features has some connection with Aethiopia, but is also based on generic views. Even though he located them toward "the interior of the eastern part" (*ab orientis parte intima*), which would imply between the Nile and the Indian Ocean, many of the peoples were said to be found elsewhere. Pomponius Mela (3.91) recorded a similar catalogue, located in West Africa. Yet there is some legitimacy to the people without noses, a Greco-Roman description of sub-Saharan Africans who came to places like Meroë for trading purposes; they are probably the people described by Diodorus who had flat noses and woolly hair, and were black in color (Diodorus 3.8.2; see also Strabo, *Geography* 16.4.11).

6.188. There is a further description of anatomically strange people, probably other sub-Saharan visitors to Meroë and elsewhere, but with racial characteristics more in the realm of fantasy. These were probably based on a misinterpretation of unfamiliar personal adornment or dress; such people were also said to be around the fringes of India, and similar people without noses were reported by Megasthenes (F27b; Strabo, *Geography* 15.1.57). Speaking by nods and gestures is nothing more than communication between those without a common language.

Ignorance of fire – which seems improbable – was another alleged characteristic of exceedingly primitive people. The mention of Ptolemy IX Lathyrus (ruled 116–107 BC) connects the local discovery of fire with the expedition of his contemporary Eudoxus of Cyzicus, who went far to the south on both the eastern and western coasts of Africa, and was the first Greek to go into some of these regions. Although the accounts of his journeys have nothing to do with fire, it is perhaps a suggestion of someone from the Mediterranean bringing civilization to these remote people (Strabo, *Geography* 2.3.4–5; *NH* 2.169, 6.198; Roller, *Through the Pillars* 107–12).

Pygmies (probably originally an ethnym, Pygmaioi) had been known to the Greek world as early as the time of Homer (*Iliad* 3.6; see also Herodotus 4.43; *NH* 6.70). The reference to marshes at the source of the Nile is an allusion to Nero's explorers, who travelled into them (Seneca, *Natural*

Questions 6.8.3–4). The mountains that appear to be on fire may be those that Strabo reported near the Red Sea which, due to the heat, give off a red glow (Strabo, *Geography* 16.4.20), or a reference to the volcano of Theon Ochema in West Africa, discovered by Hanno (Hanno 16; *NH* 6.197).

6.189. The account mentions a route from Napata to the Red Sea. The passage seems confused, because Napata is in the great bend of the Nile, just below the Fourth Cataract, and to head from there to the Red Sea would mean approaching and crossing the river again after about 150 km. Pliny may have meant Meroë, which would have had a route to the sea, or it may be that Napata, the largest city for hundreds of kilometers, was the staging point for a route that went along the river shore (avoiding the cataract) and then headed across the desert to the sea. But any approach to the sea would have been easier from Meroë, and this may be why it replaced Napata as the Meroitic capital, perhaps in the sixth or fifth century BC (Shinnie, *Meroe* 31).

Nevertheless from either Napata or Meroë to the Red Sea would be far more than 3 days; from Meroë it is about 400 km. over rough territory: the 15 days reported by Strabo (*Geography* 16.4.8) is far more reasonable. Gold was a common product of this region, and there is physical evidence of its mining and processing in Hellenistic and Roman times (J. Vercoutter, "The Gold of Kush," *Kush* 7 [1959] 120–53). The Atabuli were probably north of Meroë (Pliny, *Histoire naturelle* 6.4, 171); the Megabarri were also in this region. A detailed account of their weaponry, preserved by Strabo, may mean that they were Ptolemaic mercenaries (Strabo, *Geography* 16.4.17). Indigenous populations that subsisted on the elephant were common in this region. Expectedly, they were called Elephantophagoi ("Elephant Eaters") by the Greeks: Strabo preserved their techniques of hunting the animal (Strabo, *Geography* 16.4.10; see also Diodorus 3.26).

6.190. Pliny listed 15 ethnic groups without providing a source. The context is the left (African, or west) bank of the Nile, presumably above and below Meroë. Two have generic Greek names: the Macrobii ("Long-Lived People") and the Gymnetes ("Naked People"). Macrobii was an epithet applied regularly to the Aethiopians from early times: they were said to live 120 years (Herodotus 3.17, 23). There was a known settlement of naked people at Endera (perhaps the home of Pliny's Anderae), seemingly between the Nile and the coast (Strabo, *Geography* 16.4.9). For the Megabarri, see *NH* 6.189. The remaining ethynms cannot be located, but there are some hints about their habits. The Hipsodores were black-skinned, and because they were ashamed of this, they covered their bodies with a red pigment, probably minium (cinnabar, or red mercuric sulphide), a known product of the region (*NH* 33.118); this is a rare qualitative statement about skin color from antiquity (Gruen, *Rethinking* 205). For the *cynoscephali* (baboons), see *NH* 6.184. The unusual height of the Syrbotae (see also *NH* 7.31) – eight cubits or about 3.5 m. – may be a reference to some of the tall peoples of central Africa.

6.191. Aristocreon (*FGrHist* #667, F3) was probably another early Ptolemaic explorer, essentially only documented by Pliny (see also *NH* 5.59), who was the author of an *Aithiopika*. He provided some information on the settlements and people located at least 17 days upriver from Meroë, as far as Esar, also known as the Island of the Fugitive Egyptians, which Herodotus placed two months from Meroë (Herodotus 2.30–1; Strabo, *Geography* 16.4.8). The king in question is probably Psammetichos II (ruled 595–589 BC), from whom mutinous soldiers fled to this remote locale. Sembobitis, Semberritae, and Sembritae all seem to based on the same word, translated as "immigrants" or "foreigners" (Eratosthenes, *Geography* F98; Bion F6). Arabia here refers to the territory on the right bank of the Nile, not the peninsula. Some of the ethnic groups (the Simbarri, Palugges, and Asachae) were in the mountainous territory east of the river, closer to the coast. A queen as ruler for the Semberritae is certainly possible, and reported elsewhere (Strabo, *Geography* 17.1.2), but there may be confusion with the queens of Meroë.

6.192. The various reports from south of Meroë and to its west into the interior of north central Africa become increasingly vague and somewhat fantastic. Legitimate ethnyms are mixed with Greek descriptive ones, and there are strange and unusual (from the Greco-Roman point of view) physical and dietary characteristics. Most of these are hearsay reports received at places like Meroë that were either misunderstood or fit into preconceptions.

The Nubians were reported to extend along the upper Nile, from (in their broadest sense) the First Cataract to regions beyond Meroë. To some extent they were seen as synonymous with the Meroitic peoples (Strabo, *Geography* 17.1.7, 53; *NH* 6.35). Pliny, however, specifically located them in a region eight days from the Island of the Fugitive Egyptians, which would be well upstream of Meroë. Yet their capital, Tenupsis, has been suggested to be near modern Gumbus, just north of the Third Cataract, thus placing the Nubians along a lengthy stretch of river (L. P. Kirwan, "Nubia and Nubian Origins," *GJ* 140 [1974] 46.

There is no obvious explanation for the lack of ears among the Sesambri, unless it is some misunderstood facial adornment, and they, the Tonoberi, and Ptoenphani are otherwise unknown. The latter were ruled by a dog, whose barking gave instruction to the populace. Since he was described as sitting in a majestic posture and was worshipped, he may have been a cult statue (Plutarch, *On Common Conceptions: Against the Stoics* 11.1064b; Aelian, *On the Characteristics of Animals* 7.40). The Archisarmi, Phalliges (perhaps the Palugges of *NH* 6.191), Marigarri, and Chasamari were peoples on the upper Nile.

6.193. Returning to the report of Bion (F6), Pliny provided a catalogue of localities on islands in the Nile, seemingly beginning at a point 20 days above Meroë, perhaps in the vicinity of modern Khartoum, where the junction of the White and Blue Nile creates the impression of an island. For the Seberritan town ruled by a queen, see *NH* 6.191 (Semberritae). This is all part of an account of the region around the Fugitive Island, and the area between it and Meroë.

6.194. Pliny has finished using Bion as a source. The account now covers a wide range of interior Aethiopia (in its broadest sense), west of the Nile and as far as the Atlantic, probably the "southern ocean" and certainly the location of the "coastal Aethiopians" (see further, *NH* 6.195). The Sirbites may be the same as the Syrbotae of *NH* 6.190.

The Nisicathae and Nisitae are otherwise unknown; the men with three or four eyes are probably from a piece of folk wisdom about their ability with arrows. Dalion (F3) was probably limited to the Nile in his explorations, but would have learned about groups far away. The Vacathi, who only used rain water, would hardly have been on the river, and lived to the south, where rain was dependable. The Longopori were five days from Oecalices, which implies a land route. If the Perusii are the same as the Perorsi (*NH* 5.10, 16, 43), this is another indication that some of the peoples described are far to the west.

6.195. The ethnyms continue to be for people west of the Nile, and south into the tropics, an area so little known that it was believed to be imaginary. The Nigroe may be "river people," from the generic word for river found in such names as "Niger," or it could be a Latin descriptive ethnym, "black people." The single eye may be another example of facial adornment.

Many of the ethnyms that follow are Greek descriptive terms. The Agriophagi ("Wild Beast Eaters") subsisted on the panther and lion. The panther was mentioned in too broad a context in Greek and Roman literature to be useful, but the African lion (*Panthera leo*) was found in these regions until the twentieth century (Kitchell, *Animals* 108, 147). The omnivorous Pamphagi and the cannibal Anthropophagi reflect habits of peoples of central Africa and even south of Rhapta, at modern Dar es-Salaam (Ptolemy, *Geographical Guide* 4.8.1). The Cynamolgi ("Dog Milkers") and Artabatitae (or Atabatitae) are otherwise unknown. The Hesperioe ("Westerners") and Perorsi were far to the west.

Locusts (*Schistocerca gregaria*) as a food item were generally the diet of ascetics, but the Locust Eaters ("Akridophagoi") were said to be the westernmost of the Aethiopians. Locusts never became part of the Greco-Roman diet; thus any population that ate them was curious, although it is unlikely that they could have subsisted only on them. Diodorus provided a detailed habit of the harvesting and processing of the insect (Diodorus 3.29; Strabo, *Geography* 16.4.12; Dalby, *Food* 199).

6.196. Nearing the conclusion of his discussion of Aethiopia – a region broadly conceived – Pliny listed measurements taken from the map of Agrippa (F29) and other sources. Agrippa's distance (east–west) from the Red Sea is 2,170 miles (3,212 km.), which would be about two-thirds of the way across the continent of Africa, perhaps to the eastern borders of the Mauretanian territory of Juba. The width (1,296 miles or 1,918 km.), from Upper Egypt to the south, would be as far as the vicinity of Meroë or the Fugitive Island. The Sirbites and Dabelli were between Meroë and the coast; the reported total journey of 33 days is at variance with other figures, but there is no reason to believe that the routes were identical. The Aethiopian Ocean is not a term used previously by Pliny and can only refer to the southern Red Sea. An amount of 625 miles (925 km., from 5,000 stadia) is about twice the direct distance but the vagaries of the actual routes are uncertain. For the distance between Meroë and Syene, see *NH* 6.184.

6.197. This section brings together material from disparate sources in order to provide an overall view of Aethiopia in its broadest sense: in other words, the entire strip of the continent of Africa from the vicinity of Meroë east to the Red Sea and west to the Atlantic. The orientation is from the summer sunrise (northeast) to the winter sunset (southwest), creating an axis roughly from Meroë to Mt. Cameroon (Theon Ochema).

Ebony had been known to the Greeks since the fourth century BC, and had medicinal as well as decorative uses. It was most famous as the wooden beams of the Temple of Artemis at Ephesus (Theophrastus, *Research on Plants* 1.5.4; Herodotus 3.97, 114; Dioskourides 1.98; *NH* 16.213).

The active volcano of Theon Ochema (Chariot of the Gods, modern Mt. Cameroon) had already been discussed by Pliny in the context of Polybius' West African exploration (Polybius 34.15.7; *NH* 2.238, 5.10). The mountain was originally identified and named by Hanno (16). There is a vestige of a *periplous* along the West African coast to the Horn of the West (Hesperu Ceras, the Hesperium Promontory of *NH* 5.10) in the land of the western Aethiopians, material taken from Hanno. The Aegipanes and Satyrs seem to be in what Pliny would call northwestern Aethiopia, between Mt. Atlas and the Niger River; mention of them comes from the expedition of Suetonius Paulinus in the AD 40s (*NH* 5.7, 44, 46).

Islands Around The Continent of Africa

6.198–9. The final topographical section of Book 6, and indeed of the entire geographical portion of the *Natural History,* is about islands thought to be around the continent of Africa (through *NH* 6.205). Several sources are cited, but the emphasis on northwest Africa indicates that the primary one was Juba. There are also vague references to Ephorus, Eudoxus, and Timosthenes. Ephorus (F30, 172), the historian of the fourth century BC, was probably the first to define Aethiopia as the entire southwest quadrant of the inhabited world. It is not certain whether Pliny meant Eudoxus of Cnidus, the mathematician and astronomer of the early fourth century BC (*NH* 2.130), or the adventurer from Cyzicus who was involved in establishing the route to India and then was lost at sea off the West African coast around 100 BC (*NH* 2.169; Posidonius F49; Strabo, *Geography* 2.3.4–5). Timosthenes, the naval chief of staff of Ptolemy II, would have been more likely to have been interested in the Red Sea than the Atlantic; Ptolemaic interests did not extend that far (F14 Roller).

In addition, Clitarchus (Kleitarchos), who late in the fourth century BC reported on the expedition of Alexander the Great (*FGrHist* #137, F29; Heckel, *Who's Who* 86), wrote about islands in the Indian Ocean, using material probably obtained from Nearchus and Onesicritus; the island rich in gold may be Chryse (*NH* 6.80), and the one with a sacred mountain and aromatic trees may be a place that produced cinnamon, perhaps Taprobane or somewhere farther east (Dalby, *Food* 87–8).

Much of Pliny's account centers on the island of Cerne (Kerne), with material from Ephorus (F172) and Polybius (34.15.9) as well as Cornelius Nepos. Pliny's description is unusually badly garbled. Cerne was first said to be somewhere in the western Indian Ocean and accessible from the Red Sea. If the "columns" are the Pillars of Hercules, this outlines the actual route to Cerne, but the island could not be approached from the Red Sea except by circumnavigating Africa. It was a Carthaginian trading post on the west coast of Africa established by Hanno (8) around 500 BC. There have been many suggestions for its location, none of them definitive

(Roller, *Through the Pillars* 37; Walbank, *Historical Commentary* 3, 637–8). After stating that neither its size or distance from the mainland had been determined, Pliny then provided the figures for both. It was an island eight stadia (1.6 km.) offshore according to Polybius – Nepos had a mile (1.5 km.), perhaps a casual conversion of stadia to miles – and two miles (three km.) in circumference. "Opposite [*adversus*] Carthage" means that it was believed to be on the same latitude as the Carthaginian city. Cerne became known to Greeks and was visited in the mid-fourth century BC, when it was described as a vigorous trading post with objects from as far away as the eastern Mediterranean (Pseudo-Skylax 112). Polybius visited two centuries later and was the last to report on it as an active post, since it had already begun to fade.

The island of Atlantis (or Atlas) is also confusing, given the prevalence of similar toponyms in the region (Mt. Atlas, Atlantic Ocean), unless it is the famous Platonic locale (*Timaios* 24–6). The remainder of the passage continues to be from Hanno's report (13–14), although it is difficult to comprehend the land turning west at the Hesperu Ceras, and this may be an error for "east," since the narrative is in the vicinity of the long east–west coast of West Africa.

6.200. Xenophon of Lampsacus was a geographer of the second or first century BC, cited several times by Pliny, who seems to have written on the extremities of the known world (Dueck, "Xenophon," *EANS* 840–1). The Gorgades Islands were near Hesperu Ceras (*NH* 6.197), and thus presumably along the east–west coast of West Africa. There are a number of possibilities for their location, from Fernando Po and São Tomé in the east to the Cape Verdes in the west. The Gorgons were mythical monstrous sisters, best known because of the confrontation between one of them, Medusa, and Perseus. As early as the time of Hesiod they were located in the far west (Hesiod, *Theogony* 270–83), and they may have originally been a western ethnic group that evolved into myth: this is certainly the context of Pliny's summary of them (Diodorus 3.54.7–55.3). Yet the account does not conform to the extant text of Hanno, where the hairy women are called the Gorillai (Hanno 18). Although he did report the removal of their skins to the Temple of Juno (Tanit) at Carthage, the note that they were still there when Carthage was taken (146 BC) cannot be from Hanno, and Pliny's source may be Juba (mentioned at *NH* 6.201), from his *Wanderings of Hanno* (F6; Athenaios 3.83d), which may also have been the source of Xenophon of Lampsacus. It is possible that Hanno's Gorillai evolved in Greek texts to the more familiar Gorgons.

6.201. Sebosus Statius was later than Juba, and may have updated some of his information (*NH* 6.183; Keyser, "Sebosus Statius," *EANS* 729). The two Hesperides ("Western [Islands]") are connected with the Gorgons (Hesiod, *Theogony* 274–5); here the implication is that they are somewhere between the coast near Mt. Atlas and Hesperu Ceras. The Gardens of the Hesperides were located around Lixus (*NH* 5.3), but it took Hanno only a few days to reach Lixus from the Pillars of Hercules (Hanno 2–6), and Pliny's reported sailing time to the Hesperides conforms more with the southern location reported at *NH* 6.200.

The Mauretanian dyeing industry was established on islands off the coast of Mauretania and produced Gaetulian purple, considered in the Augustan period to be one of the better varieties (Horace, *Epistles* 2.2.181–2; Tibullus 2.3.58; Ovid, *Fasti* 2.319). Pliny elsewhere described its production in detail (*NH* 9.125–41). The islands are thought to be the islets at modern Essaouira in Morocco, where there is archaeological evidence for the industry, which may have been started in Carthaginian times (Roller, *World of Juba* 115–17).

6.202–5. Using in part Sebosus' recension of Juba's material, Pliny described the Fortunate Islands (Antonio Santana Santana and Trinidad Arcos Pereira, "The Canary Islands in Pliny the Elder's *Naturalis Historia*," *AfrRom* 16 [2004] 297–312). These – also known as the Islands of the Blessed – were the mythical paradise that was the home of the dead (Homer, *Odyssey* 4.561–8; Hesiod, *Works and Days* 166–73). Stories of Hercules' last labor (the Apples of the Hesperides) were also mingled with tales about the islands (Hesiod, *Theogony* 215–16; *NH* 5.3). Juba commissioned an expedition to discover them (F44), and found islands off the coast of southern Mauretania that he identified as the Fortunate Islands, which are the modern Canaries.

Yet the distances and sailing directions reported by Pliny are confused. The most accurate report is that the islands are 750 miles (1,110 km., from 6,000 stadia) from Gades, but the Canary group is no more than 350 km. across, and the subsequent figures are in error. But there is no difficulty in identifying the individual islands. Pliny tended to use Latin translations of Juba's Greek names, perhaps obtained from Sebosus. All the names are descriptive.

Some of the islands are visible from the Mauretanian coast, and Juba's explorers approached the group from the northeast, coming first to Ombrios (Sebosus' Pluvalia), the only one where Juba's Greek name is preserved. This is modern Lanzarote, the northeasternmost of the group and the first one that the explorers would encounter. The tree that produces water is the giant fennel, *Ferula lancerotensis*, still conspicuous on the island today.

The explorers then went to Junonia (probably Heraia to Juba), modern Fuerteventura and immediately to the southwest of Ombrios, where there was evidence of previous habitation, presumably Carthaginian. The smaller island also named Junonia may be modern Graciosa, just north of Lanzarote. From Junonia they continued on to Capraria, seemingly Goat Island (modern Gran Canaria), whose name, if properly interpreted, also suggests previous habitation, since the goats were probably the domesticated variety (*Capra hircus*). Capraria was also noted for its lizards.

Juba's people do not seem to have visited Ninguaria ("Snowy," modern Tenerife), to the northwest of Capraria. Its snowy peak, Mt. Teide (3,715 m.), is the highest location on any island in the Atlantic. But the expedition passed the island and continued on to Canaria, or Dog Island (modern Gomera), where the local dogs – some of which were presented to Juba – gave the entire group its name. It too had signs of earlier habitation. Two other islands were only mentioned in Sebosus' version: Invallis ("Mounded," modern Palma), and Planasia ("Flat," modern Hierro). These are the two westernmost and may only have been visited briefly, if at all. Juba's report also included details about the natural history and material culture of the islands, and local commodities, especially noticeable on Canaria. As usual, there is the incredulity at encountering large ocean creatures, thought to be sea monsters (*belvae*).

Knowledge of the islands was lost to Europeans in late antiquity and only reacquired in the fourteenth century, at which time Pliny's names were reattached. These were later changed to the modern ones except for Canaria, which was applied to the wrong island (John Mercer, *The Canary Islanders: Their Prehistory, Conquest, and Survival* [London 1980] 155–9).

Measurements and Parallels of the Earth

6.206. To conclude his geography, Pliny devoted the last sections to a variety of measurements and a catalogue of the parallels that crossed the inhabited world. The first source mentioned is Polybius (34.15.2). All distances have been converted from stadia, and vary both in the manuscripts of Pliny and citations by Polybius (Walbank, *Historical Commentary* 3, 631–2). From the Pillars of Hercules ("Gades Strait") to the mouth of the Maeotis was 3,437.5 miles (5,088 km., from 27,500 stadia), probably a sailing distance through the Aegean rather than any coastal measurement. There is also another series of distances, clearly a sailing route that has been specifically laid out: from the Pillars to Sicily (1,250 miles or 1,850 km., from 10,000 stadia), continuing to Crete (375 miles or 555 km., from 3,000 stadia), Rhodes (187.5 miles or 278 km., from 1,500 stadia), and the same distance to the Chelidonian Islands off the south coast of Asia Minor. Then it is 225 miles (333 km., from 1,800 stadia) to Cyprus, and another 115 miles (170 km. or 920 stadia) to the terminal point, Seleucia Pieria at the mouth of the Orontes. The total is 2,340 miles (3,463 km., from 18,720 stadia). All the figures are even hundreds of stadia except the last (perhaps a copying error), suggesting that the numbers have been rounded. Yet the distance from one end of the Mediterranean to the other is quite accurate.

6.207. Pliny then turned to Agrippa (F66) as his next source, some of whose figures are questionable, as Pliny himself noted. Agrippa's distance from the Pillars to the Issic Bay (the modern Gulf of Alexandretta at the northeastern corner of the Mediterranean: see *NH* 5.91) should hardly vary from that to Seleucia Pieria, as the two locations are only about 100 km. apart. But Agrippa's distance (3,440 miles or 5,091 km.) is half again as large as that of Polybius, and Pliny was correct to suspect an error, either in the map of Agrippa itself or any accompanying text. Agrippa's distance from Sicily to Alexandria (1,350 miles or 1,998 km.) is more reasonable, because it would reflect a sailing route that did not go directly to Alexandria but headed southeast to the closest part of the African coast (near Cyrene) and then east.

Agrippa also provided a circuit of the Mediterranean system, not including the Maeotis (15,509 miles, or 22,953 km.). This figure is also based on sailing routes from specific ports rather than an actual measurement of the coast, and may ultimately derive from Artemidorus, whose figure was an additional 756 miles (1,119 km.), although this may be a matter of a conversion factor. Artemidorus also provided the circuit of the Maeotis (1,125 miles or 1,665 km.), which conforms to the 9,000 stadia reported by Strabo (*Geography* 2.5.23).

6.208–9. On occasion Pliny would give credit to the heroic work of those who determined the measurements and distances and thus made it possible to record the size and extent of the inhabited earth. This passage is similar to his comments regarding the explorers of the Nile and its sources, who went far while unarmed and for reasons other than war, as well as his remarks on the peaceful activities of the Roman military (*NH* 5.51, 6.182). Elsewhere, in his section on linen, the essential component of sails (*NH* 19.4–6), he noted how a single plant – flax (the material from which linen is made) – reduced the size of the world, a metaphor for the progress of civilization through navigation.

Pliny's next topic is a discussion of the size of the three continents, and thus of the entire inhabited earth. The only source mentioned is the map of Agrippa (F34). Because in antiquity the full extent of the continents was not known, the numbers are highly speculative. The problem is compounded by a gap in the text of *NH* 6.208 so that some of the figures for Europe are not extant. The single surviving number, 8,714 miles (12,897 km.), vastly exceeds any possible dimension of the continent other than, perhaps, its circumference. Elsewhere Pliny reported that it was 3,437½ miles from the Pillars of Hercules to the mouth of the Maeotis (*NH* 6.206); Agrippa's significantly larger figure assumes a much longer distance from the Maeotis to the External Ocean, probably around the north of Europe and thus through regions unlikely to have been seen by any Greek or Roman. Nothing was known about the assumed eastern coast of Europe, from the Baltic to the upper Tanais, or whatever theoretical point was used in that direction to divide Europe from Asia.

Africa was the only continent where it was believed that there was solid evidence for its full circuit, but this was based on the idea that its coast ran directly from a point south of the mouth of the Red Sea to somewhere in West Africa, perhaps near Mt. Cameroon (Strabo, *Geography* 2.3.3–4; *NH* 6.175). The stated east–west length of 3,698 miles (5,473 km.) is a reasonable figure for the northern part of the continent, but Pliny seemed to be confused by Agrippa's 910 miles (1,347 km.), taking it as an additional east–west dimension, providing a total of 4,608 (6,820 km.), which is excessive. In fact Agrippa meant that because of the expedition of L. Cornelius Balbus in 21–20 BC (*NH* 5.35–7), the interior of North Africa was known as far as 910 miles inland, rather than only the 250 miles (370 km.) of the inhabited coastal strip.

Pliny's length of Asia (5,013¾ miles or 7,420 km., perhaps 40,000 stadia) has no end points. The width (north–south) is calculated at 1,875 miles (2,775 km., from 15,000 stadia), said to be from the ocean east of Meroë to Alexandria. This is certainly reasonable for these endpoints, but it is a peculiar figure for the width of the entire continent and is actually the width of Arabia, not Asia as a whole. The number does not take into account the extent of Asia to the north across the Black Sea to the Tanais.

6.210. In a complex series of calculations, Pliny determined the relative sizes of the three continents. This was a matter of interest ever since the concept of continents had been established in the fifth century BC. Herodotus believed that there were significant differences in their size, with Europe by far the largest (Herodotus 4.36, 42, 45). By the first century BC this was the generally held opinion, with Libya (Africa) thought to be the smallest (Strabo, *Geography* 17.3.1). Elsewhere Pliny noted that Europe was as large as the other two put together (*NH* 3.5), a point of view perhaps as much ideological as geographical, and oddly not sustained by his calculations.

The figures for the relative sizes are complex to modern readers because of the Greek method of handling fractions. Yet when reduced to a common denominator they show that Europe was thought to be about 46% of the inhabited earth, Asia 32%, and Africa 21%. These figures have little meaning beyond a largely theoretical conception of the size of the parts of the inhabited world. Much of Europe was vaguely understood (except for northern Scandinavia and most of western Russia), but there was an awareness of only a small portion of Asia and less than half of Africa, although it was believed otherwise. As a matter of curiosity, Europe is by far the smallest of the three and Asia is larger than the other two put together.

6.211. To conclude his geographical account, Pliny recorded the various latitude parallels of the inhabited world (illustrated in Pierre Schneider, "Les 'climats' de Pline l'Ancien (*N. H.* 6, 212–220): essais de transcriptions graphiques," *GA* 20–1 [2011–12] 179–204). Previously he had credited Eratosthenes with the sublime ability to grasp the size of the earth (*NH* 6.171). Here he noted that the concept of parallels – a line created by connecting points of the earth at the same latitude – was another subtle and meticulous Greek invention. Yet rather than use the Greek world *parallelos*, he preferred the Latin *circuli* ("circles"). Pliny justified this survey of the "circles" by stating that it was necessary so that nothing be lacking in his overview of the world, a visual image that recalls the map of Agrippa (Carey, *Pliny's Catalogue* 70–1).

The concept of parallels was developed in the fourth century BC by Eudoxus of Cnidus, who began to theorize that the length of day at the solstice could be used to determine latitude (Roller, *Ancient Geography* 75–6, 105–8). This idea was further refined by Dicaearchus of Messana, who created a latitude line from the Pillars of Hercules to the Himalayas (Dicaearchus F123; Keyser, "Geographical Work" 367). But it was Eratosthenes who established a system of parallels, essentially from the latitude of Meroë to that of Thule in the Arctic (Eratosthenes, *Geography* F57–63). His parallels always connected known points, sometimes even following shipping lanes, and thus tended to wobble, but were intended to provide a concept of latitude, based on the hours of the day and other astronomical phenomena, and thus could be used to locate the relative position of places on the surface of the earth.

Pliny's system originated with Eratosthenes, but was revised and refined at a later date, with some adjustments as to the position of the parallels and the addition of many toponyms, especially in the west, that reflected the greater knowledge of his era. The intermediate source between Eratosthenes and Pliny is not identified, but because it demonstrates a more Roman environment it is probably P. Nigidius Figulus, the only one mentioned in these sections (*NH* 6.217). He lived in the first century BC and was a polymath who included astronomy among his various writings, presumably producing a Latin treatise on the parallels (B. Centrone, "P. Nigidius Figulus," *EANS* 572–3).

6.212. Pliny's first parallel is where the longest day is 14 hours, which Eratosthenes placed 400 stadia (80 km.) south of Alexandria and Cyrene (Eratosthenes, *Geography* F60). In Pliny's recension, it extended from India to the Pillars of Hercules. Although most of Eratosthenes' datum points are listed, quite a few have been added that reflect the Roman world rather than the third century BC, especially in Africa. The focus of Pliny's version is no longer on Alexandria but all the way to the Atlantic.

Pliny also mentioned the essential tool for these calculations (using solar shadows), the gnomon, a pole used originally for land surveying, perhaps an Egyptian invention (Herodotus 2.109; Strabo, *Geography* 1.1.20). By the third century BC it had come to be a basic implement of the geographer in the field. Its length was variable, depending on the size of the shadow needed, and Pliny reported one over 35 feet long.

6.213. The next one is where the day at the solstice was 14⅖ hours long, perhaps Nigidius' adjustment of Eratosthenes' parallel of 14¼ hours, centered on Ptolemais in Phoenicia. Pliny's version extends from the Parthians through Mesopotamia, Judaea, Ptolemais, and west as far as points in North Africa and Sicily.

6.214. The next one (14½ hours and two minutes) corresponds to Eratosthenes' parallel of 14½ hours, centered on Lycia, Rhodes, and the Peloponnesus. Pliny's list includes 40 locations (as opposed to Eratosthenes' four), from the Caspian Sea through the places recorded by Eratosthenes and west to the Iberian peninsula and the Pillars of Hercules.

6.215. The next circle is where the longest day is 14⅔ hours. This does not correspond to any of Eratosthenes' parallels and is probably the creation of Nigidius. Its extremities are based on recent knowledge: in the east there is a vague reference to "what is on the other side of Imavus [Himalayas]," reflecting the trade information that was available in Roman times about a world north of the mountains. The parallel passes through 27 places, mostly in the heartland of Asia Minor and Greece, extending as far west as Gallia Narbonensis, coastal Hispania, and New Carthage, and then to unspecified points beyond. Moreover, it shows particularly well the inherent wobble in these parallels, since the latitudes of coastal Gallia and New Carthage differ by about 5°.

6.216. The next parallel is where the maximum length of the day is 15 hours. Eratosthenes placed it across the northern part of Asia Minor and the Greek peninsula, and west across Italy between Rome and Naples. Pliny listed 46 localities in addition to the four cited by Eratosthenes, beginning with Bactria in the east and continuing as far west as central Hispania. Yet, in a further indication of how the parallels connected known places at the expense of creating a straight line, these 50 toponyms are hardly at the same latitude: both Thasos (the northern Aegean island) and Regium in southern Italy are included, whose parallels are 4° apart.

6.217. The sixth parallel, at $15^{1}/_{9}$ hours, was not created by Eratosthenes, but by Nigidius (who had 15½ hours) or his sources. Pliny pointed out that this line passed through the city of Rome, and it may have been established in order to place it on a parallel. Eratosthenes hardly knew about Rome: the fragments of his *Geography* cite it only twice, one of which is suspect (F155; see also F60).

Pliny's parallel begins at the Caspian Sea and continues across Armenia, the northern Greek peninsula, north central Italy, the Ligurian coast, central Hispania, and Lusitania. Again the need to connect known points is excessive, as it swings north from Pisa in Italy and follows the coast of the western Mediterranean through Massilia and then southwest into Hispania.

6.218. The next parallel is at 15⅗ hours, close to the 15½ hour line of Eratosthenes, for which Pliny provided no specifics other than to note that it crossed the Black Sea and was at a point halfway between the equator and the north pole (Hipparchus, *Against the Geography of Eratosthenes* F56). This parallel begins on the coast of the Caspian Sea and goes across Illyria and the head of the Adriatic, northern Italy, and the Alps. It continues through Vienna in Gaul and then curves south to cross the Pyrenees. Only with this parallel did Pliny note any anomalies in the data, drawing attention to the fact that in parts of Venetia (at the head of the Adriatic) the shadow of the gnomon varied between 35 and 36 feet (about 10–11 m.). This suggests that actual measurements were taken in this region, but not at all places along the parallel, since Pliny and his sources seemed oblivious to the southern curve at the west end.

6.219. These seven parallels were based on those of Eratosthenes, with refinements by Hipparchus, Nigidius, and probably others. Pliny made a distinction between "past writers" (those who were responsible for the seven parallels) and "their successors," who calculated the three additional ones in this section. Yet he did not state who these successors were, or which group Nigidius belonged to. Although the three new parallels were, to some extent, products of the Roman world, they are also based on received data from Greek calculations. The failure to provide the length of the gnomon or its shadow demonstrates that Pliny was indeed using another source.

The first of the three additional parallels is the one with a 16-hour day, extending from the mouth of the Tanais across into the Dacian and German territory, and Gaul. This is essentially a Roman parallel, passing through points little known to the Greek world. Then there is the 17-hour parallel through the Hyperboreans and Britannia. The Hyperboreans were the semi-mythical people of far northeastern regions (*NH* 6.34, 55), so this parallel is more speculative than based on actual data, and could be assumed because the size and shape of the world was known. Finally there is a farthest parallel, passing through another mythical locale, the Ripaean Mountains (*NH* 6.15, 19, 33–4), in the land of the midnight sun. Yet there was some measured data used in both these parallels, since Britannia and Thule were places visited by Pytheas in the late fourth century BC, where he took latitude calculations and also reported on the continuous day and night (Pytheas T4–5, F6; Hipparchus, *Against the Geography of Eratosthenes* F57–8; Strabo, *Geography* 2.1.18, 2.5.8; Roller, *Through the Pillars* 71–2, 81–7). Pliny may not have consulted Pytheas directly at this point, but obtained his information through a romanized recension.

6.220. Returning to the southern extremes of the known world, Pliny reported on two additional parallels, both of which were originally calculated by Eratosthenes but perhaps accessed by Pliny through later sources, since he noted a revision to the circles.

The first is the parallel through Meroë and Ptolemais For the Hunt (at 12½ hours, adjusted from Eratosthenes' 13 hours), and the second is through Syene, at 13 hours (adjusted from 13½). Pliny added no toponyms beyond those reported by Eratosthenes and in fact provided fewer than the earlier scholar, who placed Berenice (on the Red Sea), Trogodytice, the Gedrosians, and India on the Syene parallel.

With a terse four-word statement (*et hactenus de terris*, or "this is sufficient about the lands") Pliny concluded the five books of his geography. His use of the plural, "terris," shows that he thought of the earth not as a single unit, but as made up of many lands.

Bibliography

Abeydeera, Ananda. "Raki's Mission to Romanukharattha. New Evidence in Favor of Pliny's Account of Taprobanê, *N. H.* 6.84–91," *Habis* 40 (2009) 145–65.

Agatharchides of Cnidus. *On the Erythraean Sea* (ed. Stanley M. Burstein, London 1989).

Alcock, Susan E. *Graecia Capta: The Landscapes of Roman Greece* (Cambridge 1993).

Alföldy, Géza. "Spain," *CAH* 11 (second edition 2000) 449–63.

Anderson, Andrew Runni. "Bucephalas and His Legend," *AJPh* 51 (1930) 1–21.

Andreu Pintado, Javier. "Entre la Literatura y la Historia: Plin. *Nat.* III 30 y la Latinización de Hispania," in *"Urbs aeterna," Actas y Collaboraciones del Coloquio Internacional "Roma entre la Literatura y la Historia"* (ed. Alonso del Real, Pamplona 2003) 187–210.

"Foederatos Tarracenses: Sobre una Comunidad de la Tarraconense en los Listados de la *Naturalis Historia* de Plino el Viejo," *Gerión* 35 (2017) 461–70.

Angelini, I. and P. Bellintani, "Archaeological Ambers From Northern Italy," *Archaeometry* 47 (2005) 441–54.

Arnaud, Pascal. "Marcus Vipsanius Agrippa and His Geographical Work," in *Brill's Companion to Ancient Geography: The Inhabited World in Greek and Roman Tradition* (ed. Serena Bianchetti *et al.*, Leiden 2016) 205–22.

"Pline, *Nat. Hist.* III, 37, et la cité de Digne: Problèmes de toponymie historique," in *Παλαιὰ Φιλία: Studi di Topografia Antica in Onore di Giovanni Uggeri* (ed. Cesare Marangio e Giovanni Laudizi, Galatina 2009) 593–600.

"Texte et carte de Marcus Agrippa: historiographie et données textuelles," *GA* 16–17 (2007–8) 73–126.

Arnott, W. Geoffrey. *Birds in the Ancient World From A to Z* (Abingdon 2007).

Arruda, Ana Margarida. "Phoenician Colonization on the Atlantic Coast of the Iberian Peninsula," in *Colonial Encounters in Ancient Iberia: Phoenician, Greek, and Indigenous Relations* (ed. Michael Dietler and Carolina López-Ruiz, Chicago, Ill. 2009) 113–30.

Ash, Rhiannon. "The Wonderful World of Mucianus," *BICS Supplement* 100 (2007) 1–17.

Audet, Jean-Paul. "Qumrân et la notice de Pline sur les Esséniens," *RBi* 68 (1961) 346–87.

Badian, E. "A King's Notebooks," *HSCP* 72 (1968) 183–204.

Beagon, Mary. *The Elder Pliny on the Human Animal: Natural History Book 7* (Oxford 2005).

Roman Nature: The Thought of Pliny the Elder (Oxford 1992).

Beltrán, A. "Celsa," *PECS* 210–11.

Béquignon, Y. "Dodona," *PECS* 279–80.

Beresford, James. *The Ancient Sailing Season* (Leiden 2013).

Bianchetti, Serena, "I Lotofagi Nella Tradizione Antica: Geografia e Simmetria," in *L'Africa Romana: Atti del* XIII *Convegno di Studio, Djerba 10–13 Dicembre 1998*, vol. 1 (ed. Mustapha Khanoussi *et al.*, Rome 2000) 219–29.

Bickerman, E. J. *Chronology of the Ancient World* (second edition, Ithaca, N.Y. 1980).

Bispham, Edward. "Pliny the Elder's Italy," *BICS Supplement* 100 (2007) 41–67.

Blackman, D. J. "Syene," *PECS* 871.

Blázquez Martínez, José M. "Siria en la *Historia Natural* de Plinio," in *"Debita verba": Estudios en Homenaje al Profesor Julio Mangas Manjarrés* (ed. Rosa María Cid López *et al.*, Oviedo 2013) 1, 413–28.

Bliujienė, Audronė. *Northern Gold: Amber in Lithuania (c. 100 to c. 1200)* (Leiden 2011).

Blomqvist, Jerker. *The Date and Origin of the Greek Version of Hanno's Periplus* (Lund 1979).

Blum, Rudolf. *Kallimachos: The Alexandrian Library and the Origins of Bibliography* (tr. Hans H. Wellisch, Madison, Wis. 1991).

Boardman, John. *The Greeks Overseas: Their Early Colonies and Trade* (New York 1999).

Boëthius, Axel. *Etruscan and Early Roman Architecture* (second edition, Harmondsworth 1978).

Bollansée, Jan, *et al.* "Kallimakhos of Kurene," *EANS* 463.

Bolton, J. D. P. *Aristeas of Proconnesus* (Oxford 1962).

Borlenghi, Aldo. "Mura Serviane e Mura Aureliane: l'Occupazione di Spazi Pubblici e di Aree Private in Occasione dell'Abbandono e della Costruzione delle Due Cinte Urbane," in *Les confiscations, le pouvoir et Rome, de la fin de la République à la mort de Néron* (ed. Clément Chillet *et al.*, Bordeaux 2016) 301–34.

Boschi, E., *et al.*, "Catalogue of Strong Italian Earthquakes From 461 B. C. to 1997," *Annals of Geophysics* 43 (2000) 609–863.

Bostock, John and H. T. Riley. *The Natural History of Pliny the Elder* (London 1855–7).

Bosworth, A. B. *Alexander and the East: The Tragedy of Triumph* (Oxford 1996).

A Historical Commentary on Arrian's History of Alexander (Oxford 1980–95).

Bowen, Alan C. "Hellenistic Astronomy," in *The Oxford Handbook of Science and Medicine in the Classical World* (ed. Paul T. Keyser with John Scarborough, Oxford 2018) 293–314.

"Some Early Hypotheses in Greco-Roman Astronomy," in *Hellenistic Astronomy: the Science in its Contexts* (ed. Alan C. Bowen and Francesca Rochberg, Leiden 2020) 95–111.

"Sosigenes (1)," *EANS* 752.

and Robert B. Todd. *Cleomedes' Lectures on Astronomy* (Berkeley, Calif. 2004).

Bradley, Guy. *Ancient Umbria* (Oxford 2000).

Braund, David. *Georgia In Antiquity* (Oxford 1994).

"Greeks and Barbarians: The Black Sea Region and Hellenism Under the Early Empire," in *The Early Roman Empire in the East* (ed. Susan E. Alcock, Oxford 1997) 121–36.

"The River Cyrus and Its Tributaries: Routes Through and Beyond the Caucasus," in *Topoi* (forthcoming).

Bredow, Iris von. "Regnum Bosporanum," *BNP* 12 (2008) 443–50.

Breton, Jean-François. *Arabia Felix from the Time of the Queen of Sheba* (tr. Albert LaFarge, Notre Dame, Ind. 1999).

Briant, Pierre. *From Cyrus to Alexander: A History of the Persian Empire* (tr. Peter T. Daniels, Winona Lake, Ind., 2002).

Bridgman, Timothy P. *Hyperboreans: Myth and History in Celtic–Hellenic Contexts* (New York 2005).

Brincken, Anna-Dorothee von den. "Antipodes," in *Trade, Travel, and Exploration in the Middle Ages* (ed. John Block Friedman *et al.*, New York 2000) 27–9.

Brodersen, Kai. "The Geographies of Pliny and His 'Ape' Solinus," in *Brill's Companion to Ancient Geography: The Inhabited World in Greek and Roman Tradition* (ed. Serena Bianchetti *et al.*, Leiden 2015) 298–310.

Broderick, George. "Some Island Names in the Former 'Kingdom of the Isles': A Reappraisal," *Journal of Scottish Name Studies* 7 (2013) 1–28.

Bromwich, James. *The Roman Remains of Southern France* (London 1993).

Broughton, T. R. S. *The Magistrates of the Roman Republic* (New York 1951–1952, 1986).

Brown, Truesdell S. *Onesicritus: A Study in Hellenistic Historiography* (Berkeley, Calif. 1949).

Bruère, Richard T. "Pliny the Elder and Virgil," *CP* 51 (1956) 228–46.

Burchard, Christoph. "Pline et les Esséniens," *RBi* 69 (1962) 533–69.

Burkert, Walter. *Lore and Science in Ancient Pythagoreanism* (tr. Edwin L. Minar, Jr., Cambridge, Mass. 1972).

Burr, Viktor. *Mare Nostrum* (Stuttgart 1932).

Burstein, Stanley M. "The War Elephants East and West," *World History Connected* 17 (2020) 1–14.

Burton, G. P. "Proconsuls, Assizes and the Administration of Justice under the Empire," *JRS* 65 (1975) 92–106.

Burzacchini, Gabriele. "Sul 'Fonte Pliniano' di Manduria (*NH* 11 103 [106], 226)," *Eikasmos* 11 (2000) 297–307.

Caldwell, Thomas Francis. "*The Career of Licinius Mucianus*" (M. A. thesis, University of Melbourne, 2015).

Carey, Sorcha. *Pliny's Catalogue of Culture: Art and Empire in the Natural History* (Oxford 2003).

Cary, M. and E. H. Warmington. *The Ancient Explorers* (revised edition, London 1963).

Casson, Lionel. *The Ancient Mariners* (second edition, Princeton, N.J. 1991).
Libraries in the Ancient World (New Haven, Conn. 2001).
The Periplus Maris Erythraei (Princeton, N.J. 1989).
Ships and Seamanship in the Ancient World (Princeton, N.J. 1971).
Celestino Pérez, S. and Carolina López-Ruiz. *Tartessos and the Phoenicians in Iberia* (Oxford 2016).
Centrone, Bruno. "C. Sulpicius Gallus," *EANS* 768.
"P. Nigidius Figulus," *EANS* 572–3.
"Ocellus Lucanus," *EANS* 586.
Chartrand, Mark R. *National Audubon Society Field Guide to the Night Sky* (New York 1991).
Ciprés, Pilar. "*Hispania Citerior* en la Geografía de la *Naturalis Historia* de Plinio," *Veleia* 31 (2014) 15–32.
Clarke, Katherine. *Making Time For the Past* (Oxford 2008).
Clayman, Dee L. *Berenice II and the Golden Age of Ptolemaic Egypt* (Oxford 2014).
Cobb, Matthew Adam. "From the Ptolemies to Augustus," in *The Indian Ocean Trade in Antiquity* (ed. Matthew Adam Cobb, Leiden 2018) 17–51.
Cohen, Getzel M. *The Hellenistic Settlements in the East from Armenia and Mesopotamia to Bactria and India* (Berkeley, Calif. 2013).
The Hellenistic Settlements in Europe, the Islands, and Asia Minor (Berkeley, Calif. 1995).
The Hellenistic Settlements in Syria, the Red Sea Basin, and North Africa (Berkeley, Calif. 2006).
Collins, Roger. *Spain: An Oxford Archaeological Guide* (Oxford 1998).
Coltelloni-Trannoy, Michèle. *Le Royaume de Maurétanie sous Juba II et Ptolémée* (Paris 1997).
Connors, Catherine. "A River Runs Through It: Waterways and Narrative in Strabo," in *The Routledge Companion to Strabo* (ed. Daniela Dueck, Abingdon 2017) 207–18.
Cook, J. M. "Pliny on Icarian Shores," *CQ* 9 (1959) 116–25.
The Troad: An Archaeological and Topographical Study (Oxford 1973).
Cornell, T. J. (ed.). *The Fragments of the Roman Historians* (Oxford 2013).
Couprie, Dirk L., *et al. Anaximander in Context: New Studies in the Origins of Greek Philosophy* (Albany, N.Y. 2003).
Courrént, Mireille. "*Innumerabiles mundos*... Pline l'Ancien et l'esprit de découverte (*Histoire Naturelle* 11, 1–3)," *Euphrosyne* 27 (1999) 51–64.
Cunliffe, Barry. *The Ancient Celts* (Oxford 1997).
The Extraordinary Voyage of Pytheas the Greek (New York 2001).
Facing the Ocean: The Atlantic and Its Peoples (Oxford 2001).
Cusset, Christophe and Philip Thibodeau. "C. Iulius Hyginus," *EANS* 454–5.
Dakaris, S. "Ephyra," *PECS* 310–1.
Dalby, Andrew. *Food In the Ancient World From A to Z* (London 2003).
D'Andria, Francesco. "Messapi e Peuceti," in *Italia: omnium terrarum alumna* (Milan 1988) 653–715.
Davies, Oliver. *Roman Mines in Europe* (Oxford 1935).

de Ligt, Luuk. "The Economy: Agrarian Change During the Second Century," in *A Companion to the Roman Republic* (ed. Nathan Rosenstein and Robert Morstein-Marx, Chichester 2007) 590–605.

de Souza, Philip. *Piracy in the Graeco-Roman World* (Cambridge 1999).

Demougin, Ségolène. *Prosopographie des chevaliers romains Julio-Claudiens* (Paris 1992).

Desanges, Jehan. "Des éléphants, ou artisans, et des quarante-cinq rois des Ethiopiens (*Pline l'Ancien,* vi, *186*)," in *Orbis Aethiopicus: Studia in honorem Stanislaus Chojnacki* (Albstadt 1992) 83–7.

"Pline l'Ancien et l'Istrie: Anomalies et hypothèses," *CRAI* (2004), 1181–203.

Desautels, Jacques. "Les monts Riphées et les Hyperboréens dans le traité hippocratique *Des airs, des eaux et des lieux*," *RÉG* 84 (1971) 289–96.

Devoto, Giacomo. "Umbri e Etruschi," *SE* 28 (1960) 263–76.

Dicks, D. R. *The Geographical Fragments of Hipparchus* (London 1960).

Dilke, O. A. W. *Greek and Roman Maps* (Ithaca, N.Y. 1985).

Diller, Aubrey. "The Ancient Measurements of the Earth," *Isis* 40 (1949) 6–9.

Dissanayake, C. B. and S. W. Nawaratne, "The Occurrence of Gold in the Gem Beds of Sri Lanka," *Economic Geology* 76 (1981) 733–8.

Dissanayake, C.B., *et al.* "The Geology, Mineralogy and Rare Element Geochemistry of the Gem Deposits of Sri Lanka," *Bulletin of the Geological Society of Finland* 72 (2000) 5–20.

Dmitriev, Sviatoslav. "Observations on the Historical Geography of Roman Lycaonia," *GRBS* 41 (2000) 349–75.

Draycott, Jane. "The Sacred Crocodile of Juba II of Mauretania," *AClass* 53 (2010) 211–17.

Dueck, Daniela. "Artemidoros of Ephesos," *EANS* 165.

"Herakleides 'Kritikos'," *EANS* 366.

"Isidoros of Kharax," *EANS* 443–4.

"Krates of Mallos," *EANS* 490.

"Xenophon of Lampsakos," *EANS* 840–1.

Dunand, Françoise and Christiane Zivie-Coche, *Gods and Men in Egypt, 3000* BCE *to 395* CE (tr. David Lorton, Ithaca, N.Y. 2004).

Edelstein, L. and I. G. Kidd. *Posidonius 1: The Fragments* (second edition, Cambridge 1989).

Eichel, Marijean H. and Joan Markley Todd, "A Note on Polybius' Voyage to Africa in 146 B. C.," *CP* 71 (1976) 237–43.

Elvers, Karl-Ludwig. "Verrius," *BNP* 15 (2010) 323–5.

Errington, R. M. "Rome and Greece to 205 BC," *CAH* 8 (second edition 1989) 81–106.

Espinosa Espinosa, David. *Plinio y los 'Oppida de Antiguo Lacio': el Proceso de Difusión del Latium en Hispania Citerior* (*BAR-IS* 2686, London 2014).

Euzennat, M. "Lixus," *PECS* 521.

"Sala," *PECS* 793–4.

"Volubilis," *PECS* 988–9.

Evans, James. *The History and Practice of Ancient Astronomy* (Oxford 1998).

Evans, Rhiannon. "The Cruel Sea? Ocean as Boundary Marker and Transgressor in Pliny's Roman Geography," *Antichthon* 39 (2005) 105–19.

Fakhry, Ahmed. *Siwa Oasis* (Cairo 1973).

Ferguson, John. "China and Rome," *ANRW* 2.9.2 (1976) 581–603.

"Classical Contacts With West Africa," in *Africa in Classical Antiquity: Nine Studies* (ed. L. A. Thompson and J. Ferguson, Ibadan 1969) 1–25.

Fernández Nieto, Francisco Javier. "Incidentes de una Correción Geográfica de Solino a Plinio: la Isla de Cos," in *Solinus. New Studies* (ed. Kai Brodersen, Heidelberg 2014) 90–5.

Finocchi, S. "Segusium," *PECS* 819–20.

Fishwick, Duncan. "The Annexation of Mauretania," *Historia* 20 (1971) 467–87.

Forbes, William T. M. "The Silkworm of Aristotle," *CP* 25 (1930) 22–6.

Forsyth, P. Y. "After the Big Bang: Eruptive Activity in the Caldera of Greco-Roman Thera," *GRBS* 33 (1992) 191–204.

Fossey, John M. "Notes on the Texts of Pliny's *Natural History*," *Euphrosyne* 3 (1969) 215–23.

Topography and Population of Ancient Boiotia (Chicago, Ill. 1988).

Fotheringham, J. K. "Cleostratus," *JHS* 39 (1919) 164–84.

Fraser, P. M. *Cities of Alexander the Great* (Oxford 1996).

Ptolemaic Alexandria (Oxford 1972).

Frazer, James G. *Pausanias's Description of Greece* (New York 1965).

Fridh, Å. J. "Les théories de l'Océan chez Pline l'Ancien," in *Göteborgs Kungl. Vetenskaps- och Vitterhets-Samhälles Handlingar* 4 (1952) 1–22.

Gabba, E. "Rome and Italy: the Social War," *CAH* 9 (second edition 1994) 104–28.

Gábli, Cecilia. "Pliny the Elder on the Etruscans," *AAntHung* 52 (2012) 137–63.

Galsterer, Hartmut. "Coloniae," *BNP* 3 [2003] 545–54.

"Municipium," *BNP* 9 (2006) 293–7.

Gayraud, Michel. *Narbonne antique des origines à la fin du* III*e siècle* (Paris 1981).

Gazzano, Francesca. "La Lidia di Plinio il Vecchio," *Maia* 70 (2018) 260–79.

Geus, Klaus. "Hellenistic Maps and Lists of Places," in *Hellenistic Astronomy: the Science in its Contexts* (ed. Alan C. Bowen and Francesca Rochberg, Leiden 2020) 232–9.

"Progress in the Sciences: Astronomy and Hipparchus," in *Brill's Companion to Ancient Geography: The Inhabited World in Greek and Roman Tradition* (ed. Serena Bianchetti *et al.*, Leiden 2016) 150–60.

Gillespie, Caitlin C. *Boudica: Warrior Woman of Roman Britain* (Oxford 2018).

Goudineau, C. "Tropaeum Alpium," *PECS* 936–7.

Graf, Fritz. "Fortuna," *BNP* 5 (2004) 505–9.

Graham, A. J. *Colony and Mother City in Ancient Greece* (Manchester 1964).

Graham, Daniel W. "Anaximandros of Miletos," *EANS* 75–6.

The Texts of Early Greek Philosophy (Cambridge 2010).

Grane, Thomas. "Pliny and the Wandering Mountain. A New Interpretation of Pliny's Account of the Northern Barbaricum With an Archaeological Context," *ARID* 37 (2012) 7–28.

Greaves, Alan M. *Miletos: A History* (London 2002).

Groom, Nigel. "The *Periplus*, Pliny and Arabia," *AAE* 6 (1995) 180–95.

Gruen, Erich S. "The Expansion of the Empire under Augustus," *CAH* 10 (second edition 1996) 147–97.

The Hellenistic World and the Coming of Rome (Berkeley, Calif. 1984).

Rethinking the Other in Antiquity (Princeton, N.J. 2011).

Habicht, Christian. "New Evidence on the Province of Asia," *JRS* 65 (1975) 64–91.

Haley, Evan W. *Baetica Felix* (Austin, Tex. 2003).

Hamilton, J. R. "Alexander and the Aral," *CQ* 21 (1971) 106–11.

Hammond, N. G. L. and L. J. Roseman, "The Construction of Xerxes' Bridge over the Hellespont," *JHS* 116 (1996) 88–107.

Hansman, John. "Charax and the Karkheh," *IrAnt* 7 (1967) 21–58.

Harrell, James A. "Discovery of the Red Sea Source of *Topazos* (Ancient Gem Peridot) on Zabargad Island, Egypt," in *Twelfth Annual Sinkankas Symposium: Peridot and Uncommon Green Gem Minerals* (ed. Lisbet Thoresen, Fallbrook, Calif. 2014) 16–30.

Harris, W. V. "Roman Expansion in the West," *CAH* 8 (second edition 1989) 107–62.

Hart, Timothy C. "Evolving Ethnographies in Pliny the Elder's Transdanubian Exegesis (*HN* 4.80–1)," *CQ* 71 (2021) 1–8.

Head, Barclay V. *Historia Numorum* (Oxford 1911).

Healy, John F. *Pliny the Elder on Science and Technology* (Oxford 1999).

Heckel, Waldemar. *Who's Who in the Age of Alexander the Great* (Malden, Mass. 2006).

Hind, J. G. F. "Greek and Barbarian Peoples on the Shores of the Black Sea," *AR* 30 (1983–4) 71–97.

Hine, Harry M. *An Edition With Commentary of Seneca, Natural Questions, Book Two* (New York 1981).

Hirt, Alfred Michael. *Imperial Mines and Quarries in the Roman World* (Oxford 2010).

Hodge, A. Trevor. "Purity of Water," in *Handbook of Ancient Water Technology* (ed. Örjan Wikander, Leiden 2000) 95–9.

Hölbl, Günther. *A History of the Ptolemaic Empire* (tr. Tina Saavedra, London 2001).

Hoover, Donald B. "Determining Garnet Composition From Magnetic Susceptibility and Other Properties," *Gems and Gemology* (2011) 272–85.

Hope Simpson, R. and J. F. Lazenby. *The Catalogue of the Ships in Homer's Iliad* (Oxford 1970).

Horn, Heinz Günter. "Die antiken Steinbrüche von Chemtou/Simitthus," in *Die Numider: Reiter und Könige nördlich der Sahara* (ed. Heinz Günter Horn and Christoph B. Rüger, Köln 1979).

Huxley, George. "Choirilos of Samos," *GRBS* 10 (1969) 12–29.

Irby, Georgia L. "Hydrology: Ocean, Rivers, and Other Waterways," in *A Companion to Science, Technology, and Medicine in Ancient Greece and Rome* (ed. Georgia L. Irby, Chichester 2016) 181–96.

"Tracing the *Orbis Terrarum* From Tingentera," in *New Directions in the Study of Ancient Geography* (ed. Duane W. Roller, University Park, Pa., 2019) 103–34.

Irby-Massie, Georgia L. "Apollodoros of Artemita," *EANS* 106–7.

"Aristokreon," *EANS* 138.

"Basilis," *EANS* 190–1.

"Dalion (Geog.)," *EANS* 222.

"P. Terentius Varro of Narbo, 'Atax'," *EANS* 774.

Isserlin, B. S. J., *et al.* "The Canal of Xerxes: Summary of Investigations 1991–2001," *BSA* 98 (2003) 369–85.

Jodin, André. "L'éléphant dans le Maroc antique," in *Actes du quatre-vingt-douzième congrès national des sociétés savantes* (Paris 1970) 51–64.

Jones, A. H. M. *Cities of the Eastern Roman Provinces* (second edition, Oxford 1971).

Jones, Alexander. "Dionusodoros," *EANS* 266.

"Timaios (Astrol.)," *EANS* 810.

Jones-Lewis, Molly. "Tribal Identity in the Roman World: The Case of the Psylloi," in *The Routledge Handbook of Identity and the Environment in the Classical and Medieval Worlds* (ed. Rebecca Futo Kennedy and Molly Jones-Lewis, Abingdon 2016) 192–209.

Kahrstedt, Ulrich. "Zwei Probleme im kaiserzeitlichen Griechenland," *SO* 28 (1950) 66–75.

Kaplan, Philip. "Ktesias of Knidos," *EANS* 496.

Karttunen, Klaus. *India in Early Greek Literature* (Helsinki 1989).

India and the Hellenistic World (Helsinki 1997).

Keddie, Anthony. *Class and Power in Roman Palestine* (Cambridge 2019).

Kent, John Harvey. *Corinth 8.3: The Inscriptions 1926–1950* (Princeton, N.J. 1966).

Kepler, Johannes. *Harmonices Mundi* (Linz 1619).

Keyser, Paul T. "Amometos (of Kurene?)," *EANS* 67.

"Aristarchos of Sikuon," *EANS* 133.

"Aristeides (of Knidos?)," *EANS* 135.

"Astunomos," *EANS* 175.

"Baiton," *EANS* 186.

"Demodamas of Miletos," *EANS* 233.

"The Geographical Work of Dikaiarchos," in *Dicaearchus of Messana: Text, Translation, and Discussion* (ed. William W. Fortenbaugh and Eckart Schütrumpf, New Brunswick, N.J. 2001) 353–72.

"Koiranos," *EANS* 485.

"The Kozy Kosmos of Early Cosmology," in *New Directions in the Study of Ancient Geography* (ed. Duane W. Roller, University Park, Pa. 2019) 5–55.

"Licinius Mucianus," *EANS* 507–8.

"Philemon," *EANS* 645.

"Sebosus Statius," *EANS* 729.

"Serapion of Antioch," *EANS* 733.

"Turranius Gracilis," *EANS* 820.

and Georgia L. Irby-Massie, "Agathokles of Atrax," *EANS* 43–4.

Kidd, I. G. *Posidonius 2: The Commentary* (Cambridge 1988).

Kirwan, L. P. "Nubia and Nubian Origins," *GJ* 140 (1974) 43–51.

Kitchell, Kenneth F., Jr. *Animals in the Ancient World From A to Z* (Abingdon 2014).

Kleiner, Diana E. E. *Roman Sculpture* (New Haven, Conn. 1992).

Klotz, Alfred. "Die geographischen commentarii des Agrippa und ihre Überreste," *Klio* 24 (1931) 38–58, 386–466.

Kokkinos, Nikos. "The City of 'Mariamme': An Unknown Herodian Connection?" *MediterrAnt* 5 (2002) 715–46.

The Herodian Dynasty: Origins, Role in Society and Eclipse (Sheffield 1998).

"The Location of Tarichaea: North or South of Tiberias?" *PEQ* 142 (2010) 7–23.

Köves-Zulauf, Thomas. "Plinius d. Ä. und die römische Religion," *ANRW* 2.16.1 (1978) 187–288.

Kyle, Donald G. *Sport and Spectacle in the Ancient World* (Malden, Mass. 2007).

Landels, John G. *Music In Ancient Greece and Rome* (London 1999).

Law, R. C. C., "The Garamantes and Trans-Saharan Enterprise in Classical Times," *JAH* 8 (1967) 181–200.

Leekley, Dorothy and Robert Noyes. *Archaeological Excavations in the Greek Islands* (Park Ridge, N. J. 1975).

Lefranc, Jean-Phillipe. "La Géologie, Pline l'Ancien et l'Itinéraire de Cornelius Balbus (20 avant Jésus-Christ): Nouvelles Identifications," in *Histoire et archéologie de l'Afrique du Nord: actes du IIIe colloque international réuni dans le cadre du 110e congrès national des sociétés savantes, Montpellier, 1–15 avril 1985* (Paris 1986) 303–15.

Lehoux, Daryn. "L. Annaeus Seneca," *EANS* 84–5.

"Panaitios of Rhodes (Lindos)," *EANS* 608.

Lieberman, Samuel. "Who Were Pliny's Blue-Eyed Chinese?" *CP* 52 (1957) 174–7.

Liverani, Mario. "The Libyan Caravan Road in Herodotus IV 183–185," *JESHO* 43 (2000) 496–520.

López-Ruiz, Carolina. "Tarshish and Tartessos Revisited: Textual Problems and Historical Implications," in *Colonial Encounters in Ancient Iberia: Phoenician, Greek, and Indigenous Relations* (ed. Michael Dietler and Carolina López-Ruiz, Chicago, Ill. 2009) 255–80.

Luque Moreno, Jesús. "Plinio, *nat.* 2.20 (§ 84): *De sideribus musica*," *RELat* 14 (2014) 53–86.

McKenzie, Judith. *The Architecture of Alexandria and Egypt c. 300* BC *to* AD *700* (New Haven, Conn. 2007).

Magie, David. *Roman Rule in Asia Minor to the End of the Third Century After Christ* (Princeton, N.J. 1950).

Malaise, Michel. "Pline l'Ancien a-t-il séjourné en Égypte?" *Latomus* 27 (1968) 852–63.

Manconi, D. "Nora," *PECS* 628.

Ma'oz, Zvi Uri. "Banias," *NEAEHL* 136–43.

Marcy, G. "Remarques sur l'habitation berbère dans l'antiquité: à propos des *mapalia*," *Hespéris* 29 (1942) 23–40.

Martín-Arroyo Sánchez, Daniel Jesús. "*Interpretatio* de la Titulatura Cívica de *Hasta Regia* a Partir del Culto a Juno Regina," *DHA* 43 (2017) 175–211.

Mayer, Marc and Isabel Rodà. "El Broccatello de Tortosa: Testimonios Arqueológicos," *Pallas* 50 (1999) 43–52.

Mederos Martín, Alfredo. "Los Periplos Atlánticos Norteafricanos de Polibio y Juba II Trasmitidos por Plinio (*N. H.*, v, 9–10)," *Gerión* 31 (2013) 239–68.

Mendell, Henry. "Oinopides of Khios" *EANS* 587.

Mercer, John. *The Canary Islanders: Their Prehistory, Conquest, and Survival* (London 1980).

Merrills, Andy. *Roman Geographies of the Nile From the Late Republic to the Early Empire* (Cambridge 2017).

Mierse, William E. *Temples and Towns in Roman Iberia* (Berkeley, Calif. 1999).

Mikula, Karol *et al.* "What Was the River Ister in the Time of Strabo? A Mathematical Approach," *Tatra Mountains Mathematical Publications* 78 (2021) 1–47.

Miles, S. B. "Pliny's Geography of the East Coast of Arabia," *JAS* 10 (1878) 157–72.

Miller, J. Innes. *The Spice Trade of the Roman Empire* (Oxford 1969).

Mitchell, Stephen. "The Galatians: Representation and Reality," in *A Companion to the Hellenistic World* (ed. Andrew Erskine, Oxford 2005) 280–93.

Mitford, Timothy Bruce. *East of Asia Minor: Rome's Hidden Frontier* (Oxford 2018).

Momigliano, A. "The Origins of Rome," *CAH* 7.2 (second edition 1989) 52–112.

Moro, Constantino. "*Liquidiore Flamma*: un Problema Lessicale in Plinio il Vecchio (*Nat. Hist.* III 93 s.)," *PP* 57 (2002) 115–20.

Murphy, Trevor. *Pliny the Elder's Natural History: The Empire in the Encyclopedia* (Oxford 2004).

Netz, Reviel. "Archimedes of Surakousai," *EANS* 125–8.

Nicolet, Claude. *Space, Geography, and Politics in the Early Roman Empire* (Ann Arbor, Mich. 1991).

Ogden, Daniel. *The Legend of Seleucus* (Cambridge 2017).

Ogilvie, R. M. and A. Drummond. "The Sources for Early Roman History," *CAH* 7.2 (1989) 1–29.

Olbrycht, Marek J. "Die Aorser, die Oberen Aorser und die Siraker bei Strabon. Zur Geschichte und Eigenart der Völker im nordostpontischen und nordkaukasischen Raum im 2. – 1. Jh. v. Chr.," *Klio* 83 (2001) 425–50.

Oleson, John Peter. "Irrigation," in *Handbook of Ancient Water Technology* (ed. Örjan Wikander, Leiden 2000) 181–215.

Oost, Stewart Irvin. "Amynander, Athamania, and Rome," *CP* 52 (1957) 1–15.

Ossendrijver, Mathieu. "Hellenistic Astronomy and the Babylonian Scribal Families," in *Hellenistic Astronomy: The Science in its Contexts* (ed. Alan C. Bowen and Francesca Rochberg, Leiden 2020) 426–39.

Pais, Ettore. *Fasti Triumphales Populi Romani* (Rome 1920).

Pandey, Nandini B. "Caesar's Comet, the Julian Star, and the Invention of Augustus," *TAPA* 143 (2013) 405–49.

Parker, Grant. *The Making of Roman India* (Cambridge 2008).

Parker, S. Thomas. "The Decapolis Reviewed," *JBL* 94 (1975) 437–41.

Pearson, Lionel. *The Greek Historians of the West: Timaeus and His Predecessors* (Atlanta, Ga. 1987).

The Lost Historians of Alexander the Great (New York 1960).
Pettegrew, David K. "The *Diolkos* of Corinth," *AJA* 115 (2011) 549–74.
Pina Polo, Francisco. "De la Ciudad Indígena *Salduie-Salduvia* a la Colonia Romana *Caesar Augusta*," *Gerión* 35 (2017) 541–50.
Pliny. *C. Plini Secundi Naturalis Historiae Libri* XXXVII (ed. Karl Friedrich Theodor Mayhoff, Leipzig 1892–1909).
Histoire naturelle 2 (ed. Jean Beaujeu, Paris 1950).
Histoire naturelle 3 (ed. Hubert Zehnacker, Paris 2004).
Histoire naturelle 4 (ed. Hubert Zehnacker and Alain Silberman, Paris 2015).
Histoire naturelle 5.1 (ed. Jehan Desanges, Paris 2003).
Histoire naturelle 6.2 (ed. Jacques André and Jean Filliozat, Paris 2003).
Histoire naturelle 6.4 (ed. Jehan Desanges, Paris 2008),
Histoire naturelle 13 (ed. A. Ernout, Paris 1956).
Natural History (vols. 1–2, ed. H. Rackham, Cambridge, Mass. 1938–42).
Plinius Naturkunde (ed. Roderich König *et al.*, Munich, 1973–96).
Podossinov, Alexander V. "The Indians in Northern Europe? On the Ancient Roman Notion of the Configuration of Eurasia," in *The Periphery of the Classical World in Ancient Geography and Cartography* (ed. Alexander V. Podossinov, Leuven 2014) 133–45.
Potts, Daniel T. *The Arabian Gulf in Antiquity* 2 (Oxford 1990).
Pytheas of Massalia. *On the Ocean* (ed. Christina Horst Roseman, Chicago, Ill. 1994).
Radt, Stefan. *Strabons Geographika* (Göttingen 2002–11).
Redmount, Carol A. "The Wadi Tumilat and the 'Canal of the Pharoahs,'" *JNES* 54 (1995) 127–35.
Rees, B. R. "Pliny, *N. H.* ii.22," *CR* 8 (1958) 213–15.
Richardson., L., Jr. *A New Topographical Dictionary of Ancient Rome* (Baltimore, Md. 1992).
Riess, Ernest. *Nechepsonis et Petosiridis Fragmenta Magica* (Bonn 1890).
Rochberg, Francesca. "Epigenes of Buzantion," *EANS* 290.
"Kidennas (Kidinnu) of Babylon," *EANS* 475.
"Kritodemos," *EANS* 493.
"Petosiris, or Nekhepso-Petosiris," *EANS* 637–8.
Roller, Duane W. *Ancient Geography* (London 2015).
The Building Program of Herod the Great (Berkeley, Calif. 1998).
"The Boiotian Pig," in *Essays in the Topography, History, and Culture of Boiotia* (*Teiresias Supplement* 3, ed. A. Schachter, Montreal 1990) 139–44.
Cleopatra: A Biography (Oxford 2010).
Cleopatra's Daughter and Other Royal Women of the Augustan Era (Oxford 2018).
Empire of the Black Sea: the Rise and Fall of the Mithridatic World (Oxford 2020).
Eratosthenes' Geography (Princeton, N.J. 2010).
A Historical and Topographical Guide to the Geography of Strabo (Cambridge 2018).
"Phoenician Exploration," in *The Oxford Handbook of the Phoenician and Punic Mediterranean* (ed. Carolina López-Ruiz and Brian Doak, Oxford 2019) 645–53.

"The Problem of the Location of Straton's Tower," *BASOR* 252 (1983) 61–6.
Review of Hirschfeld, *Qumran in Context*, *BASOR* 340 (2005) 94–5.
Scholarly Kings: The Writings of Juba II of Mauretania, Archelaos of Kappadokia, Herod the Great and the Emperor Claudius (Chicago, Ill. 2004).
"Seleukos of Seleukeia," *AntCl* 74 (2005) 111–18.
Through the Pillars of Herakles (New York 2006).
"Timosthenes of Rhodes," in *New Directions in the Study of Ancient Geography* (ed. Duane W. Roller, University Park, Pa., 2019) 56–79.
The World of Juba II and Cleopatra Selene: Royal Scholarship on Rome's African Frontier (London 2003).
Romanelli, Pietro. "La Campagna di Cornelio Balbo nel Sud Africano," in *Mélanges offerts à Léopold Sédar Senghor* (Dakar 1977) 429–38.
Roncaglia, Carolynn E. *Northern Italy in the Roman World* (Baltimore, Md. 2018).
Rose, Charles Brian. *The Archaeology of Greek and Roman Troy* (Cambridge 2014).
Roseman, Christina Horst. *Pytheas of Massalia: On the Ocean* (Chicago, Ill. 1994).
Ross, Micah T. "Hellenistic Egyptian Planetary Theory," in *Hellenistic Astronomy: The Science in its Contexts* (ed. Alan C. Bowen and Francesca Rochberg, Leiden 2020) 160–5.
Rostovtzeff, M. *Social and Economic History of the Hellenistic World* (Oxford 1941).
Sader, H. "Tell Kazel and Ancient Simyra," *Berytus* 38 (1990) 16–22.
Sallmann, Klaus Günther. *Die Geographie des älteren Plinius in ihrem Verhältnis zu Varro: Versuch einer Quellenanalyse* (Berlin 1971).
Salmon, Edward Togo. "The Iron Age: the Peoples of Italy," *CAH* 4 (second edition 1988) 676–719.
Salway, Benet. "Putting the World in Order: Mapping in Roman Texts," in *Ancient Perspectives* (ed. Richard J. A. Talbert, Chicago, Ill. 2012) 193–234.
Santana Santana, Antonio and Trinidad Arcos Pereira. "The Canary Islands in Pliny the Elder's *Naturalis Historia*," *AfrRom* 16 (2004) 297–312.
"La Descripción de África en Plino (*Nat.* v y vi)," in *Libyae Lustrare Extrema: Realidad y Literatura en la Visión Greco-romana de África* (Seville 2008) 347–60.
Santiago, Rosa-Araceli. "Enigmas en Torno a Saguntum y Rhoda," *Faventia* 16 (1994) 55–64).
Sauer, Eberhard. *Dariali: The 'Caspian Gates' in the Caucasus From Antiquity to the Age of the Huns and the Middle Ages* (Oxford 2016).
Scarborough, John. "L. Aelius Gallus," *EANS* 34–5.
Schiemann, Gottfried. "Ius," *BNP* 6 (2005) 1122–34.
Schmitt, P. "A la recherche du Char des Dieux," in *Littérature gréco-romaine et géographie historique: Mélanges offerts à Roger Dion (Caesarodunum* 9bis, Paris 1974) 473–9.
Schneider, Pierre. "Les 'climats' de Pline l'Ancien (*N. H.* 6, 212–220): essais de transcriptions graphiques," *GA* 20–1 (2011–12) 179–204.
"Erythraean Pearls in the Roman World," in *The Indian Ocean Trade in Antiquity* (ed. Matthew Adam Cobb, Leiden 2018) 135–56.
"*Fauces Rubri Maris*: The Greco-Roman Bab al-Mandab (5th Century B. C. – 2nd Century A. D.)," *OT* 12 (2014) 193–272.

Schwartz, Daniel R. "One Temple and Many Synagogues: On Religion and State in Herodian Judaea and Augustan Rome," in *Herod and Augustus* (ed. David M. Jacobson and Nikos Kokkinos, Leiden 2009) 385–98.

Scopacasa, Rafael. *Ancient Samnium: Settlement, Culture, and Identity Between History and Archaeology* (Oxford 2015).

Scott, David A. *Copper and Bronze in Art* (Los Angeles, Calif. 2002).

Scott, Kenneth. "The *Sidus Iulium* and the Apotheosis of Caesar," *CP* 36 (1941) 257–72.

Scullard, H. H. *The Elephant in the Greek and Roman World* (Cambridge 1974).

Shaw, Brent D. "The Elder Pliny's African Geography," *Historia* 30 (1981) 424–71.

Shinnie, P. L. *Meroe: A Civilization of the Sudan* (London 1967).

Shipley, D. Graham J. *The Early Hellenistic Peloponnese* (Cambridge 2018).

Sidebotham, Steven E. *Berenike and the Ancient Maritime Spice Route* (Berkeley, Calif. 2011).

Smith, Andrew M., II. *Roman Palmyra: Identity, Community, and State Formation* (Oxford 2013).

Speke, John Hanning. *Journal of the Discovery of the Source of the Nile* (New York 1868).

Standish, J. F. "The Caspian Gates," *G&R* second series 17 (1970) 17–24.

Starr, Chester G. "The Roman Emperor and the King of Ceylon," *CP* (1956) 27–30.

Stern, E. Marianne. "Glass Production," in *The Oxford Handbook of Engineering and Technology in the Classical World* (ed. John Peter Oleson, Oxford 2008) 520–47.

Stern, Sacha. *Calendars in Antiquity* (Oxford 2012).

Stiehle, R. "Der Geograph Artemidoros von Ephesos," *Philologus* 11 (1856) 193–244.

Stoneman, Richard. *The Greek Experience of India: From Alexander to the Indo-Greeks* (Princeton, N.J. 2019).

Strabo. *Géographie* 8 (ed. François Lasserre, Paris 2003).

Sullivan, Richard D. *Near Eastern Royalty and Rome, 100–30* BC (Toronto 1990),

Svennung, J. *Belt und Baltisch. Ostseeische Namenstudien mit besonderer Rücksicht auf Adam von Bremen* (Uppsala 1953).

Syme, Ronald. "Pliny the Procurator," *HSCP* 73 (1969) 201–36.

Székely, Melinda. "Tra Coptos e Berenice (Plin. *Nat. Hist.* 6. 26.100–106)," *AAntHung* 50 (2010) 63–9.

Tarn, W. W. *Alexander the Great 2: Sources and Studies* (Cambridge 1948).

The Greeks in Bactria and India (revised third edition, Chicago, Ill. 1997).

Tarrant, Harold. "Thrasullus, Ti. Claudius (of Mendes?)," *EANS* 806–7.

Tausend, Klaus. "Inder in Germanien," *OT* 5 (1999) 115–25.

Tautz, Burkhard. "Ein Komet im Jahre 76 n. Chr.?" *Gymnasium* 111 (2004) 15–23.

Thibodeau, Philip. "Agathokles of Khios," *EANS* 44.

"Bion of Soloi," *EANS* 193.

"Epigenes of Rhodes," *EANS* 291.

"M. Terentius Varro of Reate," *EANS* 774–8.

"Traditionalism and Originality in Roman Science," in *The Oxford Handbook of Science and Medicine in the Classical World* (ed. Paul T. Keyser with John Scarborough, Oxford 2018) 593–613.

Thomsen, Rudi. *The Italic Regions from Augustus to the Lombard Invasion* (tr. Anne Monrad and Paul Monrad, Copenhagen 1947).

Thomson, J. Oliver. *History of Ancient Geography* (Cambridge 1948).

Thoresen, Lisbet. "Archeogemmology and Ancient Literary Sources on Gems and Their Origins," in *Gemstones in the First Millennium* AD*: Mines, Trade, Workshops and Symbolism* (ed. Alexandra Hilgner *et al.*, Mainz 2017) 155–217.

and James A. Harrell. "Archaeogemmology of Peridot," in *Twelfth Annual Sinkankas Symposium: Peridot and Uncommon Green Gem Minerals* (ed. Lisbet Thoresen, Fallbrook, Calif. 2014) 31–51.

Tierney, J. J. "The Map of Agrippa," *Proceedings of the Royal Irish Academy* 63 (1962–4) 151–66.

Torelli, M. "Sublaqueum," *PECS* 864.

Travlos, John. *Pictorial Dictionary of Ancient Athens* (New York 1971).

Trofimova, Anna. A. (ed.) *Greeks on the Black Sea: Ancient Art From the Hermitage* (Los Angeles, Calif. 2007).

Turner, Brian, and Richard J. A. Talbert. *Pliny the Elder's World: Natural History Books 2–6* (Cambridge 2022).

Untermann, Jürgen. "Hispania, Iberia: Languages," *BNP* 6 (2005) 392–6.

Van der Meer, L. B. *The Bronze Liver of Piacenza: Analysis of a Polytheistic Structure* (Amsterdam 1987).

Van der Waerden, B. L. "History of the Zodiac," *AOF* 16 (1952–3) 216–30.

Van Son, D. W. L. "Zur Deutung von Plinius *N. H.* II 167," *Mnemosyne* 15 (1962) 146–52.

Vanhove, Doris, *Roman Marble Quarries in Southern Euboea* (Leiden 1995).

Vercoutter, J. "The Gold of Kush," *Kush* 7 (1959) 120–53.

Von Oppolzer, Theodor. *Canon der Finsternisse* (Vienna 1887).

Wachter, Rudolf. "Alphabet: The Greek Alphabet," *BNP* 1 (2002) 530–41.

Wade-Gery. H. T. "The Islands of Peisistratos," *AJPh* 59 (1938) 470–5.

Walbank, F. W. *A Historical Commentary on Polybius* (Oxford 1957–79).

Wallace, Paul W. *Strabo's Description of Boiotia* (Heidelberg 1979).

Ward-Perkins, J. B. *Roman Imperial Architecture* (second edition, Harmondsworth 1981).

Weerakkody, D. P. M. *Taprobanê: Ancient Sri Lanka as Known to Greeks and Romans* (Turnhout 1997).

Weinstock, Stefan. "Libri Fvlgvrales," *PBSR* 19 (1951) 122–53.

West, Stephanie. "'The Most Marvellous of All Seas': the Greek Encounter with the Euxine," *G&R* 50 (2003) 151–67.

Wheeler, Mortimer. *Rome Beyond the Imperial Frontiers* (Harmondsworth 1955).

Wiedemann, T. E. J. "From Nero to Vespasian," *CAH* 10 (second edition 1996) 256–82.

Wiegand, Theodor and Hans Schrader. *Priene* (Berlin 1904).

Wilkes, J. J. "The Danubian and Balkan Provinces," *CAH* 10 (second edition 1996) 545–85.

Winkler, Gerhard. "Geographie bei den Römern: Mela, Seneca, Plinius," in *Geographie und verwandte Wissenschaften* 2 (ed. Wolfgang Hübner and Germaine Aujac, Stuttgart 2000) 141–61.

Wiseman, James and Konstantinos Zachos. "The Nikopolis Project: Concept, Aims, and Organization," in *Landscape Archaeology in Southern Epirus, Greece* (ed. James Wiseman and Konstantinos Zachos, *Hesperia Supplement* 32, 2003) 1–22.

Woolf, Greg. *Becoming Roman: The Origins of Provincial Civilization in Gaul* (Cambridge 1998).

Woźniak, Marek A., *et al.* "Ptolemaic Berenike: Resources, Logistics, and Daily Life in a Hellenistic Fortress on the Red Sea Coast of Egypt," *AJA* 125 (2021) 247–81.

Yalouris, N. "Epidauros," *PECS* 311–14.

Young, Gary K. *Rome's Eastern Trade* (London 2001).

Zaninović, M. "Pola," *PECS* 720–1.

"Salona," *PECS* 799.

Zehnacker, Hubert. "La description de Rome dans le livre 3 de la *NH*," in *Pline l'Ancien, témoin de son temps: Conventus Pliniani internationalis Namneti 22–26 Oct. 1985* (ed. Jackie Pigeaud and José Oroz Reta, Salamanca 1987) 307–20.

"L'Europe du Nord dans l'*Histoire Naturelle* de Pline l'Ancien (*N. H.* IV, 88–104)," *RÉL* 82 (2004) 167–86.

Zucker, Arnaud. "Papirius Fabianus," *EANS* 610–11.

Index of Passages Cited

Greek and Latin Authors

Italicized numbers are citations in ancient texts; romanized numbers are pages in this volume. References to Pliny the Elder are limited to citations in the preface and Books 7–37.

Abydenos (*FGrHist* #685)
  F2, 49
Acilius, C. (*FGrHist* #813)
  F1, 201
Aelian
  On the Characteristics of Animals
    7.2, 253
    7.8, 61
    7.40, 397
    16.18, 354
Aeschylus
  Agamemnon
    256–7, 195
    281–316, 82
  Eumenides
    681, 202
  Persians
    306, 339
    318, 339
    732, 339
    865, 325
  Prometheus Bound
    349–52, 42
    365–74, 157
    422, 336
    714–16, 326
    717–21, 336
    729–35, 323
  Seven Against Thebes
    273–4, 203
  Suppliant Women
    260–7, 195
  F402, 157
Agatharchides of Cnidus
  5.62, 370, 386
  F73, 388
Agrippa, M. Vipsanius
  F4, 117
  F7, 239
  F8, 237
  F9, 132
  F13, 157
  F16, 183
  F17, 235
  F21, 227
  F22, 232
  F23, 335
  F25–6, 303
  F29, 398
  F30–1, 375
  F32, 344
  F34, 404
  F37, 112
  F42, 159
  F47, 183
  F50, 212
  F51, 324
  F53, 336
  F55, 384
  F58, 272
  F66, 403
Airs, Waters, and Places
  5, 394
  5–7, 85
  13, 76, 340, 394
  13–24, 85
Aithiopis
  Argument 4, 232
Akousilaos of Argos (*FGrHist* #2)
  F25a, 201

Alexander Polyhistor (Cornelius Alexander, *FGrHist* #273)
  F104, 170
Alkaios
  F34, 59
Ammianus Marcellinus
  15.10.2–7, 176
  22.8.10, 225
Anaxagoras of Clazomenae (ed. Graham)
  F3, 55
  F9, 71
  F37, 47
  F45, 49
  F50, 53
  F50–1, 62
Anaximander of Miletus (ed. Graham)
  F22–5, 53
  F23, 46
  F25, 46
  F31, 52
  F34-6, 94
Anaximander the Younger (*FGrHist* #9)
  F2, 217–18
Anaximenes of Miletus (ed. Graham)
  F29–31, 63
Androsthenes of Thasos (*FGrHist* #5)
  F2, 366
  F5, 357
Antiochos of Syracuse (*FGrHist* #555)
  F1, 158
  F13, 161
Apollodorus, *Bibliotheke*
  Book 1
    9.3, 196
  Book 2
    5.11, 251, 254
  Book 3
    10.3, 52
    15.2–4, 315
Apollonios of Rhodes, *Argonautika*
  Book 1
    151–5, 52
    601–4, 223
  Book 4
    269, 281
    289, 172
    325–6, 172
    480–1, 184
    569–71, 185
    634, 131
    659–63, 143
    1232–49, 94
    1477–80, 52
    1717–20, 88
Appian
  Civil War
    Book 1
      111, 266–7
    Book 2
      86, 288
      89, 314
      98–9, 261–2
      154, 43–4
    Book 3
      49–76, 57
    Book 4
      53, 263
    Book 5
      75, 180
  Iberike
    50–7, 242
    53, 125
    56–60, 113
    76–98, 125
  Illyrike
    2–3, 183
    10, 172
    16, 180
    16–21, 170
  Libyka
    10–28, 261
    28, 259–60
    75, 263
  Mithridateios
    28, 221
    30, 203
    89, 326
    95, 161
    101, 327
    101–3, 332
    103, 342
  Syriaka
    11, 319
    57, 369
Aratos of Soloi, *Phainomena*
    454–9, 36
    454–61, 51
    544–52, 35
    926–32, 41
Archimedes
  On Floating Bodies 1, Proposition 2, 74
  Sand Reckoner 1.9, 46
Aristides (*FGrHist* #444)
  F5, 220
Aristides, P. Aelius
  36.85–95, 255

Aristocreon (FGrHist #667)
  F3, 396
Aristophanes
  Ploutos 210, 52
Aristotle
  Metaphysics
    12.8, 51
  Meteorologika
    Book 1
      4, 41
      5, 57
      5.342b, 58
      6.342b, 36
      6.343b, 55–6
      7, 55
      8, 35
      9, 60, 62
      10–12, 72
      13, 60
      13.349a, 62
      13.350b, 226, 236, 301
      14, 202
    Book 2
      2, 95
      3.359a, 96, 290
      4–5, 64
      5, 65
      5.362a, 65
      6, 63, 66
      6.365a, 66
      8, 85
      8.366a, 94, 204
      9.369b, 62, 69
    Book 3
      1, 67
      2.372a, 58, 72
      2–4, 58
      4.375a, 72
  On the Heavens
    1.9.278b, 33
    2.3, 42
    2.4, 74
    2.9.290b, 34
    2.13.293, 36
  Parts of Animals
    3.2.663a, 388
  Politics
    2.6.6–7, 330
    5.9.6, 330
    7.9, 150
  Research on Animals
    2.1.499b, 388
    5.19.551b, 219, 343
    6.18.572a, 243
  F287, 94
  F488, 221
  F570, 315
  F595, 221
Arrian
  Anabasis
    Book 1
      1–6, 209, 226
      1–7, 183
      11–12, 224
      23.6–7, 305
    Book 2
      13.8, 295
      16–24, 292
    Book 3
      1.5, 283
      3–4, 267
      8–15, 81
      18.11–2, 368
      19–20, 338
      27.4–5, 360
      29.2, 76
    Book 4
      1.3, 340
    Book 5
      1.1–6, 353
      3, 346
      25–9, 347
      26.1–2, 76
      29.1, 340
    Book 6
      1–18, 351
      5–6, 348
      21–2, 84
      21.5, 361
    Book 7
      3, 349
      16, 76
      16.3, 336
      20.7, 366
  Indika
    4.2–7, 348
    4.13, 346
    6, 354
    6.9, 350
    8, 365
    8.10, 359
    18–42, 361
    21.6, 354
    21.11, 354
    23.5, 352
    30, 362
    31, 361
    37.1–2, 362
    39, 362

Periplous of the Euxine Sea
9, 336
11, 328
Artemidorus of Ephesus
F8, 245
F14, 245
F81, 277
F86, 272
F94, 392
F103, 384
F108, 350
F110, 335
F117, 313
Athenaios
3.83c, 280
3.83d, 400
4.170, 280
5.201c, 388
10.442b, 347
14.624b, 304
Aufidius Bassus (*FRH* #78)
F3, 332
Augustus (emperor)
Res gestae
8, 126
26, 75, 125, 235, 391
29, 367
30, 182
31, 357
32–3, 367
Aulus Gellius
13.9, 60
Avienus, *Ora maritima*
112, 237
117, 77
171–2, 241
174–7, 243
248, 120
383, 77
412–15, 77
559–60, 122

Baeton (*FGrHist* #119)
F2a, 347
F4, 350
Basilis (*FGrHist* #178)
F2, 392
Bion of Soloi (*FGrHist* #668)
F4, 390
F6, 396–7

Caelius Antipater (*FRH* #15)
F51, 173
Caesar, C. Julius
Civil War
1.60, 121
Gallic War
Book 1
10, 176
36, 227
37, 78, 227
54, 110
60, 121
Book 3
8–10, 240
41, 180
Book 4
10, 175
20–36, 237
Book 5
6–23, 237
13, 237
Book 6
24, 6
24–5, 236
25, 171, 226, 235
44, 110
Callimachus
Aitia
F11, 172
Hymn
1.21, 200
4.28–40, 88
F107, 178
F548, 266
F579, 185
F581, 222
Calpurnius Piso Frugi, L. (*FRH* #9)
F12, 68
F15a, 68
F37, 173
Cato, M. Porcius (the Elder)
Origines (*FRH* #5)
F49, 167
F53, 140
F57, 170
F58, 173
F59–60, 175
F68, 159
Catullus
29.19–20, 242
66, 80
Censorinus
de die natali
13.5, 34
Cicero, M. Tullius
Against Verres
2.3.12, 111
Catalina
1.11, 69

Cicero, M. Tullius (cont.)
- *de divinatione*
  - *Book 1*
    - *18–19*, 67
    - *35*, 68, 88
  - *Book 2*
    - *40–1*, 40
    - *42–3*, 68
    - *82*, 69
    - *89*, 60
- *de finibus*
  - *4.11*, 39
- *de natura deorum*
  - *Book 1*
    - *81–2*, 38
    - *101*, 38
    - *120–1*, 37
  - *Book 2*
    - *14*, 56
    - *51–2*, 42, 45
    - *53*, 43
    - *56*, 60
    - *58*, 33
    - *60–5*, 37
    - *62*, 39
    - *84*, 36
    - *98*, 73
    - *115*, 34, 73
    - *115–16*, 34
    - *116*, 36
    - *117*, 60
    - *130*, 298
  - *Book 3*
    - *39*, 38
    - *44*, 38
    - *63*, 37
- *de oratore*
  - *3.43*, 147
- *de provinciis consularibus*
  - *4*, 212
- *de senectute*
  - *49–50*, 48
- *Letters to Atticus*
  - *#24*, 6, 106
  - *#26*, 6
  - *#176*, 314
  - *#359*, 58
  - *#366*, 157
  - *#374*, 58
  - *#414*, 65
- *Letters to His Friends*
  - *3.8.6*, 110
  - *6.7–8*, 68
  - *6.7–9*, 18
  - *9.15.2*, 129
  - *10.15*, 131
  - *10.17*, 131
  - *12.15.2*, 355
- *Philippics*
  - *6.5*, 166
- *pro Fonteio*
  - *33*, 129
  - *36*, 131
- *Republic*
  - *2.11*, 147
- *Timaeus*
  - *1*, 20
- *Tusculan Disputations*
  - *1.45*, 323
  - *5.68–9*, 45

Claudius (emperor, *FRH* #75)
- *F4*, 332
- *F5*, 333
- *F6*, 373

Clitarchus (Kleitarchos, *FGrHist* #137)
- *F12*, 335
- *F29*, 399
- *F31*, 143

Cornelius Nepos (*FRH* #45)
- *F7*, 325
- *F10*, 171

Ctesias (*FGrHist* #688)
- *F1b*, 340
- *F1b.16–19*, 359
- *F13.21*, 224
- *F45.6*, 92
- *F45eβ*, 99

Daimachos (*FGrHist* #716)
- *F3*, 350
- *F5*, 350

Dalion (*FGrHist* #666)
- *F3*, 397

de bello africo
- *3*, 259
- *47*, 59
- *88–93*, 262
- *91–7*, 261
- *97*, 263

de bello alexandrino
- *59*, 260
- *62*, 260
- *72–7*, 326

de bello hispaniensi
- *28–31*, 116

Demodamas (*FGrHist* #428)
- *F2*, 340

Demosthenes
- *Third Philippic*
  - *26*, 292

Dicaearchus
- *F118–20*, 206

F123, 405
F125, 107
Dio, Cassius
Book 36
12–13, 326
54.2–4, 78
Book 39
54, 119
Book 40
25, 297
27, 339
Book 41
42.7, 260
45, 181
Book 43
9, 263
36–8, 260
50.3–5, 196
Book 45
7.1–2, 56
17, 58
Book 46
29–38, 57
55, 171
Book 48
20.4, 263
Book 49
35, 170
43.7, 260
Book 50
11.3, 260
Book 51
20–5, 119
23, 183
23–7, 190
25.5, 210
Book 52
22, 133
Book 53
22.6, 110
25.8, 126
26, 250
Book 54
5.4–5, 393
8.4, 148
9.8–9, 357
20, 175, 210
20.3, 190
24.3, 137
Book 55
8.4, 118
8.7, 148
28.5, 75
29–34, 179
32.2, 154
Book 56
18–24, 236
Book 57
19.6, 148
Book 60
8–9, 250
8.2, 329
19–23, 237
24.4, 176
Diodorus of Sicily
Book 1
33.9–12, 385
36.11–12, 283
37.3, 307
38–41, 281
40.5, 81
55.7–9, 389
57, 385
66.3–6, 283
Book 2
4–20, 326
29–31, 370
45, 330
48.6–9, 96
51.3–4, 394
Book 3
8.2, 394
15–17, 360
26, 395
29, 398
35.4, 388
43.5, 364
54.7–55.3, 400
68.2, 257
Book 4
18.5, 108
Book 5
13.1, 139
17.1, 153
21, 335
22, 238
25.4, 171
32, 237
38, 243
64.1, 218
Book 6
1–2, 39
Book 11
54.1, 198
Book 12
31, 145
31.1, 134
Book 13
35.3, 157
Book 14

Diodorus of Sicily (cont.)
82.7, 193
83.4–7, 58
100, 159
101, 150
Book 15
29, 288
48.3, 90
48–9, 197
Book 16
3.4, 207
8.2–3, 208
42.4, 313
53–5, 58
Book 17
7.4–7, 47
40.4, 292
49–51, 267
52, 283
72, 368
81.1, 360
93.2, 348
95, 340
95.1, 347
98, 348
113, 263
104.7, 362
105.1, 84
106.6–7, 362
107, 348
107.6, 363
110.4, 373
Book 18
22, 299
Book 19
52, 71
95–8, 286, 377
98–9, 290
Book 20
27.3, 346
55.4, 263
74.3, 385
110.3, 205
Book 31
8.8, 208
Book 33
14, 213
Diogenes Laertius
Book 1
99, 196
119, 86
Book 2
1–2, 85
10, 71
Book 7
139, 37
140, 33
Book 8
48, 34–5, 73
Book 9
21, 73
23, 44
Diognetus (*FGrHist* #120)
F1, 347
Dionysius of Halicarnassus
On Literary Composition
25, 4
Roman Antiquities
Book 1
11.2, 150
11.3, 160
15, 165
16, 165
20.5, 139
24, 165
27, 139
30, 139
34.1, 149
35.1–2, 150
35.3, 142
37, 134
53.3, 154
72.6, 220
Book 2
1.2, 150
5, 69
70, 92
Book 3
35, 68
49, 150
Book 4
54, 148
Book 5
24–31, 92
61.3, 149
Book 7
3, 145
Book 10
2, 70
Dionysios Skytobrachion (*FGrHist* #32)
F8, 257
Dioskourides
1.73, 362
1.98, 398
3.82, 58
4.76–7, 325
5.94, 362
5.105, 362
5.124, 93

Domitius Corbulo, Cn. (*FRH* #82)
F3, 296
Dosiades (*FGrHist* #458)
F4, 217

Empedokles of Akragas (ed. Graham)
F26–7, 36
F29, 100
F100, 95
Ennius
Annals
F145, 141
Ephorus of Cyme (*FGrHist* #70)
F30, 399
F30a, 211, 228, 255, 302, 334
F113, 195, 201
F151, 220
F162, 325
F172, 399
Eratosthenes
FGrHist #241
F1a, 316
Geography
F8, 301
F10, 312
F11, 81
F15, 95, 202, 268, 288
F16, 172, 196, 385
F17–18, 290
F22, 345
F23, 346
F25, 73
F27, 296
F30, 73, 78, 101, 102
F34, 101, 102
F37, 101, 338
F40, 83, 387
F41, 84, 357
F41–3, 83
F45, 47
F47, 300, 345
F48, 333
F50, 345
F54, 83
F56, 407
F57, 389
F57–8, 84, 408
F57–63, 405
F57–66, 101
F59–60, 84
F60, 406, 407
F66, 118, 135, 195, 286
F67, 356
F69, 344, 345
F70, 344
F73, 345
F74, 355
F75, 345
F76, 355
F78, 339
F83, 368
F84, 286
F89, 372
F90, 99
F93, 366
F94, 378, 380
F95, 76, 285
F97, 385
F98, 393, 396
F103, 273
F105, 266, 273
F106, 107–8
F108, 134, 339
F110–11, 332
F111, 335
F114, 225
F115, 225, 277, 324
F115-16, 225
F116, 324
F126, 312
F127, 152
F131, 274
F134, 108
F135, 186
F139, 90, 197
F140, 96, 280
F148, 226
F150, 236
F152, 120
F155, 407
Euclid
Optics
19, 80
Euripides
Andromache
795, 232
Archelaos
F228, 201
Bacchants
1–3, 204
16, 285, 383
274–83, 144
Hecuba
1270–3, 214
Helen
1673, 202
Herakles
395–402, 251
Hippolytus
735–41, 168

Euripides (cont.)
Orestes
1635–7, 59
Phrixos B
F819, 293
Trojan Women
1079, 393
Eutropius
6.10, 212

Fabius Pictor (*FRH #1*)
F10, 177
Florus
1.41.10, 161
2.33.48, 125
2.34.62, 357
Frontinus
de acquis
93, 164

Galen
On the Combining of Drugs According to Place
1, 259
Geminos
5.16, 79
16.32, 47, 255
Greek Anthology
8.177, 370
9.58, 370

Hanno of Carthage
1, 263
1–5, 254
2–6, 401
4, 252
5–7, 254
7, 253, 255, 393
8, 399
11, 255, 393
13–14, 400
14, 253, 255
16, 255, 395, 398
18, 400
Hecataeus of Abdera (*FGrHist* #264)
F7–14, 231
F12, 315
F14, 234
T6b, 343
Hecataeus of Miletus (*FGrHist* #1)
F39, 104
F41, 104
F93–101, 178
F106, 181
F170, 228
F279, 295
F290, 335
F346, 249
F356, 104
F357, 249
Hellanicus of Lesbos (*FGrHist* #4)
F52, 292
F77, 215e
Heraclides of Pontus
F26a, 90
Herakleitos of Ephesus (ed. Graham)
F97, 44
Herodotus
Book 1
6, 309, 325
6–94, 306
14, 319
15, 261, 335
26, 54
28, 324, 326
47–56, 194
56, 205
72, 325
74, 45, 48, 71
75, 325
92, 306
94, 139, 165
95–122, 337, 367
101, 368
104, 333
125, 339
144, 304
148, 307
153, 339
160, 316
163, 131
165-7, 153
173, 311
180, 295
181–3, 370
183, 285
196, 178
201, 230
201–6, 341
202–4, 335
203, 241
204, 75
214, 340
Book 2
6, 284, 285, 289, 371
10, 88, 192
11, 366
12, 79
13, 277
16, 107
19, 80, 282

20–2, 281
29, 83, 390
30–1, 396
32, 258, 264, 269, 275
33, 129
34, 252, 325
49, 293
56, 188
71, 388
102–4, 389
109, 406
135, 316
137, 385
144, 56
145, 316
148, 283
158, 385
166, 278
172–82, 283
184, 325
Book 3
6, 385
17, 395
23, 395
70, 374
90, 191
91, 339
93, 339, 340
94, 275, 333, 351
97, 398
102–5, 349
106, 351
114, 398
115, 167, 184, 243
116, 335, 341
Book 4
6, 230, 341
7, 231
8, 107, 245
13–14, 230
17, 230
19, 228
20, 229
20–1, 330
21, 329
31, 231
32–6, 231
36, 404
41, 249
42, 106, 404
43, 113, 276, 394
44, 63
45, 76, 107, 226, 327, 340, 390, 404
48, 171
49, 178, 190, 228
53, 228
55, 228
83–7, 224
92, 210
93–143, 226
100, 230
108–9, 230
110–17, 227, 230
117, 227
144, 319
150-8, 267
152, 64, 75, 107, 112, 114
170, 269
171, 268
172, 269, 276
173, 264
174, 276
176, 276
177, 263
178, 266
179, 281
180, 276, 295
180–8, 275
181, 267
181–5, 270
183, 264, 269, 276, 285
188, 276
191, 262, 276, 281
191–9, 249
196, 77
204, 268
Book 5
3, 345
3–8, 209
11, 210
30, 221
42, 269
58, 287
92, 188
124, 210
Book 6
9, 353
36, 211
76, 96
98, 221
Book 7
22, 208
22–4, 208
33, 214
59–83, 211
61, 332
64, 341
66, 340
70, 275
73, 210

Herodotus (cont.)
78, 333
91, 300
95, 195
112, 210
115, 208
123, 208
123–4, 337
129–30, 206
Book 8
46, 189
127, 210
Book 9
93, 181
Hesiod
Catalogue of Women
F2–3, 205
F111, 201
F98, 231
Theogony
124, 393
123–5, 38
215–16, 401
270–83, 400
270–94, 254
274–5, 401
287–94, 175, 244
338, 167, 280
339, 171, 226
340, 327
344, 324
507–19, 253
517–19, 42
820–52, 66
Works and Days
166–73, 244, 323, 401
503, 65
587, 61
609, 61
618–23, 66
618–94, 65
636, 309
Hesychios
"Kerketai", 329
Hipparchus
Against the Geography of Eratosthenes
F4, 376
F5, 354
F6, 233
F11, 81
F11–15, 103
F43–60, 101
F56, 407
F57–8, 84, 408
Hippias of Elis (*FGrHist* #6)
F2, 198
Hippolytos
Refutations
1.2.2, 34
Hipponax
F2, 327
Homer
Iliad
Book 1
423, 255
452, 316
Book 2
475, 278
494–510, 203
511, 205
571, 198
575, 197
612-14, 200
632, 217
649, 218
683–5, 205
684, 205
698–9, 214
749, 189
750, 188
751–4, 96
756, 206
783, 154
844–5, 209
845, 224
848–50, 207
851–5, 324
852, 173
856, 317
862–3, 318
864–6, 306
Book 3
1–6, 212
6, 350, 394
187, 324
275–7, 36
Book 4
75–8, 41
141, 256, 388
Book 5
39, 317
234, 53
487, 51
Book 6
130–5, 353
152–205, 196
178–83, 302
Book 8
133–4, 52
Book 9

5, 191
129, 316
383, 283
529, 192
589, 192
Book 10
429, 204, 311
Book 11
25, *243*
Book 12
30, 224
128, 206
Book 14
187, 35
226, 207
Book 16
148–51, 243
234–5, 189
719, 324
Book 18
541, 207
Book 20
219–22, 243
Book 23
205–6, 255
Odyssey
Book 1
22–7, 391
23–4, 275
51–4, 107, 241
246, 216
Book 4
354–7, 88
354–62, 312
561–8, 401
Book 5
246–8, 53
270–5, 43
272, 60
273, 80
273–5, 356
295, 216
295–6, 64
Book 9
22, 217
Book 10
133–574, 143
Book 11
14, 335
284, 205
582–92, 90
107, 156
Book 12
37–54, 146
97, 288
101–14, 93
165–200, 146
260–70, 94
Book 13
161–4, 215
Book 14
314–20, 188
Book 17
526, 188
Book 21
145, 139
Book 22
267, 216
Book 24
377–8, 192
Homeric Hymns
Apollo 370–4, 91
Dionysos 7.29, 231
Herakles 1–3, 203
Hermes 68–141, 39
Horace
Epistles
2.2.181–2, 256, 401
2.2.184, 289
Odes
1.12.46–8, 56
1.29, 285
4.4, 174
4.13.13, 220
4.14, 174
Satires
1.2.101–3, 343
1.5.97–100, 100
1.8.14–16, 149

in C. Sallustium Crispum oratio
19–20, 263
Isidorus of Charax (*FGrHist* #781)
Parthian Stations
1, 284
F6, 101
F7, 102
F8, 245
F9, 272
F10, 277
F11, 237
F12, 195
F13, 313
F19, 312
T1, 376

John of Lydia
On Portents
43–52, 67

Josephus
Jewish Antiquities
Book 16
17–19, 310
Book 18
372, 370
Jewish War
Book 1
419–21, 290
Book 2
94, 292
98, 289
119–61, 291
124, 291
567, 291
Book 3
416, 63
420, 289
459–75, 289
Book 4
459–75, 290
Book 6
289, 55
Book 7
244, 227
Juba II of Mauretania (*FGrHist* #275)
F1, 376
F3, 362
F5, 162
F6, 400
F14, 280
F28, 361
F29, 371
F30–3, 365, 375, 376
F34, 384
F35–7, 389
F37, 390
F38, 279
F38a, 269, 275, 279
F42, 258
F43, 256
F44, 258, 401
F47–54, 253
F66, 285
F74–5, 386
Julius Obsequens
16, 69
21, 169
Justin
Epitome
42.3.5, 336
Juvenal
3.305–8, 143
7.130, 388
13.86–9, 40
King Nikomedes Periplous (Pseudo-Skymnos)
268, 156
Kleomedes
1.5.30, 81
2.1, 52
2.3, 46
2.5, 49
2.6, 45, 49

Licinius Mucianus, C. (*FRH A27*)
F1, 98
F6, 278
Livy
Book 1
1.3, 173, 174
3.5, 140
33.6, 149
33.9, 142
35.7, 149
38.6, 148
44.4, 148
Book 2
12–13, 92
16.8, 147
16–17, 142
33–40, 142
Book 3
10.6, 70
26–9, 163
Book 4
11.4, 142
Book 5
34–5, 167
Book 8
15, 142
24, 160
Book 9
16, 162
27.14, 163
Book 10
21.8, 144
Book 21
5.5, 119
6–15, 120
23, 122
31, 131
41.7, 175
55, 173
60–1, 121
61.5, 121
Book 22
1.8, 59
4–7, 88
21.6, 121
22.45–50, 163

Book 24
20.15, 162
42.1–8, 116
48–9, 266
Book 25
34–6, 100
36.13–16, 114
Book 28
20, 124
30.3, 112
Book 29
37.2, 39
Book 30
12.15, 260
Book 32
28.11, 109
29–31, 167
Book 34
17.2, 113
Book 35
1.1–12, 243
7–8, 119
Book 36
36.3–5, 39
38, 167
Book 37
3.2, 69
Book 38
28–30, 216
Book 39
51, 319
55, 173
Book 40
21.2, 210
38, 163
41.10, 109
50.1, 123
Book 42
19.3–6, 331
Book 43
1.8, 170
5, 174
9.7, 180
Book 44
31, 180
37.5–9, 48
Book 45
26.14, 190
34, 209
Book 91
F18, 123
F59, 107
Epitome
114, 261
Summary
20, 137
41, 124
54–5, 242
59, 172
117–18, 57
Lucan, *Civil War*
Book 1
524–9, 56
Book 3
187, 181
Book 5
650, 185
Book 6
331, 180
Book 9
411–13, 108
Book 10
141–3, 219
144–6, 257
201–3, 51
239–47, 282
247–8, 63
Lucian
On the Syrian Goddess
21–4, 294
Lucretius
Book 1
1–20, 44
25, 4
Book 2
991–8, 72
Book 4
134, 35
134–6, 33, 35
Book 5
203, 108
Book 6
167–70, 69
225–38, 67
423–50, 67
712–18, 281
840–7, 99
848–78, 268
879–89, 97
890–9, 312
Lykophron
Alexandra
594–632, 162
712–37, 146

Macrobius
Saturnalia
1.10.7, 147

Macrobius (cont.)
3.7.2, 21, 68
3.9.3, 147
3.17.14, 358
Manilius
Astronomica
1.206–14, 73
5.600, 288
Martial
10.103, 122
13.54, 121
14.1.7, 162
Martianus Capella
6.596–8, 283
Megasthenes (*FGrHist* #715)
F4, 348
F7, 350
F7b, 350
F8, 345
F9a, 346, 348
F9b, 348
F13a, 358
F14, 346
F19, 348
F20, 349
F23b, 349
F26, 354
F27, 350
F27b, 351, 394
F34, 348
Menaechmus (*FGrHist* #131)
F7, 220
Myrsilus (*FGrHist* #477)
F11, 156

Naevius
Bellum Punicum
F17, 154
Nearchus (*FGrHist* #133)
F1b, 371
F13, 361
F14, 365
F15, 366
F27, 380
Nechepso
F2, 54
Nikolaos of Damascus (*FGrHist* #90)
F57, 192
Nonnos
41.379–81, 45

On Marvellous Things Heard
58, 320
Onesicritus (*FGrHist* #134)
F6, 346
F9, 83
F10, 84
F12, 355
F13, 354
F21, 349
F28, 361
F29, 366
F30, 371
Orosius
1.2.32, 274
4.11.5–9, 141
Ovid
Ars Armatoria
2.298, 220
Epistulae ex Ponto
1.2.77, 277
Fasti
2.319, 256, 401
4.75–6, 162
5.165–82, 60
5.635–7, 140
5.725, 100
6.31, 149
Metamorphoses
2.171, 79
4.604–803, 288
6.382–400, 304
6.424–674, 213
8.304, 52
13.565–75, 214
14.458–9, 162
15.237–51, 36
15.293–5, 90
15.307–36, 98
15.313–14, 96

Parmenides (ed. Graham)
F43, 43, 78
Patrocles (*FGrHist* #712)
F4a, 336
F4b, 334
Pausanias
Book 1
9.8, 213
12.3–4, 349
41.7, 134
44.3, 217
Book 2
1.5, 308
5.3, 98
5.7, 195
28.2, 200
37.4, 199

Book 3
13.1, 52
23, 221
Book 6
10.8, 181
16.5, 82
Book 7
3.9, 308
5.1–3, 308
7, 197
17.3–4, 202
17.5–6, 197
18.7, 196
25.8–9, 90, 197
Book 8
3.5, 160
30.8, 254
Book 9
29–31, 203
30.6, 188
36.6, 205
38.2, 205
Book 10
10.6, 160
11, 158
17.5, 156
17.9, 155
19.5–12, 318
24.2, 90
24.3, 222
31.3, 191
38.1–3, 194
Periplous of the Erythraian Sea
4, 388
5, 388
9, 336
11, 328
13, 388
14, 389
15–16, 389
20, 364
23, 382
25, 364
27, 381
33, 388
36, 388
39, 364
49, 219–20, 364
53, 363
53–4, 365
53–5, 365
54, 365
56, 364
57, 65, 363
59, 358, 365
61, 354, 359
63, 353
64, 76, 343
65, 358
Petronius
Satryicon
16–18, 38
Pherecydes of Syros (FGrHist #3)
F17, 251
F74, 167–8
F156, 160
Philistides of Mallos (FGrHist #11)
F2, 218
F3, 244
Philon
Every Good Man is Free
75–91, 291
Hypothetika
11.1, 291
Philonides (FGrHist #121)
F1, 313
Philostratos
On Heroes
8.1, 311
Life of Apollonios
5.2, 94
5.41, 202
Pindar
Isthmean
4.52–7, 251
Nemean
1.1–2, 95
3.19–29, 108, 157
6.40, 200
Olympian
1.54–71, 90
2.68–72, 323
6.87–90, 204
7.54–70, 88
Paian
9, 48
Pythian
1.19–28, 157
F173, 326
Plato
Kratylos
24, 45
Phaidon
58.108d–110b, 73
Theaitetos
174a, 50
Timaios
24–6, 89, 400

Plato (cont.)
28b, 33
30ab, 34
33bc, 35
Plautus
Curculio
613, 162
Poenulus
1034, 249
Pliny the Elder
Preface
12, 4
14, 4
17, 4
28, 13
33, 15
Book 7
11–12, 342
15, 24
206, 366
30, 359, 360
95–9, 119
96, 119
115–16, 161
125, 283
136, 104
160, 26
192–8, 19
193, 25
197, 243
205, 307
206, 366, 380
208–9, 287
Book 8
32, 252
74, 388
115, 378
152, 61
166, 243
217–18, 153
Book 9
11, 107, 288
35–6, 366
35–9, 388
46, 20
62–3, 145
106, 351
106–23, 358
119–21, 358
125–38, 292
125–41, 401
127, 256
168–9, 145
Book 10
127, 184
Book 11
76–8, 343
86–91, 274
109, 61
149, 61
Book 12
19, 392
51–72, 380
51–84, 365
52, 381
53, 296
54–65, 285
63–8, 380
69, 382
69–70, 389
78, 332
84, 385, 364
Book 13
13, 222
34, 285
62, 388
104–6, 266
Book 14
56, 2
60–1, 170
73–82, 221
Book 15
8, 236
Book 16
1–2, 236
82, 44
178, 261
213, 398
Book 17
31, 145
41, 262
193, 68
Book 18
1, 72
75, 21
112–14, 145
207–17, 41
210–12, 30, 43
269–71, 65
311–12, 61
321–5, 94
Book 19
4–6, 103
10, 126
38–45, 269
101, 38
Book 25
77–9, 258
Book 27
4–5, 324

Book 29
74, 72
Book 31
8, 91
10, 141, 165
17, 390
27, 91
80, 124
Book 32
10, 77
Book 33
2–3, 72
41, 39
66, 242
78, 177
111–22, 128
118, 6, 39
Book 34
144, 123
156, 243
Book 35
39, 386
202, 153, 274
Book 36
47, 320
48, 139, 220
49, 262
49–50, 315
125, 19
130, 99
131, 92
163, 91
165, 109
190–3, 292
Book 37
32, 184
32–5, 168
42, 235
45, 75
107, 387
107–8, 386
108, 83
202, 177
204, 256
Pliny the Younger, *Letters*
Book 3
5.1–8, 87
5.3–6, 1
5.6, 9
5.7–9, 1
5.8, 1
5.17, 16, 109
Book 4
30, 98
Book 6
16, 1
16.4, 1
20, 1
Plutarch
Aemilius Paullus
29, 209
Alexander
25.4–5, 285
26, 283
31–3, 81
35, 99, 362-3
38, 368
62.2, 348
69.3, 349
70, 363
Antonius
60.2–3, 197
63, 210–11
Caesar
3.1, 314
57.8, 196
58, 144
69, 58
69.3, 57
Camillus
22, 143
Cato the Younger
65–70, 261
71, 262
Cicero
4.4, 314
Concerning the Face that Appears in the Globe of the Moon
6.923a, 37
19.931e, 48
Crassus
19–33, 297
21, 296
Demetrios
52, 196
Flamininus
20, 319
Galba
15, 329
24.4, 148
Greek Questions
15, 194
Lucullus
35, 326
Lysandros
12.1–2, 71
Marius
17, 71
Nikias
23, 49
Numa
1, 198

Plutarch (cont.)
On Common Conceptions Against the Stoics
11.1064b, 397
On the Cleverness of Animals
21.974f, 61
On the Generation of the Soul in the Timaios
31, 50
On Isis and Osiris
41, 56, 95
On the Virtue of Alexander
5.328c, 360
Otho
4.5, 141
Perikles
19.1, 211
Pompeius
34–6, 327
35, 342
80, 288
Roman Questions
61, 16, 147
74, 74
Sayings of Kings and Commanders
179f, 285
Sertorius
9, 251
14, 123
21, 121
Sulla
5.4–6, 369
9, 69
Theseus
27, 135
Polybius
Book 1
19.2, 261
39.1–6, 264
39.3–4, 93
Book 2
10–11, 170
11.11, 180
14.4–6, 134
14.5, 158
17, 167
23–31, 177
39.1, 134
41, 197
Book 3
14.1–3, 119
33.9, 119
47, 130
59.7, 254
76.5, 121
91, 144
113–17, 163
Book 5
5.8, 194
5.12, 192
45.4, 370
108.2, 180, 190
Book 6
55.1, 140
Book 10
34.5, 122
38.7, 124
Book 12
4d, 95
Book 13
9, 378, 380
Book 21
41.7, 299
Book 23
8, 209
8.4, 190
Book 24
4, 210
Book 29
16, 48
Book 30
15, 189, 209
Book 34
1.5, 106
1.7, 81
9.5–7, 245
9.8–11, 128
10.5, 130
10.6, 246
12.2–8, 208
12.2a–8, 54
15.1, 246
15.2, 402
15.3, 244
15.4, 152
15.6, 272
15.7, 398
15.7–9, 6
15.8, 264
15.9, 399
Book 35
1–3, 199
F96, 117
Polykleitos of Larisa (*FGrHist* #128)
F7, 340
Pomponius Mela
Book 1
1, 33, 34
6, 107
13, 339, 340
21–3, 249

23, 274, 275, 276
25, 104
26, 251
27, 108
30, 260
33–8, 263
35, 264
38, 265
39, 62, 267
40, 249, 272
46, 276
47, 276
48, 253
53, 281
55, 278
67, 293
88, 307
94–5, 47
109, 302
111, 327
116, 227
Book 2
78, 130
84, 129
85, 127
102, 313
110, 215
123, 155
Book 3
13, 241
15, 125, 127
30–2, 75
33, 227
39, 302, 340, 241
45, 77–8
47, 245
54, 234, 236, 238
56, 234
58, 342
59, 342
60, 342, 343
61, 344
67, 275
71, 361
73, 366
79, 366, 380
88, 388, 390
90–5, 77
91, 394
95, 253
96, 275
103, 276
104, 256
Posidonius
F8, 34
F10, 95
F49, 47, 77, 89, 309
F85, 33
F117, 34
F120, 54, 59
F122, 34, 47
F204–5, 80
F206, 333
F208, 47
F212, 345
F215, 93
F217, 94, 94, 245
F221, 95
F227, 89, 158
F235, 181
F236, 99
F244, 288
F277a, 190
F279, 96
T22, 64
Priscian
Institutiones grammaticae
6.83, 106
Prokopios
On Buildings
4.9.17, 213
Propertius
1.2.2, 220
4.5.57, 220
Pseudo-Skylax
10, 133
24, 179
98.2, 214
108, 268
109.3, 265
112, 77, 257, 400
Ptolemy, *Geographical Guide*
Book 1
1, 106
4, 81
Book 2
6.24, 126
6.38, 126
6.70, 122
6.76, 244
16.5, 180
Book 3
3.3, 156
3.6, 155
14.7, 188
Book 4
1.3, 257
3.24, 266
4.9, 268
8.1, 398

Ptolemy, *Geographical Guide* (cont.)
  Book 5
    9.12–21, 336
    9.22, 333
  Book 6
    9.8, 342
    11.6, 339
    14.1–9, 336
    14.10, 341
  Book 7
    4.1, 354
    4.10, 357
Pytheas of Massilia (ed. Roseman)
  F2, 84
  F5, 84, 227
  T3, 239
  T4–5, 408
  T18a, 44
  T19, 93
  T19–25, 335
  T21, 233
  T22, 237
  T23, 238

Quintilian,
  Institutes
    1.10.47–8, 49
    10.1.124, 20
Quintus Curtius
  6.4.18–19, 76
  9.2.3, 349

Sallust
  Histories
    2.39, 266, 267
    2.80, 121
    3.63, 119
    4.60.19, 370
  Jugurtha
    17.3, 108, 219
    17.4, 249
    17.5, 249
    18.4, 276
    18.8, 261
    79, 265
    89.5, 264
    110.8, 259
Sappho
  F199, 45
Seneca the Elder
  Suasoriae
    1.15, 235
Seneca the Younger
  Consolation to Marcia
    17.4–5, 72
  de clementia
    1.26.5, 39
  de consolatione ad Helviam Matrem
    7.2, 307
  Hercules Furens
    235–8, 108
  Letters
    16.4, 39
    31.9, 180
    110.1, 38
    114.17–19, 17
  Medea
    375–9, 78
    817–19, 99
  Natural Questions
    Book 1
      1.5–6, 57
      1.15, 59
      3–14, 72
    Book 2
      4.1, 60
      11.2, 43, 60
      13.1, 53
      18, 52
      31.1, 67
      32, 68
      32–50, 40
      39–51, 67
      45–6, 41
    Book 3
      3–7, 75
      25, 98
      25.8–10, 165
    Book 4a
      2.7, 392
      2.20–2, 281
      2.22, 281
    Book 5
      1–18, 62
      4–5, 63
      12.1, 66
      15, 73
    Book 6
      5, 50
      5–20, 86
      8.3–4, 392, 394–5
      24.3, 86
      26, 86
    Book 7
      3.2, 66
      17.2, 56
      21, 56
      21.3–4, 55
Silenos (*FGrHist* #175)
  F7, 244

Silius Italicus
3.303, 259
3.340–3, 116
3.420–41, 114
4.1.5, 175
4.232, 169
11.134–8, 175
16.368–9, 241
17.88–9, 261
Simonides the Younger (*FGrHist* #669)
T1, 392
Solinus
11.7, 47
11.15, 24
Sophocles
Antigone
104–5, 203
Oedipus at Colonus
1303, 195
Women of Trachis
1099–1100, 251
F776, 223
Sosibios (*FGrHist* #595)
F20, 59
Stephanos of Byzantion
Elixoia, 334
Kattousa, 212
Methouriades, 218
Minya, 206
Stesichorus
F271, 48
Strabo of Amaseia, *Geography*
Book 1
1.6, 80
1.20, 406
2.6, 307
2.10, 335
2.21, 217
2.23, 312
2.25, 390
2.28, 255
2.36, 93–4
2.39, 327
3.2, 77, 172
3.4, 202, 267
3.4–5, 233
3.11, 74, 385
3.11–12, 204
3.15, 172
3.16, 222
3.17, 90
3.19, 219
3.20, 89
4.2, 84, 238
Book 2
1.16, 229, 235
1.17, 329, 336
1.18, 84, 231, 408
1.20, 83, 387
1.22, 134
1.41, 274
2.2, 47
3.3–4, 404
3.4, 77, 363
3.4–5, 289, 394, 399
3.6, 78, 89-90
3.7, 85
4.1, 94, 227, 233, 245
4.8, 186
5.5–6, 78
5.7, 227
5.8, 84, 231, 408
5.13, 78
5.14, 80
5.15, 243
5.19, 249
5.20, 142
5.22, 225
5.23, 403
5.24, 85
5.26, 227
5.33, 262, 269
5.35, 389
5.42, 84
Book 3
1.3, 127
1.6, 109, 113
1.8, 251
2.3, 110, 115, 128
2.5, 64, 115
2.6, 153
2.7, 112
2.8–10, 128
2.10, 124
2.11, 116, 245
3.1–4, 127
3.2, 119
3.3, 243
3.4, 74, 242
3.7, 127
3.7–8, 125
4.1, 119, 122
4.5, 116
4.6, 124
4.8, 121
4.10, 123
4.11, 121
4.12, 121, 125
4.13, 125
4.20, 109

Strabo of Amaseia, *Geography* (cont.)
5.1, 152
5.5, 108, 340
5.7, 94, 245
Book 4
1.3, 122, 176
1.4–5, 131
1.7, 130
1.8, 130
1.11, 131
4.1, 239, 240
5.1, 237
5.1–5, 237
6.1, 138
6.5, 167
6.6, 130, 175
6.7, 175, 177
6.8, 175
6.9, 170, 174
6.10, 170, 174
Book 5
1.1, 150
1.2, 134
1.3, 138
1.4, 173, 325
1.6, 1, 78, 167
1.8, 171
1.9, 168, 184
1.10, 165, 166
1.11, 166
2.2–5, 139
2.4, 195, 200
2.5, 139
2.6, 139, 154
2.7, 155
2.8, 155
3.1, 164
3.2, 142, 163
3.3, 201
3.4, 91, 142, 149
3.6, 88, 142, 143
3.7, 148
3.7–8, 147
3.13, 164
4.2, 162, 163
4.3, 144, 146
4.5, 189
4.6, 91, 145
4.7, 146
4.8, 146
4.9, 89, 154
4.12, 55, 160, 162, 165
4.13, 150
Book 6
1.1, 89, 150, 155, 165
1.2, 134
1.2–3, 150
1.4, 150
1.5, 157, 160
1.6, 150
1.7, 72, 108
1.10, 159
2.3, 157
2.4, 95, 98, 150
2.10–11, 158
2.11, 89, 185
3.1, 160, 161
3.2, 161
3.5, 161
3.8, 160
3.8–9, 162
3.9, 92, 162
4.1, 133
4.2, 150
Book 7
1.3, 227, 236
1.3–4, 75
1.5, 167, 226
2.4, 227
3.1, 226, 301
3.2, 183, 227
3.2–3, 190
3.8, 183, 190, 226
3.9, 225, 230, 342
3.11, 167
3.13, 171
3.15, 226
3.16, 232
3.17, 229
3.18, 228, 229
3.19, 228
4.3, 229
4.6, 183
5.1, 191, 210
5.2–3, 182
5.3, 170, 178–9, 182
5.5, 179, 180
5.6, 180
5.7, 190
5.8, 181, 362
5.12, 180, 190, 211, 227
6.1, 210, 211, 212, 232
6.2, 319
7.2, 205, 210
7.3, 209
7.4, 208
7.5, 188, 189, 191
7.6, 192
7.8, 178, 180, 193, 208
7.9, 208

7.10, 189
7.11, 188, 189
F6, 207
F7, 207
F15, 208, 223, 294
F16, 208, 210
F20, 210
F21, 212, 214, 223
F22, 214
Book 8
1.1, 195
2.1, 195
3.5, 193
3.6, 205
3.7, 197, 198
3.9, 198
3.10, 197
3.19, 205
3.23, 217
4.3, 260
6.8, 199
6.13, 189
6.16, 205
6.17, 197
6.19, 197, 198
6.23, 196
7.1, 197
7.2, 90, 197
7.5, 90, 197
8.4, 96, 280
Book 9
1.4, 217
1.20, 203
2.3, 204
2.8, 94, 204
2.13, 160
2.18, 90
2.24, 203
2.25, 203
2.40, 205
3.13, 213
4.2, 204
4.4, 204, 219
4.8, 193, 194
4.11, 193
5.3, 205
5.5, 139, 193
5.6, 202, 203, 205
5.10, 189, 205
5.11, 193
5.15, 206
5.16, 223
5.17, 193, 205
5.19, 189
5.23, 205
Book 10
1.9, 221
2.1, 193
2.8, 192
2.13, 216
2.21, 193
3.1–18, 192
3.5, 106
3.15, 304
5.2–4, 221
5.5, 381
5.14, 221
Book 11
1.5, 333, 335
2.2, 330
2.3, 76
2.9, 228
2.12, 328
2.14, 328, 333
2.16, 328
2.18, 327
3.1–6, 332
4.1–8, 332
5.1, 339
5.8, 227, 228, 336
6.1, 339
7.1, 341
7.4, 340
8.2, 343
8.4, 333
10.2, 339
11.1–2, 339
11.2, 340
11.5, 296, 371
11.6, 334
12.4–5, 333
12.5, 338
13.2, 339
13.3, 339
14.2, 295
14.8, 96
14.14, 339
Book 12
2.7, 298
2.10, 128
3.3, 311
3.4, 324
3.7, 324
3.8, 324
3.9, 326
3.12, 325
3.17, 326
3.19, 326
3.20–1, 317
3.21, 308

Strabo of Amaseia, *Geography* (cont.)
4.3, 319
4.6, 190
8.6, 308
8.10, 337
8.11, 317
8.16, 96
8.17, 91
8.18, 98
Book 13
1.8, 309
1.26–7, 311
1.30, 311
1.31, 214
1.58, 223
4.6, 306
4.14, 91, 318
Book 14
1.3, 306
1.5, 306
1.20, 307
1.27, 300
1.30, 316
1.37, 308
1.43–4, 305
1.44, 91
2.13, 314
2.23–4, 306
3.5, 302
3.10, 311
4.3, 300
5.23, 325
Book 15
1.3, 346
1.4, 365
1.5, 340, 359
1.11, 347
1.15, 355
1.20, 343
1.20–2, 345
1.25, 275, 352
1.26, 340
1.27, 353
1.32, 351
1.33, 346
1.37, 351
1.44, 349
1.57, 234, 349
1.68, 348–9
1.72, 348
1.73, 349, 394
2.1, 361, 362
2.2, 360
2.8, 338
2.10, 360
2.12, 362
2.14, 365
3.4, 365
3.6, 368
Book 16
1.6, 44, 370
1.8, 369
1.15, 99
1.16, 370
1.17, 374
1.18, 374
1.23, 337
1.26–7, 285, 297
2.6, 293
2.10, 294
2.11, 373
2.13, 97, 293, 312
2.24, 287
2.25, 292
2.33, 285, 287, 288
2.41, 289
2.41–4, 290
2.42, 290
3.1, 245
3.2, 378
3.3, 379
3.4, 379
3.5, 366, 380
3.7, 362
4.2, 285
4.2–3, 381
4.4, 76, 360, 380, 389
4.5, 83, 386
4.6, 386
4.7, 83, 97, 384
4.8, 395, 396
4.9, 395
4.10, 395
4.11, 394
4.12, 398
4.17, 370, 380, 386, 395
4.18, 250
4.19, 381
4.20, 380, 395
4.21, 386, 377
4.22–4, 383
Book 17
1.2, 276, 393, 397
1.4, 282
1.5, 391
1.6–10, 284
1.7, 278, 396
1.8, 284
1.10, 364
1.13, 268

1.14, 250
1.18, 277
1.24, 284, 371
1.25, 385
1.35, 278
1.37, 283
1.42, 283
1.43, 267, 283
1.44, 264
1.48, 83, 282
1.50, 282, 283
1.53, 396
1.54, 391, 393
2.1, 85
2.2, 280, 393
2.4, 279
3.1, 405
3.2, 253, 257
3.3, 113, 251, 254, 257, 276
3.4, 258
3.5, 260, 275
3.7, 113, 258, 259, 274, 276
3.9, 259
3.12, 266
3.16, 262
3.17, 263
3.19, 263, 265, 267
3.20, 264, 265
3.22, 272
3.23, 264
3.25, 197
Suetonius Tranquillus, C.
Augustus
14, 41
21, 174, 177
21.3, 357
23, 236
29, 70
30.1, 148
49.1, 1
60, 328
90, 70
92, 41
Claudius
1.2, 75
17, 237
25, 302
35, 250
46, 56
Divine Julius
6.1, 67
37, 326
42, 196
44, 164
44.2, 143
52, 260
88, 56
Gaius
35, 250
Grammarians
17, 22
Nero
18, 176
22, 215
Otho
6.2, 148
8.3, 141
Tiberius
20, 179
37.1, 148
69, 70
Titus
4, 90

Tacitus
Agricola
30–5, 391
Annals
Book 1
3, 154
55–6, 236
Book 2
23–5, 235
41.2, 57
47, 88
52, 266, 267
54, 98, 318
Book 4
2, 148
5, 131
23, 250
24, 266
Book 6
21, 40
22, 40
Book 12
15, 329
15–16, 228
Book 13
8, 81
Book 14
12, 81
22, 164
26, 331
29, 85
31–7, 257
Book 15
44, 39
47, 56

Tacitus (cont.)
  Germania
    2, 234, 236
    16, 190
    28, 182
    34, 237
    35–6, 236
    38, 227
    45, 235
    46, 234
  Histories
    Book 1
      27, 148
      86, 141
    Book 2
      3, 92
    Book 4
      56, 237
      68, 164
Tatian
  Oration to the Greeks
    4, 378
Thales of Miletus (ed. Graham)
  F24, 52
  F25, 44
Theokritos
  1.67, 207
  7.114, 276
  17.58–76, 346
  20.37–9, 45
Theophrastus
  Lithika
    25, 360
    58, 363
  Research on Plants
    Book 1
      5.4, 398
      10.1, 62
    Book 4
      2.7, 388
      7.7–8, 97
    Book 5
      4.7–8, 357
      7.6, 169
      8.3, 143
    Book 6
      3.1–6, 269
    Book 9
      4, 285
Theopompus (*FGrHist* #115)
  F181b, 324
  F208, 292
  F316, 99
  F317, 143
  F318, 160
  F319, 189
Thucydides
  Book 1
    3.2, 195
    24, 180
    29.3, 192
    46, 189, 191
  Book 2
    8, 221
    80–1, 188
    96, 183, 190
    96.4, 207
    98, 190
    99.4, 207
  Book 3
    15, 195
    51.1, 217
    88, 158
    89, 89, 219
  Book 4
    24.4, 157
    102, 210
  Book 6
    2, 156
  Book 7
    33.4, 160
    50, 49
Tibullus
  2.3.58, 256, 401
  2.4.29-30, 220
Timaeus of Tauromenium (*FGrHist* #566)
  F13, 134
  F41c, 95
  F63, 156
  F67, 244
  F74, 238
  F75a, 233
Timagenes (*FGrHist* #88)
  F7, 174
Timosthenes of Rhodes (ed. Roller)
  F6, 64
  F6–8, 64
  F13, 384
  F15, 277
  F16, 313
  F27, 328

Valerius Antias (*FRH* #25)
  F19, 149
Valerius Maximus
  1.6.5, 70
  1.8.ext., *18*, 98
  8.7.ext., *2*, 134
Varro, M. Terentius
  de lingua latina

Book 5
18, 35
25, 91
30, 140
31, 106, 108
41–54, 147
42, 149
71, 165
85, 92
143–65, 147
Book 6
20, 100
23, 147
Book 7
7, 69
14, 35
22, 215
Book 9
60, 150
de re rustica
Book 1
2.1, 135
Book 2
Preface, *6*, 161
Velleius Paterculus
Book 1
15.5, 130
Book 2
49.4, 166
96, 182
106, 75
110–16, 179
112.3, 182
Vergil
Aeneid
Book 1
108–12, 273
242–53, 173
744, 60, 79
Book 2
21, 316
Book 3
384–7, 143
500, 140
516, 60
Book 4
473, 206
Book 7
10, 143
378–81, 45
410–13, 142
563–71, 91
Book 8
332, 140
681, 57
Book 9
173, 39
680, 168
Book 10
179, 139
198–203, 173
Book 11
225–90, 162
532–835, 142
Book 12
701–3, 139
Eclogues
9.47, 56–7
Georgics
Book 1
5, 35
5–6, 33
204–7, 60
240–6, 81
365–7, 41
365–9, 59
Book 2
136–76, 133
217–25, 144
490, 4
Book 3
271–9, 243
Book 4
317, 206
563–6, 146
Vitruvius
Book 1
4.12, 143
6.10, 62
Book 2
Preface 1–4, 284
8.10, 320
Book 6
1.3–4, 85
1.9–11, 85
1.11, 43
Book 8
2.6, 257
3.2, 141
3.20, 98
3.24, 153
5.1, 54
Book 9
1.11, 51

Xenophanes of Colophon (ed. Graham)
F29–30, 38
F29–42, 37
F32–3, 37
F59, 79, 94
F70, 55

Xenophon
Anabasis
1.5.1, 296
4.1.8–11, 338
5.4.11, 365
6.2.2, 324
Hellenika
3.2.10, 211
4.3.11–14, 58
6.1.7, 193
Kyropaideia
1.1.4, 340
Memorabilia
4.7.5, 36

Greek and Latin Inscriptions
CIL
2.2638, 126
2.3267–8, 124
6.3828, 213
9.3019, 87
IG
7.570, 204
9.2.521, 206
ILS
8885, 172
OGIS
586, 277
SIG
495.104, 235

Biblical References
2 Chronicles
9:21, 112
Daniel
3:19–27, 99
Exodus
14:2, 285
14:9, 285
Ezekiel
38:6, 335
Genesis
2:14, 331
8:4, 331
8:6–12, 356
10:2, 335
11:31, 297
19, 290
37:25, 285
Joshua
17:11, 291
Judges
1:27, 291
1 Kings
4:12, 291
10, 380
1 Maccabees
11:28, 289
Numbers
21:25–30, 285
Psalms
120:5, 285

Other Ancient Literature
Aitareyabrahmana
7.18, 349
Mahabharata
2.48.4, 349

General Index

Ethnyms may be indexed with their toponyms. Romans are generally listed under their nomen. Many toponyms appear in a number of variants, which often indicate different sources or even errors, but in the interests of thoroughness and clarity, many of these are listed in this index. It is not always possible to determine whether repeated mentions of the same toponym refer to identical or different places. Ancient authors only appear in this index when discussed as a part of the commentary; a full list of ancient authors cited appears in the Index of Passages Cited.

Abdera, 25, 223, 225
Abgar II of Osrhoene, 296
Abila, Mt., 107, 259
Aborigines, 142
Absyrtides (Apsyrtides) Islands, 184
Abydos, Egyptian city, 283
Abydos, Hellespontine city, 71, 213
Acamas, Cape, 313
Acarnania, 191–2
Achaeans (Black Sea region), 328
Achaeans, Anchorage and Tower of, 214
Achaeans, Homeric ethnym, 214, 216
Achaei, 204–5
Achaia (province), 205
Achaia (region), 90, 197–8, 201, 215
Achaian Federation, 197
Achaton, 194
Achelous River, 192
Acheron River, 191
Acherusia Cavern, 324
Acherusia Lake, 191
Achilles, 205, 232, 228, 311, 316
Acila, 364, 379
Acilius Balbus, M'., 58, 70
Acone, 324
Acoreus, 51
Acra Iapygia, 161
Acrauceles, 269
Acroceranium, 159
Acroceraunian Cape, Mountains, or Promontory, 180, 183–8, 191, 214
Acta Triumphorum, 17, 70, 270–1
Acte, 202
Actium, 191–2; Battle of, 131, 196–7, 210
Ada of Halicarnassus (sister of Mausolos), 305
Addiris, Mt., 253
Adiabeni, 332
Adiectum Latium, 142
Adoni, 240
Adramytteos, 310, 316
Adramyttium Gulf, 310
Adriatic Sea, 10, 87, 104, 135–6, 146, 149, 164–5, 177, 334, 407–8; and amber, 184–5, 232; and Argonauts, 180; and Danube, 171–2, 181–5; Gulf or Bay, 108, 159, 161, 184, 18; head of, 82, 93, 96, 133, 172–3, 324; oysters from, 145; southern end of, 102. *See also* Hadriatic.
Adulis, 387
Adunicates, 131
Aea, 327, 384
Aedemon, 256
Aedi, 211
Aegae, 307
Aegean Sea, 99, 102, 186, 208–12, 212–14, 217, 221, 268, 200–1, 304, 307, 317–19, 325, 402; Bay, 108; currents in, 90, 94, 224–5, 232, 311, 323; etymology of, 215; islands in, 189, 215, 218–19, 222–3, 303, 315; and Phoenicians, 287; routes around, 199–200, 307; winds in, 65
Aegeus, King, 215
Aegialos, 197
Aegila, 217
Aegina, 205, 217
Aegipani, 253, 276
Aegium, 202
Aegos River, 71

Aegosthena, 202
Aela (Aelana), 381
Aelamitae, 381
Aele, 270
Aelius Gallus, L., 383, 391
Aemilia (Italian region), 166
Aemilia, Via, 87, 167
Aemilius Lepidus, M., triumvir, 58
Aemilius Lepidus, M., (187 BC), 87
Aemilius Papus, L., 177
Aemilius Paullus, L., (cos. 168 BC), 70, 209
Aemilius Paullus, L., (cos. 50 BC), 70
Aemilius Scaurus, M., (cos. 115 BC), 70
Aemilius Scaurus, M., (aedile 58 BC), 288
Aeminius, 242
Aenaria, 154
Aeneas, 142, 147, 154
Aenienses, 193
Aeningia, 234
Aenona, 178
Aenuscabales, 382
Aeolia (Aeolis), Aeolians, 308–10, 325
Aeolian Islands, 89–90, 158
Aequicoli, 164
Aequiculani, 163–4
Aeschylus, 82, 130, 336
Aesculapius, 38–9; Spring of, 96
Aesepus River, 317
Aesonenes, 122
Aether, 36, 46
Aetheria, 393
Aethiopia, Aethiopians, 56, 85, 270, 280–3, 350, 389–99, 35; as generic term, 78, 255, 259, 267, 274
Aethiopian Aroteres, 388
Aethiopian Ocean or Sea, 102, 398
Aethiopian Tarraelii, 275
Aethiops, 393
Aetna (Etna), Mt., 98, 100, 157
Aetolia, Aitolians, 192–3, 215
Aetolian Federation, 194
Aex Rock, 214–15
Afranius, L., 77
Africa (continent), 82, 119, 136, 249–73, 389, 406; as continent, 106–8, 390, 398, 404; circumnavigation of, 77, 113, 399–400, 404; East, 77, 102, 384, 389; interior, 274–6; North, 218, 253–4, 261, 278; sailing routes of, 93, 136, 155, 157, 314, 321, 403; size of, 404–5; Sub-Saharan, 77, 212, 257, 280–2, 305, 394, 396–8; West, 77, 81, 99–100, 112–14, 275, 279, 398–400
Africa (Roman province), 261–7
Africa Nova, 261, 263
Africa Vetus, 263
African wind, 222
Aga, Mt., 295
Agamemnon, 37, 162, 200, 307–8
Agatha, 130
Agatharchides of Cnidus, 381
Agathocles, people so named, 22
Agathyrsi, 231
Ager Gallicus, 167
Aglaosthenes of Naxos, 22
Agra, 381
agraian issues, 79, 231, 269
Agriophagi, 397
Agrippa Postumus, 154
Agrippina the Younger, 56, 81
Aiaia, 143
Aidoneus, 35
Aietes, 327
Aigaion (mountain), 218
Aigospotamoi, 71
Ailana, 383
Ajax, Tomb of, 311
Akran Iapygiam, 12
Akridophagoi, 398
Alabanda, 305
Alabanenses, 125
Alabaster, 358
Alabeta, 279
Alani, 227
Albania in Caspian region, 332, 336, 209
Albanus, Mt., 149
Albenses, 163
Albion, name for Great Britain, 237
Albiones, Hispanic people, 241
Albula River, 140
Album, Levantine cape, 292
Album, Mauretanian cape, 107
Albuminganum, 137
Alcyons, 65–6
Aldebaran, 60
Aleian Plains, 298–9
Alexander (Paris), 36
Alexander I of Epiros, 160
Alexander Polyhistor, 22, 170
Alexander the Great, 22–3, 47, 82, 160, 183, 188, 207–9, 224, 268, 283–4, 290, 294–5; and Arabia, 76, 97, 367; eastern expedition of, 24–6, 63–4, 83, 224, 292, 305, 308, 317, 333, 337–9, 360; and India, 11, 83–4, 275, 345–54; and Mesopotamia, 100, 263, 362–3, 369–83; and Persians, 337, 367; and Rome, 143; successors of, 72, 213; topographical manipulation by, 76, 329, 336–7, 340, 351; writings on, 23, 25, 28–9, 143, 336, 347–8, 350
Alexander, King, 315
Alexandria (Egypt), 81, 84, 88, 273, 278, 283–4, 388, 404, 406; library at, 4, 16; location and

routes to and from, 272, 276–7, 314, 364, 392, 403
Alexandria at Charax, 375
Alexandria by Issos, 101
Alexandria Eschate, 340
Alexandria in Aria, 359
Alexandria in Gedrosia, 362
Alexandria Troas, 92
Alexandropolis in Nisiaea, 367
Alica, 145
Alinda, 305
Allobroges, 131–2
Alopeconnesus, 223
Alpes Maritimae (province), 137
Alphabet, 287
Alpheus River, 95, 198
Alps, 74, 86, 95, 129, 174–7, 181–2, 235–6, 254, 301, 407; Cottian, 167, 170; inhabitants of, 137–8; Maritime, 71, 131, 137; passes and routes through, 170, 245–6; Romans and, 246
Alsium, 136
Altanus wind, 62
Altars (Arae, islands), 273
Altinum, 168, 170
Amalcius, 233
Amanirenas, Meroitic queen, 393
Amantis, Amantes, 181
Amanus, Mt., 294, 298
Amasis, Egyptian king, 283
Amazonius Mt., 301
Amazons, 134, 227, 301, 308, 330, 335
Amber, 168, 184–5, 232–3, 235
Amber Islands, 184
Ambilatri, 239
Ambisontes, 176–7
Ambitouti, 318
Ambracian (Ambracius) Bay or Gulf, 90, 188–9, 191–2
Amenemes III of Egypt, 278
Ameria, 71, 166
America, 78
Amisum, 326
Ammon Oracle, 62, 267–8, 278–9. *See also* Hammon.
Amometus, 23
Ampelome, 382–3
Ampelusia, Cape, 251
Amphipolis, 209
Ampitus, 328
Ampsaga River, 260, 263, 266
Ampsanctus, 91
Amymone Spring, 199
Amyntas of Galatia, 300
Anactorica, 191
Anahita, 326
Anakreon, 316
Anapauomenon Spring, 97
Anaphe, 88
Anas River, 109, 111–12, 116–18, 242
Anatili, 130
Anaxagoras of Clazomenae, 31, 45, 71
Anaxikrates, 76
Anaximander of Miletus, 23, 34, 42, 85–6
Anaximander of Miletus, the younger, 23, 217–8
Anaximenes of Miletus, 31, 85
Anazaberi, 299
Anchialum, 212
Ancona, 82, 136, 165–7, 172–3
Ancus Marcius, Roman king, 142
Andelonenses, 123
Anderae, 395
Andromeda (mythic personality), 35, 288, 312, 291
Andros, 98, 221
Androsthenes of Thasos, 357, 366, 378
Anemurium, 313
Angerona, 147
Anio Novus (aqueduct), 141
Anio River, 164
Annius Milo, Titius, 70
Annius Plocamus, Popilius, 354–6
Antaeus, 251
Antandros, 215
antarctic, 79
Antenor, 173
Anthedon, 288
Anthi, 335
Anthropophagi, 230, 342, 397–8
Antias, Valerius, 17, 21, 100
Antichthones, 354–6
Anticlides, 23
Antigenes, 23
Antigoneia, 188
Antigonenses, in Epirus, 188
Antilibanus, Mt., 293
Antioch, cities so named, 369
Antioch-on-the Orontes, 293–7, 331–2
Antiochos of Syracuse, 151, 158, 161
Antiochus I, 295, 339, 368, 373
Antiochus III, 360, 378, 380
Antiochus IV, 375, 378, 380
Antiochus V, 375
Antipodes, 354–5
Antipolis (in Gallia), 153
Antipolis (locality near Rome), 149
antipous, antipodes, 73
Antirhium, 90, 193, 197
Antissa, 90
Antium, 154
Antoniopolis, 306

Antonius, M., 57–8, 192, 196, 210, 250, 260, 300, 306, 364
ants, gold digging, 349
Aornos, 188–9, 191
Aorsi, 227
Aous (Aoos) River, 181, 188
Apame, 312
Apamea in Caria, 305
Apamea in Mesopotamia, 373, 378
Apamea, towns so named, 312
Apameia in Phrygia, 90
Apeliotes wind, 164
Apennines, 67, 138, 140, 146, 160, 162–8
Aphas River, 191
Aphrodisias in Caria, 305
Aphrodisias in Cilicia, 299
Aphrodision in Hispania, 244
Aphrodite (goddess), 44, 80, 92, 122, 144
Aphrodite (planet), 43
Apia, 195
Apina, 162
Apiolae, 149
Apis, Egyptian town, 272
Apollo (god), 38, 88–90, 134, 192, 221, 231–2, 304, 306, 317, 340
Apollo (planet), 44
Apollo Clarius, 98
Apollo, Cape, 262
Apollodorus of Artemita, 23
Apollodorus of Athens, 23
Apollonatae, 232
Apollonia in the Cyrenaica, 268
Apollonia in Epirus, 99, 136, 161, 180–1
Apollonia in Lydia, 306
Apollonia in Thrace, 232
Apollonihieritae, 306
Apollonios of Perge, 50, 52, 130, 143
Apollonios of Rhodes, 172
Apollophanes, satrap, 352
Appenninus Mountains, 138. *See also* Apennines
Appia (aqueduct), 141
Appia, Via, 160, 176
Appian, 288
Apples, 251, 254, 401
Apros, 213
Apsis, 50
Apsyrtus, 184
Apulia, 159–62
Apulia et Calabria, 159–60
Aquae Albulae, 141
Aquae Curtiliae, 91
Aqueducts, 141
Aquicaldenses, 122
Aquileia, 96, 170–3, 183
Aquilo wind, 64, 231, 334
Aquilones, 65
Aquitania, Aquitani, 132, 239
Arabia, 10, 97, 246–7, 282–9, 296–8, 362–3, 375, 382–4, 404; Aelius Gallus and, 383, 391, 396; aromatics from, 286, 364, 380, 388–9; and C. Caesar, 76–7, 376–7, 381, 390; coasts of, 357, 366, 378–9; Felix, 377; interior of, 249, 286; and Seleucids, 380; trade with, 364, 377
Arabia, region in Osrhoene, 296–7
Arabian Gulf, 366
Arabian Mardani, 369
Arabian Sea (Bay), 79
Arabian Scenitae, 285, 297, 372
Arabians in Mesopotamia, 369, 390
Aracelitani, 123
Arconesus, 223
Arachosia, 359
Aracynthus, Mt., 194
Arados in Phoenicia, 277–8, 293, 295, 297, 312–13
Arados on Persian Gulf, 97, 397
Araethyrea, 198
Aramii, 341
Arasmi, 339
Aratos of Soloi, 35, 36, 51
Araxus, 196, 216
Arbela, 81
Arbii, 361
Arbilitis, 337
Arbis River, 361
Arcades, Italian population, 142
Arcadia, region in Peloponnesus, 96–7, 142, 160, 188, 195–202, 334
Archilochos, 210
Archimedes of Syracuse, 23, 74
Archisarmi, 397
Archous River, 373
Arcobrigenses, 123
Arconesus, 223
Arctic, 29, 79, 84, 92, 99, 231, 238, 405
Arcturus, 60, 65, 362
Ardaeans, 179–80
Ardea, 142
Ares (planet), 43
Aresene Lake, 96, 372
Arethusa Spring, 95
Aretissa Lake, 96, 372
Arevaci, 119, 126, 241
Argestes wind, 64
Argive Plain, 199
Argo, 52, 93, 336
Argolid, 96, 195, 197, 199, 217
Argonauts, 88, 171–2, 184, 232, 326–7, 336
Argos, 199, 202
Argos, Gulf of, 199
Argyra, 354

Ariane, 360
Ariarathes V of Cappadocia, 331
Arii, 359, 367
Arimaspi, 230
Ariminium, 87, 165
Arimphaei, 335
Aristarchus of Samos, 46, 53
Aristarchus of Sicyon, 23
Aristeas of Proconnesus, 230–1
Aristides, 23
Aristocreon, 24, 391, 396
Aristocritus, 23
Aristotle, 24, 30, 72, 86, 236
Armenia, 81–2, 96, 227, 301–3, 323, 331, 369, 373, 407; Greater, 295, 331, 337; Lesser, 331, 372
Armenochalybes, 332
Arminius, 236
Arnistae, 180
Arnus River, 140
Aromatics, 286, 380–1, 383
Aroteres, peoples so named, 211, 336, 388
Arpi, 92
Arrathus River, 191
Arretini, 140
Arrhabaios, king of the Lynchestae, 208
Arrian, 328, 352, 361, 363
Arroni, 240
Arrotrebae, 240
Arruntius, L., 18
Arsanias River, 373
Arsenic, 362
Arsia River, 135–6, 174, 183
Arsinoë Kleopatris, 286, 385
Artabatitae, 398
Artabres, 241
Artabrum, Cape, 101, 241
Artacabene, 360
Artaxata, 337
Artaxerxes II, 25
Artemidorus of Ephesus, 24, 101–2, 272, 381, 384, 403
Artemis, 88, 124, 221, 308, 374, 399
Artemita in Acarnania, 192
Asachae, 396
Asana River, 257
Asbytae, 269
Ascitae Arabians, 389
Ascius, 84
Asculapium at Epidaurum, 200
Asia, continent and region, 28, 66, 82, 117, 277, 308, 377, 404–8; and Alexander the Great, 224; boundaries of, 76, 225–6, 228, 294, 323, 327–30, 340, 390, 404; central, 23, 25, 230, 343–4, 358; coasts of, 224, 231; as continent, 106–7, 276, 310; eastern, 7, 76, 364; southern, 11, 13
Asia, Roman province, 22, 59, 78, 91, 303, 314, 318
Asia Minor, 8, 11, 19, 91, 96–8, 107, 111, 162, 303, 335; coast of, 89, 302, 324, 403; interior, 330–1; northern, 190, 229, 311, 317–20, 324–5; routes around, 102; southern, 92, 97, 100, 298–300; western, 204, 304
Asido Caesarina, 115
Asinaeus Bay, 198–9, 217
Asoka, Mauryan king, 349
Asopis River, 198
Aspagani, 353
Asphalt, 99, 363
Asphaltitis Lake, 96, 290
Asprutani, 159
Assian stone, 92
Assos, 92
Assyrians, 325–6
Astacenus Bay, 319
Astapus River, 280
Astasapes River, 280
Astigi, 110, 115
Astrology, 25–6, 28–9, 41
Asturga, 126
Asturia, Astures, 125–6, 243
Astynomus, 24
Astypalaea, 28, 222
Atabuli, 395
Atalante, 89
Atargatis, 294
Atbana River, 280
Ateius Capito, 17–18
Ateius Philologus, 17–18
Ateius, L., 17–18
Ater, Mt., 270
Aternum, 87
Aternus River, 87, 135, 163–5
Athamanes, 193
Athena, 41, 92
Athenaion Promontory (Campania), 146
Athenian Stoa (philosophical school), 28–9
Athens, Athenians, 31, 48, 71, 64, 96, 202–3, 349
Athos, Mt., 208, 223, 283, 294
Atilius Regulus, C., 177
Atina, Plain of, 96
Atinates, 162
Atlantes, 275
Atlantia, 39
Atlantic Ocean, 21, 27, 43, 64, 75, 118–19, 152, 232–3, 250–6, 289, 393, 397–8, 406; and Carthaginians, 113; coasts of, 101, 233–4; currents in, 106, 108; islands in, 237–40, 243–6, 402; name of, 241, 400; and Persians, 276; rivers emptying into, 109–110, 114, 242, 275; writers on, 78
Atlantis, 32, 90–1

Atlantis Island, 400
Atlas (divinity and toponym), 42
Atlas Island, 400
Atlas, Mt., 21, 252–60, 274–5, 393, 399–401
Atlas, Pillars of, 107, 241, 259
Atraces, 193
Atramitai, 381
Atrax, town and river, 193
Attacians, 231
Attacori, 343
Attali, 372
Attalids, the, 311
Attana, 379
Attica, 64, 200, 202–3, 217, 219, 222
Atticus, 106
Auchatai, 341
Auchetae, 230
Aufidius Bassus, 2, 87, 332
Augila, Augilae, Augiles, 264–5, 269, 276
Augury, 69
Augusta Emerita, 243
Augusta Firma, 115
Augusta Praetoria, 135
Augusta Taurinorum, 169
Augusta Vindelicum, 174
Augustani, the, 122
Augustus (emperor), 1, 18–19, 30, 126, 153, 176–7, 292, 314–5, 318, 326; and Arabia, 383; building program of, 7, 69, 119, 262; and geography, 27; and Hispania, 109–10, 122–5, 128, 244; and Indians, 349, 357, 365; and Italy, 133, 136, 145, 154; and Mauretania, 250, 252; and natural phenomena, 40–1, 56–7, 70; and northern Europe, 6, 75, 103; and Rome, 148; and wine, 144; writings of, 57
Aulis, 204
Aulocrene, Lake, 307
aulon, 67
Aulon Cilicius, 313
Aunios, 241
Aurunci, 142
Ausetani, 121
Auson, 142, 152
Ausones, 142
Ausonian Sea, 152
Autaei, 382, 385
Autoleles, 259
Automate, 88–9, 222
Automula, 352
Autrigones, 126
Auximum, 146
Avalitae, 382
Avantici, 132
Avens River, 164
Axenus, 225. *See also* Black Sea
Axiacae, 228
Axius River, 207
Azali, 182
Azanian Sea, 387
Azetae, 294
Azores, 77

Baaethaemi, 294
Baal, 114, 285, 370
Babylon, Babylonia, Babylonians, 86, 286, 298, 366, 369–70, 374; astronomical research and, 42, 44, 49, 370–1; Alexander the Great and, 99, 263, 283, 347, 362; writers on, 25–6, 52, 61, 67–8, 371
Bacchus, 38, 98, 113, 144, 340, 353
Bactria, 100, 339–40, 342–4, 347, 353, 374, 407
Baebius Macer, Q., 2
Baebius Tamphilus, M., 163
Baeculonenses, 122
Baestulians, 112
Baetica, 109–22, 242–3, 251
Baetis River, 94, 110, 114–15, 119, 124, 127
Baeton, 24, 338, 347, 350
Baeturia, 116
Bagienni, 176
Baiae, 145
Baianian Bay, 97
Balari, 155
Balcia, 233–4
Baliares, 152–3
Baltic, 75–8, 103, 174, 185, 232–8, 404
Bambyce, 294
Banasa, 252
Baniurae, 259
Baragaza, 389
Barbatia, 378
Barbitace, 374
Barea, 119
Bargylia, 304
Barium, 161
Barsalium, 296
Basilia, 233
Basilidae, 229–30
Basilis, 24
Basta, 161
Bastarnae (Basternae), 227, 236
Bastitani, 124
Bastuli, 119
Batavi, 237
Bateni, 339
Baton, brothers so named, 179
Baunonia, 233
Bear (constellation), 35, 60, 62, 80, 345, 350, 355, 362
Bebrykes, 311

Bebryx, King, 114
Bedesis River, 166
Belbina, 217
Belgica, 166
Belion River, 242
Benacus Lake, 173
Beneventum, 160, 163
Berenice in Cyrenaica, 266–9
Berenice on the Red Sea, 83, 356, 364, 385–6, 389, 408
Berenice, places so named, 386–7
Berenice I, 386
Berenice II, 80, 386
Bergi, 238
Bergomum, 170
Beroia, 295
Berossos, 25, 52, 68
Berrice, 238
Besbicus, 89
Bessi, 210
Bibaga, 353–4
Biballi, 127
Bilbiltani, 122
Bilibis, 122
Billis River, 324
Bindusara, 349
Bion of Soloi, 24, 390–1, 397
Birds, 65–6, 86, 188–9, 334, 356
Bisaltae, 308, 210
Bisgargitani, 122
Bithia, 156
Bithynia, 317–19, 324, 337
Bitienses, 156
Bitumen, 99, 290, 362, 390
Bizone, 211
Bizye, 213
Black Sea, 10–11, 81, 98, 103, 172, 181–3, 209, 214, 224, 307, 333, 336; as bay, 108; coasts of, 95, 321–31, 323; currents of, 90, 94, 311; distances and measurements of, 79, 84, 212, 229, 277, 303, 404, 407; islands in, 232–3, 334; meaning of name of, 225, 227; populations around, 210–11, 291, 301–2; regions north of, 230–2; rivers of, 228; routes around, 102, 212, 224–5, 342; size of, 335. *See also* Axenus, Euxinus, Pontus.
Blascon, 153
Blemmyae, 276
Blessed, Islands of the, 244, 321, 401
boats, 115, 208, 355, 364–5
Bocchus II, 251, 260
Bodiontici, 132
Boea, 199
Boeotia, 90, 178, 194, 203–4, 219–20
Bogudes II, 260
Boians, Boii, 78, 167, 170, 181
bolides, 57
Bononia, 167
Boreas (mythological figure and wind), 64, 315
Borion, 266
Borysthenes River and region, 228, 230, 235
Bosporanian Kingdom, 228–9, 329–30, 342
Bosporus, Cimmerian, 20, 90, 224–5, 228–31, 277, 323–30, 333, 335
Bosporus, Thracian, 90, 212–14, 224–5, 231, 247, 277, 298, 303, 319–21, 323–5, 334
Bottiaei, 210
Boudica, 21, 257
Boukephalos, 352–3
Bovianum, 164
bracae, 129
Bracares, 127
Bracata, 129
Bragmanae, 348
Branchidae, 306
brettioi, 150
Breuci, 182
Brigae, 210
Brigiani, 177
Brisari, 344
Britannia, 21, 84, 237–8, 257, 408
British Isles, 27–9, 77, 84, 134, 168, 186, 335
Brixia, 173
Bronze Age, 16, 69, 184–5, 193, 200, 256, 287, 374, 384–5, 388
Brundisium, 98, 102, 160–1, 215
Bruttium, Bruttians, 133, 150–1
Bruttium et Lucania, 150, 159
Bruttium, Cape, 108–9
Bruttius Gulf, 135
Bucephala, 352–3
Budini, 230
Bulensi, 194
Bulini, 178
Bulis, 194
Bull (constellation), 35, 60
Bullidenses, 208
Bullis, 181, 208
Bura, 90, 197
Burcana, 235
Burgodiones, 123
Bursaonenses, 123
Buthrotum, 191, 215
Butrium, 166
By the Head of the Rock, 271
Byllis, 178, 181
Byzacium, 262–3
Byzantium, 212–13, 225, 319

Cabalia, 302
Cadi, 306

Cadistus, Mt., 218
Cadmus of Miletus, 307
Cadmus, mythological personality, 178, 292–3
Cadmus, Mt., 309
Caecilius Metellus Celer, Q., 77
Caecilius Metellus, C., 59
Caecina, Aulus, 18
Caecinus, 68
Caecubian wine, 144
Caecubus, 144
Caelius (Coelius) Antipater, L., 18, 77, 173
Caelobothras, 365
caelum, 33, 35, 59
Caenus, 151
Caere, 139
Caesar, Gaius, 27, 76, 376–8, 381–3, 389–90
Caesar, Throne of, 80–1
Caesar, Victory of, 56
Caesaraugusta, 122–3, 128
Caesarea in Mauretania, 120, 250, 260, 280
Caesarea in Armenia, 331–2
Caesarea Maritima, 289, 292
Caesarii Iuvenales, 124
Caicus River, 311
Calabria, 138, 150–1, 159–60, 163
Calaguris, Calagurritani, 123
Calaminae Islands, 91
Calauria, 217
Calchadon, 319
Calchas, 162
calendars, 30, 43–4, 65, 365
Calenus, 144
Calidonia, 237
Caligula, Gaius, 196, 250, 359
Callaeci, 127
Callenses Aeneanici, 116
Callicrates, 24
Callidemus, 24
Callimachus of Cyrene, 4, 16, 24, 80, 266
Calliphanes, 24
Callipolis, 161
Calpe, 107–8, 111
Calpetanus Rantius Quirinalis Valerius Festus, C., 271
Calydon, 193–4
Calydonian Boar Hunt, 52, 193
Camacae, 341
Camactulici, 131
Camalodunum, 84
Camels, 378, 381–2
Cameroon, Mt., 99, 255, 398, 404
Camilla, 142
Camirus, 304
Campania, 70, 82, 91, 97, 100, 133–4, 142–50
Campanian Bay, 89
Campi Lapidei, 130
Camunni, 174–5
canals, 115, 168, 229, 244; and Alexander the Great, 308; at Athos, 208; between Black Sea and Caspian, 333–4; at Corinth, 195–6, 385; in Egypt, 79, 244, 278, 284–6, 384–5; at Leukas, 192; in Mesopotamia, 298, 372
Canaria, 258, 402
Canarii, 258
Canary Islands, 244, 321, 401
Cancer (constellation), 80
Cancer, Tropic of, 80, 83, 345, 350, 356
Canchlei, 285
Candace, 393
Candaei, 386
Candavia, Mt., 180
Canicula, 60, 65
Cannae, Cannenses, 163
Cannenefates, 237
cannibals, 230, 342, 397
Canopus (star), 357, 380
Cantabria, Cantabrians, 69–70, 125–6, 241
Cantabrian War, 57, 125
Canusium, 161
Capella (star), 60
Caphereus, 220
Caphrena, 369
Capillati, 137, 176
Capisene, 359
Capitalia, Mt., 351
Capotes, Mt., 295
Cappadocia, 91, 325–6, 331
Capraria, 402
Capreae, 154–5
Capricorn (constallation), 80
Capricorn, Tropic of, 80
Capsitani, 266
Caralis, Cape, 155
Caralis (city), 155–6
Carambis Cape or Promontory, 102, 229, 325
Carambucis River, 334
Caranitis, 295
Carbilesi, 210
Carbula, 115
carbuncles, 270
Carcinites Bay, 228–9
Cardia, 213
Carduchi, 338
Cardytenses, 295
Carecini, 163
Caria, 88, 92, 303–5
Carice, 90
Carietes, 125
Carmania, Cape, 362
Carmania, region, 359, 361–2, 365–6, 379

Carmel, Mt., 289
Carni, 170, 173–4, 181
Carpathos, 218, 222, 314
Carpetani, 119
Carrhae, 296–7, 339
Carteia, 112, 118, 245
Carthage, Carthaginians, 136, 249–52, 254–6, 262–3, 269, 401; distances to and from, 262, 272–3; explorations from, 75, 77, 112, 159, 279–80, 400; foundation of, 292; outposts of, 113, 153–6, 257, 274, 399–400, 402; and Romans, 6, 81, 88, 108, 114–16, 120–1, 261–4; writings on, 18
Carthageniensis, 119
Carthaginian Gulf, 262
Carthago Nova (New Carthage), 109, 117–20, 124, 128, 407
Caryanda, 315
Carystus, 220
Cascantenses, 123
Cascantum, 123
Casilinum, 149
Casinum, 97
Casiri, 344
Casius, Mt., (Egypt), 285, 287–8, 385
Casius, Mt., (Syria), 293–4
Casos, 219, 222, 314
Caspian Gates, 333, 336–8, 347, 352
Caspian (Hyrcanian) Sea, 11, 79, 209, 326–7, 333, 337, 342, 373, 406–7; exploration of, 29, 76, 345; location of, 332–3; name of, 301, 335, 339; and Ocean, 75–6, 79, 329, 334–6, 341, 344; peoples around, 339–41, 367; size of, 335
Cassandria, 71
Cassiopaei, 189
Cassiterides Islands, 243–4
Castabala, 299
Castellum Pucinum, 170
Castra Hannibalis, 159
Castrum Novum, 135–6
Castulo, Castulonenses, 119, 124, 127
Catabanes, 285
Catabathmos, 249, 268, 272
Catali, 174
Catalogue of Ships, 203
cataracts, 83, 281–3, 389–1, 395–6
Cato, people so named. *See Porcius*
Catoslugi, 239
Catullus, 18, 80, 242
Caturiges, 170
Caucasian Gates, 333, 336
Caucasus Mtns., 11, 301–2, 326–37; Alexander the Great and, 336–7, 346, 351, 353; exploration of, 253, 341–2; Lesser, 301
Caucenses, 125
Cauchian Plains, 373
Caui, 180
Caulon, 159
Cavares, 131–2
Cea (Ceos, Keos), 90, 219, 304, 345–6
Cebenna Mtns., 129
Cedar, Mt. Atlas, 257
Cedrei, 285
Celegeri, 183
Celsa, Celsenses, 122
Celtiberians, 116, 119, 125
Celtic Lusitanians, 113
Celtic Turduli, 242
Celtica, 116, 334
Celticum, Cape (Celtic Promontory), 240–3
Celts, 1, 6, 10, 126, 170, 255; in Galatia, 318; in Hispania; and Lusitania, 113, 116, 127; in Italy, 165, 167, 169, 173
Cenchreae, 200
Cenomani, 173
census, Roman, 126
Cephallania, 216
Cephalotomi, 328
Cepheni, 332
Cepheus, king, 391
Cephisus River, 204–5
Ceranius (Ceraunian) Mtns., 302
Cerauni, 179
Ceraunian Mtns., 181, 188, 330
Ceraunian Promontory, 181
Cerberus, 324
Cercei, 88, 141, 144
Cercetae, 328
Cercina, 273
Cercinitis, 273
Ceres, 144
Ceribobus, 368
Cerne, 252, 399–400
Cerretani, 121–2
Cessetania, 121
Cestrini, 189
Ceto, 288–9
Ceutrones, 175–6
Chabrias of Athens, 287
Chalceritis, 334
Chalcidean Peninsula, 208, 210
Chalcidice, 58, 71, 211, 223
Chalcis on Euboea, 146, 330
Chalcodontis, 220
Chaldaic people, 270
Chalonitis, 373
Chalybes, 326, 332–3
Chandragupta, 28, 344, 346, 349, 353
Chaones, Chaonians, 188–9
Charax, 26–7, 363, 371–2, 374–8, 381
Charibael, king, 382

Charini, 235
Chariot of the Gods, 99, 398. *See also* Theon Ochema
Chariots, 168, 227, 276, 326–8
Charonea, 91
Charybdis, 94, 157
Chasamari, 397
chasms, meterological phenomena, 58
Chatramotitae, 380, 383
Chatti, 236
Chauci, 236–7
Cheimerion, 191
Chelidoniae Islands, 97, 313, 403
Chelidoniae wind, 65
Chelidonium, Cape, 300
Chelonates Bay, 196, 198
Chelonitis Islands, 379
Chelonophagi, 366
Chersonesus in Cyrenaica, 268
Chersonesus, Tauric, 92, 228–30
Chersonesus in Thrace, 211–13, 223
Chersonesus Wall, 212
Cherusci, 236
Chimaera, Mt., 99, 100
Chimaera, 100, 302
Chimera Fortress, 191
China, Chinese, 220, 342–3, 358
Chione, 315
Chios, 22, 215, 310, 315–16
Choaspes River, 373
Choirilos of Samos, 225
Chorasari, 341
chorography, 106
Christians, 39
Chryse, Cape, 343
Chryse, Indian island, 353, 399
Chryseon Ceras, Cape, 212
chrysokolles, 320
Chytrophoria Islands, 308
Cibarci, 240
Cibotus, 90
Cibyratic Jurisdiction, 304
Cicero, M. Tullius, 16, 18, 21, 32–3, 60, 67, 314
Ciconae, 244
Ciconians, 96
Cidamus, 271
Cidenas of Babylon, 31, 44
Cileni, 241
Cilicia, 286, 294, 298–9, 313, 325, 331
Cilician Gates, 298, 300
Cimbri, 75, 236
Cimbrian Cape or Promontory, 75, 235
Cimbrian Wars, 70–1
Cimician Woods, 92
Cimmerians, 231, 335
Cinaedopolis, 315
Cincibilus, 174
Cingulum, 146
Cinithi, 267
cinnabar, 128, 362, 395–6
cinnamon, 389
Cinnamon-Bearer Territory, 389, 399
Cios, 318–19
Circe, 88, 142–3, 146, 152, 361
Circei, 88, 141–6
circuli, 405
Cirrhaean Plain, 194
Cirta Sittianorum, 261, 266–7, 274
Cisippades, 265
Cithaeron, Mt., 203
Citis (Cytis), 387
citrus wood, 256
Clariae, 211
Classis, 168
Claudia Valeria, Via, 87
Claudius (emperor), 19–20, 25, 56–8, 145, 176, 228, 250, 302, 325, 329, 359, 372; and Britannia, 168; and hydraulics, 141, 333; writings of, 16, 18, 237, 284, 332
Claudius Marcellus, C., (cos. 50 BC), 70
Claudius Marcellus, M., (cos. 183 BC), 173
Claudius, Mt., 182
Clazomenae, 308
Clementia, 37
Cleobulus, 24
Cleonae, 197–8, 201
Cleopatra Selene, 120, 250, 160–1, 379
Cleopatra VII, 43, 192, 196, 219, 250, 256, 284–6, 358, 385
Cleostratus of Tenedos, 32, 42
Clidae, Cape, 313
Climax Megale, 368
Clitarchus, 25
Clitorium, 201
clouds, 49, 54, 61–2, 86
Clunia, 124–6
Clupea, 274
Cnemis, 204
Cnidus, 58, 80, 304, 314
Cocceius Nerva, M., 204
Cocynthos (Cocynthum), 135, 158
Codanus Gulf, 234
Coelaletae, 210
Coele Syria, 127, 293
Coelerni, 127
Coeranus the Philosopher, 25
Colapiani, 182
Colapis River, 182
Colchinium, 180
Colchis, 96, 171, 301, 326–7, 333, 336, 341–2

Colentum, 178
coloniae, 110–11, 132, 166, 173, 196–7, 252, 260
Colophon, 98
Colubraria, 152
columna, meterological phenomenon, 67
Columna Regia, 151
Colycantii, 311–12
Coma Berenice, 80
comets, 21, 55–8, 80
Commagene, 99, 331
Compsa, 70
Comum, 1, 86, 98, 167, 170
Concordia, 37
Conplutenses, 123
Consaburrenses, 124
Consentia, 151
Constellations, 34–5, 42, 51–2, 60–1, 65. *See also* specific names
Consuarani, 129
continents, 8–9, 78, 106, 231, 321, 354, 404–5. *See also* Africa, Asia, Europe
conventus juridici, 110, 124, 304–5, 309–10
Copae, 203–4
Copaic Basin, 90, 205
Copori, 241
copper, 128, 220, 320
Coptus, 283, 356, 364
Cora, 146–7
coracinus, 279
Coracum, 182
Corambis, 390
Corax, Mt., 301–2
Coraxians, 95, 301, 327, 336
Corcyra, 102, 136, 191, 215–16
Corcyra Melaena, 185
Corduba, 110, 114–16, 128
Cordueni, 338
Coretus Bay, 22
Corinth, Corinthia, 157, 192, 195–8, 200
Corinth, Isthmus of, 102, 213, 385
Corinthian Gulf, 90, 188, 191–7, 202, 215–16
Cornacates, 182
Cornelius Balbus, L., 104, 264–5, 270, 276, 404
Cornelius Cethegus, P., 163
Cornelius Dolabella, P., 58
Cornelius Nepos, 18, 77, 107, 252, 399
Cornelius Orfitus, Ser., 58
Cornelius Scipio Aemilianus, P., 6, 254, 264
Cornelius Scipio Africanus, P., 122
Cornelius Scipio, Cn., 100, 114, 116
Cornelius Scipio, P., 100, 114
Cornelius Sulla, L., 56, 69, 369
Cornelius Sulla, P., 153
Coronaeus Bay, 198–9
coronas, meteorological phenomenon, 58, 72
Corpili, 210
Corseae, 315
Corsi, 155
Corsiae, 195
Corsica, 153–5
Corticata, 241
Corycus, Mt., 218
Coryphasium, places so named, 199–200
Cos, 23, 304, 314, 343–6
cosmology, 25, 29, 32–4, 72
cosmos, 31–3, 35–7, 40, 42, 50, 53, 60, 72, 100
Cossiaei, 374
Cossora, 373
Cotieri, 341
Cottian Alps, 167, 170, 176
Cottius, M. Julius (father and son), 176–7
Cottonara, 365
Cotys I of Bosporus, 329
cows, 70, 323
cranes, 212
Crania, Mt., 194
Crassus, M. Licinius (cos. 70 BC, 55 BC, triumvir), 70, 296–7, 339, 369
Crassus, M. Licinius (cos. 30 BC, grandson of triumvir), 190
Crates of Mallos, 16, 25, 218
Cremona, 169–70, 173
Crete, 192, 218–9, 222, 243
Crete (nymph), 217
Crinis Berenices. *See Coma Berenices*
Crisaeus Bay, 194
Critodemus, 25
Criu Metopon, on Crete, 218, 268
Criu Metopon, in Taurica, 229
Crobyzi, 228
Crocala, 353–4
crocodiles, 39, 254–5, 275, 279–80, 352, 394
Crocodilon, 292
Croesus of Lydia, 194, 305–6
Cronium, 238
Croton, 92, 135
Croucasis, 341
Crustuminium, 92
Ctesias of Cnidus, 25
Ctesiphon, 370
cuci, 388
Cucios Spring, 388
Cunicularia e, 155
Curetes, 192, 218
Curetis, 191
Curio the Elder, 18–19
Cutilia Lake, 164
Cybele, 39
Cyclades, 221–3

Cyme in Aeolis, 308
Cynamolgi, 398
cynocephalus, 393
Cynossema, 214, 224
Cyparissus Bay, 198
Cyprus, 89, 92, 156, 299, 313–14, 403
Cyrenaica, 62, 218, 249, 263–9, 272
Cyrene, 249–50, 265–8, 272, 403, 406
Cyrus River, 332–3, 338, 341
Cyrus the Great of Persia, 75, 230, 340–1, 359–60, 368
Cythera, 217
Cyzicene marble, 320
Cyzicus, 27, 317, 320, 399

Daans (Dahae), 341
Dabelli, 398
Daci, 227
Dacia, 23, 408
Dahae, 341
Daimachos, 349
Dalion, 25, 391
Dalmatia (Delmatia), 12, 62, 178–9
Damascus, 286, 291, 293, 297–8
Damastes of Sigeion, 25
Damea, 312
Danaë, 142
Daniel, biblical figure, 99
Dantheletae, 190
Danube (Danubius, Danuvius) River, 8, 174, 181, 190–1, 209, 214, 226, 334, 346; and Adriatic, 171–2; peoples along, 210–12; rivers empyting into, 170, 227; Romans and, 19, 171, 174, 226, 235, 266
Daorizians, 179
Daphne, 293, 349, 357
Darat River, 2, 254–5
Dardae, 349
Dardani, Moesian peoples, 183, 190, 207
Dardani, Trojan allies, 190
Dardanian Cape, 311
Dardanus, 147
Dardi, 162
darekh, 96
Daritis, 360
Darius I of Persia, 63, 112, 210, 224–6, 353, 374, 385
Darius III of Persia, 81, 337, 363
Dascusa, 332
Dassaretae, 180, 190
Daunii, 162
Daversi, 179
Dead Sea, 96, 285, 288–91. *See also* Asphaltitis
Decapolis, in Syria, 291, 293
Deciates, 131, 137
Deiotaros of Galatia, 300
Delmatia. *See* Dalmatia
Delos, 23, 88, 97, 221–2, 231–2, 381
Delphi, 91, 168, 194
Delta, Egyptian, 277–9, 284–5, 385, 390
Deltoton (constellation), 35
Demeter, 144
Demetrius Poliorcetes, 195–6
Democritus of Abdera, 25, 37, 55
Democritus, lapidary writer, 25
Demodamas of Miletus, 25, 340–1
Demonnesus, 320
Denda, 180
Denselatae, 190, 210
Derasidae Islands, 308
Dertona, 138
Dertosa, Dertosani, 122, 128
Desitiates, 179
Detumo, 115
Deuri, 172
deus, 32, 37
Develton, 212
Dexivates, 130
Dianium, Dianenses, 124, 152
Dicaearchus of Messana, 74
Didani, 333
Didyma, 306, 340
Diegylis, 213
Dienses, 208
Diggeri, 210
Dindari, 179
Dinochares, 283
Dinocrates, 283
Diobessi, 210
Diognetus, 24, 26, 338
Diolkos, 195
Diomedes of Argos, 162, 334
Diomedes, Bistonian king, 211
Diomedia, 184
Dionysius I of Syracuse, 159
Dionysius II of Syracuse, 94
Dionysius of Chalkis, 220
Dionysius of Charax, 26–7
Dionysius of Halicarnassus, 26
Dionysius, people so named, 26, 220
Dionysius, Selucid envoy, 348–9
Dionysodorus of Melos, 102–3
Dionysopolis, 212
Dionysos, 113, 144, 203, 209, 291, 340, 346, 353, 359
dioptra, 57, 80
Dioryctos, 192
Dioscuri, 38, 59, 327–8
Dioscurias, 327–30
Dioscuridu, 380
Dirae, 206
Ditiones, 179

Dium (Dion), 208
divinity, 32–3, 37–41
Docleatae, 180
Dodona, 97, 188–9, 202, 232
Dog Star, 60, 168, 281, 364
dogs, 61, 185, 214, 258, 381, 397, 402
dokoi, 58
Dolopes, 193
Dolus, 38
Domitian (emperor), 39
Domitius Ahenobarbus, Cn., 59
Domitius Corbulo, C., 1, 19, 81, 323, 331, 337
Dora Baltea River, 169
Dorakta, 362
Dorians, 53, 204, 304, 325
Dorieus, 269
Dorion, 198
Doris, Central Greek region, 205
Doris Bay, in Caria, 304
Doriscum, 211
Dorylaeum, 309
Dosiades, 26
Draco, Mt., 309
Dramasa, 350
Draus River, 182
Drepane, 215
Drepanum, 389
Dribices, 339
Drinium River, 182
Drion, 162
Druentia River, 130–2
Drusus (brother of Tiberius), 2, 75, 174, 226, 233–5
Drymodes, 200
Dryopes, 189
Dryopis, 205
Dumatha, 378
Durius River, 119, 125, 241
dye, purple, 256, 292, 401
Dyme, 197
Dyris, Mt., 253
Dyrrachium, 161, 180–1

earth, 6, 53, 72–99, 224, 233, 251, 356, 392, 408; beneath, 52, 63, 66, 68, 280; changes to, 202, 267; circumference and size of, 26, 53, 83, 101–2, 282, 387, 405; creatures of, 34; inhabited portion of, 11, 51, 34, 195, 403–5; location of, 43–4, 52, 80; moisture of, 62; and moon, 46, 49, 54; parallels across, 402–7; peoples of, 345; phenomena of, 99–101; and planets, 51; as primeval element, 35–6; zones of, 47
earthquakes, 85–90, 197, 211, 219, 221
ebony, 398
Ebusus, 152–3
Ecbatana, 337, 368
Echinades Islands, 192, 215
eclipses, 45–9, 58, 71, 81, 86
ecnephias, 66
Edeko, 122
Editania, 120
Edoni, 210
Egelestani, 124
eggs, 34, 38
Egi, 240
Egnatia, Via, 208, 211–13
Egypt, 67, 82, 86, 249, 267–9, 276–88, 356, 371, 377, 389, 398, 406; and astrology, 28, 54; and astronomy, 43–4, 51, 56, 66, 95; economy of, 385; explorations from, 396; gods of, 38–9; and Levant, 294; and Mediterranean, 272–6; and Rome, 20–1, 383, 391; routes to, 314, 364–5
Egyptian Sea, 79
ekleiptikos, 51
Elam, Elamites, 374
Elaphonnesus, 320
Elatium, 297
Elatos (suitor), 216
Elatus, Mt., 216
Eldamari, 369
Electrides Islands, 184
Elegea, 296
Eleia, Elii, 198
Elephantine, 282
Elephantophagoi, 395
Elephants, 83, 252–3, 259, 348–9, 354, 384, 388–90, 394–5
Eleusis, in Attica, 202
Eleusis in Boeotia, 90
Eleutherae, 202
Elis, 82, 198, 202
Elymais, 374
Emathia, 207
Emodus, Mt., 301
Empedokles of Akragas, 62
Emporiae, in Hispania, 121
Encheleae, 178
encyclopedia, Greek and Roman, 2–4
Endera, 395
Endymion, 45
Enetians, 173
Engada, 291
Enoecadoe, 228
Eoan Ocean or Sea, 334, 344, 355
Epagerritae, 328
Epetium, 179
Ephesus, 88, 102, 308–9, 315, 398
Ephorus of Cyme, 26, 308
Ephyra, at Corinth, 196
Ephyra, in Thesprotia, 189
Ephyra, places so named, 193

Epians, 198
Epidamnum, 180–1
Epidaurum in Argolid, 200, 202
Epidaurum in Epirus, 180
Epigenes of Byzantion, 26
Epigenes of Rhodes, 26
epikyklos, 50
Epiros (Epirus), 100, 136, 181, 186–91, 202, 208–9
Epopos, 89
Eporedia, 169
Equaesi, 127
equator, 47, 55, 80–1, 103, 250, 253, 355, 407
equinox, 52, 79–80
Erasinus River, 96
Eratosthenes of Cyrene, 6, 11, 26, 73, 103, 127, 188, 109, 224, 272–3, 282, 286, 292, 335, 405–8
Eravisci, 182
Erebus, 38
Ergavicenses, 123
Eridanus River, 167–8
Erinyes, 206
Erythea, places so named, 244–5
Erythrai, 26
Erythras, King, 366, 380
Erythrean Sea, 245
Esar, 396
Esbonitae, 385
Essedones, 230, 341
Essenes, 291
Etesian winds, 281, 364
ethnology, 19, 269, 325, 352
Etruria, 92, 100, 138–40, 155, 163, 177
Etrusca disciplina, 67–8
Etruscans, 18–19, 21, 53, 67–9, 87–8, 139, 145–6, 150, 166, 173
Euandros (Evander), 201
Euboea, 24, 89, 204, 220–22, 232; islands near, 219, 221, 223; settlements from, 145–6, 151
Euboean Strait, 94, 204
Euchatae, 341
Euclides (Euclid) of Alexandria, 26
Eudoxus of Cnidus, 26–7, 46, 51, 66, 80–1, 399, 405
Eudoxus of Cyzicus, 27, 77, 394, 399
Euergetae, 360
Euganians, 173–4
Euhemeros of Messene, 39
Eulaeus River, 374–5
Eunoe Maura, 260
Eupalimna, 193
Euphorbos, 258
Euphrates River, 285–7, 294–8, 303, 332, 367–71, 374, 377, 383; and Babylon, 370; as boundary, 369; length of, 372–3; source of, 331
Euripus, 94, 204, 220, 222
Europe, 77–8, 101, 174–6, 181, 212, 214, 228, 318; and Alexander the Great, 340; boundaries of, 76, 224–6, 277, 317, 323, 329–30, 340; central, 167, 170–1, 226; coasts of, 8, 11, 104, 240–1, 249; as continent, 106–8; gulfs of, 151–2, 158–9, 186, 224; northern, 6, 29, 130, 211, 227, 233–8, 243; rivers of, 130–1
Eurotas River, 199
Eurus wind, 216
Eurydice, 188
Euthymenes of Massilia, 255, 275, 279, 281
Euxinus Pontus, 320–1, 323
Evenus River, 196
eye disease, 61

Fabianus, Papirius, 19, 95
Fabius Maximus Aemilianus, Q., 131
Fabius, Q., 58, 131
Fabius Pictor, Q., 177
Fabricius Tuscus, 19
faces, 57
Falernian wine, 144
Fannius, C., 59
Favonius wind, 243, 345
Febris, 37
Fecusses, 174
Fenni, 234
Feronia, temple of, 70
Fertor River, 162
Fibularienses, 123
Fides, 37
fire, 53, 82, 141, 221, 253, 368, 393–5; as primeval element, 35–6, 100; as meterological phenomenon, 52, 62, 67; as volcanic phenomenon, 97, 99, 181, 302
fish, 89, 96, 279, 288, 360–1
fishing industry, 145
fish salting, 290
Flanatic Bay, 172
flatus, 63
Flavian ideology, 4
floods, 90, 168, 196, 348, 351, 373–6; controlling of, 191, 375–6, 378; of Euphrates, 298; of Nile, 30, 44, 279–82; of Tiber, 141
flora, 72–3, 97, 143, 228, 257–8, 275, 345, 353, 393
Florentini, 140
Fonteius Capito, C., 81
Forat, 378
Formiae, 144
formica, 61
Formio River, 171
Fortuna, 38, 40
Fortunate Islands, 11, 244, 321, 401
Forum Iuli, 131
Fossa, 155

Fossa Augusti, 168
Fossae Marianae, 130
frankincense, 285, 364, 380
Frentanians, 162
Frigidus River, 171
Frisi, 237
Frisiauones, 237
Fucinus Lake, 95, 163–4, 372–3
Fugitive Egyptians, Island of, 396–8
fulmina, 52, 67
Fulvius Flaccus, Ser., 179
Fulvius Nobilior, M., 216
Fut River, 257

Gabii, 91
Gabuleni, 294
Gades (Gadir, Gadeira), 75–7, 104–7, 110, 116–17, 242–5, 270; distances and routes to, 101–2, 118, 155, 233, 245, 401; Hercules and, 107, 245; Phoenicians and, 75, 192; Posidonius and, 80, 93–4
Gades Strait, 402
Gaditanian Strait, 106–7
Gaeli, 339
Gaetulia, 266–7
Gaetulian purple, 256, 401
Gaetulians, 255, 259, 274
Gaia, king of the Massili, 266
Galata, 158, 273–4
Galatia, 299–300, 318
Galba, Ser. Sulpicius, 128, 132, 329
Galene, 90
Galilee, Sea of, 90, 289–92
Gallaecia, 243
Gallia, 65, 75, 86, 104, 127, 138, 153, 165–6, 170, 183, 239, 40, 346, 407
Gallia Comata, 171
Gallia Narbonensis, 1, 65, 71, 129–132, 137, 175–6, 407
Gallia, Lugdenensian, 239
Gallic invasion, 143
Gallic Ocean, 94, 109, 241
Galloi, 318
Gamale, 90
Gamphasantes, 276
Gandari, 339
Gangaridae Calingae, 348
Ganges River, 101–2, 343–51, 355
Garamantes, 264, 269–71, 276
Garganus Peninsula, 160–3
garlic, 38
Garunna River, 238, 240
Gaugamela, 337
Gaul, 78, 110, 407–8; Cisalpine, 77. *See also* Gallia.
Gaulos, 157, 273
Gaza, 284–7, 377, 380–1
Gebbanitae, 380
Gedrosia, 360, 408
Gedrusi, 360
Gellianus (Gnaeus Gellius), 19
Geloni, 230
Gelonos, 230
gems, 56, 72, 92, 270–1, 358, 386
Genesara Lake, 290
Genetae, 326
Genua, 138
geocentric system, 42, 50, 79
geometry, 26, 54
Georgi, 335
Ger River, 257–8
Gerama, 264
Gerania, 212
Germani, 124, 227, 236
Germania, 1, 2, 11, 19, 22, 75, 86, 103, 118, 235, 246
Germanian Oretanians, 124
Germanicopolis, 318
Germanicus, 57, 75, 98, 235, 318
Gerovacium, 237
Gerrha, in Arabia, 378–80
Gerrha (Gerrhum) in Egypt, 385
Gerunda, 122
Geryon, 244–5; Cattle of, 175, 245
Geryones, 245
Getae, 210, 227
Gethone, 223
Gigurri, 126
Gindareni, 294
Gir River, 280
Giri, Mt., 271
Glaesaria, 235, 238
Glanis River, 141
glass, 292
Glinditiones, 179
Gnatia, 100
gnomon, 52, 82, 85, 406–8
Gobares, 396
gold, 35, 128, 175–7, 238, 242, 349–54, 358, 362, 365, 374, 383, 395, 399
Golden Apples, 251, 254, 401
Golden Fleece, 326–7
Gordyaea, 369
Gordyaean Mtns., 373
Gorgades Islands, 406
Gorgons, 254, 400–1
Gorillai, 401
Grabaei, 180
Graccurritani, 123
Graecia, 202
Graecus, ethnym, 162
Graecus, son of Zeus, 205

Graiae Pass, 169
grain, 121, 144–5, 359
Graius, 162
Granicus River, 317
Graucomen, 12
Great Year, 44
Greek fire, 99
Grovi, 241
Gutones, 235
Gyani, 362
Gymnasiae, 153
Gymnetes, 274, 395
Gynaecocratumenians, 330
Gytheates Bay, 199
Gythium, 199

Hadriatic Sea 184; *see also* Adria, Adriatic
Haedi (constellation), 60
Haemasi, 180
Haemodae, 238
Haemon, personalities so named, 205
Haemonia, 205
Haemus, Mt., 191, 209–10, 212
halcyons, 65–6
Haleti, 210
Haliacmon River, 207
Halicarna, 193
Halicarnassus, 26–7, 304–5, 314–15, 320
Halizonians, 317
Halmon, or Halmonia, 206
Halone, 88
Halonesos, 223
Halys River, 325
Ham, 121
Hamaxobi, 227
Hammon Oracle, 267, 278–9. *See also* Ammon
Hammoniac Nome, 278
Hannibal of Carthage, 59, 119–21, 130–1, 149, 159, 162–3; in Alps, 169, 175; in Bithynia, 319
Hannibal, Towers of, 82
Hanno of Carthage, 27; explorations of, 13, 77, 99, 254–5, 280; writings of, 99, 112, 250, 252–4, 274–5
Harmonia, mythological personality, 178
harmonia, cosmic concept, 53
Harmozae, 366
Harpasa, 92
Harpokration, 39
Hasta Regia, 115
Hebrus River, 210–12
Hebudes, 238
Hecataeus of Abdera, 27, 225
Hecataeus of Miletus, 27, 106, 112, 225
Hecatompylos, 338
Hecuba, 214
Hekatomnids, 303, 305, 320
helena, meteorological phenomenon, 59
Helene (island), 219, 222
Helice, 90, 197
Helicon, Mt., 194, 203
heliocentric system, 42–3, 50
Helios (divinity), 88, 94, 168
heliotrope, 61
Hellanicus of Lesbos, 27
Hellas (town and region), 202, 205
Hellen, 205
Helleni in Hispania, 241
Hellespont, 71, 90, 186–8, 211–14, 220, 224, 310–11, 317–19, 337
Hellopes, 189
Helveti, 239
Hemesa (Emesa), Hemeseni, 294
Hemodi Mtns., 300, 344, 347, 357
Heniochi, 326–7
Hephaisteion, companion of Alexander the Great, 283
Hephaistos, 393
Hera, 35, 150
Heraclea in Chalcidice, 211, 324
Heracleotic Mouth, 284
Heraclides, 27
Herakleion, in southern Italy, 108
Herakles, hero. *See Hercules*
herbs, 10
Hercules, hero, 38, 113–14, 130, 169, 175, 203–5, 251, 254, 276, 352, 358; and Alexander the Great, 209, 340; Labors of, 211, 244–5, 251, 268, 334, 401
Hercules, or Herakles, planet, 43
Hercules, Grottoes of, 251
Hercules, Pillars or Columns of, 11, 75, 101, 107, 113, 241–50, 255, 259–60, 323, 340, 399, 401–6
Hercuniates, 182
Hercynian Mtns., 236
Herennius, M., 67
Hermes (planet), 38, 44
Hermiones, 236
Hermunduri, 236
Hermus River, 309
Herod the Great, 289, 292, 310
Herodion, 289–90
Herodotus, 16, 27, 32, 112, 244, 368
Heroonopolis Bay (Heroonpolitic Gulf), 381, 384
Heroum, 381
Hesperides, divinities, 217, 251; Apples of, 254, 401; Gardens of, 401
Hesperides Islands, 254, 401
Hesperides, locale in Cyrenaica, 268
Hesperioe, 398

Hesperu Ceras, Cape (Hesperium Promontory), 255, 399–400
Hiberia, Caspian region, 209, 326, 332
Hiberian Sea, 109, 152
Hibernia, 237–8
Hibernians, 113
Hiberus River, in northern Hispania, 8–9, 113, 116, 120–4, 126, 8, 240
Hiberus River in Baetica, 120
Hiera, Aegean island, 88–9, 222
Hiera, Aeolian island, 158
Hierapolis in Phrygia, 91
Hiericous (Jericho), 289
Hilleviones, 234
Himani, 178
Himantopodes, 276
Himilco of Carthage, 27, 77, 237
Himyarites, 382
Hippalos, 363, 387
Hippalum, Cape, 387
Hippalus wind, 387
Hipparchus of Nicaea, 27, 48–9, 51, 54, 57, 61, 81, 85, 407
Hipparenum, 370
Hippo Dirutus, 262
Hippokrates of Cos, 315
Hippolyte, Amazon, 134
Hipponiensian Bay, 262
Hippopodes, 234
hippopotami, 255, 388, 394
Hippuros, 356
Hippus of Tyre, 287
Hipsodores, 395
hirci, 55
Hirpini, 91, 160, 162
Hirri, 235
hirundo, 65
Hispalensian district, 115
Hispalis (Hispal), 94, 110, 115
Hispania, 66, 76, 82, 98, 100, 104, 108, 109–29, 177, 240–3, 260, 304, 307
Hispanic Sea. 152
Hister River, 171, 181, 211–12, 214, 226–8, 232. *See also* Danube River
Histria, 16, 31, 170–4, 177–8
Histropolis, 226
Hittites, 299, 326
Homer, 190, 218
Homeritae, 382–3
Homna, 379
Honor (divinity), 37
Horcos River, 206
Hormiae, 144
horologium, 85
horses, 234, 243
Hyades, 60
Hyampolis, 204
Hyantes, 204
Hybanda, 89
Hyctanis River, 362
Hydaspes River, 352
Hydata Therma, 122
Hyde, 306
Hydra of Lerna, 199
Hydraulics, 31, 95, 115, 144, 168, 206, 333, 372–3
Hydreuma, places so named, 364
Hydruntum, 136, 161, 184
Hydrusa, 219, 221
Hyginus, C. Julius, 16, 19
Hylaei, 229
Hylatae, 294
Hypaea, 153
Hypanis River (modern Kuban), 229
Hypanis River (modern Bug), 228–30
Hypasis River, 83, 347
Hyperboreans, 231, 331, 343, 408
Hyrcanian Sea, 301. *See also* Caspian
Hyrcanians, 339, 367
Hyrie, 216
Hysiae, 197
Hyspaosines, 375

Iadar, 178
Ialysus, 304
Iamnea, 228
Iapudes, 170, 178, 181
Iapudia, 172, 178
Iapygian Promontory or Cape, 161–3, 181
Iapyx, 161
Iasian Bay, 304
Iaxartes River, 25, 340
Iazygae, 227
Iber River, 113, 120
Iberia, Caspian region, 209, 326, 332
Iberia (Hispania), 11, 21, 27, 64, 112–14, 119–20, 122, 127–8, 130, 136, 241, 245, 321
Icaros, 222, 315
Ichthyophagi, 360–1, 379, 389
Iconium, 299
Icosium, 260
Ictis, 238
Ida, Mt., on Crete, 47, 218
Ida, Mt., in Troad, 47, 223
Idumaea, 288–9
Iesses, 122
Iessonienes, 122
Ilergaones, 121, 123
Ilgertes, 121
Ilienses, 155
Ilium, 88, 92, 310–11

Illyrian Bay, 193
Illyricum (Illyria), 97, 136, 161, 177–84, 407
Ilorcitani, 124
Ilva, 154
Imavus (Imaus) Mtns., 310, 347, 406
Imbrus, 223
Imeni, 297
India, 7–8, 11, 77–8, 83, 92, 101, 231, 234, 255, 275, 300–1, 328, 339–40, 344–59, 406–8; coasts of, 75–6, 334; trade with, 63–4, 82, 283, 342–3, 358, 359; routes to and from, 101–2, 330, 342, 361–3, 380, 384–8, 399; writings on, 20, 24–8, 99, 392
Indian Ocean or Sea, 63, 249, 379, 387, 394, 399
Indigetes, 121
Indo-Europeans, 358
Indus River, 63, 84, 339, 344, 350–3, 359–61, 366, 375; and Alexander the Great, 28, 83, 275, 345–6, 354; flooding of, 298; mouth of, 112, 344, 355, 362–3, 366
Inferum Sea, 152, 174
Ingauni, 137
Inguaeones, 234, 236
Inopus, spring and river, 98–9
Insubres, 170
intercalary year, 43, 66
Intercatienses, 125
Internal Sea, 106–7
Invallis, 402
Ionia, Ionians, 98, 131, 305–10, 316, 325–7, 353
Ionian Islands, 216
Ionian monism, 23, 31, 34, 42, 85
Ionian Sea, 184–6, 193, 200
Ios, 222
Iovis Cassius, Temple of, 215
Iphigeneia, 229
Ireland, 77, 237
iron, 70, 123, 154, 326
Iron Age, 137, 139, 165
Isauria, 299
Ishtar (planet), 43
Isidorus of Charax, 26–7, 101, 196, 199, 272, 296, 372, 375–6
Isis, 39, 280
Ismenus River, 203
Issa, 136, 179, 185
Issic Bay, 403
Issidones, 230
Isthmia, 200
Isthmian Games, 200
Isthmos/Isthmus, 12, 102, 195–6, 199, 200, 203, 213, 385
Istria, 133, 135, 170, 172
Istros River, 171–2, 226, 346
Istuaeones, 236
Italy (Italia), 43, 46, 58, 64, 72, 80, 86–7, 92–6, 100, 108–11, 129, 131–78, 201, 321, 373–4, 377; Diomedes and, 184, 334; Hannibal and, 121; ideology and, 108; parallels across, 407–8; rivers of, 98; routes to and from, 109, 185, 196, 215, 246; trade with, 115, 126–8, 185; weather and, 43, 65–7, 82, 85; writings on, 19, 110
Ithaca, 215–6
Ithome, 198
Ituraeans, 294
Iuliopolis (Juliopolis), 317, 319
Iures Mtns., 129
Ivor River, 257, 388
Ivory, 256

Jason (Argonaut), 150, 196, 326–7, 336
Jason of Pherai, 193, 206
Jerusalem, 290–1
Jewish War, 90
Jews, 40
Johann of Speyer, 9
Jomanes River, 351
Jope, 288–9, 391
Jordan River, 289–91
Juba I of Numidia, 261, 263
Juba II of Mauretania, 6, 15, 28, 76, 102, 120, 250, 259–61, 276, 280–2, 286, 361–6, 371–2, 376–83, 384–90, 398–402
Judaea, 247, 289–91, 406
Julia Augusta, Via, 176
Julia Gaditana, 244
Julian Teari, 122
Juliobriga, 126
Julio-Claudian period, 2, 8, 20, 30, 154, 219, 261, 310, 359, 392
Juliopolis in Bithynia, 319
Juliopolis in Egypt, 364
Julius Agricola, C., 327
Julius Aquila, 19, 68
Julius Caesar, C., 7, 30, 43–4, 51, 55–6, 67, 80, 110, 116–22, 130–1, 157, 171, 175, 196, 226–7, 237, 260, 270–1, 388, 314, 326
Julius Caesar, L., 58
Julius Caesar, Sex., 87
Junius Brutus, D., (cos. 138 BC), 57, 69, 127, 242
Junius Silanus, M., 88
Juno, 69, 115, 150, 244, 400
Junonia, islands so named, 402
Jupiter (god), 41, 52, 68–9, 370
Jupiter (planet), 43, 50, 52
Jupiter (Zeus) Spring, 97

Kalanos of Taxila, 348
Kalypso, 53, 60, 159, 361
Kane, 364

Kanobos, 80
Kardytos, 295
karst topography, 96, 171, 180
Kelaina, 304
Kepler, Johannes, 53
Kerkyra, 185, 191, 215
Kestros River, 300
Kinys River, 269
Kleanthes of Assos, 36–7
Kleomedes, 46, 52
Kolaios of Samos, 64
Konon the astronomer, 80
Kophen River, 359
Korbilon, 246
Koryphon, 200
Kouretians, 192
Kourion Mt., 192
Kronos (divinity), 114
Kronos (planet), 36, 42
kyanos, 320
Kyrnos, 153

Labeateae, 180
labyrinth, in Egypt, 283
Lacedaemon, 85
Lacetani, 121, 123
Lacinian Promontory, 158
Lacinium, 159–60, 135
Laconic Bay, 217
Laconica, 160, 198–9
Laeanitic Bay, 384
Laeetani, 121
Laevi, 170
Laminitani, 124
Laminium, 109
lampades, meterologial phenomena, 57
Lamponia, 223
Lancia, Lancienses, 126
Laodicea in Persis, 368
Laodikeia, 96
lapis specularis, 128
Lapithae, 206
Larinates, Larinum, 163
Larius, Lake, 198
Latium, 70, 88, 91, 97, 136, 140–9, 165
Latium et Campania, 141
Latovici, 182
Leborinian Fields, 145
Lecheae, 197
Lectum, 310
Legio XI Claudia, 164
Lelanton Plain, 220
Lelegians, 311
Lemannus Lake, 130
Lemavi, 126
Lemnos, 88, 223
Leo (constellation), 80, 281
Leonicenses, 123
Leonnatus, 361
Lepadusa, 273
Lepcis Magna, 265, 268, 271–2
Lepontii, 175
Lepta Acra, 389
Lerina, 153
Lerna, 199
Lero, 153
Lesbos, 27, 316–17
Leto, 88
Leucas (Leucadia), 90, 192, 196
Leucatas Cape, 319
Leucoe Aethiopians, 274
Leucopetra, 151
Leucosia, 89, 156
Leucosyri, 325–6
Leucothea, 155–6
Leuka, Mt., 218
Leuke, 232
Leukosia (siren), 156
Leuni, 241
Levant, 385–92, 314, 391
Libanus, Mt., 393
Liber, Father, 98, 113, 144, 203, 209, 291, 340, 353, 359
libraries, 4, 16, 19, 25, 31, 284
Liburnia, 133, 135, 166, 172, 177–8
Libya, 60, 93, 218, 262–4, 272, 281; as continent, 106, 404; name, 249
Libyan Egyptians, 274
Libycan Sea, 249
Libyes, 249
Libyphoenician, 263
Libyssa, 319
Lichades Islands, 219
Licinius Lucullus, C., 212
Licinius Lucullus, L., (cos. 151 BC), 125
Licinius Lucullus, L., (cos. 74 BC), 99, 125, 232, 315
Licinius Mucianus, C., 20, 95, 144, 221, 295
Ligauni, 131
Liger River, 239–40, 246
lightning, 40, 62, 68–70
Ligures, Ligurians, 133, 137–8, 152, 163, 166, 407
Ligurian Alps, 174–5
Ligyans, 130, 137
Lilybaeum, 157, 262
Limaea River, 242
Limici, 127
Lindos, 28, 304
linen, 403
lions, 100, 302, 397

Lipara, 158
Liparaean Islands, 158
Liris River, 141–2, 144
Lisa, 302
Lissum, 180
Livia Drusilla, 326, 370
Liviopolis, 326
Livius, T., the son, 20
Livy (T. Livius), 16, 20
Lixus, city and river, 251–2, 401
Lochias, Cape, 312
Locri, Italian town, 72, 92, 109, 151, 158
Locris, 188, 192–3; Epicnemidian, 194, 204; Opuntian, 205; Ozolian, 194
locusts, 398
Longi Muri (Macron Tichos), 213
longitude, 81, 200, 262
Longopori, 397
Lotophages (Lotus Eaters), 265
Lucani (Leukanoi), 150, 162
Lucania, 70, 89, 96–7, 150, 159
Lucanian Thebes, 159
Lucensis Jurisdiction, 240
Luceria, 162
Lucifer (planet), 43
Lucius, Lucanian leader, 150
Lucretius, 4, 33, 35, 37, 67, 69, 72, 97, 98, 268
Lucus Augusti (Hispania), 126
Lucus Feroniae, 70
Lugudunum, 239
Luna, Eturian town and quarries, 139, 176
Luna (planet), 44
Lupeni, 333
Lusitania, 109, 113, 116, 119, 127–8, 241–5, 407
Lycaon of Arcadia, 201
Lycaon, Mt., 201
Lycaonia, 299–300
Lycia, 92, 298–304, 311, 313, 319, 406
Lycomedes, Lake, 265
Lycus River, 311
Lydia, 88, 139, 164, 306–8, 311
Lykophron, 146
Lykos, rivers so named, 96
Lynceus of Messene, 51–2
Lynchestae, 207–8
Lysas, freedman, 356
Lysimachea, in Thrace, 213
Lysimachos, 213
Lytharmis, Cape, 334

Macae, 269, 362
Macedonia, Macedonians, 71, 166, 177–80, 183, 190, 207–9, 214, 284–5, 294, 363, 369–70; Roman province of, 19, 47–8, 190, 209; settlers elsewhere, 306, 309
Macerones, 333
Machroae, 265
Macra River, 137–8
Macris, 89
Macrobii, 395
maculae, lunar phenomenon, 45
Macurebi, 260
Macynium, Mt., 194
Maeander River, 88–9, 305–9
Maedi, 190, 210
Maenuba River, 115–6
Maeonia, Maeonii, 306
Maeotic Marsh or Lake, 13, 76, 90, 224–30, 277, 301, 323, 329–30, 336, 340, 402–4
Maezei, 179
Magadalensberg, 174
Magi, 368
Magium, 275
Magna Graecia, 134, 158
Magna Mater, 39
Magnesia (Magnetis), Greek region, 190, 206
Magnesia near Sipylus, 88, 90
Magnesia on the Maeander, 307
magnetism, 92
Magnus Annus, 44
Maior (island), 153
Makarioi Nesoi, 321–2
Makros Plain, 288
malachite, 320
Malaia, Mt., 84
Malay Peninsula, 7, 353
Malea, Cape, 199–200, 202, 217–18
Maleus, Mt., 83–4, 350
Maleventum, 163
Malli, 348
maltha, 99
Malvane River, 259
Mamertini, 157
Manduria, 96
mangroves, 97, 352
Manilius, M., 40
Mantinea, 199, 201
Mantua, 173
mapalia (*magalia*), 261
maps, 7–8, 17, 23, 107, 135, 191–2, 337, 386, 392
Maraces, 193
marble, 128, 139, 176, 220, 261, 315, 320, 358
Marchubi, 267
Marcia, 67
Marcius Philippus, L., 87
Marcius Rex, Q., 58
Marcius Septimus, L., 100–1
Mardani, 369
Mardi, 328, 374
Mardians, 339

Marduk (planet), 43
Mareotis Lake, 278, 283
Marethesium, 307
Margiane, 339
Margus River, 339
Mariamme, 294–5
Mariamnitani, 294
Mariandyni, 324
Maribba, 382
Marici, 170
Marigarri, 397
Marius, C., 59, 69–71, 130, 153
Marmaridae, 268
Marotiani, 339
Marrebbata (Mariaba), 381
Marrucini, 163
Mars (god), 157
Mars (planet), 43, 50–1, 68
Marsaci, 237
Marsi, 163–4
Marsic (Social) War, 70, 87, 149, 163
Marsyas, 164, 305
Marsyas River, 304
Martial, 122
Marus River, 149, 164
Masada, 291
Masaesyli, 259
Masei, 369
Masica, 298
Masinissa, 259, 261, 266
Massagetai, 230, 340–1
Massili, 266
Massilia, 29, 131, 137, 246, 407
Masthitae, 180
Mastusia, Mt., 309
Matiani, 399
Mauretania, 6, 28, 120, 249–51, 260, 389, 398; Hercules and, 251; and Nile, 269, 279–282; peoples of, 255; products of, 256, 401; Romans and, 250, 256–9, 261
Mauretania Caesariensis, 250, 259
Mauretania Tingitana, 250, 256, 259
Mauri (Marusii), 259
Mauryans, 83, 344–9
Mausolos, 303, 305
Medea, 99, 184, 327
Media, 227, 328, 337, 339, 367
Mediolanum, 170
Mediterranean Sea, 11, 86, 106–7, 135, 151–8, 249, 286–7; coasts of, 117, 247, 249, 330, 407; distances in, 118, 333, 403; exploration of, 64; extent of, 76, 107, 328, 402; and piracy, 9, 63; rivers emptying into, 140, 277, 281; routes from, 101, 199, 211, 218, 385; tides in, 93, 376; winds in, 64–5
Medusa, 254, 400
Mega Tichos, 12
Megabarri, 395
Megale Hellas. *See* Magna Graecia
Megara, Megaris, 64, 195, 200, 202, 204, 217, 219
Megasthenes, 28, 93, 344–5, 347, 349, 354–5
Melanoscopium, 302
Melas Gulf, 214, 223
Melcumani, 179
Melita, 157, 185, 273
Melite, 185
Melitene, 296
Mellaria, 107
Melos, 27
Melpum, 170
Memphis, 277–9, 282
Menaechmus of Sikyon, 28
Menelaus, 36, 80, 195, 312
Meninx, 265, 267, 273
Menobardi, 332
Menocaleni, 174
Mens, divinity, 37
Mentesa, Mentesani, 114
Mentores, 178
Mercurii (Mercury), Cape, 157, 262
Mercury (god), 38
Mercury (planet), 43–4, 50–1
meridians, 84
Meroitic Kingdom, 83, 390, 393, 395–6
Meroë, 83, 102, 280, 389–98, 404–5, 408
Meros, Mt., 353
Merv (oasis), 339
Mesala, 382
Mesammones, 269
Mesene, 373, 380
Mesopotamia, 16, 27, 102, 249, 285–6, 321–5, 337–9, 367–75, 406; astronomy and astrology of, 40, 43–5; rivers through, 295–7
Messana, 94; Strait of, 136, 151
Messapia, 160
Messene, in Messenia, 51, 191
Messenia, 198, 200, 216
metallurgy, 116
Metapontum, 159–60
meteorites, 47, 57–9, 71
meteorology, 19, 29
Methone, 198
Methurides, 217
Methymna, 316
Metina, 153
Metrodorus of Scepsis, 28, 168–9
Metubarbis, 182
Midacritus, 243
Midas of Gordium, 318
Miletus, 306–8, 315, 340, 382

Milky Way, 35
Miltiades, elder, 211
Mimas, Mt., 308
Minaei, 382–3
Minerva, Cape, 146
mining, 73, 96, 116, 124, 127, 175–7, 320, 326, 386, 395
minium, 128, 395–6
Minius River, 241–2
Minoecians, 138
Minor (island), 153
Minos, King, 217, 382
Minturnae, 142, 144
Minya, in Thessaly, 206
Minyans, 205
Minyas, Treasury of, 205
Misenum, 1, 146, 154
Mithridates of Bosporus, 329
Mithridates VI, 28, 91, 99, 169, 203, 221, 229, 232, 310, 326–7, 331, 342, 369
Mithridatic Wars, 310–11, 319
Mizaei, 374
Moci, 339
Modogalinga, 349
Moeris, Lake, 278, 283
Moesia, Moesi, 182–3, 190, 210
Molossia, 160, 209
Mona, 84–5
Monadi, 162
Monaedes, 350
monoxylis, 365
monsoons, 64, 365, 356
moon, 31–6, 44–9, 52–4, 59–61, 72, 93–5, 253, 357. *See also* Luna
Mopsopia, 300
Morgantion, Morgetes, 150
Morini, 237, 246
Morogi, 240
Mosa River, 236–7
Moscheni, Moschi, 301, 332–3
Mossylitic Cape, 389
mountains, cultural significance of, 8
Mucian Altars, 92
Mucius Scaevola, C., 92.
Mucius Scaevola, Q., 58
Muluccha River, 259
Munatius Plancus, L., 58
Munda, 115–6
mundus, as cosmic concept, 4, 33, 35, 37
municipia civium Romanorum, 111
murex, 292
Murgi, Murgitania, 109, 118
musculus aquaticum, 97
Muses, Sanctuary of, 203
music, 53, 253, 304
Musulami, 266–7
Musuni, 267
Mutina, 57, 87, 100, 167
Muza, 364
Muziris, 365
Mygdonia, places so named, 337
Mykale, Mt., 305
Mylae, 94
Mylasa, 305
Myos Hormos, 385
Myriandrus, 101
Myrina, 223
Myrmidones, 205
myrrh, 285, 380, 382
Myrsilus of Methymna, 28
Myrtoan Sea, 200, 217
Mysia, Mysians, 190, 309–11
Mysotimolitae, 306
mystery cults, 30
Mytilene, 316–17

Nababes, 260
Nabataeans, 286, 296, 382
Nabu (planet), 44
Namarinian Varri, 240
Namnetes, 239
Napata, 391, 395
naphtha, 97, 99
Nar River, 164, 166
Narbo Martius, 130
Naresi, 179
Naro River, 179–80
Narona, 179–80
Nasamonians, 268–9
Nasican Calagurritani, 123
Nattabutes, 266
Naucratitic Mouth, 284
Nauportus River, 171
Navia River, 240
navigation, 60, 80, 114–15, 120, 141, 169, 282, 287, 296, 308, 312, 334, 355, 362, 403
Naxus, 221–2
Nea in Troad, 92
Neacyndes, 365
Neae, 88
Neapolis in Campania, 86, 146
Neapolis in Judaea, 289
Neapolitani, 156
Nearchus of Crete, 28, 63, 354, 361, 374–5, 399
Nechepsus, 28–9
Necho II, king, 385
Nelkynda, 365
Nemausus, 132
Nemea, 201
Nemean Games, 200

Neminie Spring, 98
Neocaesarea in Cappadocia, 325
Neolithic Period, 311
Neoptolemos, Mithridatic commander, 229
Nergal (planet), 43
Neri, 240–1
Neritis, 192
Neritus (Neriton), Mt., 261
Nero (emperor), 1–2, 19–20, 25, 39, 55–6, 87, 145, 165, 176, 196, 201, 215, 269, 331, 391–2
Nesactium, 178
Nesibis, 337
Nesimi, 259
Nestis, 35–6
Nestor, 198
Neuroe, 230
New Carthage, 407
New Pleuron, 193
Nicaea in Bithynia, 27
Nicaea in Gallia, 137
Nicasia, 222
Nicephorion, 369
Nicias of Athens, 48
Nicives, 266–7
Nicomedia in Bithynia, 319
Nicopolis, near Actium, 192
Niger, rivers so named, 257–8, 264, 267–9, 275, 280, 397–9
Nigidius Figulus, P., 20, 406–8
Nigris River, 275
Nigris Spring, 280
Nigritae, 274
Nigroe, 397
Nikomedes IV, 319
Nikopolis in Egypt, 264
Nile, 80, 83, 97–8, 102, 107, 275–85, 346, 357, 364, 380, 385–6, 389–97; expeditions up, 25, 385–7, 391–5, 403; flooding of, 30, 44, 281–2, 298; and Indus, 352; mouths of, 168, 272–3, 276–8, 285, 312; source of, 269, 279–80, 295; underground course of, 167, 280, 372, 374
Nilometer (Niloscope), 282
Ninguaria, 402
Ninos (Ninevah), 337
Ninurta (planet), 42
Nisicathae, 397
Nisitae, 397
Nola, 146, 149
Nomads, 227–8, 258, 274, 285, 297, 326–8, 336, 344, 372
nomos, 278
Nora, Norenses, 156
Norbanus Balbus, L., 88–9
Northern Ocean (North Sea), 233
Notian Sea, 152
Nubians, 396
Numa, 68
Numantia, 125
Numantine Wars, 125
Numenius, 380
Numidia, 6, 259–66
nuraghi, 156
Nymphaeum, at Apollonia in Epirus, 100, 181
Nymphaeum, places so named, 92
Nymphaeum, Mt., 206
Nymphaeus, Crater of, 100
Nymphodorus of Syracuse, 28
Nymphorum Cubile, 361
Nysa, Decapolis city, 291
Nysa in Caria, 305
Nysa in India, 353
Nyx, 38

Oblivio River, 242
Ocean, External, 11, 74–5, 282–3, 327–9, 334–6, 341, 362, 376, 404
Ocelis, 364, 379
Ocellus Lucanus, 33
Ocra, Mt., 174
Octavian, C., 56–9, 170, 180, 192, 196, 364. *See also* Augustus
Octavius, Cn, cos. 76 BC, 59, 87
Octodurus, Octodurenses, 175
Odomantians, 210
Odysians, 210
Odysseus, 53, 60, 64, 80, 94, 142–6, 152, 195, 215
Oea, 271–2
Oecalices, 197
Oechalichae, 275
Oeensi, 271
Oenoe, 202
Oenotri, 150, 156
Oeonae, 234
Oeta, Mt., 193, 205
Oglasa, 154
Ogygia, 159
Ogyris, 380
Oinopides of Chios, 42, 44
Oinotros, 150
Olarson, 127
Olbia, 228, 235
Olisipo, 241, 243
Olisiponensian Promontory, 241
olives, 61, 97, 145, 229, 355, 365
Olta, 68
Olympia, 82, 139, 198, 202
olympiads, 198
Olympian Games, 200
Olympias, mother of Alexander the Great, 188
Olympias wind, 64

Olympus, Mt., in northern Greece, 64, 41
Olympus, Mt., in Lydia, 309
Olynthos, 58, 211
Omani, 378–9
Ombrios, 401–2
omens, 40–1
Omonandes, 299
Onesicritus of Astypalaia, 28, 63, 346, 354–5, 361–3, 374–5, 399
onions, 38
Ophiodes, 386
Ophiophagi, 386
Ophiusa, places so named, 152–3, 221
Opimus, L., 58
Opus, 204
oracles, 62, 98, 169, 181, 188–9, 194, 267, 278–9, 305–6
Oracta, 362
Oratae, 352
orbis, cosmological term, 34
Orbona, shrine of, 38
Orcades, 238
Orcheni, 371, 373
Orchomenus, places so named, 205–6
Oretani, 119, 124
Oretians, 83–4
Orgenomesci, 240
Oricum, 180–1
ornamentum, cosmological term, 35
Ornithias wind, 65
Oroandes, 299
Orodes II of Parthia, 339
Oromobii, 170
Orontes River, 95, 287, 293–7, 331, 372, 403
Oropus, 203
Orospeda Mtns., 114
Orpheus, Orphic cosmology, 34, 188
Orroean Arabians, 331
Oruros, 369
Osca, Oscenses, 123
Osci (Oscans), 142, 145, 157
Oseriates, 182
Osicerdenses, 123
Osrhoene, 296
Ossigetania, 114
Ossismians, 239
Osteodes, 158
Ostracine, 288
Ostia, 136, 140–2, 168, 372
Otho, M. Salvius, 128, 141
Othryonei, 208
Our Sea, 107
Ovid, 227, 288
oxen, 94
Oxii, 374
Oxubii, 131
Oxus River, 342
Oxyttagae, 339
oysters, 145
Ozolae, 194
Ozuaei, 180

Pacensis, 243
Pactolus River, 306
pados (*pedos*) tree, 169
Padus River, 137–8, 165–9, 169–70, 177, 184
Padus, personal name, 169
Paeligni, 163
Paeonia, 207
Paesici, 126
Paestum, 151, 156
Paesuri, 241
Pagae, 194, 202
Palaeogoni, 354
Palaesimundum, 354, 357
Palaetyros, 292
Palantia, 125
Palestine, 285, 288–9, 293
Palinurus Promontory, 151
Pallantias, 266
Pallantium, 201
palm trees, 388, 393
Palmyra, 297, 377
Palugges, 396–7
Pamphagi, 397
Pamphile, 219
Pamphylia, 399–300, 302
Pan (divinity), 113, 290
Panaetius of Lindos, 28–9
Panaetolium, 194
Panda, 340, 352
Pandae, 352
Pandaia, 352
Pandateria, 154
Pandion, 365
Pandosia in Epirus, 191
Pandosia in Italy, 159–60
Pandya kingdom, 365
Paneas, 290
Panionia, 307
Pannonia, 181–2
Panticapaeum, 225, 229–30, 329–30
Panticapes River, 228
Paphlagonia, 317, 324–5
Papirius Carbo, Cn., 59
Paraetaceni, 368
Paraetonium, 268, 272
parallels, 402, 405–8
Parapotamia, 373
parasang, 13, 101, 371

paraselene, 58–9
Parasinus, 92
parhelion, 58
Paria Island, 312
Pariades, Mt., 301
Pariani, 339
Parihedrians, 301
Parisii, 239
Parma, 167
parma (shield), 134
Parmenides of Elea, 35, 43, 73
Parnassus, Mt., 194
Paropanisidae, 360
Parra, 170
Parthenios, 23
Parthenon (Athens), 206, 284
Parthenope, 146
Parthenopolis, 212
Parthia, 70, 81, 296–8, 337, 339, 367–72, 376, 378–80, 406
Parthini, 180
Parthyaea, 337, 367
Pasargadai, 348, 362, 368
Pasini, 178
Pasitigris River, 363, 373–4
Passala, 315
Patala, 351
Pataliputra, 28, 83, 247–50, 355
Patalis, 84
Patmus, 222
Patras (Patrae), 102, 196
Patrocles, Seleukid explorer, 29, 76, 335, 339–41, 345
Pausanias, 188
Paxos, 216
pearls, 351, 354, 358
Pelagones, 207
Pelasgia, 195, 200–1
Pelasgians, 139, 150, 195
Pelasgikos, 139
Pelasgis, 200, 205
Pelasgos, mythological figure, 201
Pelion, Mt., 75
Pella, 375
Pellene, 197
Pelondones, 125
Peloponesian War, 30, 188, 191, 221
Peloponnesus, 47, 82, 90, 139, 142, 157, 193–202, 205, 216–17, 268, 328, 406
Pelops, 139
Pelorum, 157
Pelusium, 278, 286–8, 385
Pendelonians, 241
Penius River, 206–8, 214
Pentapolis, in Cyrenaica, 267
pepper, 365
Peraea, 289
Pergamum, 25, 311
Periander of Corinth, 188, 195
peridots, 386–7
Perimula, Cape, 351
periplous, periploi, 8, 111–12, 120–1, 129–33, 137–8, 159, 178–80, 199–203, 207, 211–14, 227–8, 250–2, 256
Perorsi, 258, 274, 397–8
perpendiculum, carpentry tool, 54
Perrhaebi, 189, 193
Persepolis, 368
Perseus (constellation), 35
Perseus (hero), 142, 254, 288, 400
Perseus of Macedonia, 48, 166, 209
Persia, Persians, 13, 25, 99, 211, 276, 307, 332–3, 339–41, 360–3, 371, 374; and Greek world, 58, 75, 308; in Iberia, 113; language of, 225, 323; and Parthians, 337, 367; royal road of, 102; topography of, 321, 330, 337
Persian Gulf or Sea, 28, 63, 83, 117, 330, 337, 354, 357, 361–9, 372–80
Persis, 339, 348, 361, 368, 373–4, 378
Perusia (Perugia), 41, 140
Perusii, 397
Pestici, 341
Petosiris, 28–9, 54, 56
Petra, 286, 298, 377–8, 381–2
Petronius, Publius, 391, 393
Petubastes IV, 379
Peucetia, 160–1
Peucini, 236
Peutinger Map, 365
Phacusa, 222
Phaeacia, 215
Phaethon, mythological figure, 168, 184
Phaethon, planet, 43
Phainon, planet, 42
Phalacrum, 215
Phalasarna, 217
Phalerum, 96
Phalliges (Palugges), 397
Pharia, 185
Pharnakes II, 326
Pharos Island, 88, 312
Pharsalus, 21, 180, 282, 287
Pharusians, 113, 258, 274, 276
Phasargis, 368
Phasis River, 226, 327, 342
Phazania, 270
Phegium, 90
phengites, 85, 91
Pherecydes of Syros, 31, 86
Phiala, 281
Phila, 153

Philae, 282
Philaeni, Altars of, 265–6
Philemon, 29, 233
Philip II of Macedonia, 30, 58, 99, 189, 207–8
Philippi, 211, 213
Philiscum, 298
Philistides of Mallos, 29
Philoktetes, 206
Philomela, 213
Philon, explorer, 83, 387
Philonides of Chersonesos, 29, 82
Phlius, 198
Phocaea, 120–1, 124, 131, 243, 309–10
Phocis, 188, 194–5, 204
Phoenicia, Phoenicians, 97, 153, 156, 238, 263, 277, 286–93, 312, 390; in Arabia, 245; in Iberia, 75, 106, 112–14, 244
Phosphoros (planet), 43
Phrygia, 52, 310, 300, 303–7, 309–11, 318
Phthiotis, 203, 205
Phthirophagi, 327
phthongos, 327
Phycus, Cape, 218, 268
Phygela, 307–8
Piacenza liver, 69
Picentines, 149, 165
Picentinus Ager, 149
Picenum, 165, 146
Pictones, 239
Pieria, 207
pigs, 60, 204
Pindar, 31
Pindus Mtns., 188–9, 193
piracy, 9, 63, 66, 161, 179, 364
Piraeus, 139, 198, 219
Pisa in Eleia, 139, 198
Pisa or Pisae in Italy, 139, 407
Pisidia, 299–300, 309
Pithecusa (Pithecussae), 154
Pityus, 328
Pityusa, 307
Pityussae, 152
Placentia, 87
Placia, 317
Planasia, island near Corsica, 154
Planasia, island off African coast, 402
plane tree, 195
planets, 36, 42–4, 50–4, 60–1. *See also* individual names.
plants, poisonous, 72–3, 324
Platanodes, Cape, 198
Plato, 31
Pleiades, 62, 66
Pleuron, 193
Plinia, 1
Pliny the Younger, 1–2
Plotae, 216
Plutarch, 48, 58, 288
Poediculi, 161
Poeninae Pass, 169
Pola, 136, 172, 178
poles: celestial, 35; north, 103, 407; south, 55, 350
Pollio, Asinius, 18
Polybius of Megalopolis, 29, 254
Pompaelo, 121
Pompeii, 67, 146
Pompeius (Pompey) Magnus, Cn., 21, 56, 118–21, 161, 180, 262, 282, 287, 369; activities east of Black Sea of, 327, 332, 341–2
Pompeius Strabo, C., 177
Pompeius, Cn., the Younger, 116
Pompeius, Sextus, 116
Pompelonenses, 123
Pomponiana, 153
Pomponius Mela, 7–8, 11, 16, 20
Pomponius Secundus, P. 2
Pomptine Marsh, 143
Pontic Sea (Pontos), 79, 211, 224–5, 277, 320, 324. *See also* Black Sea.
Pontus (province), 81, 303
Pontus (region), 144, 301, 305, 324–7, 331, 342
poplar trees, 46
Populonium, 139, 154
Porcius Cato, C., 70
Porcius Cato, M., the Censor, 18, 167, 173
Porcius Cato, M., consul 118 BC, 58
Porcius Cato, M., the Younger, 262
Porsina (Porsenna) of Clusium, 68
Portae (Taurus), 301
Portus Herculis, 138
Portus Veneris, 122
Poseidon, 88, 200, 251
Poseidonia (in southern Italy), 151
Posidonius of Apameia, 33, 34, 46–48, 51, 81, 29, 239, 243–5, 292–3, 361–3
Postumius Albinus, Sp. 58
Potheinos, 287
Potidaea, 71
Potinusa, 245
pottery, Greek, 330
Praeneste, 40. 388
Praestamarci, 241
Praetavi, 297
Prasii, 349
Pratitae, peoples so named, 338, 367
prester, meterological term, 67
Prettanike, 84
Priam, 214, 391
Priapus (town), 224
Priene, 89, 305

Probalinthos, 202
Prochyta, 89, 154
Proconnesus, 230, 320
prodromi, 65
Prokopios, 213
Prometheus, 205, 221
Propontis, 88–90, 213–14, 224, 310, 317–20
Proserpina, 38
Protesilaus, 214
Prusa, 319
Psammetichos II, 396
psittacum, 392
Psylli, 264–5
Pterophoros, 230
Ptoenphani, 397
Ptolemaic period, 30, 39, 250, 266, 276–9, 284–6, 291, 313, 364, 382–7, 390–2, 395–6, 399
Ptolemais in Phoenicia, 292, 406
Ptolemais on the Red Sea (epi Theras), 83, 384, 408
Ptolemy Ceraunus, 333
Ptolemy I, 284, 333, 346, 386
Ptolemy II, 30, 83, 277, 392, 345–6, 385–8, 277, 392
Ptolemy III, 80, 386
Ptolemy IX, 77, 394
Ptolemy XIII, 287
Ptolemy of Ituraea, 294
Ptolemy of Mauretania, 250, 256
Pudicitia, 37
puleium, 61
pumice, 99
Punic Wars: First, 17; Second, 18, 88, 101, 109, 112, 119–20, 124, 173, 261, 266; Third, 261–2
Puteoli, 91, 356
Pydna, 47, 209
Pygmaeans, 212
Pylos, places so named, 198
Pyraei, 180
Pyrenaea Venus, 121–2
Pyrene, city, 129
Pyrene, person, 114
Pyrenees, 102, 109, 114, 119, 121, 125, 127–32, 239–43, 407
Pyroeis (planet), 43
Pyrpyle, 2, 21
Pyrrander, 29
Pyrrha, 90
Pyrrhus of Epirus, 161–4, 188
Pythagoras of Kroton, 29, 31, 43, 86
Pythagoreans, 29, 34–5, 53, 135
Pytheas of Massilia, 29, 75, 84, 99, 102, 231–4, 327–8, 245
Pythian Games, 200
Pythodoris of Pontus, 305, 326–8
python, 91

Quariates, 131
quarries, 139, 220, 261–2
Querquerni, 127
Quinctius Cincinnatus, L., 163

rabbits, 153
Rachias, 354, 356–60
Radata, 390
Raetians, 173–4, 181, 235
rainbows, 72
Raurici, 239
Ravenna, 166, 168, 171, 320
Reate, 91, 97–8, 164
rector, 39, 41
Red Sea, 79–80, 83, 97, 107, 112, 270, 276, 285–6, 297–8, 333, 345, 356, 360–5, 377–87, 390, 395, 398–9, 404, 408
Red Sea, alternative name for Persian Gulf, 365–6, 372
reeds, 261, 275
Regium, 135, 151, 156, 407
Resisthos (Rhaidestos), 213
Rha River, 76, 336, 341
Rhadamaei, 382
Rhadamanthus, 382
Rhambakia, 361
Rhapta, 398
Rhenus River, 75, 175, 236, 239
Rhesos, king of the Edoni, 210
rhinoceroses, 388
Rhinocolura, 288
Rhium, 193, 196
Rhoali, 297
Rhoda, town in Rhodanus delta, 130
Rhodanus River, 129–31, 153, 168, 175
Rhode, 121, 130
Rhodes, 130, 222, 303–4, 314, 382, 403, 406
Rhodope, Mt., 191
Rhoeteum, 311
Rhosian Mtns., 294
Rhoxolani, 227
Rhyndacus River, 310, 317
rings, meteorological phenomenon, 58
Ripaei Mtns., 226
Riphearina, 382
rivers: cultural significance of, 8–9; underground, 95–6, 167, 269, 280, 372–4, 383. *See also* individual names.
roads, Roman, 46, 117, 148. *See also* individual names.
Roadside Pratitae, 388
Rome (city), 147–50; secret names of, 147

Rome (monuments and features in city): *agger*, 148; Arch of Augustus, 17; Arch of Titus, 38, 69, 147; Aurelian Wall, 148; Capitol (hill and temple), 149, 232; Carinae, 235; Castra Praetoria, 148, 159; Circus Maximus, 149; Clivus Suburanus, 38; Esquiline Hill, 38; Flavian Amphitheater, 135; Forum Boarium, 140; gates of, 148–9; Golden Milestone, 148; Palatine Hill, 37, 39, 201; Pons Sublicus, 141; Porticus Vipsania, 17, 118, 376; Quirinal, 118, 376; Saturnia, 149; Servian Wall, 148; Shrine of Febris, 37; Shrines and temples of Juno, 69; Shrines and Temples of Jupiter, 68–9; Shrine of Lares, 38; Shrine of Mala Fortuna, 38; Shrine of Volupia, 147; synagogues in, 39; Tarpeian Rock, 149; Temple of Antoninus and Faustina, 220; Temple of Concordia, 37; Temple of Divine Julius, 57; Temple of Saturn, 148; Temple of Tellus, 135; Temple of Venus Genetrix, 57; Velabrum, 147
Romulus, 69, 147
Rubens, Peter Paul, 144
Rubico River, 166
Rubrensis Lake, 129–30
Rubuta River, 138
Rudini, 180
Rumnici, 341
Rutilius, P., 58
Rutubis Harbor, 250
Rutuli, 142

Saba, Sabaei, 380
Sabbatha (Sabota), 380–1
Sabines, 164–6
Sabrata, 262–3
Sacae, 341
Sacasani, 333
Sacred Promontory (Hispania), 101, 242–3
Sacred Spring, 165
Saetabitani, 124
Saevo Mtns., 234
Sagrus River, 163
Saguntum, 120
Sahara Desert, 167, 253, 257–8, 267–71, 274–5; routes across, 268–9, 280, 264
sailing season, 65–6
Saitae, 374
Sala, 252–3, 256
Salamis in Greece, 219; Battle of, 55
Salapia, 162
Salassi, 175, 177
Salduba, 124
Saliares, 92
Sallentine Promontory, 181
Sallentinus, Sallentini, 96, 160
Sallustius Crispus, C., 18, 263, 265
Sallvi, 137, 169
Salmani, 369
Salon Island, 161
Salona, 179
salt, 124, 142, 270, 379
Samaria/Sebaste, 289
Same, 216
Samiramis (Semiramis), 325
Samnites, 150, 160, 164–5
Samnium, 140, 163
Samonium Cape or Promontory, 218
Samos, 23, 222, 315
Samosata, 99, 296–7
Samothrace, 223
Sangaris River, 324
Sapaei, 210
Sappho, 45
Saraparae, 339
Sarbabares, 266
sarcophagus stone, 92
Sardeates, 179
Sardi, 228
Sardinia, 101, 136, 155–8, 262, 273, 313
Sardis, 88, 306, 308–9
Sardoan Sea, 152
Sarmadium, 160–1
Sarmatae, 211, 227, 230, 234, 328–30, 336
Sarmatian Iazygae, 227
Sarmatian Wilderness, 227
Saronic Gulf, 196, 199–200, 202, 217
saronis, 200
saros cycle, 49
Sarunetes, 175
Sasaei, 180
Sasine, 160
Sasonis (Sanos), 185
Sataspes, 113, 276
Saturn, divinity, 38, 114, 148
Saturn, planet, 36, 42–5, 50–1, 60, 68
Saturn, Cape, 120
satyrs, 253, 276, 399
Saunitians, 150, 160
Sauromatians, 211, 227, 230, 328–9, 336
Saus River, 182
Savus River, 171
Scalabis, 243
Scaldis River, 235, 238–9
Scamander River, 92
Scandiae, 238
Scandinavia, 103, 233–4, 236, 334, 405
Scantiae Spring, 100
Scardona, 178–9
scarus, 145
Scatebra Spring, 97

Scatinavia. *See* Scandinavia
Scaugdae, 211
Scenitae, 285, 297, 372
Scheria, 215
Schinusa, 222
Schoenos, 202
schoenus, 13, 284, 296, 371
Sciathos, 223
scintilla, meterological phenomenon, 59
Scioessa, Mt., 197
sciothericon, meterological instrument, 59
Scipio, Tomb of, 114
Sciri, 235
Sciron, wind and person, 64
Scirtari, 179
Scodra, 180
Scolacium, 158
Scordisci, 182
scorpions, 274
Scotusa, 211
Scribonius Curio, C., 18, 59
Scylace, 317
Scylla, 94, 157
Scyllaeum, Cape, 199
Scyllaeus Gulf, 135
Scyros, 222
Scythia, Scythae (Scythians), 67, 75, 118, 183, 210–12, 226–9, 232, 255, 261, 291, 302, 333, 335–6, 341
Scythian bow, 225, 229
Scythian Cape, 342
Scythian Ocean, 335
Scythians, Degenerate, 227
Scythicus, Mt., 302
Scythopolis, 291
Scythotaurians, 229
sea depth, 75, 93, 95, 97, 312–13
seafaring cults, 288
seal, monk, 70
seamen, 30, 65, 74, 106, 355, 378, 381
seashells, 267
seasons, 36, 47, 52, 61, 65–6, 80
Sebaste, 289
Sebastopolis, in Colchis region, 327–8
Sebastos, 328. *See also* Augustus
Seberritans, 397
Sebosus Statius, 20, 392, 401–2
Sebubus River, 252
Seduni, 177
Segasame, 125
Segienses, 123
Segisamonenses, 125
Segismaiulienses, 125
Segovellauni, 131
Segusium, 176–7
Seirenes Islets, 146
Selene (divinity), 45
Selene (planet), 44, 139
selenite, 128
Seleucia Pieria, 293–4, 297, 372, 403
Seleucia, on the Eulaus, 374
Seleucia, on the Euphrates, 297
Seleucia on the Tigris, 297–8, 337–8, 370, 372–3
Seleucid period, 25, 29, 76, 209, 293–7, 340–5, 312, 332–3, 340, 347–9, 360, 369–70, 375, 378, 380
Seleucus I, 76, 297, 312, 333, 337–8, 344, 347–8, 353, 369
Selloe, 189
Semberritae, 396
Sembobitis, 396
Sembritae, 396
Sempronius Gracchus, Ti. (agrarian reformer), 79, 123
Sempronius Gracchus, Ti. (cos. 177 BC), 123
Sempronius Tuditanus, C., 172
Sena River, 167
Senate, Roman, 173, 259, 369
Seneca the Younger, 16, 20, 41, 346
Senedi, 180
Senones, 167, 170
Senta Cave, 62
Separi, 179
Septem Maria, 168
Septentrionales Ocean, 75, 233
Septentriones, 62, 80–1, 350, 355, 362
Seraci, 328
Serapilli, 182
Serapion of Antioch, 30
Serapion the *gnomicus*, 30
Serapis, 39
Seres, 219, 335, 342–3, 357–8, 364
Sergius Catilina, Q., 67
Sergius Orata, M., 145
Sergius Plautus, 20–1
Sericus Ocean, 335
Serretes, 182
Sertorius, Q., 119, 123, 251
Servius Tullius, 40, 100, 148
Sesambri, 396–7
Sesostris, 385; Pillars of, 389
Sestius Quirinalis, L., 241
Sestos, 213
Setae, 349
Seth, 56
Setinian wine, 144
Seurbi, 241
Seven Wonders of the World, 312, 370
Seven-Stadia Strait, 213–14

shadows, 45–7, 49, 82, 85, 223, 406–8; absence of in tropics, 83–4, 282, 387; reversal of, 83, 357; Venusian, 44
Sibyl, 169
Sicambri, 236
Sicilia (Sicily), 103, 134, 136, 157–9, 185, 273, 313, 406; Athenian expedition to, 48; and Hannibal, 59; peoples of, 142, 150–1
Sicilian Strait, 93, 262
Sicoris River, 123
Siculi, peoples so named, 142, 166
Sicyon, 82
Sideris River, 339
Sidon, 287, 292
Sidus Iulium, 56
Siga, 259–60
Sigeum, 310, 317, 319
Sila Forest, 151
Silanus, people so named, 59
Silerus River, 95, 150–1
silk, 219–30, 342–3, 358
Silk People, 342
Silk Road, 342, 344
silphium, 269
siltation, 88–9, 117, 130, 153, 192, 295, 307–9, 316, 363, 371–6
Silures, 237
siluris, 279
silver, 35, 124, 127–8, 349, 351, 353, 358
Silvi, 333
Simbarri, 396
Simonides the Younger, 30
Singara, 297
Singidunum, 182
Singilis River, 115
Sinope in Pontus, 325
Sinope, towns so named, 144
Sinuessa, 91, 144–5
Siphae, 194
Sipylus, Mt., 90
Siraci, 228
Sirbites, 397–8
Sirbonis, Lake, 288
Sirens, 144, 146
Sirens, Promontory of, 89
Sirius (Seirios), 60–1, 65, 168, 281
Sirmium, 182
Sitioganus River, 362
Sittace, Sittacene, 373
Six Gods (islands), 244
Skylax of Karyanda, 63, 112
Skyllaion Promontory, 157, 199
Skythia, Little, 211–12
slaves, 70, 150, 209, 221, 388
slingshots, 152
snakes, 100, 153, 264, 302
snow, 66, 72, 100, 168, 231, 257, 280, 342, 347, 402
Snow, mythological figure, 315
Sobek, 280
Socrates, 36
Sodi, 333
Sogdiana, 340
Solentia, 179
Solis Island, 361
Solomon of Israel, 112
Solorius Mtns., 110
solstices, 52, 61, 80–3, 147, 282, 356, 387, 405–6
Solymi, 311
Sophonisba, 259–61
Sora, 146
Soracte, Mt., 91
Sordones, 129
sorex, 61
Sosigenes, 30, 43–4
Sotades, 30
Soti, 138
Spaosines, 375
Spaosinou Charax, 375–6
Sparta, Spartans, 52, 58–9, 85, 160, 188, 199, 269
Speke, John Hanning, 279
Spes, divinity, 37
sphingia, 388
spikes, meterological phenomenon, 58
Spinetian treasury, 168
spiracula, volcanic phenomena, 91
Sporades, 221–3
Stabiae, 2, 149
Stactares, 229
stadia, 13, 54, 152, 195, 371
Staphylus, 30
Staures, 339
Stesichorus, 31, 175, 245
Stilbides, 48
Stilbon (planet), 44
stipendaria, 111
Stobi, 207
Stoechades, 153
Stoeni, 175
Stoicism, 20, 32–4, 40, 45, 50, 63, 72
Stonians, 175
Strabo of Amaseia, 8, 11, 16
Strato's Tower, 289
Straton of Arados, 294–5
Straton of Lampsakos, 202
Straton I of Sidon, 289
Stratonike, 294
Strongyle, 158
Strophades Islands, 215, 217
Strymon River, 190, 208–11
Sturi, 237

Sturium, 153
Stymphalus Lake, 97, 334
Styx River, 206
Suari, peoples so named, 83, 350, 360
Sublaqueum, 164
Subocrini, 174
Subsolanus wind, 64
Sucro River, 152
Suculae, constellation, 60, 62
Suebi, 78, 227, 236
Suebri, 131
Suelteri, 131
Suetonius Paulinus, C., 21, 85, 253, 256–8, 275–6, 399
Sulcitani, 155
Sulpicius Galba, Ser., 128, 132, 329
Sulpicius Gallus, Ser., 21, 47–9, 53
Sulpicius Camerinus, Ser., 70
Summanus, 68
sun, as planet, 36–7, 43
sundials, 30
Sunium, Cape, 203
Supertamarci, 240–1
Sura, 297–8
Surdaonian Ilerdenses, 123
Surius River, 96
Surrentum, 146
Susa, Susiane, 176, 363, 373–4
Syagrum, Cape, 363, 380
Syene, 83–4, 282, 387–92, 398, 408
Symi, 314
Symplegades, 232
Synnas Jurisdiction, 304
Syphax, 259–61
Syracuse, 22, 26, 72
Syrbotae, 396–7
Syria, 19, 38, 285, 293–6, 326, 370, 373, 377, 385; borders of, 289; rivers of, 95; Roman province of, 291, 293; routes to and from, 298, 300, 325; writers on, 31
Syrian Gates, 294
Syrie (island), 308
Syrmatae, 339
Syros, 98–9, 221
Syrtes, 93, 264–6

Tabis, Mt., 342
Tabora, 379
Tacfarinas, 267
Tacitus, Cornelius, 2, 98, 234, 236–7, 391
tacking, 66
Tader River, 114, 120
Taenarum, Cape, 198–9, 268
Tagus River, 117, 119, 242–3
Talge (Talka), 342
Tamil poetry, 365
Tamna, 285
Tanager River, 96
Tanagra, 204
Tanais (town and river), 76, 107, 226–7, 233, 245, 277, 329–30, 340, 404, 408
Tanit, 244, 400
Tantalis, 90
Tantalos, 90
Taphios, 155
Taphrae, 230
Taprobane, 11, 321, 353, 354–9, 389, 399
Tarantinus Bay, 160
Tarbelli, 239
Tarentum, Tarantines, 160–1
Tarichaea, 290
Tariotae, 179
Tarnis River, 240
Tarquinius Superbus, 148
Tarquinsian Lake, 92
Tarquitius, 21, 68
Tarracenses, 123
Tarracina, 70
Tarraco, Tarracon, 109, 120–2, 127
Tarraconensis, 1, 109–10, 122, 128
Tarsos, 299
Tartaros, 66
Tartessos (Tartesos), 112–13, 116, 245
Taulanti, 180
Taurania, 149
Taurian (Tauric) Chersonesos, 92, 229
Taurianum Promontory, 151
Tauric Range, 229
Taurica, 229
Taurisci, 174–5, 182
Tauromenitanian Strait, 93
Tauromenium, 93
Taurus (constellation), 35, 60, 62, 80
Taurus Mtns., 286, 294, 296, 300–2, 309, 338, 373–7
Taxila, 28, 348, 353
Taÿgetus, Mt., 85–6
Teani, 162
Teanum Apulium, 162
Teate, 87
Tectosages, 130
Telamon, in Etruria, 177
Telandrus, 302
Telavium River, 178
Telchius, 328
telinum, 222
Telmesus, city and bay or gulf, 302–3
Telos, 222
Tempe, Vale of, 206
Tenedos, 316–17

Tenos (Tenus), 214, 221
Tenupsis, 396
Teos, 316–17
Teredon, 378
Tereus, 213
Tergeste, 171, 173–4
Termetis, Mt., 309
tetrarchy, 291–4, 300
Teura, 159
Teuthrania, 311
Teutoni, 236
Teutosages, 318
Teutria, 184
textiles, 219, 292
Thales of Miletus, 31, 42, 45, 48, 50, 71, 281
Thasos, 223, 407
Thebae also called the Corsiae, 194
Thebai in Ionia, 305
Thebais in Asia Minor, 305
Thebais in Egypt, 278, 311
Thebata, 369
Thebes in Boeotia, 178, 194, 203–4, 292, 305
Thebes in Egypt, 203, 278, 283
Thebes in Lucania, 159
Thebes in Phthiotis, 203
Themiscyra River, 301
Thenae, 264
Theodosia, 229
Theodosia festival, 98
Theon Ochema, 99, 255, 395, 398
Theophanes of Mytilene, 342
Theophrastus of Eresos, 30, 143
Theopompus of Chios, 30, 207
Thera, 88, 215, 222
Therasia, 88, 222
Thermaic Bay or Gulf, 208, 223
Thermon, 194
Thermopylai, 301
Therothoae, 389
Thespiae, 194
Thesprotia, 188
Thessalonica, 208
Thessalos, 205
Thessaly, 74, 139, 189–90, 193, 195, 205–6, 292
Thia, 215, 222
Thina, 343, 358
Thomna (Thomala), 380
Thracia (Thrace), Thracians, 17, 19, 90, 183, 190–1, 207–14, 223, 337, 389
Thracian Chersonesos, 71, 211, 213, 223
Thrasyllus of Mendes, 30
Thrinakia, 156
Thucydides, 16, 30, 89, 191
Thule, 84–5, 102, 238, 405, 408
thunderbolts, 52, 66–70
Thuni, 343
Thyamis River, 189
Thyatira, 308
Thybris River, 140
Thyni, 210
Thynias, 212
Thyssagetae, 230
Tiberis (Tiber) River, 140–1
Tiberius (emperor), 19, 28, 30, 40, 70, 75, 88, 103, 148, 154, 182, 267, 326; Alpine expedition of, 174, 226, 235
Tibur, 141, 164, 259
Ticer River, 121–2
Ticinum, 170
tides, 6, 10, 93–5, 245, 376, 380
Tifernus River, 163
tigers, 351
Tigranes II, 331–2, 377
Tigranes VI, 331
Tigranocerta, 332
Tigris River, 298, 338, 363, 368–78
Tiligul River, 228
Timachi, 183
Timaeus of Tauromenion, 30, 95, 150, 156, 168, 198, 233, 238, 244
Timaeus the mathematician, 30, 44, 281–2
Timagenes of Alexandria, 31
Timavus River, 184
Timosthenes of Rhodes, 30–1, 65, 152, 214, 313–14
tin, 238, 243
Tingi, 251–2, 259, 273
Tinia River, 141
Titaresios River, 206
Titius River, 172, 178
Titus (emperor), 1, 4, 9, 13–16, 21, 39, 49, 55, 87, 90, 128, 147
Tmolus, Mt., 306, 309
Toarci (Tochari), 343
Togata Gallia, 166
Toletum, 124
Tolosani, 240
Tolostobogii, 318
Tomarus, Mt., 194
Tomi (Tomis), 212, 227
Tomouroi, 189
Tonoberi, 396–7
toparchy, 289
topazos, 386–7
Topazos Island, 386–7
Toronaei, 208
tortoise shell, 358–9, 379, 387–8
Tractus Umbriae, 140
Trageae, 315
Trajan (emperor), 141

Trallis, 305
Transmontani, 126
Transpadana, 136, 139
Trapeza, Cape, 311, 317
Trapezus, 326
Trasimenus, Lake, 88, 100
Treres, 207
Triballi, 183, 190, 207
Trica, 162
Tricores, 131
Trinacria, 156–7
Triones Septem: *see* Septentriones
Tripsedi, 311–12
Tritolli, 131
Triton River, 280
Tritonis Lake, 280
Troas (Troad), 309–11, 315–16
Trocmi, 318
Troezen, 217
Troglodytae, 270, 276
Trogodytae, peoples so named, 227, 270, 276, 280, 386, 388
Trogodytice, 80, 408
Trojan War, 150, 162, 173, 190, 201, 214, 309, 311, 316
Tropaeum Alpium, 176–7
Trumplini, 174
Tubero, Q. Aelius, 21
Tuder, 71
Tullius Tiro, M., 21, 60
Tullus Hostilius, 68
turbo, celestial phenomenon, 45, 67
Turdulians, 112, 116, 241–2
Turiaso, Turiassonenses, 123
Turmogidi, 125
Turranius Gracilis, 21, 107
Turri, 176
turtles, 97, 359, 366
Tusca River, 261–2
Tuscan Bay, 89
Tuscan Sea, 177
Tusci, 139, 145
Tusculum, 92
Twelve Gods, Altar of, 347
Tyde, places so named, 241
Tyle. *See* Thule
Tylos on the Persian Gulf, 379
Tylos in Phoenicia. *See* Tyre
Tymphaei, 193, 208
Tyndaris, 90
Typhon, mythological creature or personality, 55–6, 66, 154
typhon, meterological phenomenon, 66
Tyra (Tyras) River, 227–8
Tyragetae, 228
Tyre (Tyros), 256, 292, 317
Tyrrhenian Sea, 108
Tyrrhenians, 139
Tyrrhenus of Lydia, 139

Uberi, 177
Udiva River, 120
Ulixes. *See* Odysseus
Ulla River, 241
Umbria, Umbrians, 71, 91, 138–40, 163, 165–6
Umbro River, 140
universe, 33–7, 40, 52–3, 79–80, 103, 106
Uranus (planet), 36
Urci, 119
Uria, 160
Uruk, 371
Ustica, 158
Utica, 93, 262–3, 272
Utica, Gulf of, 262

Vacathi, 397
Vaccaei, 119, 125, 241
Vada Volaterrana, 153
Vadimonis, Lake, 91
Vagia River, 241
Valentia, 120
Valentini, 155
Valerianus, 21
Valerienses, 124
Valerius Flaccus, L., 59
Valerius Soranus, Q., 16, 147
Valerius Triarius, C., 326
Vamacures, 267
Vandili, 235
Vannius, 227
Varciani, 182
Vardaeans, 179
Varduli, 125, 240
Vareia, 120–1
Varinnae, 235
Varro, M. Terentius, 21, 35, 113, 135–7, 179, 212, 221, 225, 242–5, 336, 341–2
Varro Atacinus, P. Terentius, 21–2
Varus River, 129, 131, 133, 135–7, 174
Vascones (Vaconi), 121
Vediantii, 137
Veii, 92, 170
Veins of the Nile, 392
Velini Lakes, 164
Velinus Lake, 72, 96–7
Venafrum, 145
Vendelians, 227
Venedi, 234–5
Veneni, 138
Venesti, 240

Venetia, Venetians, 82, 173, 240, 407
Venetia et Histria, 170
Venneses, 125
Vennonenses, 175
Venus (planet), 43–4, 46, 50–2, 144
Venus, Cape, in Cilicia, 299
Verbannus Lake, 173
Veretini, 163
Vergil, 4, 22, 33, 35, 59, 91, 133, 138, 142, 168, 173, 243, 273, 316
Vergiliae, 62, 65
Verrius Flaccus, M., 22
Verucini, 131
Vespasian (emperor), 1, 9, 15, 20, 39, 49, 63, 72, 87, 128, 147, 164, 201–2, 212, 260, 263, 289, 321, 325
Vesper (planet), 43, 52
Vestini, 164
Vesulus, Mt., 167
Vesuvius, Mt., 1, 89, 133, 144, 146, 149, 157
Veteres Turduli, 241
Vettius Marcellus, 87
Vettones, 119, 241–2
Vetus, L. Antistius, 22
Vibi Forum, 167
Victoria Iuliobrigensium, 240
Vienna in Gaul, 407
Vindelici, 174
viniculture, 170, 254
Vipsania Polla, 118
Vipsanius Agrippa, M., 7, 17, 112, 118, 135–7, 190, 201, 207, 214, 218, 233, 242, 250, 252, 255, 260, 265, 268, 277, 286–90, 297–9, 306, 323, 333, 367, 372, 376–7, 381, 405
Vipstanus Apronianus, C., 81
Virgilienses, 124
Vistula River, 75, 167, 227, 234–6
Vitellius, A., 128
Vitruvius, 51, 62, 64, 85
Vocontii, 71, 131–2, 169
Volcae, 130
Vologescerta, 370
Vologesus I, 370
Volsci, 142
Volsinii, 68
Volta, 68
Volubili (Volubilis), 252
Volumnius Amintinus, P., 70
vortex, 66
Voturi, 318
Vulcan, 33, 100, 393

weather, 22, 40, 48, 58–62, 66–7, 70, 229, 342
whales, 288, 362
wheat, 145
wind rose, 64, 69
winds, 42, 59–60, 62–6, 86, 158, 243, 365, 386. *See also* individual names
wine, 67, 98, 144–5, 170, 222, 355. *See also* individual names
wolves, 55, 162
woodpeckers, 165
wool, 327
writing: development of, 287; unknown forms of, 379

Xanthos of Lydia, 202
Xanthus River, 98
Xenagoras, 31
Xenophanes of Colophon, 37–8, 55, 94
Xenophon of Athens, 324, 338, 371
Xenagoras, 31
Xenophon of Lampsacus, 31, 400
Xerxes, Persian king, 113, 206, 208, 211, 224, 373

Zacynthus, 216–17
Zamnes, 390
Zanganae, 389
Zarangae, 339
Zarmanochegas, 349
Zazata, 342
Zelil, 251
Zeugma, 287, 297–8, 369, 372
Zeus (god), 22, 35, 52, 60, 66, 97, 154, 189, 192, 198, 201, 205, 218, 267
Zeus (planet), 43
Zeus, as primeval element, 36
Zeus Baal, 370
Ziela (Zela), 326
Zigerus, Cape, 363
Zimara, 296
Zimyra, 293
Zmyrna (Smyrna), 12, 308–9, 316
zodiac, 35, 41–2, 51, 53, 80, 85
Zoelae, 126
zones, climate, 47, 49